Promising Detoxification Strategies to Mitigate Mycotoxins in Food and Feed

Promising Detoxification Strategies to Mitigate Mycotoxins in Food and Feed

Special Issue Editor

Ting Zhou

MDPI • Basel • Beijing • Wuhan • Barcelona • Belgrade

MDPI

Special Issue Editor
Ting Zhou
Guelph Research and Development
Center (AAFC)
Canada

Editorial Office
MDPI
St. Alban-Anlage 66
Basel, Switzerland

This is a reprint of articles from the Special Issue published online in the open access journal *Toxins* (ISSN 2072-6651) from 2016 to 2018 (available at: http://www.mdpi.com/journal/toxins/special_issues/promising_detoxification)

For citation purposes, cite each article independently as indicated on the article page online and as indicated below:

LastName, A.A.; LastName, B.B.; LastName, C.C. Article Title. *Journal Name* **Year**, *Article Number*, Page Range.

ISBN 978-3-03897-027-9 (Pbk)
ISBN 978-3-03897-028-6 (PDF)

Cover image courtesy of Yousef I. Hassan.

Contents

About the Special Issue Editor

Ting Zhou is a research scientist with Agriculture and Agri-Food Canada, Guelph Research and Development Center, Guelph, Ontario, Canada, and a member of associated graduate faculty at the University of Guelph. He is a leader of a well-established food safety research program specialized in controlling fungal hazards in food value chains and is involved with several international and national projects on mycotoxin mitigations using physical, chemical and biological strategies. Dr. Zhou is the author or coauthor of over 120 peer reviewed publications of scientific journals and book chapters. In particular, he has made significant contributions to the microbial/enzymatic detoxifications of Fusarium mycotoxins, and was awarded the most prestigious award of Agriculture and Agri-Food Canada in 2016, the Prize for Outstanding Achievement in Science for his research in mycotoxin biodetoxification. Dr. Zhou received his Ph.D. degree (1991) from McGill University, Canada.

Preface to "Promising Detoxification Strategies to Mitigate Mycotoxins in Food and Feed"

Mycotoxins have been a threat to mankind for thousands of years. Contaminations of mycotoxins in food and feed are responsible for many different acute and chronic toxicities, including induction of cancer, mutagenicity, and many other toxic effects ranging from discomfort to death. With government regulations and routine monitoring of the food value chain, the risks of mycotoxins entering our food supplies have significantly reduced. However, due to the ubiquitous nature of mycotoxin producing fungi and our inability to prevent conditions that favor growth and mycotoxin production of mycotoxigenic fungi, contaminations of agricultural produce with mycotoxins are still inevitable under the current agri-food production system. On the other hand, mycotoxins are generally tolerant to common food cooking and processing, thus, the existing food processes are not effective in mitigating mycotoxins in food and feed. Annual costs related to the occurrence of mycotoxins in food and feed are continuing to rise, causing the international economy to lose billions of dollars. Even more challenging, climate change may alter populations of mycotoxigenic fungi and increase levels of mycotoxins in crop production. Therefore, innovative strategies and techniques are critically needed to address the worldwide threat of mycotoxins.

One of such innovations is to mitigate mycotoxins by detoxifications, i.e., to inactivate the toxicity or to reduce the adverse effects of mycotoxins. After several decades of research on mycotoxin detoxifications, our understanding started to reach a pinnacle. Detoxifications by biological and enzymatic means have been intensively researched for almost all major mycotoxins, resulting in a great number of discoveries, technologies and even some commercial products that can be applied to reduce the adverse effects of mycotoxins. Moreover, advances in food and feed processing techniques, coupled with state-of-the-art molecular research tools, are leading the way for optimized empirical and feasible solutions.

This book correlates papers from a Special Issue of the open access journal, Toxins—Promising Detoxification Strategies to Mitigate Mycotoxins in Food and Feed. The focus of this book is to look into the most recent advances related to mitigation of mycotoxin contamination in food and feed through detoxifications. Collectively, the authors have provided many insights in the development of mycotoxin detoxifications and addressed certain critical challenges in the applications of such strategies. The book also provides comprehensive strategies with state-of-the-art tools for the future research and development in the field of mycotoxin detoxifications. It is my hope that this book will further stimulate research interest in this field and speed up the development of mycotoxin detoxifications.

Ting Zhou
Special Issue Editor

![toxins logo] *toxins*

MDPI

Editorial

Promising Detoxification Strategies to Mitigate Mycotoxins in Food and Feed

Yousef I. Hassan and Ting Zhou *

Guelph Research and Development Centre, Agriculture and Agri-Food Canada, 93 Stone Road West, Guelph, ON N1G 5C9, Canada; Yousef.Hassan@AGR.GC.CA
* Correspondence: Ting.Zhou@agr.gc.ca

Received: 19 February 2018; Accepted: 7 March 2018; Published: 9 March 2018

Mycotoxins are secondary fungal metabolites associated with adverse human health and animal productivity consequences. Annual costs connected with mycotoxin occurrences in food/feed are continuing to rise. It is estimated that close to five billion dollars are lost yearly in association with fungal infections and crop contamination with mycotoxins within the North American region alone. More recent evaluations valued losses associated with aflatoxin (AF) contaminations within the corn industry to reach as high as US$1.68 billion annually in the United States [1]. Similarly, the U.S. swine industry was reported to face current losses (in the form of weight gain reduction) due to fumonisins contamination in dried distillers' grain and solubles (DDGS) of $9 million annually [2]. This value represents only those losses attributable to one mycotoxin on one adverse outcome in one species. In Europe, deoxynivalenol (DON) is typically found in more than 50% of investigated samples [3]. When 18,884 samples collected between 2007 and 2012 from member-states of the European Union (EU) and Norway were investigated for mycotoxins, DON was found in 44.6%, 43.5%, and 75.2% of unprocessed grains, food, and feed samples, respectively [4]. The same pattern is also encountered in North Asia, with DON being the main contaminant (present in 92% of all tested samples) with average levels of 1154 ppb (part per billion) [5]. The latest multi-city survey conducted in China on the occurrence of DON in different cereal-based products indicated that more than 80% of the analyzed samples were positive with DON levels ranging between 0.1 and 2511.7 µg/kg [6].

After more than five decades of continuous mycotoxin-mitigation research, our understanding started to reach a pinnacle point where biological and enzymatic means can be used to address such toxins. Moreover, advances in food and feed processing techniques (such as cold atmospheric pressure plasma, hot air and infrared rays roasting, neutral electrolyzed water, etc.) coupled with state-of-the-art molecular research tools are leading the way for optimizing empirical and feasible solutions.

In light of the above facts, the focus of this special issue of *Toxins* was to look into the most recent advances related to mitigating mycotoxin contamination in food and feed. Multiple recent microbial and enzymatic investigations are included and many novel and promising techniques for food/feed applications are covered.

Wilson et al. [7] reported the screening of plant and soil samples for microorganisms capable of degrading trichothecenes and eventually identified two mixed cultures consistently decreasing DON levels through oxidation to 3-keto-DON. Another study screened 43 bacterial isolates and identified *Bacillus shackletonii* L7, which is capable of reducing aflatoxin B1 (AFB$_1$), AFB$_2$, and AFM$_1$ [8] where a thermostable-enzyme enriched in the culture's supernatant was purified with an estimated molecular mass of 22 kDa. Moreover, a separate study focused on elucidating how one bio-control agent, *Sporobolomyces* sp., targets and degrades patulin (PAT), a commonly encountered mycotoxin that contaminates apple and cider products [9]. The involved mechanism behind this microorganism's ability to degrade PAT was shown to be inducible with a rapid degradation of PAT, especially when the cells of this agent are exposed to low concentrations of PAT ahead of time. Furthermore, the mechanism(s) behind the degradation of PAT by another yeast isolate, *Pichia caribbica*, was examined. The collected results indicated the involvement of an enzymatic mechanism [10], while the rigorous proteomics analysis (with two-dimensional gel electrophoresis) revealed the upregulation of multiple proteins involved in the cellular metabolism and/or stress response which could be responsible for PAT degradation at the same time.

The presented special issue additionally reported on expanding the current empirical utilization of innovative mitigation strategies to control mycotoxins in actual farm settings. The use of neutral electrolyzed water to prevent aflatoxicosis in Turkey poults was among the promising studies that were shared by Gómez-Espinosa et al. [11]. As reported, alterations of serum biochemical constituents, enzyme activities, relative organs weights, and morphological changes associated with AF(s) were all mitigated by using the described neutral electrolyzed water detoxification procedure. Another novel investigation led by Bosch et al. scrutinized the use of cold atmospheric pressure plasma for the degradation of multiple mycotoxins including AAL toxin, enniatin A, enniatin B, fumonisin B_1, sterigmatocystin, DON, T2-toxin, and zearalenone (ZEA) [12]. The results reflected a significant influence of the involved mycotoxin's structure in addition to the matrix on the overall degradation rates. The results collectively indicated the suitability of the introduced approach for the decontamination of mycotoxins in food commodities where mycotoxins are confined to or enriched on surfaces such as cereal grains. Roasting with the use of infrared or static hot air was investigated for its ability to decontaminate AFs in hazelnuts [13]. Both traditional static hot-air roasting and infrared rays roasting methods were effective (85–95% reduction) when temperatures of 140 °C for 40 min were maintained, but infrared rays proved to be slightly better in this regard. More importantly, the nutritional quality and lipid profile of all tested hazelnut varieties were not affected after such roasting. Ultraviolet irradiation was also suggested to reduce AF(s) genotoxicity and carcinogenicity [14]. In order to define the final by-products of this non-specific degradation method, especially in edible oils, an Ultra Performance Liquid Chromatograph-Thermo Quadrupole Exactive Focus Mass Spectrometry/Mass Spectrometry (UPLC-TQEF-MS/MS) approach was used. The obtained high-resolution mass spectra reflected two main products while the toxicological evaluations conducted using human embryo hepatocytes indicated that these products had much lower toxicity than the parental compound, AFB_1.

The aqueous extract of hyssop, *Micromeria graeca*, was shown to halt AFB_1 production in *Aspergillus flavus*. The observed inhibitory effect was attributed to the downregulation of specific transcripts within the AF biosynthesis pathway [15]. The proposed approach falls well into green farming practices aiming at reducing the use of fungicides. Similarly, the ability of curcumin to prevent AFB_1 hepatoxicity was reported. The alleviation in the typical symptoms associated with AFB_1-induced hepatotoxicity due to curcumin inclusion/supplementation was attributed mainly to the pivotal inhibition of CYP450 isozyme-mediated activation of AFB_1 to toxic AFBO [16].

The detailed *in silico* analysis of a laccase (and two different isoforms) capable of degrading AFB_1 and AFM_1 was reported in this special issue [17]. This interesting investigation helped in pinpointing the structural differences among the three studied isoforms and highlighting the most suitable isoform for future protein engineering approaches.

An exciting report about using *Bacillus subtilis* ANSB060 to ameliorate the negative effects of AFs in ducks is presented [18]. The bacterium was originally isolated from fish gut and showed the ability to protect the growth performance of Cherry Valley ducks fed with moldy maize naturally contaminated with AFs. In a parallel fashion, the ability of *Bacillus licheniformis* CK1 to protect post-weaning gilts from ZEA-contaminated feed was demonstrated. The capability of this bacterium to degrade ZEA was associated with the reported protection mechanism [19]. Finally, the ability of sporoderm-broken spores of *Ganoderma lucidum* to enhance the immune function and maintain the growth performance of broiler chickens exposed to AFB_1 was detailed. The results showed that diets contaminated with a low level of AFB_1 can be consumed without any negative consequences as long as they are supplemented with the sporoderm-broken spores of *G. lucidum* [20]. Moreover, the introduced treatment was able to restore the normal levels of IgA and IgG in the serum of chickens exposed to AFB_1.

The enzymatic modifications of DON occupied a considerable part of this issue, particularly the C3 chemical group modifications. First, Tian et al. suggested that the glycosylation of this group is part of the self-protection mechanism(s) possessed by multiple *Trichoderma* strains serving as antagonists towards *Fusarium graminearum* growth [21]. Similarly, Hassan et al. [22] explored the epimerization of the above group (C3) and its influence on the molecular interactions of DON and its C3 stereoisomer (3-*epi*-DON) with well-defined enzymes such as Tri101 acetyltransferases to conclude that the associated changes within the involved –OH group not only influence DON's toxicity but also increase the overall polarity of this toxin as well as changing its acetylation patterns [22].

This issue also encompasses some excellent in-depth reviews. The review shared by Loi et al. covered mycotoxins biotransformation by native and commercial enzymes [23], covering purified enzymes isolated from bacteria, fungi, and plants with validated potentialities using *in vitro* and *in vivo* methods and setting examples for applications in food, feed, biogas, and biofuel industries. Zhu et al. brought attention to the most recent strategies and methodologies for developing microbial detoxification systems to mitigate mycotoxins [24], highlighting the tremendous and unexpected challenges facing any progress in this regard including the isolation of single colonies harboring the reported biotransformation activity and the assessment of the cellular toxicity of final biotransformation by-products. The review prepared by Hojnik et al. was dedicated to explore the use of cold atmospheric pressure plasma to decontaminate mycotoxins [25], presenting the advantages of this approach (cost efficiency, ecologically-friendly, negligible influence on food quality and attributes) which may overcome many weaknesses associated with the conventional/classical methods of inactivation. Finally, the mitigation of PAT in fresh and processed food commodities including beverages was discussed by Ioi et al. in a separate review [26] that covered the pre-processing stage (storage conditions, use of fungicides, and the physical removal of fungi and infected tissues). The review further detailed the effects of common processing techniques (including pasteurization, filtration, and fermentation) on PAT and reviewed non-thermal methods (such as high hydrostatic pressure, UV radiation, enzymatic degradations, and binding to microorganisms) to remove or detoxify PAT.

Overall, we are thrilled to present the above genuine contributions with the diligent work of many involved teams that collectively aimed at addressing some of the main challenges that remain within the mycotoxin-mitigation arena using both integrative and innovative approaches. The promising outcomes of this research focus create a foundation to use recombinant enzymes/proteins for the detoxification of many agriculturally-important mycotoxins, including AFs, PAT, and DON. Moreover, the use of innovative processing techniques (such as infrared roasting, non-ionizing radiations, cold atmospheric pressure plasma, and neutral electrolyzed water) will greatly enhance the safety of numerous food/feed commodities with diverse physical attributes/chemical compositions.

Acknowledgments: Appreciation is due to all the authors who shared their cutting-edge research findings with the readers of this special issue. The rigorous and in-depth evaluation of the submitted manuscripts carried out by our expert peer-reviewers made this special issue possible. The valuable contributions, organization, and editorial support of the MDPI management team and staff cannot be ignored for the overall success of this humble effort to push the boundaries of human knowledge in the correct direction in regard to fighting mycotoxins and guaranteeing safer and more wholesome food/feed commodities of future generations.

Conflicts of Interest: The authors declare no conflict of interest.

References

1. Mitchell, N.J.; Bowers, E.; Hurburgh, C.; Wu, F. Potential economic losses to the US corn industry from aflatoxin contamination. *Food Addit. Contam. Part A Chem. Anal. Control Expo. Risk Assess.* **2016**, *33*, 540–550. [CrossRef] [PubMed]
2. Wu, F.; Munkvold, G.P. Mycotoxins in ethanol co-products: Modeling economic impacts on the livestock industry and management strategies. *J. Agric. Food Chem.* **2008**, *56*, 3900–3911. [CrossRef] [PubMed]
3. Streit, E.; Schatzmayr, G.; Tassis, P.; Tzika, E.; Marin, D.; Taranu, I.; Tabuc, C.; Nicolau, A.; Aprodu, I.; Puel, O.; et al. Current situation of mycotoxin contamination and co-occurrence in animal feed—focus on Europe. *Toxins* **2012**, *4*, 788–809. [CrossRef] [PubMed]

4. European Food Safety Authority. Deoxynivalenol in food and feed: Occurrence and exposure. *EFSA J.* **2013**, *11*, 3379.
5. Rodrigues, I.; Naehrer, K. A three-year survey on the worldwide occurrence of mycotoxins in feedstuffs and feed. *Toxins* **2012**, *4*, 663–675. [CrossRef] [PubMed]
6. Woo, C.S.J.; El-Nezami, H. Mycotoxins in Asia: Is China in danger? *Qual. Assur. Saf. Crop. Foods* **2015**, *7*, 3–25. [CrossRef]
7. Wilson, N.M.; McMaster, N.; Gantulga, D.; Soyars, C.; McCormick, S.P.; Knott, K.; Senger, R.S.; Schmale, D.G. Modification of the Mycotoxin Deoxynivalenol Using Microorganisms Isolated from Environmental Samples. *Toxins* **2017**, *9*, 141. [CrossRef] [PubMed]
8. Xu, L.; Eisa Ahmed, M.F.; Sangare, L.; Zhao, Y.; Selvaraj, J.N.; Xing, F.; Wang, Y.; Yang, H.; Liu, Y. Novel Aflatoxin-Degrading Enzyme from Bacillus shackletonii L7. *Toxins* **2017**, *9*, 36. [CrossRef] [PubMed]
9. Ianiri, G.; Pinedo, C.; Fratianni, A.; Panfili, G.; Castoria, R. Patulin Degradation by the Biocontrol Yeast Sporobolomyces sp. Is an Inducible Process. *Toxins* **2017**, *9*, 61. [CrossRef] [PubMed]
10. Zheng, X.; Yang, Q.; Zhang, H.; Cao, J.; Zhang, X.; Apaliya, M.T. The Possible Mechanisms Involved in Degradation of Patulin by Pichia caribbica. *Toxins* **2016**, *8*, 289. [CrossRef] [PubMed]
11. Gomez-Espinosa, D.; Cervantes-Aguilar, F.J.; Del Río-García, J.C.; Villarreal-Barajas, T.; Vázquez-Durán, A.; Méndez-Albores, A. Ameliorative Effects of Neutral Electrolyzed Water on Growth Performance, Biochemical Constituents, and Histopathological Changes in Turkey Poults during Aflatoxicosis. *Toxins* **2017**, *9*, 104. [CrossRef] [PubMed]
12. Ten Bosch, L.; Pfohl, K.; Avramidis, G.; Wieneke, S.; Viöl, W.; Karlovsky, P. Plasma-Based Degradation of Mycotoxins Produced by Fusarium, Aspergillus and Alternaria Species. *Toxins* **2017**, *9*, 97. [CrossRef] [PubMed]
13. Siciliano, I.; Dal Bello, B.; Zeppa, G.; Spadaro, D.; Gullino, M.L. Static Hot Air and Infrared Rays Roasting are Efficient Methods for Aflatoxin Decontamination on Hazelnuts. *Toxins* **2017**, *9*, 72. [CrossRef] [PubMed]
14. Mao, J.; He, B.; Zhang, L.; Li, P.; Zhang, Q.; Ding, X.; Zhang, W. A Structure Identification and Toxicity Assessment of the Degradation Products of Aflatoxin B(1) in Peanut Oil under UV Irradiation. *Toxins* **2016**, *8*, 332. [CrossRef] [PubMed]
15. El Khoury, R.; Caceres, I.; Puel, O.; Bailly, S.; Atoui, A.; Oswald, I.P.; El Khoury, A.; Bailly, J.D. Identification of the Anti-Aflatoxinogenic Activity of Micromeria graeca and Elucidation of Its Molecular Mechanism in Aspergillus flavus. *Toxins* **2017**, *9*, 87. [CrossRef] [PubMed]
16. Zhang, N.Y.; Qi, M.; Zhao, L.; Zhu, M.K.; Guo, J.; Liu, J.; Gu, C.Q.; Rajput, S.A.; Krumm, C.S.; Qi, D.S.; et al. Curcumin Prevents Aflatoxin B(1) Hepatoxicity by Inhibition of Cytochrome P450 Isozymes in Chick Liver. *Toxins* **2016**, *8*, 327. [CrossRef] [PubMed]
17. Dellafiora, L.; Galaverna, G.; Reverberi, M.; Dall'Asta, C. Degradation of AflaToxins by Means of Laccases from Trametes versicolor: An. In Silico Insight. *Toxins* **2017**, *9*, 17. [CrossRef] [PubMed]
18. Zhang, L.; Ma, Q.; Ma, S.; Zhang, J.; Jia, R.; Ji, C.; Zhao, L. Ameliorating Effects of Bacillus subtilis ANSB060 on Growth Performance, Antioxidant Functions, and Aflatoxin Residues in Ducks Fed Diets Contaminated with Aflatoxins. *Toxins* **2016**, *9*, 1. [CrossRef] [PubMed]
19. Fu, G.; Ma, J.; Wang, L.; Yang, X.; Liu, J.; Zhao, X. Effect of Degradation of Zearalenone-Contaminated Feed by Bacillus licheniformis CK1 on Postweaning Female Piglets. *Toxins* **2016**, *8*, 300. [CrossRef] [PubMed]
20. Liu, T.; Ma, Q.; Zhao, L.; Jia, R.; Zhang, J.; Ji, C.; Wang, X. Protective Effects of Sporoderm-Broken Spores of Ganderma lucidum on Growth Performance, Antioxidant Capacity and Immune Function of Broiler Chickens Exposed to Low Level of Aflatoxin B(1). *Toxins* **2016**, *8*, 278. [CrossRef] [PubMed]
21. Tian, Y.; Tan, Y.; Liu, N.; Yan, Z.; Liao, Y.; Chen, J.; de Saeger, S.; Yang, H.; Zhang, Q.; Wu, A. Detoxification of Deoxynivalenol via Glycosylation Represents Novel Insights on Antagonistic Activities of Trichoderma when Confronted with Fusarium graminearum. *Toxins* **2016**, *8*, 335. [CrossRef] [PubMed]
22. Hassan, Y.I.; Zhu, H.; Zhu, Y.; Zhou, T. Beyond Ribosomal Binding: The Increased Polarity and Aberrant Molecular Interactions of 3-epi-deoxynivalenol. *Toxins* **2016**, *8*, 261. [CrossRef] [PubMed]
23. Loi, M.; Fanelli, F.; Liuzzi, V.C.; Logrieco, A.F.; Mulè, G. Mycotoxin Biotransformation by Native and Commercial Enzymes: Present and Future Perspectives. *Toxins* **2017**, *9*, 111. [CrossRef] [PubMed]
24. Zhu, Y.; Hassan, Y.I.; Lepp, D.; Shao, S.; Zhou, T. Strategies and Methodologies for Developing Microbial Detoxification Systems to Mitigate MycoToxins. *Toxins* **2017**, *9*, 130. [CrossRef] [PubMed]

25. Hojnik, N.; Cvelbar, U.; Tavčar-Kalcher, G.; Walsh, J.L.; Križaj, I. Mycotoxin Decontamination of Food: Cold Atmospheric Pressure Plasma versus "Classic" Decontamination. *Toxins* **2017**, *9*, 151. [CrossRef] [PubMed]
26. Ioi, J.D.; Zhou, T.; Tsao, R.; Marcone, M.F. Mitigation of Patulin in Fresh and Processed Foods and Beverages. *Toxins* **2017**, *9*, 157. [CrossRef] [PubMed]

Article

Modification of the Mycotoxin Deoxynivalenol Using Microorganisms Isolated from Environmental Samples

Nina M. Wilson [1], Nicole McMaster [1], Dash Gantulga [1], Cara Soyars [2], Susan P. McCormick [3], Ken Knott [4], Ryan S. Senger [5,6] and David G. Schmale [1,*]

[1] Department of Plant Pathology, Physiology, and Weed Science, Virginia Tech, Blacksburg, VA 24061, USA; nina09@vt.edu (N.M.W.); niki@vt.edu (N.M.); gantulga@vt.edu (D.G.)
[2] Biology Department, The University of North Carolina at Chapel Hill, Chapel Hill, NC 27599, USA; csoyars@live.unc.edu
[3] USDA-ARS, Mycotoxin Prevention and Applied Microbiology, Peoria, IL 61604, USA; susan.mccormick@ars.usda.gov
[4] Department of Chemistry, Virginia Tech, Blacksburg, VA 24061, USA; kknott@vt.edu
[5] Department of Biological Systems Engineering, Virginia Tech, Blacksburg, VA 24061, USA; senger@vt.edu
[6] Department of Chemical Engineering, Virginia Tech, Blacksburg, VA 24061, USA
* Correspondence: dschmale@vt.edu; Tel.: +1-540-231-6943

Academic Editor: Ting Zhou
Received: 2 March 2017; Accepted: 11 April 2017; Published: 15 April 2017

Abstract: The trichothecene mycotoxin deoxynivalenol (DON) is a common contaminant of wheat, barley, and maize. New strategies are needed to reduce or eliminate DON in feed and food products. Microorganisms from plant and soil samples collected in Blacksburg, VA, USA, were screened by incubation in a mineral salt media containing 100 µg/mL DON and analysis by gas chromatography mass spectrometry (GC/MS). Two mixed cultures derived from soil samples consistently decreased DON levels in assays using DON as the sole carbon source. Nuclear magnetic resonance (NMR) analysis indicated that 3-keto-4-deoxynivalenol was the major by-product of DON. Via 16S rRNA sequencing, these mixed cultures, including mostly members of the genera *Acinetobacter*, *Leadbetterella*, and *Gemmata*, were revealed. Incubation of one of these mixed cultures with wheat samples naturally contaminated with 7.1 µg/mL DON indicated nearly complete conversion of DON to the less toxic 3-epimer-DON (3-epi-DON). Our work extends previous studies that have demonstrated the potential for bioprospecting for microorganisms from the environment to remediate or modify mycotoxins for commercial applications, such as the reduction of mycotoxins in fuel ethanol co-products.

Keywords: mycotoxin; trichothecene; deoxynivalenol; bioprospecting; detoxification; *Fusarium*

1. Introduction

Mycotoxins are toxic secondary metabolites produced by fungi that are a threat to the health of humans and domestic animals [1]. This diverse class of compounds can contaminate commercial foods (e.g., wheat, maize, peanuts, cottonseed, and coffee) and animal feedstocks. Mycotoxins can be harmful even at small concentrations, creating significant food safety concerns [1,2]. The Food and Agriculture Organization estimated that approximately 1 billion metric tons of food is lost each year due to mycotoxin contamination [3]. Economic losses include yield loss from mycotoxin contamination [4], reduced value of crops [4], loss of animal productivity from health issues related to mycotoxin consumption [5], and even animal death [6,7].

The trichothecenes are a major class of mycotoxins containing over 150 toxic compounds and are toxic inhibitors of protein synthesis [8,9]. Trichothecenes are produced by several different fungi in the genus *Fusarium* [9,10]. One of the most economically important trichothecenes is deoxynivalenol (DON), which contaminates wheat, barley, and maize worldwide [11]. DON causes feed refusal, skin disorder, diarrhea, reduced growth, and vomiting in domestic animals [12]. Depending on the dose and exposure time of DON, there is also evidence that DON acts as an immunosuppressive [1]. It is among the most closely monitored mycotoxins in the US, and DON contaminations have resulted in estimated annual losses of up to $1.6 billion [13].

While there is structural variety, all trichothecenes share a core structure that includes the C-12,13 epoxide that is important to toxicity and protein inhibition [14,15]. DON is a type B trichothecene characterized by the presence of a keto group on C-8 [16]. There are mechanisms the fungus *Fusarium* implements during the biosynthesis of DON to alter the structure, making it less toxic, e.g., acetylating the C-3 position [16].

Microbial detoxification of mycotoxins has previously been reported [17,18]. Fuchs et al. [19] were able to isolate an anaerobic eubacterium that converted DON to de-epoxy-DON. A few years later, Völkl and colleagues [20] reported that a mixed culture of organisms from soil samples converted DON to 3-keto-4-deoxynivalenol (3-keto-DON), but they were unable to identify the causal microorganisms responsible for the modification. The product 3-keto-DON is approximately 90% less toxic than DON, and represents a suitable detoxified product [21]. Shima et al. [21] discovered a single organism in aerobic conditions from an environmental sample that converted DON into 3-keto-DON, and He et al. [22,23] isolated an aerobic organism, from the genus *Devosia*, converting DON to 3-epimer-DON (3-epi-DON). Ikunaga et al. [24] identified a bacterium from the genus *Nocardioides* that converts DON to 3-epi-DON. Recently, He et al. [25] discovered an aerobic culture of microorganisms converting DON to de-epoxy-DON. The current study extends these prior investigations to a series of studies to isolate additional microorganisms from the environment that modify and remediate DON. While others have shown that soil bacteria can detoxify DON, the functional enzyme(s) responsible for conversion to 3-keto-DON remains elusive. Once the enzymatic mechanism(s) and genetic element(s) responsible are identified, yeast can be engineered to remediate DON during a fermentation process involving mycotoxin-contaminated feedstocks.

Based on previous work [21–25], we hypothesized that mixed cultures of microorganisms isolated from natural soil environments incubated with a mineral salt media using 100 μg/mL DON as the sole carbon source will detoxify DON. The specific objectives of this research were as follows: (1) identify microbes isolated from plant and soil samples taken in Blacksburg, VA, that modify DON; (2) characterize DON metabolites using thin layer chromatography (TLC), gas chromatography mass spectrometry (GC/MS), and nuclear magnetic resonance (NMR); (3) identify bacterial components of mixed cultures with DON modification activity; and (4) determine if these microorganisms can modify DON in naturally contaminated wheat samples. Our work extends previous studies that have demonstrated the potential for bioprospecting for microbes that modify toxic secondary metabolites from grains and/or grain products, such as the reduction of mycotoxins in fuel ethanol co-products.

2. Results

2.1. Selection of Microbes in the Presence of High Concentrations of DON

An initial screen of 11 plant and soil environmental samples incubated in mineral media containing 100 μg/mL DON as the sole carbon source identified five cultures in which no DON remained after 7 days. These five mixed culture samples that eliminated DON from the culture media (below the limit of quantification (<LOQ), which was 0.2 μg/mL) came from soil samples taken from a landscape plot, vineyard, and peach orchard and from plant samples taken in a small grain field and a vineyard. With further subculturing, three mixed culture samples had decreased DON levels in the culture media (Table S1), all of which were derived from the landscape plot. Only two samples from the landscape plot, Mixed Cultures 1 and 2 (Figure S1), consistently removed/modified DON in the culture media. Further assays with Mixed Cultures 1 and 2 (Table 1) suggested that the glycerol stocks

were heterogeneous and likely contained mixtures of culturable and unculturable microbes; four sample replicates from Mixed Cultures 1 and 2 did not perform the same and had varying amounts of DON modification based on percentage of DON modified in each culture (Table 1). Mixed Culture 3 replicates did not modify DON and thus were not studied further.

Table 1. DON modification by soil mixed cultures. Four sample replicates from Mixed Culture 1, 2, and 3 following incubation with 5 µg/mL DON.

Culture Sample	Replicate	DON (µg/mL) Analytical Rep 1	DON (µg/mL) Analytical Rep 2	Mean DON (µg/mL)
Mixed Culture 1-R1	1	0.16	0.16	0.16
Mixed Culture 1-R2	2	2.48	2.72	2.6
Mixed Culture 1-R3	3	3.12	3.04	3.08
Mixed Culture 1-R4	4	2.72	2.68	2.7
Mixed Culture 2-R1	1	<0.2	<0.2	<0.2
Mixed Culture 2-R2	2	4.32	3.8	4.06
Mixed Culture 2-R3	3	<0.2	<0.2	<0.2
Mixed Culture 2-R4	4	<0.2	<0.2	<0.2
Mixed Culture 3-R1	1	3.92	3.8	3.86
Mixed Culture 3-R2	2	3.92	3.76	3.84
Mixed Culture 3-R3	3	4.08	3.8	3.94
Mixed Culture 3-R4	4	3.28	3.64	3.46
Control-R1	1	4.46	4.48	4.47
Control-R2	2	4.48	4.28	4.38
Control-R3	3	3.84	4.04	3.94

2.2. Isolation of Individual DON Modifying Microbes

There were two pure cultures, Pure Cultures 1 and 2 from Table S1, of bacteria that were initially associated with decreased levels of DON within culture media. Pure Culture 1 originated from the small grain field, and Pure Culture 2 originated from the landscape plot. Via 16S rRNA sequencing, it was revealed that Pure Culture 1 was from the genus *Achromobacter*, and the Pure Culture 2 was from the genus *Pseudomonas*. However, additional DON assays indicated that both Pure Cultures 1 and 2 did not consistently modify DON (data not shown). The two pure cultures were inconsistent in their ability to eliminate DON from culture media, although preparation and culture conditions remained the same.

2.3. Identification of Mixed Cultures Using 16S Ribosomal Sequencing

Sequencing of 16S of Mixed Cultures 1 and 2, which consistently modified DON, indicated that they contained mostly members of the genera *Acinetobacter*, *Leadbetterella*, and *Gemmata* (Figure 1). Mixed Culture 1 consisted of members of the genera *Acinetobacter*, while Mixed Culture 2 was composed mostly of the genera *Leadbetterella*, and *Gemmata*. Mixed Culture 1-R1 was the only culture able to modify DON in culture media (Figure 1a). Mixed Culture 1-R1 was composed mostly of the genera *Acinetobacter* and *Candidatus*. Mixed Culture 2-R1, Mixed Culture 2-R3, and Mixed Culture 2-R4 modified DON (Figure 1b). Mixed Culture 2-R2 was unable to modify DON in culture media and was the only culture that contained a large amount of *Burkholderia*.

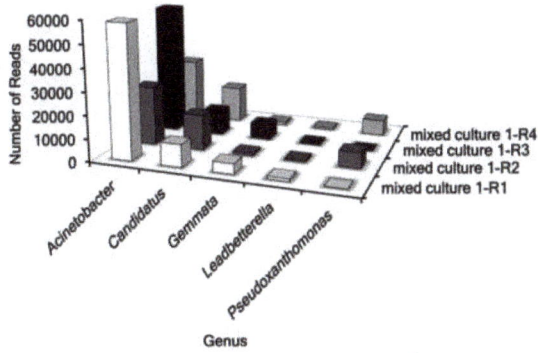

(a)

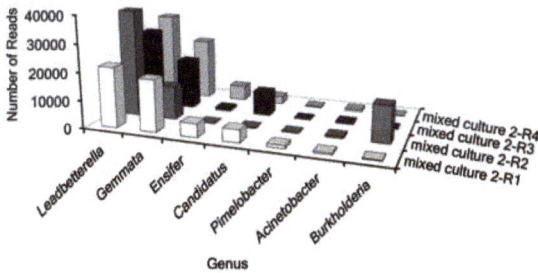

(b)

Figure 1. 16S rRNA sequencing data for within sample repetitions of (**a**) Mixed Culture 1 and (**b**) Mixed Culture 2. Each plot graph illustrates the major genera represented in each sample in accordance with number of reads resolved from sequencing. In (**a**) Mixed Culture 1-R1 was the only culture that was able to modify DON from the culture media. In (**b**) Mixed Culture 2-R1, Mixed Culture 2-R3, and Mixed Culture 2-R4 were able to modify DON from the culture media.

2.4. Thin Layer Chromatography to Identify DON Derivatives

TLC analysis of extracts of Mixed Cultures 1 and 2 showed a DON by-product that was less polar than DON, 15-ADON, and 3-A-DON (Figure 2). Mixed Cultures 1 and 2 by-products showed similar properties to 3-keto-DON, which is nonpolar and migrated further up the TLC plate. Mixed Cultures 1 and 2 by-products were also dissimilar to the acetylated versions of DON, 15-A-DON, and 3-A-DON. The product turned blue with NBP/TEPA, indicating that it contained an epoxide, and had a similar Rf (retention factor) as 3-keto DON.

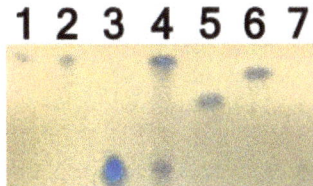

Figure 2. TLC analysis of DON by-products and controls. Lane 1: Mixed Culture 1; Lane 2: Mixed Culture 2; Lane 3: DON; Lane 4: 3-keto-DON and DON mixture; Lane 5: 15-A-DON; Lane 6: 3-A-DON: and Lane 7: De-epoxy-DON.

2.5. Nuclear Magnetic Resonance to Identify Structure of DON Derivatives

NMR analysis showed that both the DON by-product in both Mixed Culture 1 and Mixed Culture 2 was 3-keto-deoxynivalenol (3-keto-DON) (Table S2). The 3-keto-DON proton closely resembled that reported by Shima et al. [21].

2.6. DON Assays with Mixed Cultures with Naturally Contaminated Wheat Samples

DON assays using Wheat Sample #13w-7 (41.0 µg/mL) did not show any DON reduction with Mixed Cultures 1 or 2. DON reduction was observed with Wheat Sample #13-v193 (7.1 µg/mL DON). In particular, two samples from Mixed Culture 1 (Sample ID 2 and Sample ID 3 in Table 2) showed nearly complete DON reduction compared to the control. Figure 3 shows a GC/MS chromatogram overlay of the DON control and Sample ID 2 from Mixed Culture 1, where significant DON reduction was observed. The DON peak had a retention time of 6.12 min, and the new peak at 6.33 min is postulated to be 3-epi-DON based on molecular weight and fragmentation. According to He et al. [22], 3-epi-DON is significantly less toxic than DON. However, Ikunaga et al. [24] suggest that 3-epi-DON may still be just as toxic as DON since the epoxide ring is still present. DON and the postulated 3-epi-DON were detected in SIM mode with a target ion with a mass/charge ratio of 512.3 and had reference ions at 422.4 and 497.3.

Table 2. Grain culture extracts from naturally contaminated Wheat Sample #13v-193 (7.1 µg/mL DON) incubated with Mixed Cultures 1 and 2 were analyzed using GC/MS. Two separate assays were performed at different times with three replicates for Mixed Cultures 1 and 2 and the control. Sample ID 2 from Mixed Culture 1 in the first assay using Wheat Sample #13v-193 showed significant DON reduction compared to the negative control (below the limit of quantification, which is 0.20 µg/mL).

Sample ID	Culture ID	Assay	Replicate	Starting DON (µg/mL) in Wheat	Final DON (µg/mL) in Wheat
1	Mixed Culture 1	1	1	7.10	5.08
2	Mixed Culture 1	1	2	7.10	<0.20
3	Mixed Culture 1	1	3	7.10	0.08
4	Mixed Culture 1	2	1	7.10	7.04
5	Mixed Culture 1	2	2	7.10	7.76
6	Mixed Culture 1	2	3	7.10	5.88
					4.3 (mean)
7	Mixed Culture 2	1	1	7.10	4.0
8	Mixed Culture 2	1	2	7.10	6.4
9	Mixed Culture 2	1	3	7.10	3.48
10	Mixed Culture 2	2	1	7.10	7.56
11	Mixed Culture 2	2	2	7.10	7.88
12	Mixed Culture 2	2	3	7.10	7.28
					6.1 (mean)
Control	Control (no cultures)				7.10 (mean)

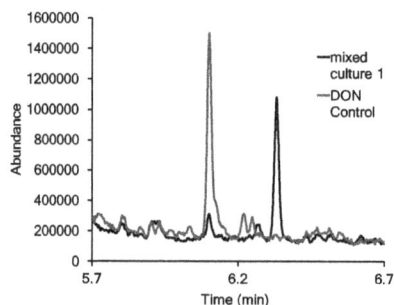

Figure 3. GC/MS chromatograph of DON derivatives in scan operating mode from Sample ID 2 in Table 2 (Mixed Culture 1, Wheat Sample #13w-7, 7.1 µg/mL DON). DON (grey line; retention time of 6.12 min) was modified to 3-epi-DON (black line; retention time of 6.33 min).

3. Discussion

New strategies are needed to reduce or eliminate DON in feed and food products. DON degrading activity restricted to anaerobic organisms limits the potential use of these microorganisms for industrial purposes as feed additives. Aerobic organisms pose their own problems, as many valuable organisms are unculturable in the lab. Even though culturing aerobic organisms from the environment can be tedious, Shima et al. [21] discovered Strain E3-39, which can convert DON into 3-keto-DON under aerobic conditions, and He et al. [25] discovered a microbial culture that converted DON to de-epoxy-DON. Here, we extend these prior studies by isolating mixed microbial cultures from the environment that modify DON, characterizing their DON derivatives, and characterizing the microorganisms present in cultures that modify DON.

Two mixed cultures that consistently decreased DON in cultures, in which DON was the sole carbon source in a minimal medium, were identified. From these mixed cultures, we were unable to isolate a pure culture that modified DON consistently in culture media. Several bacteria and fungal colonies were initially selected and screened for DON modification, but only two bacteria eliminated DON from cultures containing DON as the sole carbon source. These two bacteria, an *Achromobacter* and *Pseudomonas* species, were not consistent in eliminating DON from cultures. This inconsistency could be due to the cultures being stored in glycerol stocks at −80 °C, since the cold temperatures may have affected their ability to modify DON. Völkl et al. [20] also were unable to identify a pure organism from the mixed culture (D107) that consistently modified DON. Isolating pure cultures remains a challenge, as multiple microorganisms could be responsible for the metabolism or conversion of DON.

According to Shima et al. [19], 3-keto-DON is significantly less toxic than DON. Proton data from Mixed Cultures 1 and 2 were similar to 3-keto-DON reported by Shima et al. [21]. There is a discrepancy in the literature regarding proton data for 3-keto-DON reported by Völkl et al. [20]; the authors appear to have inadvertently switched some of the proton data for DON with the proton data for 3-keto-DON.

Results from 16S rRNA sequencing of Mixed Cultures 1 and 2, which consistently modified DON, indicated that they contained mostly members of the genera *Acinetobacter*, *Leadbetterella*, and *Gemmata*. To our knowledge, these genera have not been reported previously to modify DON in culture. Strains of *Acinetobacter* have been associated with the modification of ochratoxin A [26]. He et al. reported *Pseudomonas* and *Achromobacter* genera in their microbial culture that converted DON to de-epoxy-DON; the two pure cultures we isolated that demonstrated activity in DON cultures could have lost functionality during storage in glycerol stocks at −80 °C. Isolating DON modifying microbes is difficult, in part due to growth and function restrictions since some microbes may be inhibited by others [27]. Several microorganisms are likely responsible for the conversion of DON to 3-keto-DON, and with additional testing and analysis it may be possible to isolate the specific bacteria responsible for DON modification.

DON was nearly eliminated in two naturally contaminated samples of wheat (7.1 µg/mL DON) inoculated with Mixed Culture 1. GC/MS scans of the two samples showed the appearance of a peak with a similar mass/charge ratio as DON, but different retention times. This was postulated to be the DON metabolite, 3-epi-DON. A reduction of DON was not observed with the samples contaminated with a higher concentration of DON (41 µg/mL). The observed differences in the modification of DON in the assay with DON as the sole carbon source and the assay using naturally contaminated sources of wheat could be attributed to the naturally contaminated sources of wheat cultures containing additional carbon sources for the microbes to utilize. He et al. [28] were able to produce 3-epi-DON with their strain of *Devosia* with different carbon sources such as corn meal broth and a mixture of yeast and glucose. 3-Epi-DON may have been produced from our naturally contaminated sources of wheat, since 3-keto-DON may be further reduced to produce 3-epi-DON [29].

Our work extends previous studies that have demonstrated the potential to use mixed cultures of microbes to detoxify DON. Future work to assess how microbial assemblages change before, during, and after screening with DON will highlight what microorganisms are selected for under the pressure of DON. Additional work needs to be done to culture specific microorganisms that are unable to grow under the test conditions to greatly increase the probability of identifying an organism that can detoxify DON (e.g., the use of the iChip to identify the new antibiotic allowing scientists to screen for new microbes that are difficult to culture or unculturable with traditional laboratory practices) [30]. However, the transformation of DON to 3-keto-DON and to 3-epi-DON with our mixed cultures demonstrates the feasibility of our approach. Future work will aim to elucidate enzymes responsible for modification of DON as Chang et al. did by modifying ochratoxin A using a carboxypeptidase enzyme from the species *Bacillus amyloliquefaciens* ASAG1 [31]. Engineering yeast that express DON-detoxifying enzymes and/or adding purified enzymes that can convert DON to less toxic by-products will be of value in the fuel ethanol industry, where such strategies could reduce mycotoxins in fuel ethanol by-products destined for feed and food [27].

4. Materials and Methods

4.1. Field Collections

Plant and soil samples were collected at Virginia Tech's Kentland Farm in Blacksburg VA, USA, on 13 September 2013. Microbial samples were collected from fresh leaves or plant debris, and from soil samples collected with a soil corer (10 × 1 in. diameter galvanized steel soil sampler, Zoro). Eleven samples were collected from six different collection sites including a field of oat (*Avena sativa*), a field of corn (*Zea mays*), a landscape plot, a vineyard (*Vitis vinifera*), a peach orchard (*Prunus persica*), and an apple orchard (*Malus domestica*).

4.2. Selection of Microbes in the Presence of High Concentrations of DON

Plant samples were ground into a fine powder using a coffee grinder (Hamilton Beach, Model 80365, Southern Pines, NC, USA). Soil samples were placed in one-gallon zip lock bags and mixed thoroughly to displace soil clumps. Aliquots of 0.1 g of each sample were suspended in 1 mL of mineral salt medium (MM) [32] containing 100 µg/mL of DON as the sole carbon source. A negative control included 1 mL MM and 100 µg/mL DON without any environmental samples. Cultures were incubated on a shaker (New Brunswick Scientific Excella E-24 Incubator Shaker, Edison, NJ, USA) for 7 days at 120 RPM and 28 °C. After 7 days, 10 µL of each culture was added to 1 mL of MM and 100 µg/mL of DON and incubated for another 7 days under the same conditions. This process of subculturing and incubation was repeated four additional times for a total of six weeks.

Resulting cultures were screened for the disappearance of DON using gas chromatography/mass spectrometry (GC/MS) following standard protocols [33]. Each sample was diluted by adding 100 µL of culture to 1.9 mL of sterile water before GC/MS analysis; 250 µL of the dilution was added to 1.7 mL of acetonitrile and filtered through Whatman 1 qualitative paper. A 1 mL portion of the flow through was dried down in a glass tube using compressed air in a nitrogen evaporator set at 55 °C. Dried samples were then derivatized at room temperature with a mixture of 99 µL of N-trimethylsilylimidazole (TMSI) and 1 µL of trimethylchlorosilane (TMCS) for 20 min. Then, 500 µL of isooctane containing 0.5 µg/g of mirex (Sigma-Aldrich, St. Louis, MO, USA) was added to the glass tube and immediately vortexed, followed by an addition of 500 µL of water to quench the reaction. From the top organic layer, 150 µL was transferred to chromatography vials for GC/MS analysis.

An Agilent 6890/5975 system was used for GC/MS analysis operating in selected ion monitoring (SIM) mode. An autosampler in splitless mode injected 1 µL of each sample onto an HP-5MS column (0.25 mm inner-diameter, 0.25 µm film thickness, 30 m length) to detect DON. The inlet temperature was set at 280 °C with a column flow rate of 1.2 mL/min using helium. The initial column temperature was held at 150 °C for 1 min, increased to 280 °C at a rate of 30 °C/min, and held constant for 3.5 min. A post-run of 325 °C for 2.5 min was used to clean the column. DON was detected in SIM mode at a mass/charge ratio of 512.3 and had reference ions at 422.4 and 497.3. Mirex (hexachloropentadiene dimer) was used as an internal standard to check the quantitative precision of the instrument and was detected in SIM mode at a mass/charge ratio of 271.8 and had a reference ion of 275.8 [34]. A linear regression model was used to quantify DON with standards (Romer Labs, Austria and Sigma-Aldrich, St. Louis, MO, USA) at concentrations of 0.05, 0.10, 0.25, 0.5, 1.0, 2.50, and 5.0 μg/mL. Mycotoxin values were quantitated using a standard curve ranging from 0.05 to 1.0 μg/mL. Values determined to be greater than 1.0 μg/mL were quantitated using a curve that included the 2.5 and 5.0 μg/mL standards. The LOQ for the method was determined to be 0.2 μg/mL, based on standard protocols [33]. All cultures showing decreased levels of DON were transferred to 25% glycerol and stored at −80 °C.

Each culture that demonstrated decreased levels of DON were then further analyzed 4X (quadruplicate) to show consistency of decreased DON in culture assays with 100 μg/mL DON. A 50 μL sample of the glycerol stocks mentioned above were cultured into four separate tubes of R2A (Reasoner's 2A; a medium for culturing slow-growing microorganisms; Sigma-Aldrich) liquid media for 2 days at 28 °C. After incubation, 100 μL of each culture was then added to a new culture tube containing 1 mL MM and 100 μg/mL DON and allowed to incubate with shaking at 120 RPM at 28 °C for 7 days. Mycotoxin extraction and GC/MS analysis described above was used to determine the amount of DON in each sample. All samples were made into a 25% glycerol stock and stored at −80 °C for future identification of microbes present.

4.3. Isolation of Individual DON Modifying Microbes

Mixed cultures that resulted in decreased levels of DON were selected and 200 μL of each culture was plated on solid R2A media and incubated for 7 days at 28 °C. After incubation, bacterial and fungal colonies of different morphologies were randomly selected and cultured in 1 mL of MM and 100 μg/mL DON for 7 days with shaking at 120 RPM at 28 °C. GC/MS preparation and analysis as described above was used to determine the concentration of DON in each sample after incubation. All pure cultures that demonstrated decreased levels of DON were made into a 25% glycerol stock and stored at −80 °C. Individual microbes that demonstrated decreased levels of DON were sequenced using 16S primers (27F and 518R bacterial 16S ribosomal primers) at the Biocomplexity Institute at Virginia Tech (Blacksburg, VA, USA).

4.4. Identification of Mixed Cultures Using 16S Ribosomal Sequencing

To assist in the identification of the microorganisms present in mixed cultures, 100 μL of frozen stock from the mineral media cultures with 100 μg/mL DON was incubated in 2 mL of R2A liquid media for 2 days at 28 °C. DNA from the mixed cultures was purified using a Thermo Scientific KingFisher mL nucleic acid purification machine (Thermo Scientific, Waltham, MA, USA) and Qiagen Puregene Yeast/Bac Kit B (Qiagen, Germantown, MD, USA). Samples of 40 ng/μL suspended in water were sent to MR DNA Laboratory in Shallowwater, Texas, for a diversity assay with 16S sequencing using the 27F primer. Illumina sequencing technology was used to generate an average of 20 K reads.

4.5. Thin Layer Chromatography to Identify DON Derivatives

Thin layer chromatography (TLC) was used to detect new DON products in mixed culture extracts (GC/MS analysis of TMS-derivatized samples run in SIM mode was used to measure the disappearance of DON). Samples were dried down using compressed air in the fume hood, and 200 μL of acetonitrile was then added and vortexed to ensure the DON derivatives were dissolved. Each sample was spotted 1 in. from the bottom of a 20×20 cm silica gel plate (60-F254, Millipore, Darmstadt, Germany). The plate was placed in a TLC tank (L × H × W 27.10 cm × 26.5 cm × 7.10 cm) using 48:92:10 hexane/ethyl acetate/methanol as the solvent. The solvent was allowed to run to approximately 3 cm away from the top of the plate. The plate was dried in a fume hood, then sprayed with NBP (nitrobenzylpyridine), heated for 30 min at 100 °C, and then lightly sprayed with TEPA (tetraethylenepentamine). Products that contained an epoxide group were stained blue [35].

4.6. Nuclear Magnetic Resonance to Identify Structure of DON Derivatives

In order to obtain sufficient product for nuclear magnetic resonance (NMR) analysis, samples were assayed in 10 mL of MM containing 100 μg/mL DON to produce enough by-product, incubated for one week with shaking at 28 °C, and were subsequently dried down using air. The residue was dissolved in 200 μL of acetonitrile and streaked on TLC plates as described above. Bands were visualized under UV light, marked with a pencil, and scraped off using a razor blade and placed in a glass vial. Deuterated chloroform was added to each vial and vortexed. ^{1}H NMR spectra were recorded on a Bruker Avance II (500 Mhz) equipped with a Prodigy cryoprobe. Chemical shifts were referenced to residual proton signal of the CDCl3 solvent. MNova 11 was used to analyze the ^{1}H data.

4.7. DON Assays with Mixed Cultures with Naturally Contaminated Wheat

Two wheat (*Triticum aestivum*) samples naturally infected by *Fusarium graminearum* and containing different concentrations of DON were used: Sample #13v-193 (7.1 μg/mL DON) and Sample #13w-7 (41.0 μg/mL DON). GC/MS methods were used to determine the concentrations of DON present in the samples. Samples were ground to a fine powder using a Stein mill (Steinlite Corp., Atchison, KS, USA).

To prepare for the assays, 50 μL of the mixed cultures were added to 2 mL of R2A liquid media and allowed to incubate for 2 days at 120 RPM and 28 °C. A 1.0 g sub-sample of each wheat sample was added to a 125 mL flask topped with a foam stopper (21–26 mm) and autoclaved on a dry cycle. Under sterile conditions, 4.5 mL of sterile water was added to each flask and 500 μL of each mixed culture was added to three flasks of each wheat sample, Sample #13v-193 and Sample #13w-7. Negative controls for each wheat sample were included without any mixed culture and only 5 mL of sterile water. All flasks were incubated for one week at 180 RPM and 28 °C. To analyze DON concentration after incubation, each sample was dried down in an oven set at 55 °C. Each 1.0 g sample was combined with 8 mL of an 84% (*v*/*v*) acetonitrile in DI water to extract DON, sonicated to release any clumps, then placed on a shaker at 200 RPM overnight at room temperature. The solvent was then cleaned by passing it through a column consisting of a 1:3 ratio of a 1.5 g mixture of C18 (40 um particle size) and aluminum oxide (active, neutral, 0.063 to 0.200 mm particle size range). An aliquot of 1 mL of the eluent was added to a glass test tube, dried and evaporated using a nitrogen evaporator set at 55 °C. Derivatization of samples was performed as described above and GC/MS analysis was performed operating in scan mode analyzing from 5.7 to 8.8 min; all other parameters of the GC/MS method was kept the same as described above.

Supplementary Materials: The following are available online at www.mdpi.com/2072-6651/9/4/141/s1, Table S1: Mixed and pure microbial samples, derived from either soil or plant material, that initially eliminated DON from cultures containing mineral media and 100 μg/mL of DON as the sole carbon source (GC/MS analysis indicated a value below the limit of quantification). Not all cultures were consistent at modifying DON in repeated assays. The percent of time each culture eliminated DON from the culture medium was based on how many times each culture eliminated DON from the culture over how many times each culture was assayed. Both mixed culture 1 and mixed culture 2 will modify DON in culture material 77% of the time they are assayed, Figure S1: GC/MS chromatogram of mixed culture 2 (7.7 min; detected in SIM mode with a target ion with a mass:charge ratio of 438.2 and reference ions at 318.2 and 303.1) after incubation in mineral media with 100 μg/mL of DON as the sole carbon source. DON is represented by the peak at 6.1 min. The DON incubated with mixed culture 2 was below the limit of detection of 0.20 μg/mL. Table S2: Proton data collected with a Bruker, Avance II, 500 MHz NMR for mixed culture 1 and DON. Proton data for 3-keto-DON was produced by Shima et al. [19]. Comparison of proton data for mixed culture 1 products with DON and 3-keto-DON confirm that mixed culture 1 contained 3-keto-DON. Proton data for mixed culture 2 was similar to mixed culture 1 (data not reported).

Acknowledgments: We thank the Griffey Lab at Virginia Tech for providing wheat samples contaminated with DON. This work was supported in part by grants to D. G. Schmale from the Virginia Small Grains Board (VT PANs #449102, #449282, #449426, 449552) and the U.S. Wheat and Barley Scab Initiative (VT PANs #422288, #422533). This material is based upon work supported by the U.S. Department of Agriculture. This is a cooperative project with the U.S. Wheat & Barley Scab Initiative. Any opinions, findings, conclusions, or recommendations expressed in this publication are those of the authors and do not necessarily reflect the view of the U.S. Department of Agriculture.

Author Contributions: N.M.W. and D.G.S. planned, coordinated, and conducted most of experiments and coordinated the writing of the manuscript. D.G.S., N.M.W., and C.S. conducted field and laboratory work to culture and identify DON-modifying microbes. K.K. conducted NMR assays and N.M. conducted GC/MS assays. S.P.M. provided materials for the work, and contributed to TLC methods. D.G. and R.S.S. provided oversight of the project. N.M.W. and D.G.S. wrote the manuscript, and all authors contributed edits to the manuscript.

Conflicts of Interest: The authors declare no conflict of interest.

References

1. Sobrova, P.; Adam, V.; Vasatkova, A.; Beklova, M.; Zeman, L.; Kizek, R. Deoxynivalenol and its toxicity. *Interdiscip. Toxicol.* **2010**, *3*, 94–99. [CrossRef] [PubMed]
2. Placinta, C.; D'Mello, J.P.; Macdonald, A.M. A review of worldwide contamination of cereal grains and animal feed with *Fusarium* mycotoxins. *Anim. Feed Sci. Technol.* **1999**, *78*, 21–37. [CrossRef]
3. Food and Agriculture Organization of the United Nations. *The Global Forum for Food Security and Nutrition*; FAO: Rome, Italy, 2013.
4. Zain, M.E. Impact of mycotoxins on humans and animals. *J. Saudi Chem. Soc.* **2011**, *15*, 129–144. [CrossRef]
5. Fink-Grernmels, J. Mycotoxins: Their implications for human and animal health. *Vet. Q.* **1999**, *21*, 115–120. [CrossRef] [PubMed]
6. Pier, A.C. Major biological consequences of aflatoxicosis in animal production. *J. Anim. Sci.* **1992**, *70*, 3964–3967. [CrossRef] [PubMed]
7. Pohland, A.E. Mycotoxins in review. *Food Addit. Contam.* **1993**, *10*, 17–28. [CrossRef] [PubMed]
8. Scott, P.M. Trichothecenes in grains. *Cereal Foods World* **1990**, *35*, 661–666.
9. Desjardins, A.E. *Fusarium Mycotoxins: Chemistry, Genetics, and Biology*; American Phytopathological Society (APS Press): Paul, MN, USA, 2006.
10. Schmale, D.G.; Munkvold, G.P. Mycotoxins in crops: A threat to human and domestic animal health. *Plant Health Instr.* **2009**, *3*, 340–353. [CrossRef]
11. Tanaka, T.; Hasegawa, A.; Yamamoto, S.; Lee, U.S.; Sugiura, Y.; Ueno, Y. Worldwide contamination of cereals by the *Fusarium* mycotoxins nivalenol, deoxynivalenol, and zearalenone. 1. Survey of 19 countries. *J. Agric. Food Chem.* **1988**, *36*, 979–983. [CrossRef]
12. Pestka, J.J. Deoxynivalenol: Mechanisms of action, human exposure, and toxicological relevance. *Arch. Toxicol.* **2010**, *84*, 663–679. [CrossRef] [PubMed]
13. Nganje, W.E.; Kaitibie, S.; Wilson, W.W.; Leistritz, F.L.; Bangsund, D.A. *Economic Impacts of Fusarium Head Blight in Wheat and Barley: 1993–2001*; Department of Agribusiness and Applied Economics, Agricultural Experiment Station, North Dakota State University: Fargo, ND, USA, 2004.
14. Sato, N.; Ueno, A. Comparative toxicities of trichothecenes. In *Mycotoxins in Human and Animal Health*; Rodricks, J.V., Hesseltine, C.W., Mehlman, M.A., Eds.; Pathotox Publishers: Chicago, IL, USA; pp. 295–307.
15. Ueno, Y.; Nakajima, M.; Sakai, K.; Ishii, K.; Sato, N. Comparative toxicology of trichothec mycotoxins: Inhibition of protein synthesis in animal cells. *J. Biochem.* **1973**, *74*, 285–296. [PubMed]
16. Kimura, M.; Tokai, T.; Takahashi-Ando, N.; Ohsato, S.; Fujimura, M. Molecular and genetic studies of *fusarium* trichothecene biosynthesis: Pathways, genes, and evolution. *Biosci. Biotechnol. Biochem.* **2007**, *71*, 2105–2123. [CrossRef] [PubMed]
17. McCormick, S.P. Microbial Detoxification of Mycotoxins. *J. Chem. Ecol.* **2013**, *39*, 907–918. [CrossRef] [PubMed]
18. Zhou, T.; He, J.; Gong, J. Microbial transformation of trichothecene mycotoxins. *World Mycotoxin J.* **2008**, *1*, 23–30. [CrossRef]
19. Fuchs, E.; Binder, E.M.; Heidler, D.; Krska, R. Structural characterization of metabolites after the microbial degradation of type A trichothecenes by the bacterial strain BBSH 797. *Food Addit. Contam.* **2002**, *19*, 379–386. [CrossRef] [PubMed]
20. Völkl, A.; Vogler, B.; Schollenberger, M.; Karlovsky, P. Microbial detoxification of mycotoxin deoxynivalenol. *J. Basic Microbiol.* **2004**, *44*, 147–156. [CrossRef] [PubMed]
21. Shima, J.; Takase, S.; Takahashi, Y.; Iwai, Y.; Fujimoto, H.; Yamazaki, M.; Ochi, K. Novel detoxification of the trichothecene mycotoxin deoxynivalenol by a soil bacterium isolated by enrichment culture. *Appl. Environ. Microbiol.* **1997**, *63*, 3825–3830. [PubMed]

22. He, J.W.; Bondy, G.S.; Zhou, T.; Caldwell, D.; Boland, G.J.; Scott, P.M. Toxicology of 3-epi-deoxynivalenol, a deoxynivalenol-transformation product by *Devosia* mutans 17-2-E-8. *Food Chem. Toxicol.* **2015**, *84*, 250–259. [CrossRef] [PubMed]

23. He, J.W.; Yang, R.; Zhou, T.; Boland, G.J.; Scott, P.M.; Bondy, G.S. An epimer of deoxynivalenol: Purification and structure identification of 3-epi-deoxynivalenol. *Food Addit. Contam. Part A* **2015**, *32*, 1523–1530. [CrossRef] [PubMed]

24. Ikunaga, Y.; Sato, I.; Grond, S.; Numaziri, N.; Yoshida, S.; Yamaya, H.; Hiradate, S.; Hasegawa, M.; Toshima, H.; Ito, M.; et al. *Nocardioides* sp. strain WSN05-2, isolated from a wheat field, degrades deoxynivalenol, producing the novel intermediate 3-epi-deoxynivalenol. *Appl. Microbiol. Biotechnol.* **2011**, *89*, 419–427. [CrossRef] [PubMed]

25. He, W.J.; Yuan, Q.S.; Zhang, Y.B.; Guo, M.W.; Gong, A.D.; Zhang, J.B.; Wu, A.B.; Huang, T.; Qu, B.; Li, H.P.; et al. Aerobic De-Epoxydation of Trichothecene Mycotoxins by a Soil Bacterial Consortium Isolated Using In Situ Soil Enrichment. *Toxins* **2016**, *8*, E277. [CrossRef] [PubMed]

26. De Bellis, P.; Tristezza, M.; Haidukowski, M.; Fanelli, F.; Sisto, A.; Mulè, G.; Grieco, F. Biodegradation of Ochratoxin A by Bacterial Strains Isolated from Vineyard Soils. *Toxins* **2015**, *7*, 5079–5093. [CrossRef] [PubMed]

27. Yu, H.; Zhou, T.; Gong, J.H.; Young, C.; Su, X.J.; Li, X.-Z.; Zhu, H.H.; Tsao, R.; Yang, R. Isolation of deoxynivalenol-transforming bacteria from the chicken intestines using the approach of PCR-DGGE guided microbial selection. *BMC Microbiol.* **2010**, *10*, 182–191. [CrossRef] [PubMed]

28. He, J.W.; Hassan, Y.I.; Perilla, N.; Li, X.Z.; Boland, G.J.; Zhou, T. Bacterial epimerization as a route for deoxynivalenol detoxification: The influence of growth and environmental conditions. *Front. Microbiol.* **2016**, *7*. [CrossRef] [PubMed]

29. Karlovsky, P. Biological detoxification of the mycotoxin deoxynivalenol and its use in genetically engineered crops and feed additives. *Appl. Microbiol. Biotechnol.* **2011**, *91*, 491–504. [CrossRef] [PubMed]

30. Piddock, L.J.V. Teixobactin, the first of a new class of antibiotics discovered by iChip technology? *J. Antimicrob. Chemother.* **2015**, *70*, 2679–2680. [CrossRef] [PubMed]

31. Chang, X.; Wu, Z.; Wu, S.; Dai, Y.; Sun, C. Degradation of ochratoxin A by *Bacillus amyloliquefaciens* ASAG1. *Food Addit. Contam. Part A* **2015**, *32*, 564–571. [CrossRef] [PubMed]

32. Sato, I.; Ito, M.; Ishizaka, M.; Ikunaga, Y.; Sato, Y.; Yoshida, S.; Koitabashi, M.; Tsushima, S. Thirteen novel deoxynivalenol-degrading bacteria are classified within two genera with distinct degradation mechanisms. *FEMS Microbiol. Lett.* **2012**, *327*, 110–117. [CrossRef] [PubMed]

33. Khatibi, P.A.; McMaster, N.J.; Musser, R.; Schmale, D.G. Survey of mycotoxins in corn distillers' dried grains with solubles from seventy-eight ethanol plants in Twelve states in the U.S. in 2011. *Toxins* **2014**, *6*, 1155–1168. [CrossRef] [PubMed]

34. Mirocha, C.J.; Kolaczkowski, E.; Xie, W.; Yu, H.; Jelen, H. Analysis of deoxynivalenol and its derivatives (batch and single kernel) using gas chromatography/mass spectrometry. *J. Agric. Food Chem.* **1998**, *46*, 1414–1418. [CrossRef]

35. Takitani, S.; Asabe, Y.; Kato, T.; Suzuki, M.; Ueno, Y. Spectrodensitometric determination of trichothecene mycotoxins with 4-(p-nitrobenzyl) pyridine on silica gel thin-layer chromatograms. *J. Chromatogr. A* **1979**, *172*, 335–342. [CrossRef]

toxins

Article

Effect of Degradation of Zearalenone-Contaminated Feed by *Bacillus licheniformis* CK1 on Postweaning Female Piglets

Guanhua Fu [1], Junfei Ma [1], Lihong Wang [1], Xin Yang [1], Jeruei Liu [2] and Xin Zhao [1,3,*]

[1] College of Animal Science and Technology, Northwest A & F University, Yangling 712100, Shaanxi, China; guanhua1220@163.com (G.F.); mamafeiabc@163.com (J.M.); 18710354317@163.com (L.W.); yangx0629@163.com (X.Y.)
[2] Institute of Biotechnology and Department of Animal Science and Technology, National Taiwan University, Taipei 10617, Taiwan; jrliu@ntu.edu.tw
[3] Department of Animal Science, McGill University, Montreal, QC H9X 3V9, Canada
*   Correspondence: xin.zhao@mcgill.ca; Tel.: +86-29-8708-0899

Academic Editor: Ting Zhou
Received: 1 September 2016; Accepted: 11 October 2016; Published: 17 October 2016

Abstract: Zearalenone (ZEA), an estrogenic mycotoxin, is mainly produced by *Fusarium* fungi. In this study, *Bacillus licheniformis* CK1 isolated from soil with the capability of degrading ZEA was evaluated for its efficacy in reducing the adverse effects of ZEA in piglets. The gilts were fed one of the following three diets for 14 days: a basic diet for the control group; the basic diet supplemented with ZEA-contaminated basic diet for the treatment 1 (T1) group; and the basic diet supplemented with fermented ZEA-contaminated basic diet by CK1 for the treatment 2 (T2) group. The actual ZEA contents (analyzed) were 0, 1.20 ± 0.11, 0.47 ± 0.22 mg/kg for the control, T1, and T2 diets, respectively. The results showed that the T1 group had significantly increased the size of vulva and the relative weight of reproductive organs compared to the control group at the end of the trial. The T1 group significantly decreased the concentration of the luteinizing hormone (LH) compared with the control and T2 groups. Expression of ERβ was significantly up-regulated in the T2 group compared with the control. In addition, expression of ERβ was not different between the control and the T1 group. In summary, our results suggest that *Bacillus licheniformis* CK1 could detoxify ZEA in feed and reduce the adverse effects of ZEA in the gilts.

Keywords: *Bacillus licheniformis* CK1; zearalenone (ZEA); serum hormones; estrogen receptor (ER); post-weaning female piglets

1. Introduction

Mycotoxins are toxic secondary metabolites produced by a range of fungi, especially from *Fusarium*, *Aspergillus*, and *Penicillium* genera [1]. Mycotoxins pose great risks to the health of animals as well as humans. The ingestion of mycotoxin-contaminated feed by animals results in mycotoxin accumulation in different organs or tissues, endangering animal health or entering into the food chain through meat, milk, or eggs [2]. Humans get directly exposed to mycotoxins as a result of eating contaminated crops, or indirectly exposed by consuming contaminated animal products. In addition, the absorption of mycotoxins can be via respiratory or dermal exposure [3–5]. Over 400 mycotoxins have been identified, but thousands of mycotoxins may exist [6]. The number of mycotoxins could be changed according to a newly proposed definition of mycotoxins [7]. The new definition states that something is a mycotoxin if and only if it is a secondary metabolite produced by microfungi, posing a health hazard to human and vertebrate animal species by exerting a toxic activity on human or vertebrate animal cells in vitro with 50% effectiveness levels <1000 μM [7]. Of these, zearalenone (ZEA) is one of the most important mycotoxins for its global incidence and toxicity [8].

Zearalenone—a phenolic resorcyclic acid lactone—has good thermal stability and low solubility in water, but is highly soluble in organic solvents. Zearalenone is particularly toxic to the reproductive system, resulting in uterine enlargement, alterations to the reproductive tract, reduced litter size, increased embryo lethal resorption, decreased fertility, and changed progesterone (PRG) and estradiol (E2) plasma levels in laboratory animals [9]. ZEA and 17β-estradiol have similar structures, and both competitively bind to estrogen receptors (ERs). ZEA can activate the transcription of estrogen-responsive genes [10,11]. The estrogenic effects of ZEA are particularly pronounced in the reproductive system of pigs [12].

The Food and Agriculture Organization of the United Nations (FAO) has estimated that approximately 25% of the world's agricultural products are contaminated with mycotoxins, resulting in significant economic loss due to their impact on human health, trade, and animal productivity [13]. Streit et al. analyzed 13,578 samples of feed and feed raw materials for contamination with ZEA from all over the world over a period of eight years (January 2004–December 2011), and found that 36% of samples were positive for ZEA [14]. Among the positive samples, the average concentration and the maximum concentration were up to 101 μg/kg and 26,728 μg/kg, respectively [14]. Thus, detoxification strategies for contaminated feeds for animals are needed to reduce or eliminate the toxic effects of ZEA in order to improve food safety, prevent economic losses, and reclaim contaminated products. Numerous physical and chemical detoxification methods have been tried, including chemical, physical, and biological approaches. Among them, biological transformations (including the use of microorganisms to breakdown ZEA) are the least studied and may provide an effective means to manage this mycotoxin. Microorganisms in the *Bacillus* genus are considered as probiotics and have been shown to effectively degrade ZEA in vitro. For example, Tinyiro et al. found that *B. subtilis* 168 and *B. natto* were efficient in the removal of more than 75% of ZEA from the liquid medium [15], whereas Cho et al. reported that a *B. subtilis* strain degraded 99% of ZEA in the liquid medium [16]. Moreover, Yi et al. isolated a strain of *Bacillus licheniformis* CK1 from soil samples and found that this strain was capable of degrading ZEA [17]. However, there was limited investigation on feeding animals with microbiologically-detoxified diets [18]. Therefore, the purpose of this study was investigate effects of *Bacillus licheniformis* CK1 on growth performance, vulva size, relative weights of organs, and serum hormone of female piglets fed feed contaminated with ZEA. In addition, we also evaluated the expression of the estrogen receptors in the vagina, uterus, and ovary of the piglets.

2. Results

2.1. Growth Performance

In the seven-day adaption period, there was no significant difference in the average daily feed intake (ADFI), average daily gain (ADG), or feed efficiency (FE, feed intake/gain) among the three groups. Similarly, during the 14-day feeding period, treatments T1 and T2 exhibited no negative effect on the ADFI, ADG, or FE in comparison with the control (Table 1).

Table 1. Growth performance of piglets fed different diets.

Groups	Average Daily Feed Intake (kg)	Average Day Gain (kg)	Feed Conversion Rate
Control	0.71 ± 0.01	0.40 ± 0.08	1.73 ± 0.09
T1	0.68 ± 0.02	0.42 ± 0.05	1.63 ± 0.05
T2	0.63 ± 0.04	0.39 ± 0.06	1.65 ± 0.09
p value	0.289	0.463	0.713

Control group: the basal diet; T1 group: Zearalenone-contaminated diet; T2 group: fermented ZEA-contaminated basic diet by *Bacillus licheniformis* CK1. Values are expressed in mean ± S.E. (standard error, *n* = 6).

2.2. Vulva Size

Figure 1 shows changes in vulva size in the piglets in the three groups of piglets. There was no significant difference in the vulva size among the three groups at the beginning of the trial (d1). At the end of the trial (d15), the size of the vulva was significantly increased in the T1 group, but not in the T2 group, in comparison with the control ($p < 0.05$). The vulvae of the piglets in the T1 group were slightly red and swollen. On the other hand, no obvious change was observed in the control and T2 groups (Figure 2).

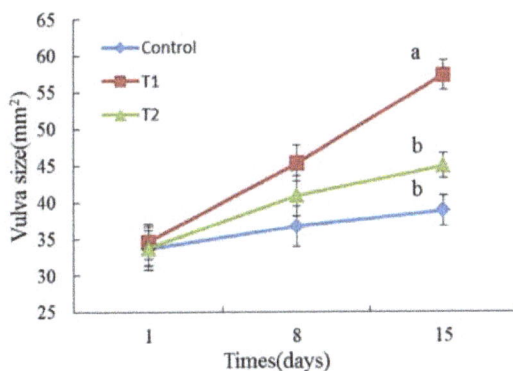

Figure 1. Effects of different diets on the vulva size of female piglets. a, b Values followed by different superscript letters differ significantly ($p < 0.05$, $n = 6$). Control group: the basal diet; T1 group: zearalenone (ZEA)-contaminated diet; T2 group: fermented ZEA-contaminated basic diet by *Bacillus licheniformis* CK1.

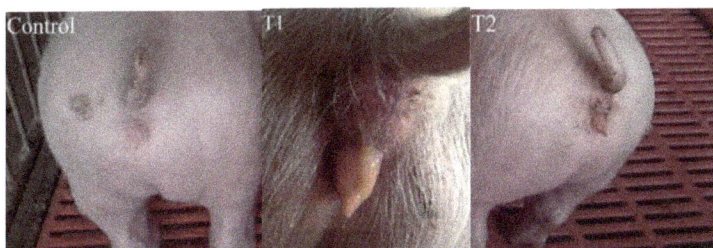

Figure 2. The representative vulva of piglets at the end of the study. (**Control group**) the basal diet; (**T1 group**) zearalenone-contaminated diet; (**T2 group**) fermented ZEA-contaminated basic diet by *Bacillus licheniformis* CK1.

2.3. Organ Weight

The relative organ weight was calculated as the weight of the organs divided by the body weight (g/kg). As shown in Table 2, the relative weight of the liver was significantly lower in piglets in the T1 group compared with the control ($p < 0.05$), while there was no significant difference between the T2 group and the control ($p > 0.05$). Piglets in the T2 group had significantly increased kidney weight, in contrast with the control and the T1 groups ($p < 0.05$). The T1 group and the T2 group had significantly heavier reproductive organs than the control ($p < 0.05$), but the relative weights of reproductive organs in T1 and T2 groups were not different ($p > 0.05$). For the other organs (heart, spleen, and lung), there were no differences among the three diet groups ($p > 0.05$).

Table 2. Relative weight of organs in weaned piglets fed different diets.

Group	Relative Weight (g/kg)					
	Heart	Liver	Spleen	Lung	Kidney	Reproductive Organs
Control	4.98 ± 0.24	27.51 ± 1.08 [a]	2.09 ± 0.30	11.68 ± 0.77	5.32 ± 0.28 [a]	0.47 ± 0.04 [a]
T1	4.87 ± 0.10	24.33 ± 0.85 [b]	1.79 ± 0.07	10.38 ± 0.20	5.24 ± 0.26 [a]	0.66 ± 0.06 [b]
T2	5.00 ± 0.17	25.53 ± 0.76 [ab]	2.04 ± 0.19	11.16 ± 0.42	6.80 ± 0.52 [b]	0.63 ± 0.04 [b]
p value	0.888	0.048	0.559	0.243	0.015	0.006

Control group: the basal diet; T1 group: zearalenone-contaminated diet; T2 group: fermented ZEA-contaminated basic diet by *Bacillus licheniformis* CK1. Values are expressed in mean $\pm$ S.E. ($n = 6$). [a, b] Means with different superscripts within same column differ significantly ($p < 0.05$).

2.4. The Level of Serum Hormones

The levels of serum hormones at the end of the test period are presented in Table 3. No significant differences were found in the level of follicle stimulating hormone (FSH), estradiol (E2), prolactin (PRL), progesterone (PRG) and testosterone (T) among the three treatments ($p > 0.05$). On the other hand, the T1 diet significantly decreased the concentration of luteinizing hormone (LH) in comparison with the control ($p < 0.05$). The T2 diet ameliorated the effect of ZEA and the levels of LH in the control and T2 groups were similar.

Table 3. The level of serum sex hormones of the female weaned piglets fed different diets on d15.

Group	Follicle Stimulating Hormone (FSH), mIU/mL	Luteinizing Hormone (LH), mIU/mL	Estradiol (E2), pg/mL	Prolactin (PRL), ng/mL	Progesterone (PRG), ng/mL	Testosterone (T), ng/dL
Control	13.33 ± 0.56	8.68 ± 0.11 [a]	21.27 ± 1.18	12.64 ± 0.48	0.80 ± 0.14	32.07 ± 2.97
T1	12.55 ± 0.56	8.02 ± 0.12 [b]	24.38 ± 1.01	12.60 ± 0.72	0.91 ± 0.2	31.85 ± 3.59
T2	12.76 ± 0.26	8.70 ± 0.13 [a]	23.32 ± 1.57	11.09 ± 0.42	1.13 ± 0.32	31.85 ± 0.96
p value	0.521	0.002	0.256	0.124	0.635	0.998

Control group: the basal diet; T1 group: zearalenone-contaminated diet; T2 group: fermented ZEA-contaminated basic diet by *Bacillus licheniformis* CK1. Values are expressed in mean $\pm$ S.E. ($n = 6$). [a, b] Means with different superscripts within same column differ significantly ($p < 0.05$).

2.5. Estrogen Receptor α (ERα) and Estrogen Receptor β (ERβ) mRNA Expression

As shown in Figure 3, mRNA expression of ERα and ERβ (two subtypes of estrogen receptor) was quantified in the reproductive organs by real-time quantitative polymerase chain reaction (RT-qPCR). No significant difference in ERα mRNA expression was found in the uterus and ovary among the three treatments, while the T2 group significantly decreased the mRNA expression of ERα in vagina ($p < 0.05$) in comparison with the control and the T1 group. The mRNA abundance of ERβ in the uterus, vagina, and ovary was significantly higher in gilts on the T1 diet. Meanwhile, ERβ mRNA expression in the uterus, vagina, and ovary in the T2 group was not significantly different from those in the control group.

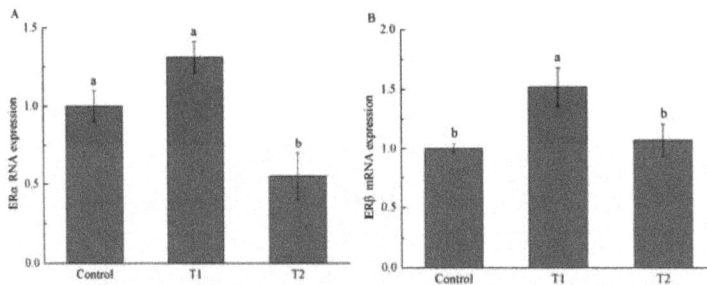

Figure 3. *Cont.*

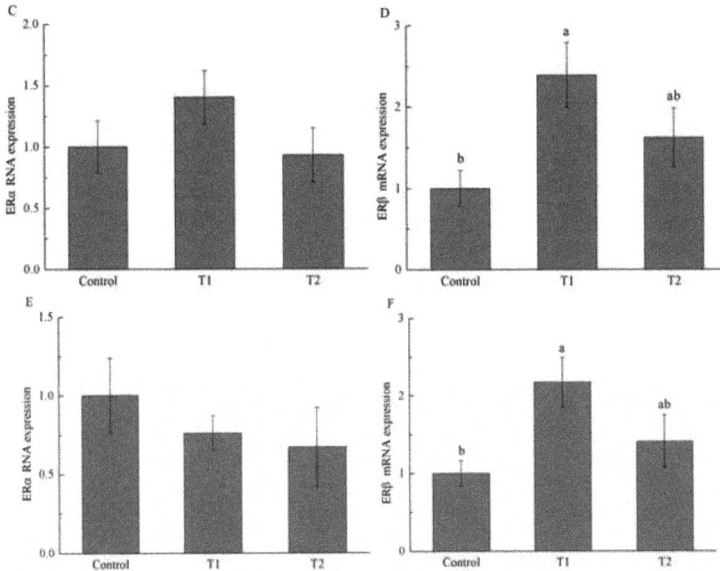

Figure 3. Effects of different diets on the mRNA expression of estrogen receptor α (ERα) ((**A**) vagina, (**C**) uterus, and (**E**) ovary) and estrogen receptor β (ERβ) ((**B**) vagina, (**D**) uterus, and (**F**) ovary) in the reproductive organs of female weaned piglets. Control group: the basal diet; T1 group: zearalenone-contaminated diet; T2 group: fermented ZEA-contaminated basic diet by *Bacillus licheniformis* CK1. [a, b] Different letters in each panel indicate significant differences ($p < 0.05$, $n = 6$).

3. Discussion

Zearalenone (ZEA) activates estrogen receptors and induces functional and morphological alteration in reproductive organs. The susceptibility to the adverse effect of ZEA is species–dependent, and pigs—especially prepubertal gilts—are very sensitive to ZEA due to its high alpha-hydroxylation activity and low glucuronidation activity [19]. ZEA is a substrate for α and β hydroxysteroid dehydrogenases, which convert ZEA into two stereoisomeric metabolites, α-zearalenol and β-zearalenol. Alpha-hydroxylation results in an increase in estrogenic potency as compared to the parent compound and beta-hydroxylation product [20]. The glucuronidation conjugates metabolites of ZEA hydroxylation and eliminates them through urine and bile fluid. Pigs have a rather low activity of glucuronidation [21].

Bacillus licheniformis (*B. licheniformis*) CK1 efficiently degraded ZEA in the basal diet. The concentration of ZEA in diets offered to the T2 group was reduced to 0.47 mg/kg from 1.20 mg/kg in the T1 group due to degradation by *B. licheniformis* CK1. The reduction of ZEA from 1.20 to 0.47 mg/kg in feed relieved red and swollen symptoms in the vulva of the piglets, and significantly decreased the vulva size of the piglets. Jiang et al. [22] reported the vulva size and relative weight of genital organs, liver, and kidney increased linearly in a ZEA-dose-dependent manner, indicating that the estrogenic effects are stronger with increasing concentration of ZEA. Thus, decreasing the concentration of ZEA in feeds could reduce the estrogenic effects of ZEA in the gilts. In addition, it has been previously shown that *B. licheniformis* CK1 decreased more than 98% of ZEA in ZEA-contaminated corn meal medium and was non-hemolytic, non-enterotoxin producing, and displayed high levels of extracellular xylanase, cellulase, and protease activities [17]. The presence of interfering substance in the basal diet might explain the lower efficiency of *B. licheniformis* CK1 to degrade ZEA in the current study, in comparison with the study of Yi et al. [17].

Whether *B. licheniformis* CK1 could reduce the adverse effects of ZEA for piglets was investigated in the feeding trial. Our results showed that feeding the ZEA-contaminated diet (T1 group) significantly increased the vulva size of gilts. In addition, organ weights were used as an index of estrogenic response to ZEA, especially for reproductive organs. In our study, we observed that the relative weight of reproductive organs in the T1 group was significantly increased compared to the control group. Our results are in agreement with previous reports. Oliver et al. reported that gilts fed ZEA-contaminated diets significantly increased vulva width and length compared with control [23]. Similarly, the vulva width, length, and area of piglets linearly increased as ZEA levels increased [24]. In addition, the relative weight of genital organs was also increased in female piglets supplemented with 1.05 mg/kg ZEA [24]. While there was no significant difference between T1 and T2 groups for the relative weight of genital organs, the T2 group significantly reduced the vulvar swelling of piglets in this study, implying that *B. licheniformis* CK1 can effectively alleviate the estrogen acting on the vulva of postweaning piglets caused by ZEA. Others have reported similar effects by using adsorbent materials or chemicals to deal with ZEA contaminations. Jiang et al. reported that clay enterosorbent at the levels of 5 or 10 g/kg was able to reduce the estrogenic effect of ZEA on vulvar swelling in postweaning female pigs [25]. Moreover, Denli et al. demonstrated that activated diatomaceous clay could effectively spare the estrogenic effect of ZEA on uterus and ovaries in rats and pigs [26]. The addition of a modified calcium montmorillonite alleviated some of the reproductive effects of ZEA on the relative weight of genital organs in postweaning piglets [24,27].

ZEA and its metabolites can be regarded as endocrine disruptors that change hormonal activity at the pre-receptor level. In the current study, ZEA decreased the level of luteinizing hormone (LH) in post-weaning gilts, but had no influence on the level of follicle stimulating hormone (FSH), estradiol (E2), prolactin (PRL), progesterone (PRG), or testosterone (T). Wang et al. observed that ZEA decreased the levels of E2 and LH in pre-pubertal gilts, but had no effect on the level of FSH [28]. Other researchers also reported that serum LH in gilts was significantly decreased by adding ZEA in feeds [29,30]. The gilts in the T2 group had similar LH concentrations to those in the control group, indicating that *B. licheniformis* CK1 has a protective effect on ZEA toxicosis symptoms in piglets.

ZEA and some of its metabolites have been shown to competitively bind to estrogen receptors (ERα and ERβ) in a number of in vitro and in vivo systems [31]. Therefore, we also investigated the mRNA expression level of estrogen receptors in different tissues of piglets among the three groups. In the present study, ERβ expression was significantly increased in the uterus, vagina, and ovary of gilts in the T1 group compared with the control group, whereas ERα was not significantly different. Our results are in agreement with a previous study by Oliver et al. [23]. ERβ could directly bind and accelerate the expression of adipogenic genes, enhancing triglyceride concentrations in ERβ-positive cells [32]. Thus, increasing the size and weight of the reproductive organs in our study—at least in part—could result from altering the expression of ERβ and the subsequent expression of other genes. Contrary to our results, Dong et al. found that the expression of ERα was significantly increased in the uterus of goats by ZEA, while the expression of ERβ was not changed [33]. The different results could be due to the species difference. Expression of ERβ in vagina, uterus, and ovary in the T2 group was similar to those in the control group, but was significantly lower than that in vagina of the T1 group.

In our study, the average daily feed intake (ADFI), average daily gain (ADG), and feed efficiency (FE) of the piglets were not different among the three groups. Likewise, Jiang et al. reported that there was no obvious difference in the growth performance of gilts between the control diet and the diet with ZEA concentration in the range of 1.1 to 3.2 mg/kg. Another study also showed that average daily feed intake did not differ between gilts consuming the control and zearalenone diets, which resulted in similar feed efficiency [23].

During the microbial transformation of ZEA, both estrogenic and non-estrogenic intermediates and by-products can be produced—for example, estrogenic α-zearalenol and β-zearalenol [34–36]. Certainly, some microbes could degrade ZEA to non-estrogenic products. For example, *Clonostachys rosea* IFO 7063 was effectively capable of converting ZEA to a non-estrogenic compound,

1-(3,5-dihydroxy-phenyl)-10′-hydroxy-1′E-undecene-6′-one, determined by 2D NMR spectroscopy [37]. The yeast strain *Trichosporon mycotoxinivorans* was also able to decarboxylate ZEA [38] and produce a compound identified as (5*S*)-5-({2,4-dihydroxy-6-[(1E)-5-hydroxypent-1-en-1-yl]benzoyl}oxy) hexanoic acid via NMR spectroscopy [39]. In addition, although the degradation product was not clear, Kriszt et al. [40] reported non-pathogenic *Rhodococcus pyridinivorans* K408 degraded 87.21% ZEA and reduced 81.75% of estrogenic effects. Our results supported that *B. licheniformis* CK1 degraded ZEA and reduced its estrogenic effects, possibly because ZEA was converted to non-estrogenic or less estrogenic compounds.

In conclusion, *B. licheniformis* CK1 could degrade the ZEA in feed and alleviated the adverse effect of ZEA for piglets. Our results support the notion that microbiological detoxification is suitable for the decontamination of mycotoxins in feed with high efficiency, strong specificity, and no environmental pollution [41].

4. Materials and Methods

4.1. Strains and Chemicals

Bacillus licheniformis CK1 was isolated from the National Taiwan University [17]. Purified zearalenone (ZEA), acetonitrile, and methanol (HPLC (high-performance liquid chromatography) grade) were purchased from Sigma-Aldrich (St. Louis, MO, USA). All other chemicals used were of analytical grade.

4.2. Preparation of the Experimental Diets

ZEA (47 mg) was dissolved in acetic ether and then mixed with talcum powder. A ZEA premix was prepared by blending ZEA-contaminated talcum powder with 3 kg of the basic diet (Table 4). We prepared two batches of ZEA premixes. One was used in mixing with the basic diet as treatment 1 (T1), which was calculated for a ZEA concentration of 1 mg/kg. The other was used for fermentation by *Bacillus licheniformis* CK1 to degrade ZEA before mixing with the basic diet as treatment 2 (T2).

Table 4. Ingredients and compositions of the basic diet.

Ingredients	Percentage, %	Nutrients	Analyzed Values
Corn	53.00	Gross energy (MJ/kg)	17.12
Wheat middling	5.00	Crude protein (%)	19.40
Whey powder	6.50	Calcium (%)	0.84
Soybean oil	2.50	Total phosphorus (%)	0.73
Soybean meal	24.76	Lysine (%)	1.36
Fish meal	5.50	Methionine (%)	0.46
L-Lysine HCl	0.30	Sulfur amino acid (%)	0.79
DL-Methionine	0.10	Threonine (%)	0.90
L-Threonine	0.04	Tryptophan (%)	0.25
Calcium phosphate	0.80	-	-
Limestone, pulverized	0.30	-	-
Sodium chloride	0.20	-	-
Premix [1]	1.00	-	-
Total	100	-	-

[1] Supplied per kg of diet: vitamin A, 3300 IU; vitamin D3, 330 IU; vitamin E, 24 IU; vitamin K3, 0.75 mg; vitamin B1, 1.50 mg; vitamin B2, 5.25 mg; vitamin B6, 2.25 mg; vitamin B12, 0.02625 mg; pantothenic acid, 15.00 mg; niacin, 22.5 mg; biotin, 0.075 mg; folic acid, 0.45 mg; Mn, 6.00 mg; Fe, 150 mg; Zn, 150 mg; Cu, 9.00 mg; I, 0.21 mg; Se, 0.45 mg.

For the fermentation, batches of 300 g of autoclaved ZEA-contaminated feed were mixed with 2700 mL sterilized water in a 5 L fermentor. The mixture was inoculated with 1% of an overnight bacterial culture of *Bacillus licheniformis* CK1 and incubated at 37 °C, 300 rpm for 36 h. The fermented feed was poured into a basin. To absorb water, the basic diet was gradually added. The mixture was dried at room temperature. A total of 3 kg ZEA-contaminated feed were fermented.

All diets were prepared at the same time and stored in covered containers before feeding.

4.3. Determination of ZEA in Feed by HPLC

The concentration of ZEA in feed was determined by high-performance liquid chromatography (HPLC) performed on HPLC instrument including a LC-20AT delivery system (Shimadzu, Kyoto, Japan), a CBM-20A system controller (Shimadzu, Kyoto, Japan), a SIL-20A autosampler (Shimadzu, Kyoto, Japan), a RF-10AXL fluorescence detector (Shimadzu, Kyoto, Japan), and an Ascentis C18 HPLC column (Sigma-Aldrich, Bellefonte, PA, USA; 5 μm particle size, L × I.D. 250 mm × 4.6 mm). The injection volume for quantifying ZEA was 20 μL. The mobile phase consisted of methanol:water 80:20 (v/v) at a flow rate of 0.5 mL·min^{-1}. The detector was set at excitation and emission wavelengths of 225 nm and 465 nm, respectively. A standard curve was established by analyzing six ZEA standard solutions (0.125, 0.25, 0.5, 1, 2, 5 μg/mL), and each concentration was determined in triplicate. The linear regression equation of the standard curve showed an R^2 value >0.99. Before the HPLC analysis, ZEN in the feed was extracted and cleaned up using the Romer Mycosep 226 column (Romer Labs Inc., Union, MO, USA) according to the manufacturer's instructions. The levels of ZEA in feeds were calculated by using the linear regression equation of the standard curve.

4.4. Experimental Design and Animals

A total of 18 post-weaning female piglets (Landrace × Yorkshire × Duroc) weaned at d30 with an average body weight (BW) of 8.19 ± 0.32 (mean ± SE) kg were used in this study. The animal protocols used in this work were evaluated and approved by Institutional Animal Care and Use Committee of Northwest A&F University (Identification code: NWAFAC2014, Date of approval: 16 August 2014). Gilts were randomly allocated to three treatments, with six gilts in each group according to BW. All animals were on the basic diet during a 7 day adaptation period after weaning. The nutrient concentrations of the basic diet met or exceeded minimal requirements according to the National Research Council (NRC) [42]. Pigs were fed the basic diet (control), treatment 1 diet (T1), or treatment 2 (T2) diet during a 14 day test period. The actual ZEA contents (analyzed) were 0, 1.20 ± 0.11, 0.47 ± 0.22 mg/kg for the control, T1, and T2 groups, respectively.

During a 14 day test period, animals were housed individually in metal pens on the Northwest A&F University farm. Throughout the study, animals had free access to feed and water, and room temperature was 26–28 °C. Body weights were measured weekly. Feed intake of each treatment was recorded daily. Vulva length and width were measured on d1, d8, and d15 after treatments started to determine the dietary ZEA estrogenic effects, and the vulva area was calculated approximately as a diamond shape ((vulva length × vulva width)/2) according to Jiang et al. [25].

4.5. Sample Collection

Pigs were fasted for 12 h at the end of the experimental period. Blood samples of approximately 10 mL were collected from the jugular vein of all animals into non-heparinized tubes, incubated at 37 °C for 2 h, centrifuged at 1500 × g for 10 min at room temperature, and the serum was separated and stored in 1.5 mL Eppendorf tubes at −20 °C for hormone analyses (described below). After collection of blood samples, piglets were immediately euthanized and genital organs (ovary + cornu uteri + vagina − vestibule), liver, kidney, heart, lung, and spleen were isolated, weighed, and examined for gross lesions. Samples of uterus, vagina, and ovary tissue were kept at −80 °C until extraction of total RNA for expression of the ERα and ERβ. Organ weights were expressed on a relative body weight basis (g/kg).

4.6. Serum Hormone Analysis

Serum samples were analyzed for follicle stimulating hormone (FSH), luteinizing hormone (LH), estradiol (E2), prolactin (PRL), progesterone (PRG), and Testosterone (T) using commercial radioimmunoassay kits obtained from Tianjin Jiuding medical bioengineering CO., Ltd. (Tianjin, China). All the samples were determined by the Yangling Demonstration Zone Hospital (Yangling, Shaanxi, China).

4.7. Total RNA Extraction and Real-Time Quantitative RT-PCR (qRT-PCR)

Total RNA was extracted from frozen tissues using a total RNA Kit from Omega (Norcross, GA, USA), according to the manufacturer's instructions. The purity of total RNA was ascertained by the A260/A280, and the integrity of total RNA was checked by agarose gel electrophoresis. Total RNA for each sample was converted into cDNA using TaKaRa PrimeScriptTM RT Reagent Kit (TaKaRa Biotechnology CO., Ltd., Dalian, China) according to the manufacturer's instructions and used for real-time quantitative polymerase chain reaction (RT-qPCR).

A SYBR® Premix Ex Taq kit (TaKaRa Biotechnology CO., Ltd., Dalian, China) was used to measure mRNA expression of estrogen receptor genes (ERα and ERβ) with glyceraldehyde-3-phosphate dehydrogenase (GAPDH) as an endogenous control. Pig-specific primers were designed from published GenBank sequences (Table 5). All of the PCR reactions were performed in triplicate. The relative gene expression levels were determined using the $2^{-\Delta\Delta Ct}$ method [43].

Table 5. Nucleotide sequences of primers for quantitative real-time polymerase chain reaction (qRT-PCR).

Gene	Forward Primer and Reverse Primer (from 5' to 3')	Size (bp)	Genbank No.
GAPDH	CCTGGCCAAGGTCATCCATG CCACCACCCTGTTGCTGTAG	500	NM_214220.1
ERα	TTGCTTAATTCTGGAGGGTAC AGGTGGATCAAGGTGTCTGTG	110	EF195769.1
ERβ	GCTCAGCCTGTACGACCAAGTGC CCTTCATCCCTGTCCAGAACGAG	138	NM_001001533.1

GADPH: glyceraldehyde-3-phosphate dehydrogenase.

4.8. Statistical Analysis

Data were analyzed through ANOVA and Duncan's multiple range tests using SPSS 16.0 statistical software (SPSS 16.0 Inc., Chicago, IL, USA, 2008). The values are expressed as mean ± S.E. Differences were considered significant at $p < 0.05$.

Acknowledgments: This work was supported by funding from an innovation project of science and technology plan project of Shaanxi Province, China (2014KTCL02-21) and a Ministry of Agriculture (No. 2013-S16), and the thousand talent program to Xin Zhao.

Author Contributions: Guanhua Fu, Junfei Ma and Xin Zhao designed the experiments, did the data analysis and wrote the paper. Lihong Wang, Xin Yang and Jeruei Liu helped the data analysis and contributed to manuscript writing. All authors read and approved the final manuscript.

Conflicts of Interest: The authors declare no conflict of interest.

References

1. Reverberi, M.; Ricelli, A.; Zjalic, S.; Fabbri, A.A.; Fanelli, C. Natural functions of mycotoxins and control of their biosynthesis in fungi. *Appl. Microbiol. Biotechnol.* **2010**, *87*, 899–911. [CrossRef] [PubMed]
2. Marin, S.; Ramos, A.J.; Cano-Sancho, G.; Sanchis, V. Mycotoxins: Occurrence, toxicology, and exposure assessment. *Food Chem. Toxicol.* **2013**, *60*, 218–237. [CrossRef] [PubMed]
3. Terr, A.I. Sick building syndrome: Is mould the cause? *Med. Mycol.* **2009**, *47*, 217–222. [CrossRef] [PubMed]
4. Hardin, B.D.; Robbins, C.A.; Fallah, P.; Kelman, B.J. The concentration of no toxicologicconcen (CoNTC) and airborne mycotoxins. *J. Toxicol. Environ. Health A* **2009**, *72*, 585–598. [CrossRef] [PubMed]

5. Boonen, J.; Malysheva, S.V.; Taevernier, L.; Di Mavungu, J.D.; De Saeger, S.; De Spiegeleer, B. Human skin penetration of selected model mycotoxins. *Toxicology* **2012**, *301*, 21–32. [CrossRef] [PubMed]

6. Nielsen, K.F.; Smedsgaard, J. Fungal metabolite screening: Database of 474 mycotoxins and fungal metabolites for dereplication by standardised liquid chromatography-UV-mass spectrometry methodology. *J. Chromatogr. A* **2003**, *1002*, 111–136. [CrossRef]

7. Taevernier, L.; Wynendaele, E.; De Vreese, L.; Burvenich, C.; De Spiegeleer, B. The mycotoxin definition reconsidered towards fungal cyclic depsipeptides. *J. Environ. Sci. Health C* **2016**, *34*, 114–135. [CrossRef] [PubMed]

8. Schatzmayr, G.; Streit, E. Global occurrence of mycotoxins in the food and feed chain: Facts and figures. *World Mycotoxin J.* **2013**, *6*, 213–222. [CrossRef]

9. Zhang, Y.; Jia, Z.; Yin, S.; Shan, A.; Gao, R.; Qu, Z.; Liu, M.; Nie, S. Toxic effects of maternal zearalenone exposure on uterine capacity and fetal development in gestation rats. *Reprod. Sci.* **2014**. [CrossRef] [PubMed]

10. Koraichi, F.; Videmann, B.; Mazallon, M.; Benahmed, M.; Prouillac, C.; Lecoeur, S. Zearalenone exposure modulates the expression of ABC transporters and nuclear receptors in pregnant rats and fetal liver. *Toxicol. Lett.* **2012**, *211*, 246–256. [CrossRef] [PubMed]

11. Turcotte, J.C.; Hunt, P.J.; Blaustein, J.D. Estrogenic effects of zearalenone on the expression of progestin receptors and sexual behavior in female rats. *Horm. Behav.* **2005**, *47*, 178–184. [CrossRef] [PubMed]

12. Songsermsakul, P.; Böhm, J.; Aurich, C.; Zentek, J.; Razzazi-Fazeli, E. The levels of zearalenone and its metabolites in plasma, urine and faeces of horses fed with naturally, fusarium toxin-contaminated oats. *J. Anim. Physiol. Anim. Nutr.* **2013**, *97*, 155–161. [CrossRef] [PubMed]

13. Mycotoxins. Available online: http://www.webcitation.org/6k9LUBU0l (accessed on 30 August 2016).

14. Streit, E.; Naehrer, K.; Rodrigues, I.; Schatzmayr, G. Mycotoxin occurrence in feed and feed raw materials worldwide: Long-term analysis with special focus on Europe and Asia. *J. Sci. Food Agric.* **2013**, *93*, 2892–2899. [CrossRef] [PubMed]

15. Tinyiro, S.E.; Yao, W.R.; Sun, X.L.; Wokadala, C.; Wang, S.T. Scavenging of zearalenone by *Bacillus* strains-in vitro. *Res. J. Microbiol.* **2011**, *6*, 304–309. [CrossRef]

16. Cho, K.J.; Kang, J.S.; Cho, W.T.; Lee, C.H.; Ha, J.K.; Song, K.B. In vitro degradation of zearalenone by *Bacillus subtilis*. *Biotechnol. Lett.* **2010**, *32*, 1921–1924. [CrossRef] [PubMed]

17. Yi, P.J.; Pai, C.K.; Liu, J.R. Isolation and characterization of a *Bacillus licheniformis* strain capable of degrading zearalenone. *World J. Microbiol. Biotechnol.* **2011**, *27*, 1035–1043. [CrossRef]

18. Zhao, L.H.; Lei, Y.P.; Bao, Y.H.; Jia, R.; Ma, Q.G.; Zhang, J.Y.; Chen, J.; Ji, C. Ameliorative effects of *Bacillus subtilis* ANSB01G on zearalenone toxicosis in pre-pubertal female gilts. *Food Addit. Contam. A* **2015**, *32*, 617–625. [CrossRef] [PubMed]

19. Fink-Gremmels, J.; Malekinejad, H. Clinical effects and biochemical mechanisms associated with exposure to the mycoestrogen zearalenone. *Anim. Feed Sci. Technol.* **2007**, *137*, 326–341. [CrossRef]

20. Shier, W.T.; Shier, A.C.; Xie, W.; Mirocha, C.J. Structure-activity relationships for human estrogenic activity in zearalenone mycotoxins. *Toxicon* **2001**, *39*, 1435–1438. [CrossRef]

21. Krishnaswamy, S.; Duan, S.X.; Von Moltke, L.L.; Greenblatt, D.J.; Sudmeier, J.L.; Bachovchin, W.W.; Court, M.H. Serotonin (5-hydroxytryptamine) glucuronidation in vitro: Assay development, human liver microsome activities and species differences. *Xenobiotica* **2003**, *33*, 169–180. [CrossRef] [PubMed]

22. Jiang, S.Z.; Yang, Z.B.; Yang, W.R.; Gao, J.; Liu, F.X.; Broomhead, J.; Chi, F. Effects of purified zearalenone on growth performance, organ size, serum metabolites, and oxidative stress in postweaning gilts. *J. Anim. Sci.* **2011**, *89*, 3008–3015. [CrossRef] [PubMed]

23. Oliver, W.T.; Miles, J.R.; Diaz, D.E.; Dibner, J.J.; Rottinghaus, G.E.; Harrell, R.J. Zearalenone enhances reproductive tract development, but does not alter skeletal muscle signaling in prepubertal gilts. *Anim. Feed Sci. Technol.* **2012**, *174*, 79–85. [CrossRef]

24. Jiang, S.Z.; Yang, Z.B.; Yang, W.R.; Wang, S.J.; Liu, F.X.; Johnston, L.A.; Chi, F.; Wang, Y. Effect of purified zearalenone with or without modified montmorillonite on nutrient availability, genital organs and serum hormones in post-weaning piglets. *Livest. Sci.* **2012**, *144*, 110–118. [CrossRef]

25. Jiang, S.Z.; Yang, Z.B.; Yang, W.R.; Yao, B.Q.; Zhao, H.; Liu, F.X.; Chen, C.C.; Chi, F. Effects of feeding purified zearalenone contaminated diets with or without clay enterosorbent on growth, nutrient availability, and genital organs in post-weaning female pigs. *Asian-Australas. J. Anim.* **2010**, *23*, 74–81. [CrossRef]

26. Denli, M.; Blandon, J.C.; Guynot, M.E.; Salado, S.; Pérez, J.F. Efficacy of activated diatomaceous clay in reducing the toxicity of zearalenone in rats and piglets. *J. Anim. Sci.* **2015**, *93*, 637–645. [CrossRef] [PubMed]
27. Wang, J.P.; Chi, F.; Kim, I.H. Effects of montmorillonite clay on growth performance, nutrient digestibility, vulva size, faecal microflora, and oxidative stress in weaning gilts challenged with zearalenone. *Anim. Feed Sci. Technol.* **2012**, *178*, 158–166. [CrossRef]
28. Wang, D.F.; Zhang, N.Y.; Peng, Y.Z.; Qi, D.S. Interaction of zearalenone and soybean isoflavone on the development of reproductive organs, reproductive hormones and estrogen receptor expression in prepubertal gilts. *Anim. Reprod. Sci.* **2010**, *122*, 317–323. [CrossRef] [PubMed]
29. Diekman, M.A.; Long, G.G. Blastocyst development on days 10 or 14 after consumption of zearalenone by sows on days 7 to 10 after breeding. *Am. J. Vet. Res.* **1989**, *50*, 1224–1227. [PubMed]
30. Green, M.L.; Diekman, M.A.; Malayer, J.R.; Scheidt, A.B.; Long, G.G. Effect of prepubertal consumption of zearalenone on puberty and subsequent reproduction of gilts. *J. Anim. Sci.* **1990**, *68*, 171–178. [CrossRef] [PubMed]
31. Kuiper, G.G.; Lemmen, J.G.; Carlsson, B.; Corton, J.C.; Safe, S.H.; van Der Saag, P.T.; van Der Burg, B.; Gustafsson, J.Å. Interaction of estrogenic chemicals and phytoestrogens with estrogen receptor β. *Endocrinology* **1998**, *139*, 4252–4263. [CrossRef] [PubMed]
32. Madak-Erdogan, Z.; Charn, T.H.; Jiang, Y.; Liu, E.T.; Katzenellenbogen, J.A.; Katzenellenbogen, B.S. Integrative genomics of gene and metabolic regulation by estrogen receptors α and β, and their coregulators. *Mol. Syst. Biol.* **2013**, *9*, 676. [CrossRef] [PubMed]
33. Dong, M.; He, X.J.; Tulayakul, P.; Li, J.Y.; Dong, K.S.; Manabe, N.; Nakayama, H.; Kumagai, S. The toxic effects and fate of intravenously administered zearalenone in goats. *Toxicon* **2010**, *55*, 523–530. [CrossRef] [PubMed]
34. Hurd, R.N. Structure activity relationships in zearalenones. In *Mycotoxins in Human and Animal Health*; Rodricks, J.V., Hesseltine, C.W., Mehlman, M.A., Eds.; Pathotox Publishers: Park Forest South, IL, USA, 1977; pp. 379–391.
35. Kiessling, K.H.; Pettersson, H.; Sandholm, K.; Olsen, M. Metabolism of aflatoxin, ochratoxin, zearalenone, and three trichothecenes by intact rumen fluid, rumen protozoa, and rumen bacteria. *Appl. Environ. Microb.* **1984**, *47*, 1070–1073.
36. Kollarczik, B.; Gareis, M.; Hanelt, M. In vitro tranformation of the *Fusarium* mycotoxins deoxynivalenol and zearalenone by normal gut microflora of pigs. *Nat. Toxins* **1994**, *2*, 105–110. [CrossRef] [PubMed]
37. Kakeya, H.; Takahashi-Ando, N.; Kimura, M.; Onose, R.; Yamaguchi, I.; Osada, H. Biotransformation of the mycotoxin, zearalenone, to a non-estrogenic compound by a fungal strain of *Clonostachys* sp. *Biosci. Biotechnol. Biochem.* **2002**, *66*, 2723–2726. [CrossRef] [PubMed]
38. Molnar, O.; Schatzmayr, G.; Fuchs, E.; Prillinger, H. *Trichosporon mycotoxinivorans* sp. nov., a new yeast species useful in biological detoxification of various mycotoxins. *Syst. Appl. Microbiol.* **2004**, *27*, 661–671. [CrossRef] [PubMed]
39. Schatzmayr, G.; Schatzmayr, D.; Pichler, E.; Taubel, M.; Loibner, A.P.; Binder, E.M. A novel approach to deactivate ochratoxin A. In *Mycotoxins and Phycotoxins: Advances in Determination, Toxicology and Exposure Management*; Njapau, H., Trujillo, S., van Egmond, H., Park, D., Eds.; Academic Publishers: Wageningen, The Netherlands, 2006; pp. 279–290.
40. Kriszt, R.; Krifaton, C.; Szoboszlay, S.; Cserháti, M.; Kriszt, B.; Kukolya, J.; Czéh, Á.; Fehér-Tóth, S.; Török, L.; Szőke, Z.; et al. A new zearalenone biodegradation strategy using non-pathogenic *Rhodococcus pyridinivorans* K408 strain. *PLoS ONE* **2012**, *7*, e43608. [CrossRef] [PubMed]
41. Hathout, A.S.; Aly, S.E. Biological detoxification of mycotoxins: A review. *Ann. Microbiol.* **2014**, *64*, 905–919. [CrossRef]
42. National Research Council. *Nutrient Requirements of Swine*, 10th ed.; National Academy Press: Washington, DC, USA, 1998.
43. Livak, K.J.; Schmittgen, T.D. Analysis of relative gene expression data using real-time quantitative PCR and the $2^{-\Delta\Delta Ct}$ method. *Methods* **2001**, *25*, 402–408. [CrossRef] [PubMed]

Article

The Possible Mechanisms Involved in Degradation of Patulin by *Pichia caribbica*

Xiangfeng Zheng, Qiya Yang, Hongyin Zhang *, Jing Cao, Xiaoyun Zhang and Maurice Tibiru Apaliya

School of Food and Biological Engineering, Jiangsu University, 301 Xuefu Road, Zhenjiang 212013, Jiangsu, China; 1879600292l@163.com (X.Z.); yangqiya1118@163.com (Q.Y.); caojing819@126.com (J.C.); zhangxiaoyungu@126.com (X.Z.); mtapaliya@yahoo.com (M.T.A.)
* Correspondence: zhanghongyin@ujs.edu.cn; Tel.: +86-511-8878-0174

Academic Editor: Ting Zhou
Received: 7 September 2016; Accepted: 30 September 2016; Published: 9 October 2016

Abstract: In this work, we examined the mechanisms involved in the degradation of patulin by *Pichia caribbica*. Our results indicate that cell-free filtrate of *P. caribbica* reduced patutlin content. The heat-killed cells could not degrade patulin. However, the live cells significantly reduced the concentration of the patulin. In furtherance to this, it was observed that patulin was not detected in the broken yeast cells and cell wall. The addition of cycloheximide to the *P. caribbica* cells decreased the capacity of degradation of patulin. Proteomics analyses revealed that patulin treatment resulted in an upregulated protein which was involved in metabolism and stress response processes. Our results suggested that the mechanism of degradation of patulin by *P. caribbica* was not absorption; the presence of patulin can induce *P. caribbica* to produce associated intracellular and extracellular enzymes, both of which have the ability to degrade patulin. The result provides a new possible method that used the enzymes produced by yeast to detoxify patulin in food and feed.

Keywords: mycotoxin; patulin; biodegradation; *Pichia caribbica*; proteomics; intracellular and extracellular enzymes

1. Introduction

Phytosanitation is critical in food safety in the globalized agribusiness, where fresh fruits have been considered as promising natural food. The Food and Agricultural Organization (FAO) estimated that 25% of the world's crops are contaminated with mycotoxins, and *Aspergillus*, *Penicillium*, and *Fusarium* genera were incriminated [1–3]. Patulin (4-hydroxy-4H-furo [3,2c] pyran, 2[6H]-one) one of these mycotoxins is an unsaturated heterocyclic lactone produced by certain fungi species (*Penicillium*, *Aspergillus*, and *Byssochlamys*). Patulin is the most common mycotoxin found in apples and its derived products [4]. *Penicillium expansum* is the most common fungus that causes blue mold and patulin contamination in stored apples [5]. Patulin was first isolated from *Aspergillus clavatus* and studied in the early 1940s [6,7]. Patulin contamination is a world-wide problem including, Portugal [8,9], Belgium [10], India [11], Northeast China [12]. Patulin has been demonstrated to induce oxidative stress and causes DNA strands to break in HepG2 cells. Oxidative damage in human cells can lead to mutagenic [13], carcinogenic [14], immunotoxic [15], neurotoxic [16], genotoxic, and teratogenic [17,18] effects.

Traditionally, to control blue mold and patulin contamination in apples, synthetic fungicides are usually relied upon. However, the development of fungicide resistance by pathogens and the public's concern over the presence of chemical residues in food have prompted an urgent need for alternative control with good efficacy, and little or no toxicity to the non-target organisms [19]. In recent years, biological control using antagonistic yeasts has emerged as a promising method to reduce synthetic fungicides [20]. In view of this, many researches have shown that some antagonist yeasts could directly inhibit the production of patulin. Through almost 30 years of research, dozens of yeasts, including *Pichia caribbica* [21], *Rhodotorula glutinis* [22], *Pichia ohmeri* [23], *Rhodosporidium kratochvilovae* [24], *Gluconobacter oxydans* [25], and several others have been shown to degrade patulin, and inhibited the growth of *P. expansum*. Coelho et al. [26] reported that patulin concentration of 223 μg was decreased over 83% by *P. ohmeri* 158 cells when incubated at 25 °C for two days at a humidity >99. Also, *R. glutinis* and *Cryptococcus laurentii* degraded patulin in vivo and significantly reduced the accumulation of patulin in apples [22]. Zhu et al. [27] reported that the yeast *Rhodosporidium paludigenum* reduced the patulin content in apples.

However, the mechanism(s) of action of yeasts is/are insufficient and remain poorly understood. A study by Coelho, Celli, Ono, Wosiacki, Hoffmann, Pagnocca, and Hirooka [26] showed that antagonistic yeast cells incubated with PAT decreased its contamination through two mechanisms: (1) PAT adsorption at the yeast cell wall, and (2) PAT absorption into yeast cells. Zhu et al. [28] pointed out that, when dead and live cells of *Rhodosporidium paludigenum* were incubated with patulin for three days, 51% of the patulin was absorbed by the dead cells while no patulin was detected in the viable yeast cells.

Genome-wide analysis of the model yeast *Saccharomyces cerevisiae* exposed to patulin indicated that there were upregulated genes which showed proteasome activity, metabolism of sulfur amino acids, and stress responses [29]. Ianiri et al. [30] reported that patulin was degraded by *Sporobolomyces* sp. strain IAM 13481 and two kinds of mesostate were produced. To determine the genes responsible for the degradation, 3000 mutants were instructed by T-DNA insertion, and some were proven to be sensitive to patulin. The genes which include *YCK2*, *PAC2*, *DAL5*, and *VPS8* were annotated to *Saccharomyces cerevisiae*. Ianiri et al. [31] described a transcriptomic approach based on RNAseq to study the changes of gene expression in *Sporobolomyces* sp. exposed to patulin. In their study, the upregulated *Sporobolomyces* genes were those involved in metabolic processes, oxidation-reduction, and transport processes. The patulin decreased the expression of genes involved in the processes of protein synthesis, modification, cell division, and cell cycle. The results provide a comprehensive analysis to identify potential mechanisms and enzymes that are involved in patulin degradation. However, there were no genes and enzymes found to be responsible for the patulin degradation.

In vitro test indicated that *P. caribbica* can degrade patulin directly [21]. However, the mechanism of degradation of patulin by the *P. caribbica* is unknown. Therefore, in the present study, we investigated the possible mechanisms involved in degradation of patulin by *Pichia caribbica* and provided a comprehensive analysis of the genes and enzymes that are involved in patulin degradation.

2. Results

2.1. Effect of Cell-Free Filtrate of P. caribbica on Patulin Degradation

The findings revealed that the cell-free filtrate of *P. caribbica* reduced the patulin concentration compared to the control (CK) during all the tested time, Figure 1. This demonstrates that that the cell-free filtrate significantly affected the patulin content. The patulin content was reduced from 18.8 μg/mL at 0 h to 11.4 μg/mL at 12 h after incubation, however, after 12 h the degradation declined steadily throughout the entire duration.

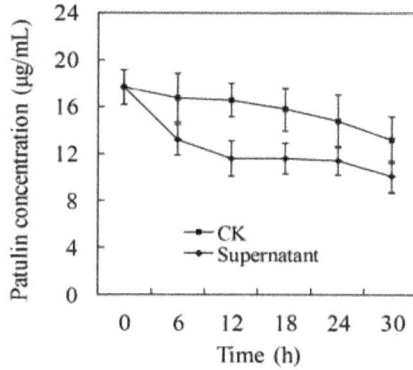

Figure 1. Efficacy of cell-free filtrate of *P. caribbica* on the degradation of patulin. The *x* axis represents the time after the addition of patulin (h: hour), the *y* axis represents the concentration of the patulin in the medium. CK: NYDB + patulin, Supernatant: *P. caribbica* cultuered in NYDB medium + patulin. Results are presented as means ± SD of triplicate experiments. The data at the same time were analyzed by the *t* test. The significant difference was assessed at the level *p* < 0.05.

2.2. Effect of Viable and Heat-Killed P. caribbica Cells on Patulin Degradation

The assay demonstrated that compared to the control, the dead yeast cells could not degrade the patulin, while the live yeast cells significantly reduced the patulin level throughout the 36 h inoculation period, Figure 2. The live *P. caribbica* cells decreased the patulin concentration to 1.4 μg/mL at 36 h of incubation. However, the patulin concentration of the control and the heat-killed cells were 13.7 μg/mL and 12.7 μg/mL at 36 h after incubation, respectively.

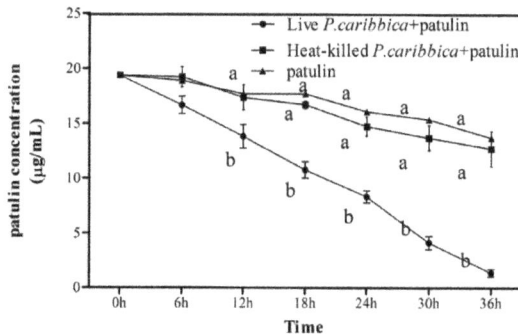

Figure 2. Efficacy of viable and heat-killed *P. caribbica* cells on degradation of patulin. The *x* axis represents the time after the addition of patulin (h: hour), the *y* axis represents the concentration of the patulin in the medium. Results are presented as means ± SD of triplicate experiments. The data at the same time were analyzed by the analysis of variance (ANOVA) in the statistical program SPSS/PC version 17.0. The significant difference was assessed at the level *p* < 0.05.

2.3. Degradation of Patulin by P. caribbica Was Inhibited by Cycloheximide

The results on the treatment without cycloheximide (*P. caribbica* alone) decreased the patulin level completely after 36 h of incubation, Figure 3. However, the cycloheximide that was added to the patulin at the beginning of the experiment decreased slightly and declined steadily. The patulin content in which the cycloheximide was added at after 6 h of incubation followed a similar trend with that of the control.

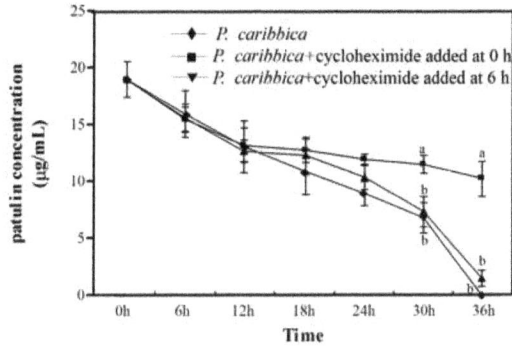

Figure 3. Effects of cycloheximide on degradation of patulin by *P. caribbica*. The *x* axis represents the time after the addition of patulin (h: hour), the *y* axis represents the concentration of the patulin in the medium. Results are presented as means ± SD of triplicate experiments. The data at the same time were analyzed by the analysis of variance (ANOVA) in the statistical program SPSS/PC version 17.0. The significant difference was assessed at the level $p < 0.05$.

2.4. Effect of P. caribbica's Supernatant on Patulin Degradation

The supernatant of *P. caribbica* significantly reduced the patulin content, regardless of whether the yeast cells were filtered after 6 h or not, Figure 4. The supernatant without yeast cells had no effect on patulin degradation. Henceforth, no significant difference was observed between the two treatments.

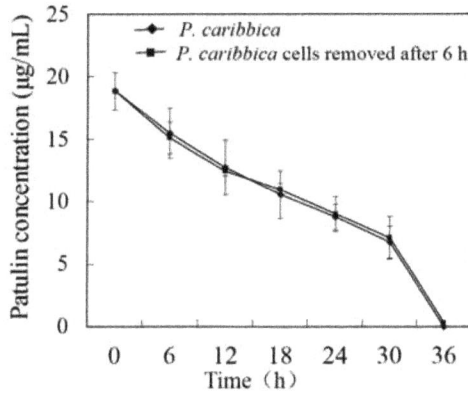

Figure 4. Efficacy of cell-free filtrate of *P. caribbica* which was induced 6 h by patulin on degradation of patulin. The *x* axis represents the time after the addition of patulin (h: hour), the *y* axis represents the concentration of the patulin in the medium. Results are presented as means ± SD of triplicate experiments. The data at the same time were analyzed by the *t* test. The significant difference was assessed at the level $p < 0.05$.

2.5. Effect of Intracellular Enzymes of P. caribbica on Patulin Degradation

The results on the intracellular enzymes demonstrates that the intracellular enzymes obtained from the *P. caribbica* that were incubated in the NYDB did not exhibit degradation effect. The patulin content remained at a high level of 8.3 µg/mL at 24 h after incubation, Figure 5C,D. However, the intracellular enzymes from *P. caribbica* incubated in the NYDB and amended with the patulin degraded the patulin to 0.25 µg/mL at 24 h after incubation, Figure 5B,D.

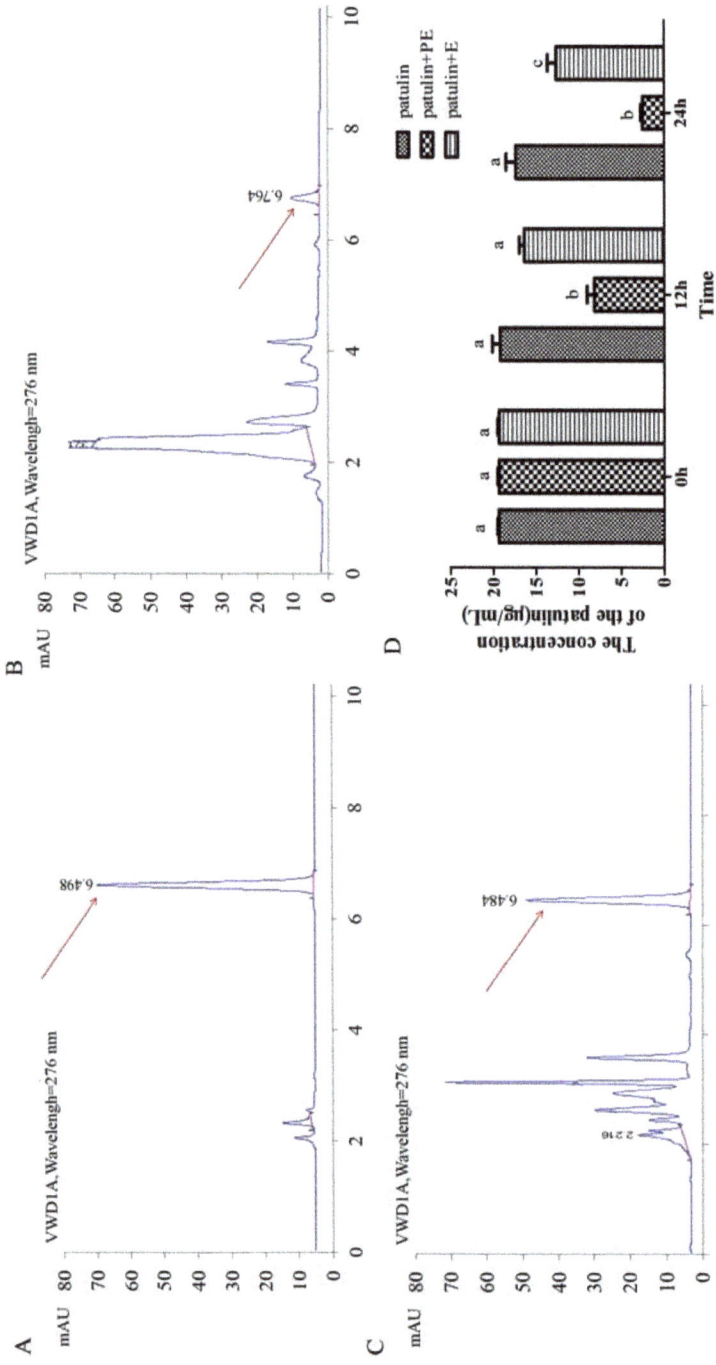

Figure 5. Effects of intracellular enzymes of *P. caribbica* on degradation of patulin. (**A**): The HPLC result of standard patulin samples in phosphate buffer at 24 h after incubation; (**B**): The HPLC result of patulin+P-E (extracted from the *P. caribbica* induced by patulin);(**C**): The HPLC result of patulin+E (extracted from *P. caribbica*); (**D**): The patulin content at 0, 12, and 24 h after treatment. The red arrows in **A**, **B**, and **C** represent the peaks of patulin. The data are presented as means ± SD of triplicate experiments. The data at the same time were analyzed by the analysis of variance (ANOVA) in the statistical program SPSS/PC version 17.0. The significant difference was assessed at the level $p < 0.05$.

2.6. Identification of Differentially Expressed Proteins

Proteins were identified by means of peptide mass fingerprints (PMF). MASCOT was used to search protein database of Viridiplantae. More than 150 protein spots were detected in each gel after ignoring very faint spots and spots with undefined shapes and areas using Image Master 2D Elite software, Figure 6A. Protein was extracted from *P. caribbica*, harvested at 24 h after incubation in NYDB, Figure 6A, and NYDB amended with patulin, Figure 6B. A total of 53 differentially expressed proteins were identified from the *P. caribbica* harvested from the NYDB amended with patulin. Out of the 53 proteins, 18 proteins were upregulated and 35 proteins were downregulated. Our test focused on 27 of them, Table 1. The differentially expressed proteins were classified with GO analysis, Figure 6C. Most of the differentially expressed proteins were related to basic metabolism such as isocitrate lyase activity, citrate (Si)-synthase activity, acetyl-CoA hydrolase activity, phosphomannomutase activity, and phosphopyruvate hydratase activity, Figure 6C. This indicated that the basic metabolism of *P. caribbica* was activated by patulin. The responses of *P. caribbica* to patulin are complex, as the differentially expressed proteins were involved in multiple metabolic pathways. Heat-shock protein 70 and Heat-shock protein SSB1 under the category of stress response may be related to the yeast cell's stress response to the patulin, Figure 6C.

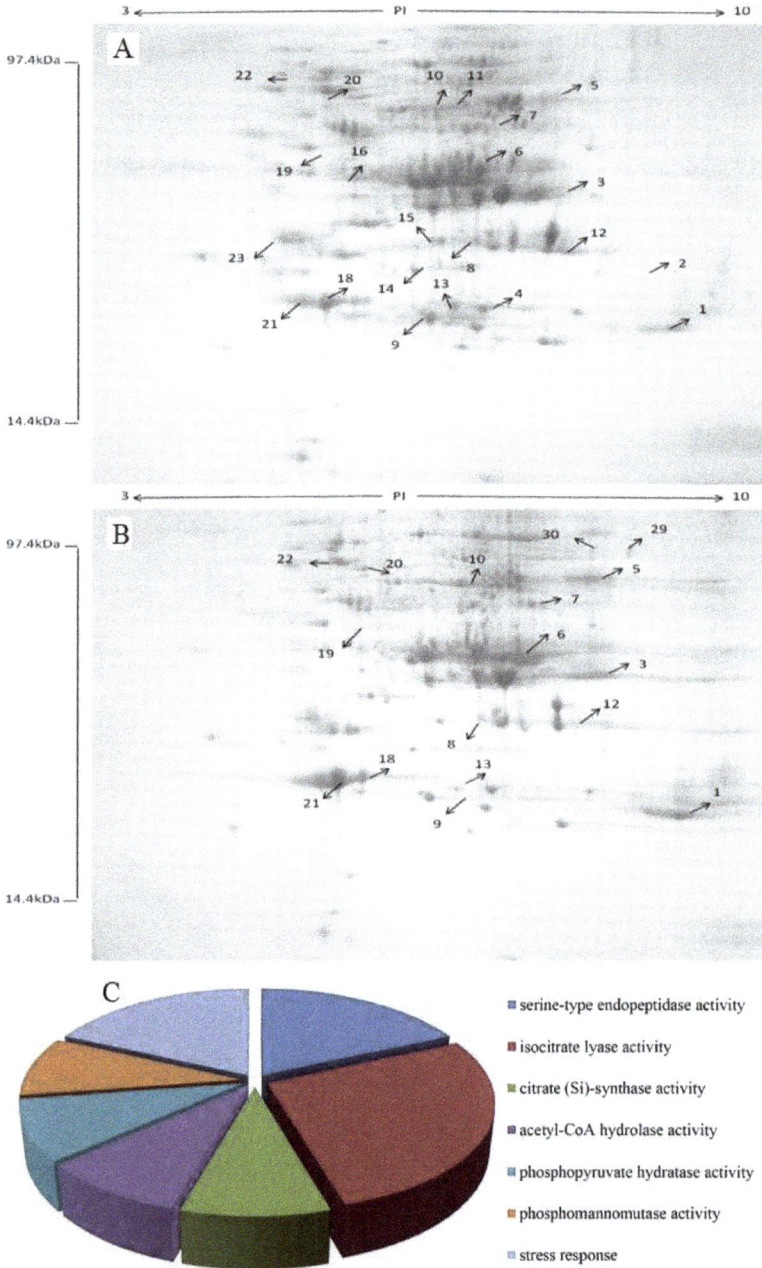

Figure 6. Two-dimensional pattern of intracellular proteins of *P. caribbica* after cultivation for 24 h in NYDB and NYDB amended with patulin. (**A**): Protein extracted from *P. caribbica* which was harvested from NYDB at 24 h after incubation; (**B**): Protein extracted from *P. caribbica* which was harvested from NYDB amended with patulin at 24 h after incubation; (**C**): Gene ontology (GO) analysis of the differentially expressed proteins of *P. caribbica* when treated with patulin.

Table 1. Proteins identified with PMF.

Molecular Function	Spot	Protein Name	NCBI Accession	Mass	PI	Species	Score	Sequence Coverage (%)	Number of Mass Values Matched
Serine-type endopeptidase activity	2	Chymotrypsinogen A	gi|117615	26,220	8.52	Bos taurus	206	16	4
Serine-type endopeptidase activity	21	Chymotrypsinogen	gi|117616	26,309	4.99	Bos taurus	110	6	1
Isocitrate lyase activity	4	Isocitrate lyase	gi|146413757	61,937	6.31	Meyerozyma guilliermondii ATCC 6260	124	5	2
Isocitrate lyase activity	8	Isocitrate lyase	gi|146413757	61,937	6.31	Meyerozyma guilliermondii ATCC 6260	429	12	5
Isocitrate lyase activity	11	Isocitrate lyase and phosphorylmutase	gi|344232420	62,282	6.78	Candida tenuis ATCC 10573	67	5	2
Citratesynthase activity	6	Citrate synthase	gi|146421975	43,995	6.25	Meyerozyma guilliermondii ATCC 6260	119	5	2
Acetyl-CoA hydrolase activity	10	Acetyl-CoA hydrolase	gi|146417797	58,385	5.96	Meyerozyma guilliermondii ATCC 6260	222	6	2
Phosphopyruvate hydratase activity	15	Enolase 1	gi|146415384	46,951	5.42	Meyerozyma guilliermondii ATCC 6260	117	3	1
Phosphomannomutase activity	17	Phosphomannomutase	gi|146423739	28,678	5.26	Meyerozyma guilliermondii ATCC 6260	361	19	5
Stress response	20	Heat shock protein SSB1	gi|146420661	66,421	5.29	Meyerozyma guilliermondii ATCC 6260	260	10	5
Stress response	22	Heat shock protein 70 2	gi|146413777	70,177	5.04	Meyerozyma guilliermondii ATCC 6260	951	17	12
Unclassified	12	DEHA2F04796p	gi|50423973	35,926	6.24	Debaryomyces hansenii CBS767	61	7	2
Unclassified	13	Conserved hypothetical protein	gi|146417765	34,916	7.17	Meyerozyma guilliermondii ATCC 6260	219	13	3
Unclassified	14	Hypothetical protein PGUG_05640	gi|146413298	65,256	5.85	Meyerozyma guilliermondii ATCC 6260	170	5	2
Unclassified	16	DEHA2G14058p	gi|50427089	47,210	5.28	Debaryomyces hansenii CBS767	197	7	2

Table 1. *Cont.*

Molecular Function	Spot	Protein Name	NCBI Accession	Mass	PI	Species	Score	Sequence Coverage (%)	Number of Mass Values Matched
Unclassified	5	Hypothetical protein PGUG_05024	gi|146414197	32,118	7.77	Meyerozyma guilliermondii ATCC 6260	122	10	2
Unclassified	9	DEHA2D06160p	gi|50420381	54,282	5.68	Debaryomyces hansenii CBS767	133	8	2
Unclassified	23	Conserved hypothetical protein	gi|146416825	44,146	5.33	Meyerozyma guilliermondii ATCC 6260	280	13	4
Unclassified	24	Hypothetical protein PGUG_00755	gi|146422888	20,317	5.45	Meyerozyma guilliermondii ATCC 6260	84	6	1
Unclassified	25	Hypothetical protein PGUG_04067	gi|146414736	34,234	6.64	Meyerozyma guilliermondii ATCC 6260	104	11	3
Unclassified	24	Hypothetical protein PGUG_03175	gi|146418962	62,015	5.21	Meyerozyma guilliermondii ATCC 6260	272	12	4
Unclassified	29	Hypothetical protein PGUG_00646	gi|190344794	78,744	8.19	Meyerozyma guilliermondii ATCC 6260	296	11	6
Unclassified	30	Hypothetical protein LOC100448380	gi|297701166	105,979	5.19	Pongo abelii	85	3	3
Unclassified	31	Hypothetical protein PGUG_03973	gi|146414548	38,139	5.53	Meyerozyma guilliermondii ATCC 6260	127	5	1
Unclassified	32	Hypothetical protein PGUG_01788	gi|146420321	37,067	5.56	Meyerozyma guilliermondii ATCC 6260	109	9	2
Unclassified	33	Hypothetical protein PGUG_00294	gi|146421948	35,991	5.22	Meyerozyma guilliermondii ATCC 6260	88	8	2
Unclassified	1	Chain Z	gi|230350	24,662	8.23	Bos taurus	316	29	6

3. Discussion

Patulin is a worldwide food mycotoxin which is present in a variety of fruit products. In recent years, attempts have been made to reduce the mycotoxin content in fruits and their derived products [32]. In view of this, biodegradation of mycotoxins has taken center stage for many researchers. Some antagonistic yeast strains used in biological control of fruits do not only control pathogens in the storage period, but also reduce patulin and other mycotoxins directly. Our research team found that the degradation of patulin treated with *P. caribbica* was higher than that of the control at all the tested time, which showed that *P. caribbica* can degrade patulin directly [21], but the degradation mechanisms are unknown. In this paper, we investigated the possible mechanisms involved in degradation of patulin by antagonistic yeast using *P. caribbica* as the model organism.

The cell-free filtrate of *P. caribbica* decreased the concentrations of the patulin at all tested times indicating that *P. caribbica* might have produced certain intracellular enzymes which degraded the patulin. These results are similar to the degradation of OTA by the cell-free supernatant of *Bacillus subtilis CW 14* [33]. The functions of these enzymes were, however, short-lived as the degradation declined after 12 h. Moreover, the patulin may have reacted with sulfhydryl containing amino acids or proteins, thereby decreasing its concentration [17].

Co-incubation of heat-killed and live cells of *P. caribbica* with patulin respectively showed that the inactive yeasts could not reduce the concentration of the patulin, while the live cells significantly reduced the concentration of the patulin (Figure 2). This finding is in line with Dong et al. [34] who reported that the live yeast *Kodameae ohmeri* could degrade patulin. In addition, significant differences in the reduction of patulin content was observed between viable and heat-killed cells. The results showed that the degradation of patulin requires the presence of live yeast cells.

Fifty-one percent of patulin was absorbed by dead yeast cells of *R. paludigenum* [28]. However, the test on the role of *P. caribbica* in patulin absorption indicated that patulin was not detected in the broken yeast cells and the cell wall. These results showed that there are other mechanisms of patulin reduction other than absorption by the yeast cells.

The test on the effects of cycloheximide on *P. caribbica* degradation of patulin, demonstrated that the *P. caribbica* without the cycloheximide decreased the patulin concentration throughout the entire duration of the experiment and after 36 h the patulin could not be detected. On the contrary, *P. caribbica* containing the cycloheximide decreased the patulin concentration sharply during the first 12 h. These results are in agreement with the findings of Zhang et al. [35] who used cycloheximide combined with *S. cerevisiae* to degrade ZEN at time 0 h, while addition of cycloheximide at 12 h significantly slowed down degradation. Moreover, cycloheximide inhibits protein (enzymes) synthesis in eukaryotes. These results suggest that enzymes produced by antagonistic yeasts play an important role in degradation of patulin.

As reported, the process of patulin degradation is an enzymatic reaction [22,24,34]. This confirmed our findings in which the supernatant from the *P. caribbica* degraded the patulin regardless of whether the yeast cells were filtered out or not. This is an indication that the yeast may have synthesized some extracellular enzymes in the supernatant which was involved in the degradation.

Additionally, Folger [36] reported that patulin can be degraded by yeast protein extract. *R. paludigenum* was observed to have degraded patulin through the activities of its intracellular enzymes, and the enzymes were induced by the patulin [28]. In this way, the intracellular enzymes of *P. caribbica* alone and the *P. caribbica* induced with patulin were extracted and their ability to degrade patulin was confirmed. Our results showed that the intracellular enzymes of the *P. caribbica* induced with patulin contained the enzymes which were responsible for the degradation of patulin.

Ianiri, Idnurm, Wright, Durán-Patrón, Mannina, Ferracane, Ritieni, and Castoria [30] in their quest to determine the relevant genes responsible for patulin degradation used *A. tumefaciens* T-DNA delivery system. They isolated 13 mutants which were affected in patulin degradation, some exhibited hypersensitivity to patulin. However, the slow patulin degradation observed for the mutants was proven not to be the inactivation of genes encoding enzymes directly involved in the patulin degradation pathway, but to the loss of functions of genes involved in resistance to patulin-induced stresses. No degradation pathway was found. In this study, the differentially expressed proteins of *P. caribbica* induced by the patulin showed that the basic metabolic proteins of *P. caribbica* were affected. Response patterns of *P. caribbica* to patulin are complex, as the differentially abundant proteins were involved in multiple metabolic pathways. The cytoplasm of eukaryotic cells, whether chemical, thermal, or in the form of misfolded or aggregated protein, triggers a complex biological response referred to as the heat shock response [37–40]. Hsp70 is a molecular chaperone belonging to heat-shock protein family. It is a stress-inducible member, in contrast to Hsc70 expressed constitutively. Hsp70 plays an important role in the folding and assembling of newly synthesized proteins, refolding of misfolded proteins, aggregation of proteins, and membrane translocation of organelles. It also secretes proteins and controls the activities of regulatory proteins [41–44]. The more concentration indicated that Hsp70 was produced during the degradation of patulin by *P. caribbica*. Phosphomannomutase (PMM, EC 5.4.2.8), catalyzing the interconversion between mannose-6-phosphate and mannose-1-phosphate, is an essential and conserved enzyme in eukaryotic organisms [45,46]. Mannose-1-phosphate is necessary for synthesizing the vital cellular metabolite GDP-mannose, which plays a crucial role in the formation of polysaccharide chains required for the glycosylation of protein and lipid molecules [47,48]. The mannan oligosaccharide is composed of mannose and glucose oligosaccharides, one of the main active ingredients of the cell wall of yeasts. It can absorb mycotoxins. So PMM may be an important enzyme in the degradation of patulin by *P. caribbica*. Most of the differentially expressed proteins are related to basic metabolism, such as acetyl-CoA hydrolase, isocitrate lyase, citrate synthase, and enolase 1.

4. Conclusions

In conclusion, the degradation of patulin by *P. caribbica* is an enzyme catalytic process, and these enzymes were induced by patulin. In the future, we will focus on the identification of these enzymes which are involved in patulin degradation and the degradation products of the patulin.

5. Materials and Methods

5.1. Antagonist and Growth Conditions

P. caribbica was isolated from soils of an unsprayed orchard (the central shoal of Yangtze River, Zhenjiang, Jiangsu Province, China) by our research team. Sequence analysis of the 5.8 S internal transcribed spacer (ITS) ribosomal DNA (rDNA) region of the yeast found that it was *P. caribbica* [49,50]. *P. caribbica* has been shown to be safe in animal testing, such as physiological, acute toxicity and Ames test [51]. *P. caribbica* isolates were maintained at 4 °C on nutrient yeast dextrose agar medium (NYDA-0.8% nutrient broth, 0.5% yeast extract, 1% glucose and 2% agar) (Sangon Biotech, Shanghai, China). Liquid cultures of the yeast were grown in 250-mL Erlenmeyer flasks containing 50 mL of nutrient yeast dextrose broth (NYDB) which had been inoculated with a loop of the culture. Flasks were incubated on a rotary shaker at 190 rpm for 48 h at 28 °C. Following incubation, cells were centrifuged at $6000 \times g$ for 10 min and washed twice with sterile distilled water. Cell pellets were re-suspended in sterile distilled water and adjusted to an initial concentration before being adjusted to the concentration required for different experiments.

5.2. Preparation of Stock Standard Solutions of Patulin

The patulin (Sigma-Aldrich, St. Louis, MO, USA) was prepared in accordance with the method described by MacDonald et al. [52], with some modifications. The patulin standard working solution (20 μg/mL) was diluted 10-fold with acetate buffer (pH 4.0). The samples were filtered through a 0.22 μm Wondadisc NY organic filter (SHIMADZU, Kyoto, Japan), and subjected to HPLC analysis. A standard curve of the patulin (μg/mL) was generated.

5.3. HPLC-UV Analysis of Patulin

Agilent 1100 series system (Agilent, Santa Clara, CA, USA) was used to analysis the patulin. The analytical column used was Zorbax, SB-C18 250 × 4.6 mm 5 µm (Agilent, Santa Clara, CA, USA). The mobile phase composed of water and CAN (9:1, *v*/*v*) that was set at 1 mL/min. The UV detection was performed at 276 nm. Data collection and subsequent processing were performed using Gilson Unipoint software 5.0 (Gilson, Inc, Middleton, WI, USA).

5.4. Efficacy of Cell-Free Culture Filtrate of P. caribbica on the Detoxification of Patulin

One milliliter culture medium of *P. caribbica* (1 × 10^8 cells/mL) was added to 50 mL NYDB in 250-mL Erlenmeyer flasks. Flasks were incubated on a rotary shaker at 180 rpm at 28 °C for 20 h. Following incubation, cells were centrifuged at 6000× *g* for 10 min. Then cell suspensions were filtered through micro pore (0.22 µm) to obtain the cell-free filtrate. Twenty-five mL cell-free filtrate of *P. caribbica* and NYDB medium (control) were each added into 150-mL Erlenmeyer flasks containing patulin at the concentration of 20 µg/mL, respectively. Samples were taken every 6 h and centrifuged at 7000× *g* for 5 min. The supernatants were filtered through a 0.22 µm filter and HPLC analysis was performed to determine the patulin contents. Every group had three replicates and the experiment was replicated twice.

5.5. Efficacy of Viable and Heat-Killed P. caribbica Cells on the Reduction of Patulin

The yeasts cells were inactivated in water at 100 °C for 15 min. Two mL of the live *P. caribbica* (1 × 10^8 cells/mL), the heat-killed cells, and sterile distilled water (control) were added to 25 mL NYDB containing 20 µg/mL of patulin in 150 mL Erlenmeyer flasks, respectively. Samples were taken at every 6 h, centrifuged at 7000× *g* for 5 min. The supernatants were filtered through a 0.22 µm filter, and subjected to HPLC analysis. Every group had three replicates and the experiment was replicated twice.

5.6. Absorption of Patulin by P. caribbica Cells

One milliliter suspension of *P. caribbica* (1 × 10^8 cells/mL) was added to 25 mL NYDB and 20 µg/mL of patulin in 150 mL Erlenmeyer flasks. The samples were incubated in a rotary shaker at 180 rpm for 20 h at 28 °C. Following incubation, cells were centrifuged at 6000× *g* for 10 min and washed twice with sterile distilled water to remove the patulin in the supernatant. The yeasts were sonicated at 1000 hz for 20 min. Fifteen mL of ethyl acetate was added to broken yeast cells, vortexed for 60 s and the upper layer was transferred into a separate funnel (125 mL). This step was repeated thrice and then 5 mL of 1.4% (*w*/*v*) Na_2CO_3 was added to the ethyl acetate layer and vigorously mixed for 2 min. The upper ethyl acetate layer was dried in a vacuum at 40 °C in a rotary evaporator. Soon after, 1 mL acetate buffer (0.2 mol/L, pH 4) was added and vortexed until it dissolved completely. The sample was filtered through a 0.22 µm Wondadisc NY organic filter (SHIMADZU, Kyoto, Japan), and analyzed in HPLC machine. Every group had three replicates and the experiment was replicated twice.

5.7. Effects of Cycloheximide on Degradation of Patulin by P. caribbica

One milliliter suspension of *P. caribbica* (1 × 10^8 cells/mL) was added to 25 mL NYDB and 20 µg/mL of patulin in 150 mL Erlenmeyer flasks. Treatments were as follows: (1) 5 µg/mL cycloheximide with patulin; (2) Addition of patulin into cycloheximide after 6 h; (3) only patulin (Control). Samples were taken at every 6 h, and centrifuged at 7000× *g* for 5 min. The supernatants were filtered through a 0.22 µm filter, and subjected to HPLC analysis to determine the patulin content. Every group had three replicates and the experiment was replicated twice.

5.8. Effects of P. caribbica Supernatant on Patulin Degradation

Two milliliter suspension of *P. caribbica* (1×10^8 cells/mL) was added to 25 mL NYDB and 20 µg/mL of patulin. After incubating for 6 h, one group was filtered twice through 0.22 µm filter to remove all the cells and the group was not filtered. Samples were taken every 6 h and centrifuged at $7000 \times g$ for 5 min. The supernatants were filtered through a 0.22 µm filter and analyzed using HPLC. Every group had three replicates and the experiment was replicated twice.

5.9. Degradation of Patulin by Intracellular Enzymes

P. caribbica (10^8 cells/mL) was incubated in NYDB and NYDB containing 20 µg/mL patulin in a rotary shaker at 180 rpm at 28 °C. After 20 h of incubation, the supernatants and the cells were centrifuged. The cells were washed with phosphate buffer (50 mM, pH 7.0) thrice. The wet cells were quickly ground in mortar using pestle with liquid nitrogen added and then suspended in 10 mL phosphate buffer. After 30 min in ice, the samples were centrifuged at $13,000 \times g$ for 10 min at 4 °C and the supernatant was collected. Afterwards, 100 µg of patulin was added to 5 mL of the yeast cells was amended with the patulin and those that were not amended. Every group had three replicates and the experiment was replicated twice.

5.10. Protein Sample Preparation

Liquid cultures of the yeasts were grown in NYDB and NYDB + 20 µg/mL patulin as described above under the section Antagonist and Growth Conditions'. The protein samples were prepared as described by Li et al. [53] with some modifications. After 24 h, the yeast cells were harvested from the NYDB and the NYDB + patulin, and centrifuged at $10,000 \times g$ for 10 min (4 °C). The cells were washed with cold distilled water each time after centrifugation to remove residual medium. Subsequently, the samples (yeast cells) were ground into fine powder with a mortar and pestle. The powder was transferred into a 50-mL tube. Thereafter, 10 mL TE buffer (10 mM Tris-HCL, pH 8.0, 1 mM EDTA, and 1 mM PMSF), 50 µg of RNase A and 200 µg of DNase I was added and incubated at 4 °C for 30 min. The samples were centrifuged at $11,000 \times g$ for 20 min (4 °C), then the supernatant was added to two volumes of 20% TCA/acetone (-20 °C pre-cooled at least 30 min), vortexed, and incubated at -20 °C for 12–16 h. Following incubation, the sample was centrifuged at $11,000 \times g$ for 20 min (4 °C) and the supernatant was discarded. The pellets were centrifuged at $11,000 \times g$ for 5 min (4 °C), washed with acetone (-20 °C pre-cooled). The pellets were air-dried at room temperature to remove residual acetone. Then solubilized in lysis buffer containing 2 M thiourea, 7 M urea, 4% (w/v) CHAPS, 65 mM DTT, 0.2% (w/v) Bio-Lyte (Bio-Rad, Hercules, CA, USA). Protein samples were kept at -80 °C until use. The protein concentration was determined according to Bradford's method using bovine serum albumin as standard [54].

5.11. 2-DE and Image Analysis

Two-dimensional electrophoresis (2-DE) and image analysis were performed as described by Wang et al. [55]. The first dimension electrophoresis was carried out on a 17-cm IPG strip (pH 3–10, Bio-Rad, Hercules, CA, USA). The strip was rehydrated for 1 h in the rehydration solution containing 2 M thiourea, 7 M urea, 4% (w/v) CHAPS, 65 mM DTT, 0. 2% (w/v) Bio-Lyte, and about 400 µg sample protein in a re-swelling trough. The rehydrated strip was then subjected to electrophoresis in the first dimension. After isoelectric focusing, the strip was equilibrated in two steps in the SDS equilibration stock solution consisting 50 mM Tris-HCl buffer, 6 M urea, 20% (v/v) glycerol and 2% (w/v) SDS supplemented with 2% (w/v) DTT, and 2.5% (w/v) iodoacetamide, respectively. The second dimension was run on a 12.5% polyacrylamide gel using the Multiphor system (Amersham Biosciences, Amersham, UK). The conditions were 1 W per strip for 1 h followed by 15 W per strip until bromophenol blue reached 0.5 cm above the bottom. To estimate the molecular weights of the protein spots, marker proteins were also separated together with the *P. caribbica* proteins. After electrophoresis, gels were visualized by Coomassie Blue stain. The stained gels were scanned and analyzed using PDQuest software (version 7.4, Bio-Rad, Hercules, CA, USA). Proteins that increased two-fold at one point after treatment, as well as exhibited the same expression pattern among the replicates, were considered as significant and reproducible change proteins. The proteins were subsequently identified. At least three biological replicates were performed for each treatment.

5.12. Protein In-Gel Digestion and Identification

Differentially expressed protein spots were excised from the gel and were washed twice by distilled H_2O, then destained with 50 mM NH_4CO_3/CAN solution. Afterwards, they were washed with 25 mM NH_4CO_3, 50% CAN until the gels became white, then vacuum drained for 5 min. Two µL trypsin (10 µg/µL) (Sigma–Aldrich, St. Louis, MO, USA) was incubated at 4 °C for 30 min, then 10 µL 25 mM NH_4CO_3 was added. The gels were incubated overnight at 37 °C. The supernatant was collected for MS analysis [56]. Proteins were identified by MALDI-TOF/TOF and database query. The peptide solution was analyzed using MALDI TOF/TOF mass spectrometer (Ultraflex III, Bruker-Daltonics, Bremen, Germany). MS/MS spectra were analyzed using the FlexAnalysis 3.0 (Bruker Daltonics GmbH, Bremen, Germany). The resulting monoisotopic peptide masses were queried against the protein database in NCBInr using MASCOT version 2.3 software (Matrix Science, Franklin, UK) with the following search parameters: all entries, trypsin, up to one missed cleavage, carbamidomethyl (C), oxidation (M), and Gln-Pyro-glu, peptide tolerance 0.3 Dal, mass value MH+, and monoisotopic [37].

5.13. Statistical Analysis

The data were analyzed by the analysis of variance (ANOVA) in the statistical program SPSS/PC version 17.0, (SPSS Inc. Chicago, IL, USA) and the Duncan's multiple range test was used for means separation. In addition, when the group of the data was two, the independent samples t test was applied for means separation. The statistical significance was assessed at the level $p < 0.05$.

Acknowledgments: This work was supported by the National Natural Science Foundation of China (31571899, 31271967), the Technology Support Plan of Jiangsu Province (BE2014372), the Technology Support Plan of Zhenjiang (NY2013004), and the Graduate Innovative Projects of Jiangsu Province (KYLX_1069).

Author Contributions: X.Z. conducted most of the experiments, analyzed the results, and wrote the initial paper. Q.Y. conducted experiments on the patulin content analysis and wrote part of the paper. H.Z. conceived the idea for the project and provided the funding. J.C. and M.T.A. revised the article. X.Z. provided experiment direction and revised the article.

Conflicts of Interest: The authors declare no conflict of interest.

References

1. Joint FAO/WHO Expert Committee on Food Additives; World Health Organization; Food and Agriculture Organization of the United Nations; International Programme on Chemical Safety. *Toxicological Evaluation of Certain Food Additives and Contaminants in Food*; Cambridge University Press: Cambridge, UK, 1996; pp. 189–197.

2. Ono, E.Y.S.; Biazon, L.; Silva, M.D.; Vizoni, É.; Sugiura, Y.; Ueno, Y.; Hirooka, E.Y. Fumonisins in corn: Correlation with *fusarium* sp. Count, damaged kernels, protein and lipid content. *Braz. Arch. Biol. Technol.* **2006**, *49*, 1967–1976. [CrossRef]

3. Pitt, J.I.; Basílico, J.C.; Abarca, M.L.; López, C. Mycotoxins and toxigenic fungi. *Med. Mycol.* **2010**, *38* (Suppl. S1), 41–46. [CrossRef]

4. Ritieni, A. Patulin in italian commercial apple products. *J. Agric. Food Chem.* **2003**, *51*, 6086–6090. [CrossRef] [PubMed]

5. Andersen, B.; Smedsgaard, J.; Frisvad, J.C. Penicillium expansum: Consistent production of patulin, chaetoglobosins, and other secondary metabolites in culture and their natural occurrence in fruit products. *J. Agric. Food Chem.* **2004**, *52*, 2421–2428. [CrossRef] [PubMed]

6. Waksman, S.A.; Horning, E.S.; Spencer, E.L. The production of two antibacterial substances, fumigacin and clavacin. *Am. Assoc. Adv. Sci.* **1942**, *96*, 202–203. [CrossRef] [PubMed]

7. Wiesner, B. Bactericidal effects of aspergillus clavatus. *Nature* **1942**, *149*, 356–357. [CrossRef]
8. Martins, M.; Gimeno, A.; Martins, H.; Bernardo, F. Co-occurrence of patulin and citrinin in portuguese apples with rotten spots. *Food Addit. Contam.* **2002**, *19*, 568–574. [CrossRef] [PubMed]
9. Barreira, M.J.; Alvito, P.C.; Almeida, C.M. Occurrence of patulin in apple-based-foods in portugal. *Food Chem.* **2010**, *121*, 653–658. [CrossRef]
10. Baert, K.; de Meulenaer, B.; Kamala, A.; Kasase, C.; Devlieghere, F. Occurrence of patulin in organic, conventional, and handcrafted apple juices marketed in belgium. *J. Food Prot.* **2006**, *69*, 1371–1378. [PubMed]
11. Saxena, N.; Dwivedi, P.D.; Ansari, K.M.; Das, M. Patulin in apple juices: Incidence and likely intake in an indian population. *Food Addit. Contam.* **2008**, *1*, 140–146. [CrossRef] [PubMed]
12. Yuan, Y.; Zhuang, H.; Zhang, T.; Liu, J. Patulin content in apple products marketed in northeast china. *Food Control* **2010**, *21*, 1488–1491. [CrossRef]
13. Schumacher, D.M.; Metzler, M.; Lehmann, L. Mutagenicity of the mycotoxin patulin in cultured chinese hamster v79 cells, and its modulation by intracellular glutathione. *Arch. Toxicol.* **2005**, *79*, 110–121. [CrossRef] [PubMed]
14. Osswald, H.; Frank, H.; Komitowski, D.; Winter, H. Long-term testing of patulin administered orally to sprague-dawley rats and swiss mice. *Food Cosmet. Toxicol.* **1976**, *16*, 243–247. [CrossRef]
15. Wichmann, G.; Herbarth, O.; Lehmann, I. The mycotoxins citrinin, gliotoxin, and patulin affect interferon-γ rather than interleukin-4 production in human blood cells. *Environ. Toxicol.* **2002**, *17*, 211–218. [CrossRef] [PubMed]
16. Devaraj, H.; Radha, S.K.; Shanmugasundaram, E. Neurotoxic effect of patulin. *Indian J. Exp. Biol.* **1982**, *20*, 230–231. [PubMed]
17. Ciegler, A.; Beckwith, A.; Jackson, L.K. Teratogenicity of patulin and patulin adducts formed with cysteine. *Appl. Environ. Microbiol.* **1976**, *31*, 664–667. [PubMed]
18. Donmez-Altuntas, H.; Gokalp-Yildiz, P.; Bitgen, N.; Hamurcu, Z. Evaluation of genotoxicity, cytotoxicity and cytostasis in human lymphocytes exposed to patulin by using the cytokinesis-block micronucleus cytome (CBMN cyt) assay. *Mycotoxin Res.* **2013**, *29*, 63–70. [CrossRef] [PubMed]
19. Nunes, C.A. Biological control of postharvest diseases of fruit. *Eur. J. Plant Pathol.* **2012**, *133*, 181–196. [CrossRef]
20. Droby, S.; Wisniewski, M.; Macarisin, D.; Wilson, C. Twenty years of postharvest biocontrol research: Is it time for a new paradigm? *Postharvest Biol. Technol.* **2009**, *52*, 137–145. [CrossRef]
21. Cao, J.; Zhang, H.; Yang, Q.; Ren, R. Efficacy of *pichia caribbica* in controlling blue mold rot and patulin degradation in apples. *Int. J. Food Microbiol.* **2013**, *162*, 167–173. [CrossRef] [PubMed]
22. Castoria, R.; Morena, V.; Caputo, L.; Panfili, G.; de Curtis, F.; De Cicco, V. Effect of the biocontrol yeast rhodotorula glutinis strain LS11 on patulin accumulation in stored apples. *Phytopathology* **2005**, *95*, 1271–1278. [CrossRef] [PubMed]
23. Coelho, A.; Celli, M.; Sataque Ono, E.; Hoffmann, F.; Pagnocca, F.; Garcia, S.; Sabino, M.; Harada, K.; Wosiacki, G.; Hirooka, E. Patulin biodegradation using *Pichia ohmeri* and *Saccharomyces cerevisiae*. *World Mycotoxin J.* **2008**, *1*, 325–331. [CrossRef]
24. Castoria, R.; Mannina, L.; Durán-Patrón, R.; Maffei, F.; Sobolev, A.P.; de Felice, D.V.; Pinedo-Rivilla, C.; Ritieni, A.; Ferracane, R.; Wright, S.A. Conversion of the mycotoxin patulin to the less toxic desoxypatulinic acid by the biocontrol yeast *Rhodosporidium kratochvilovae* strain LS11. *J. Agric. Food Chem.* **2011**, *59*, 11571–11578. [CrossRef] [PubMed]
25. Bevardi, M.; Frece, J.; Mesarek, D.; Bošnir, J.; Mrvčić, J.; Delaš, F.; Markov, K. Antifungal and antipatulin activity of *Gluconobacter* oxydans isolated from apple surface. *Arch. Ind. Hyg. Toxicol.* **2013**, *64*, 279–284. [CrossRef] [PubMed]
26. Coelho, A.R.; Celli, M.G.; Ono, E.Y.S.; Wosiacki, G.; Hoffmann, F.L.; Pagnocca, F.C.; Hirooka, E.Y. *Penicillium expansum* versus antagonist yeasts and patulin degradation in vitro. *Braz. Arch. Biol. Technol.* **2007**, *50*, 725–733. [CrossRef]
27. Zhu, R.; Yu, T.; Guo, S.; Hu, H.; Zheng, X.; Karlovsky, P. Effect of the yeast *Rhodosporidium paludigenum* on postharvest decay and patulin accumulation in apples and pears. *J. Food Prot.* **2015**, *78*, 157–163. [CrossRef] [PubMed]
28. Zhu, R.; Feussner, K.; Wu, T.; Yan, F.; Karlovsky, P.; Zheng, X. Detoxification of mycotoxin patulin by the yeast *Rhodosporidium paludigenum*. *Food Chem.* **2015**, *179*, 1–5. [CrossRef] [PubMed]

29. Iwahashi, Y.; Hosoda, H.; Park, J.-H.; Lee, J.-H.; Suzuki, Y.; Kitagawa, E.; Murata, S.M.; Jwa, N.-S.; Gu, M.-B.; Iwahashi, H. Mechanisms of patulin toxicity under conditions that inhibit yeast growth. *J. Agric. Food Chem.* **2006**, *54*, 1936–1942. [CrossRef] [PubMed]

30. Ianiri, G.; Idnurm, A.; Wright, S.A.; Durán-Patrón, R.; Mannina, L.; Ferracane, R.; Ritieni, A.; Castoria, R. Searching for genes responsible for patulin degradation in a biocontrol yeast provides insight into the basis for resistance to this mycotoxin. *Appl. Environ. Microbiol.* **2013**, *79*, 3101–3115. [CrossRef] [PubMed]

31. Ianiri, G.; Idnurm, A.; Castoria, R. Transcriptomic responses of the basidiomycete yeast *sporobolomyces* sp. To the mycotoxin patulin. *BMC Genom.* **2016**, *17*, 1–15. [CrossRef] [PubMed]

32. Di, S.V.; Pitonzo, R.; Cicero, N.; D'Oca, M.C. Mycotoxin contamination of animal feedingstuff: Detoxification by gamma-irradiation and reduction of aflatoxins and ochratoxin A concentrations. *Food Addit. Contam.* **2014**, *31*, 2034–2039.

33. Shi, L.; Liang, Z.; Li, J.; Hao, J.; Xu, Y.; Huang, K.; Tian, J.; He, X.; Xu, W. Ochratoxin A biocontrol and biodegradation by *Bacillus subtilis* cw 14. *J. Sci. Food Agric.* **2014**, *94*, 1879–1885. [CrossRef] [PubMed]

34. Dong, X.; Jiang, W.; Li, C.; Ma, N.; Xu, Y.; Meng, X. Patulin biodegradation by marine yeast kodameae ohmeri. *Food Additives & Contaminants Part A Chemistry Analysis Control Exposure & Risk Assessment* **2015**, *32*, 352–360.

35. Zhang, H.; Dong, M.; Yang, Q.; Tibiru, A.M.; Li, J.; Zhang, X. Biodegradation of zearalenone by *Saccharomyces cerevisiae*: Possible involvement of zen responsive proteins of the yeast. *J. Proteomics* **2016**, *143*, 416–423. [CrossRef] [PubMed]

36. Folger, B.C. Patulin degradation by yeast protein extract. Master's Theses, University of Minnesota, Minneapolis, MN, 2014.

37. Majoul, T.; Bancel, E.; Triboï, E.; Ben Hamida, J.; Branlard, G. Proteomic analysis of the effect of heat stress on hexaploid wheat grain: Characterization of heat-responsive proteins from total endosperm. *Proteomics* **2003**, *3*, 175–183. [CrossRef] [PubMed]

38. Calderwood, S.K.; Murshid, A.; Prince, T. The shock of aging: Molecular chaperones and the heat shock response in longevity and aging–a mini-review. *Gerontology* **2009**, *55*, 550–558. [CrossRef] [PubMed]

39. Morimoto, R.I. Proteotoxic stress and inducible chaperone networks in neurodegenerative disease and aging. *Genes Dev.* **2008**, *22*, 1427–1438. [CrossRef] [PubMed]

40. Voellmy, R.; Boellmann, F. Chaperone regulation of the heat shock protein response. *Adv. Exp. Med. Biol.* **2007**, *23*, 89–99.

41. Bukau, B.; Deuerling, E.; Pfund, C.; Craig, E.A. Getting newly synthesized proteins into shape. *Cell* **2000**, *101*, 119–122. [CrossRef]

42. Hartl, F.U.; Hayer-Hartl, M. Molecular chaperones in the cytosol: From nascent chain to folded protein. *Science* **2002**, *295*, 1852–1858. [CrossRef] [PubMed]

43. Young, J.C.; Barral, J.M.; Hartl, F.U. More than folding: Localized functions of cytosolic chaperones. *Trends Biochem. Sci.* **2003**, *28*, 541–547. [CrossRef] [PubMed]

44. Pratt, W.B.; Toft, D.O. Regulation of signaling protein function and trafficking by the hsp90/hsp70-based chaperone machinery. *Exp. Biol. Med.* **2003**, *228*, 111–133.

45. Kepes, F.; Schekman, R. The yeast sec53 gene encodes phosphomannomutase. *J. Biol. Chem.* **1988**, *263*, 9155–9161. [PubMed]

46. Seifert, G.J. Nucleotide sugar interconversions and cell wall biosynthesis: How to bring the inside to the outside. *Curr. Opin. Plant Biol.* **2004**, *7*, 277–284. [CrossRef] [PubMed]

47. Lerouge, P.; Cabanes-Macheteau, M.; Rayon, C.; Fischette-Lainé, A.-C.; Gomord, V.; Faye, L. N-glycoprotein biosynthesis in plants: Recent developments and future trends. *Plant Mol. Biol.* **1998**, *38*, 31–48. [CrossRef] [PubMed]

48. Spiro, R.G. Protein glycosylation: Nature, distribution, enzymatic formation, and disease implications of glycopeptide bonds. *Glycobiology* **2002**, *12*, 43R–56R. [CrossRef] [PubMed]

49. Kurtzman, C.; Fell, J.; Boekhout, T. Definition, classification and nomenclature of the yeasts. *Yeasts Taxon. Study* **2011**, *1*, 3–5.

50. Li, S.-S.; Cheng, C.; Li, Z.; Chen, J.-Y.; Yan, B.; Han, B.-Z.; Reeves, M. Yeast species associated with wine grapes in china. *Int. J. Food Microbiol.* **2010**, *138*, 85–90. [CrossRef] [PubMed]

51. Zheng, X.F. *P. caribbica* shown to be safe in physiological, acute toxicity and Ames test in animal. Jiangsu University, Zhenjiang, China. Unpublished work, 2016.

52. MacDonald, S.; Long, M.; Gilbert, J.; Felgueiras, I. Liquid chromatographic method for determination of patulin in clear and cloudy apple juices and apple puree: Collaborative study. *J. AOAC Int.* **2000**, *83*, 1387–1394. [PubMed]

53. Li, B.; Lai, T.; Qin, G.; Tian, S. Ambient ph stress inhibits spore germination of penicillium expansum by impairing protein synthesis and folding: A proteomic-based study. *J. Proteome Res.* **2010**, *9*, 298–307. [CrossRef] [PubMed]

54. Bradford, M.M. A rapid and sensitive method for the quantitation of microgram quantities of protein utilizing the principle of protein-dye binding. *Anal. Biochem.* **1976**, *72*, 248–254. [CrossRef]

55. Wang, Y.; Yang, L.H.; Li, Q.; Ma, Z.; Chu, C. Differential proteomic analysis of proteins in wheat spikes induced by fusarium graminearum. *Proteomics* **2005**, *5*, 4496–4503. [CrossRef] [PubMed]

56. Zhang, L.; Yu, Z.; Jiang, L.; Jiang, J.; Luo, H.; Fu, L. Effect of post-harvest heat treatment on proteome change of peach fruit during ripening. *J. Proteom.* **2011**, *74*, 1135–1149. [CrossRef] [PubMed]

toxins

Article

Patulin Degradation by the Biocontrol Yeast *Sporobolomyces* sp. Is an Inducible Process

Giuseppe Ianiri [1,*,†], Cristina Pinedo [1,2], Alessandra Fratianni [1], Gianfranco Panfili [1] and Raffaello Castoria [1,*]

1 Department of Agricultural, Environmental and Food Sciences, Università degli Studi del Molise, via Francesco de Sanctis, 86100 Campobasso, Italy; cristina.pinedo@uca.es (C.P.); fratianni@unimol.it (A.F.); panfili@unimol.it (G.P.)
2 Department of Organic Chemistry, Universidad de Cádiz, 11510 Puerto Real (Cádiz), Spain
* Correspondences: giuseppe.ianiri@duke.edu (G.I.); castoria@unimol.it (R.C.); Tel.: +39-0874-404-698 (G.I. & R.C.)
† Present address: Department of Molecular Genetics and Microbiology, Duke University Medical Center, Durham, NC 27710, USA.

Academic Editor: Ting Zhou
Received: 12 January 2017; Accepted: 7 February 2017; Published: 10 February 2017

Abstract: Patulin is a mycotoxin produced by *Penicillium expansum* and a common contaminant of pome fruits and their derived products worldwide. It is considered to be mutagenic, genotoxic, immunotoxic, teratogenic and cytotoxic, and the development of strategies to reduce this contamination is an active field of research. We previously reported that *Sporobolomyces* sp. is able to degrade patulin and convert it into the breakdown products desoxypatulinic acid and ascladiol, both of which were found to be less toxic than patulin. The specific aim of this study was the evaluation of the triggering of the mechanisms involved in patulin resistance and degradation by *Sporobolomyces* sp. Cells pre-incubated in the presence of a low patulin concentration showed a higher resistance to patulin toxicity and a faster kinetics of degradation. Similarly, patulin degradation was faster when crude intracellular protein extracts of *Sporobolomyces* sp. were prepared from cells pre-treated with the mycotoxin, indicating the induction of the mechanisms involved in the resistance and degradation of the mycotoxin by *Sporobolomyces* sp. This study contributes to the understanding of the mechanisms of patulin resistance and degradation by *Sporobolomyces* sp., which is an essential prerequisite for developing an industrial approach aiming at the production of patulin-free products.

Keywords: *Sporobolomyces* sp. IAM 13481; microbial patulin degradation; desoxypatulinic acid; ascladiol

1. Introduction

Mycotoxins are secondary metabolites produced by fungi belonging to several genera, such as *Aspergillus, Penicillium*, and *Alternaria*. Besides their economical importance due to crops infection and product losses, their toxic secondary metabolites represent a health risk to humans and animals. The maximum content of some mycotoxins in harvested commodities and derived products has been established by national and international organizations, and the reduction of mycotoxin contamination is an important research focus.

Patulin (PAT) is a toxic fungal metabolite produced mainly by species of *Penicillium* and *Aspergillus*. *P. expansum*, the causative agent of the blue mould disease of stored apples, is the main PAT producer and its infections result in PAT contamination of apples, pears, and their derived products [1]. The carcinogenic risk of PAT is classified in group 3 by the International Agency for Research on Cancer [2], and this has led to the establishment of a maximum tolerable daily intake for PAT of 0.4 mg/kg body weight/day [3]. Moreover, although there are legislative regulations in Europe and

USA that set the highest tolerable levels of PAT in fruit-based products and juices at 50 µg/kg, and for baby food at 10 µg/kg (EC Regulation 1881/2006), recent surveys in Europe and USA revealed that PAT contamination is still a common issue [4–6].

The mechanisms of PAT toxicity and its effects on living cells have been long investigated [1,7]. Patulin was found to have antibacterial and antifungal activities [7]. In mammals, the primary target organs of PAT toxicity are the gastrointestinal tract, kidney, liver and the immune system. Continued exposure to high concentrations of PAT may include mutagenicity, genotoxicity, embryotoxicity, immunotoxicity and neurotoxicity [1,7,8].

The ecological role of PAT has not been fully elucidated, and it is likely to help in competition against other microbes that PAT-producing fungi encounter in their niche. Mutation in the gene encoding enzymes involved in PAT biosynthesis [7,9] has allowed to investigate the role of PAT in the pathogenicity of *P. expansum*, which is still controversial and it seems to be dependent on factors of the host fruits [10–13].

Limiting *P. expansum* infections in postharvest settings is crucial to prevent PAT accumulation. The use of chemical fungicides is still the most effective approach. However, ethical, technical and health issues, foster an increasing demand for alternative methods to reduce the use of chemicals, and the use of biocontrol agents and/or controlled atmosphere are very promising strategies [6]. Moreover, their combination with tools aiming at the detoxification of PAT has the potential to further reduce PAT contamination, thus providing safer fruit juices; an important focus because their contamination poses a major risk for children, who consume great quantities of fruit juices.

The influence of biocontrol agents (BCAs) that are effective against *P. expansum* on PAT accumulation is an emerging and attractive field of study, and major findings have been elegantly reviewed [14]. The ability of PAT degradation seems to be a widespread feature of the *Pucciniomycotina* red yeasts. We found that *Sporobolomyces* sp. IAM 13481 was able to convert PAT to DPA and ascladiols, with (*E*) ascladiol being a transient metabolite, and DPA and (*Z*) ascladiol the two final breakdown products [15]. In the red yeast BCAs *Rhodotorula kratochvilovae* and *R. paludigenum*, DPA was found as the major PAT degradation products [16–18], although at least in the case of *R. kratochvilovae*, the isomers (*E*) and (*Z*) of ascladiol were found as transient products (our unpublished data). Using *Sporobolomyces* sp. IAM 13481 as a model, we explored the molecular mechanisms involved in PAT degradation by *Pucciniomycotina* yeasts through a random insertional mutagenesis approach based on *Agrobacterium tumefaciens*-mediated transformation (AMT) [15]) as well as a transcriptomic approach based on RNAseq analysis of yeast cells grown in the presence and absence of this mycotoxin [19].

Our data indicated that PAT toxicity is exerted through the generation of ROS (reactive oxygen species), which leads to cellular damage but also function as signalling molecules that activate cellular components required for oxidative stress resistance and PAT degradation. Based on these findings, in the present work, we investigated whether the biochemical mechanisms that lead to PAT resistance and degradation in *Sporobolomyces* sp. IAM13481 can be induced by exposition to low doses of this mycotoxin.

2. Results

2.1. Pre-Treatment of Sporobolomyces sp. IAM 13481 Cells with Patulin Induces Mechanisms Involved in Its Resistance and Degradation

Sporobolomyces sp. IAM 13481 is able to degrade in vitro the mycotoxin PAT under aerobic conditions. The products of PAT degradation have been identified as (*E*)-ascladiol, (*Z*)-ascladiol and desoxypatulinic acid (DPA); (*E*)-ascladiol is a transient or less stable product, while DPA and (*Z*)-ascladiol are the final metabolites of PAT degradation [15] (Figure 1).

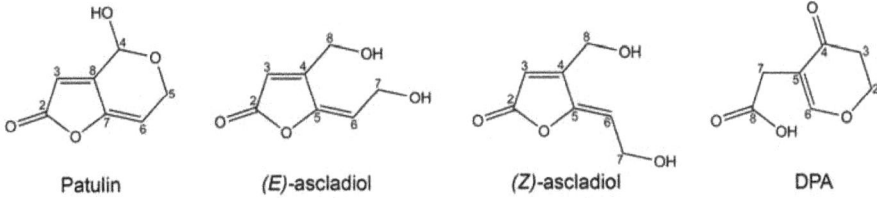

Figure 1. Chemical structures of patulin and its breakdown products (*E*) and (*Z*) ascladiol, and desoxypatulinic acid (DPA).

In the present work, the induction of the mechanisms that underlie PAT resistance and degradation by *Sporobolomyces* sp. IAM 13481 was assessed by incubating pretreated cells (PC, cells preincubated in the presence of 15 µg/mL of PAT) and non-preincubated cells (NPC, cells incubated in the absence of PAT) in the presence of two different concentrations of PAT, 50 and 100 µg/mL.

In Figure 2 the growth curves of NPC and PC of *Sporobolomyces* sp. IAM 13481 in the absence and in the presence of 50 and 100 µg/mL of PAT are reported. While in LiBa without PAT the growth kinetics of NPC and PC followed a similar trend, there were clear differences between their growth in the presence of the mycotoxin. In particular, with 50 µg/mL of PAT, PC suffered from the initial toxicity of PAT only partially and reached the highest OD_{595} value after 3 days of incubation, whereas the growth of NPC was characterized by a 1-day lag phase followed by exponential growth until the fourth day of incubation. In the presence of 100 µg/mL of PAT, there were more marked differences between the growth of PC and NPC, with PC showing a 1-day lag phase followed by active growth until the fourth day of incubation, a trend that reflected that of NPC grown with PAT 50 µg/mL, and NPC characterized by a 4-day lag phase followed by exponential growth until the seventh day of incubation.

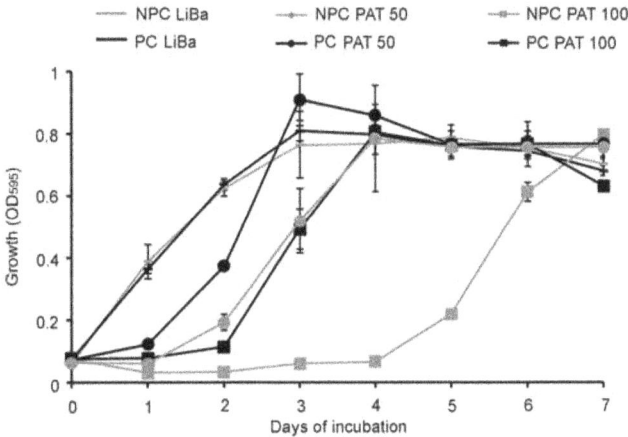

Figure 2. Growth curves of *Sporobolomyces* sp. IAM 13481 cells that were preincubated (PC, black lines) and non-preincubated (NPC, grey lines) with PAT in the presence (50 and 100 µg/mL) or in the absence (i.e., LiBa alone) of the mycotoxin. Values are expressed as OD_{595} and represent the means ± standard deviation (*n* = 9).

In accordance with our previous report [15], the kinetics of PAT degradation reflected the growth curves both for PC and NPC. As reported in Figure 3, with 50 µg/mL of PAT, after 24 h of incubation there was a significant difference between the PAT recovery of PC and NPC, with values of 33.40 ± 2.16 µg/mL and 43.91 ± 0.11 µg/mL, respectively (Figure 3A–C). At the same time point, PC produced 2.73 ± 0.14 µg/mL of (*E*)-ascladiol, 0.62 ± 0.15 µg/mL of (*Z*)-ascladiol and 1.09 ± 0.04 µg/mL of DPA (Figure 3A), while only traces of (*E*)-ascladiol were detected for NPC (Figure 3B). Also, at 48 h of incubation there was a significant difference in PAT reduction by *Sporobolomyces* PC and NPC (Figure 3C). PC led to a complete degradation of 50 µg/mL of PAT after three days of incubation, in correspondence of which they produced 6.90 ± 1.56 µg/mL of (*E*)-ascladiol, 4.47 ± 1.04 µg/mL of (*Z*)-ascladiol and 5.70 ± 0.04 µg/mL of DPA. In the same culture, at the fourth day of incubation, the (*E*)-ascladiol decreased to 3.75 ± 1.27 µg/mL, whereas the (*Z*)-ascladiol and DPA increased respectively to 5.09 ± 1.61 µg/mL and 6.06 ± 1.43 µg/mL, the highest values achieved; at the last time points, the amount of the breakdown products slightly decreased (Figure 3A) probably due to their instability or other unknown enzymatic activities. In NPC of *Sporobolomyces*, after four days of incubation, PAT was almost totally degraded (PAT was only 1.81 ± 2.52 µg/mL), and this paralleled the production of 5.36 ± 3.44 µg/mL of (*E*)-ascladiol, 4.70 ± 1.75 µg/mL of (*Z*)-ascladiol and 5.15 ± 1.30 µg/mL of DPA; at the fifth day, the (*E*)-ascladiol decreased, while (*Z*)-ascladiol and DPA slightly increased (Figure 3B).

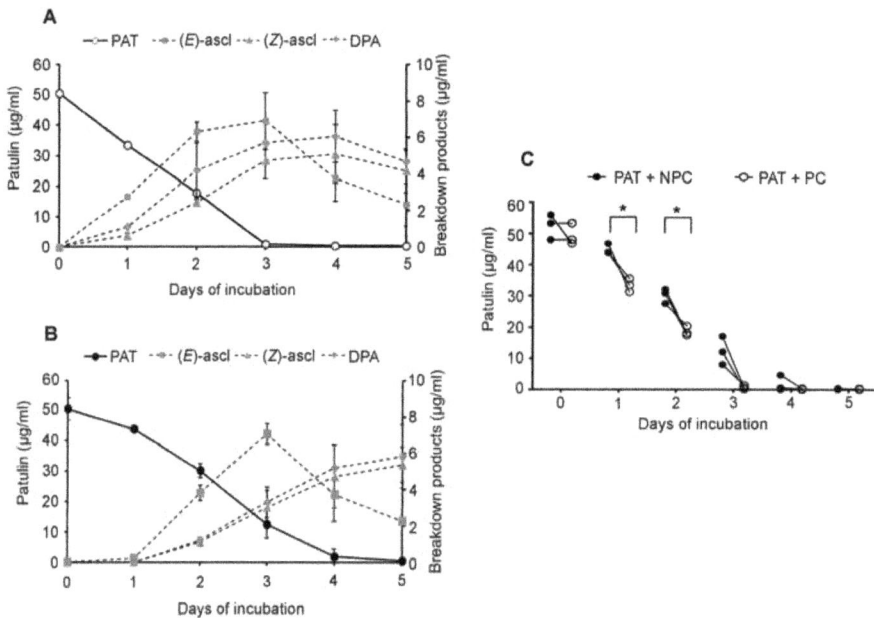

Figure 3. Time course of PAT decrease (full line) and onset of breakdown products (dotted lines) of *Sporobolomyces* PC (**A**) and NPC (**B**) grown in the presence of 50 µg/mL of mycotoxin. PAT is patulin, (*E*)-ascl and (*Z*)-ascl are (*E*)-ascladiol and (*Z*)-ascladiol, respectively. Values are expressed as µg/mL and represent the means ± standard deviation (*n* = 9, three measurements in three experiments); (**C**) Fate of 50 µg/mL of PAT in the presence of *Sporobolomyces* PC and NPC. Asterisk indicates significant differences ($p < 0.05$) when analyzed by Student's *t* test.

With 100 µg/mL of PAT (Figure 4), PC of *Sporobolomyces* started to degrade the mycotoxin since the first day of incubation (PAT recovery of 89.18 ± 0.38 µg/mL), and this paralleled the production of about 2 µg/mL of (*E*)-ascladiol and DPA, and traces of (*Z*)-ascladiol (Figure 4A). At the third day of incubation, more than 50% of PAT was degraded, and the highest production of (*E*)-ascladiol (8.00 ± 0.35 µg/mL) was also achieved, which, as expected, decreased to 1.54 ± 0.41 µg/mL on the last day of incubation. PAT was completely degraded after four days of incubation, and, at the same time point, the amount of the final breakdown products (Z)-ascladiol and DPA were 6.91 ± 1.75 µg/mL and 8.64 ± 0.55 µg/mL, respectively. Even though no more PAT was present in the medium, the amounts of these two degradation products still increased to reach the highest values at the sixth day of incubation, 11.00 ± 1.66 µg/mL and 11.34 ± 0.44 µg/mL for (Z)-ascladiol and DPA, respectively (Figure 4A). For NPC instead (Figure 4B), a considerable PAT reduction of about 40% was observed only at the fifth day of incubation, and at the same time point, (E)-ascladiol reached its production peak (4.42 ± 0.62 µg/mL) and decreased in the following days, while the amounts of (Z)-ascladiol and DPA were still low (traces for (Z)-ascladiol and 2.99 ± 0.11 µg/mL for DPA). PAT was completely degraded at six days of incubation, and at the seventh day there was the highest production of (Z)-ascladiol and DPA (12.36 ± 1.49 µg/mL and 13.36 ± 0.96 µg/mL, respectively) (Figure 4B). As shown in Figure 4C, after 1, 3, 4 and 5 days of incubation, there were highly significant differences between the degradation of 100 µg/mL of PAT by PC and NPC. With 200 µg/mL of PAT, there were even more remarkable differences between PC and NPC of *Sporobolomyces*, and when there was a lower initial concentration of cells (i.e., 1×10^5 CFU/mL), only PC were able to grow and degrade PAT, whereas NPC did not (data not shown).

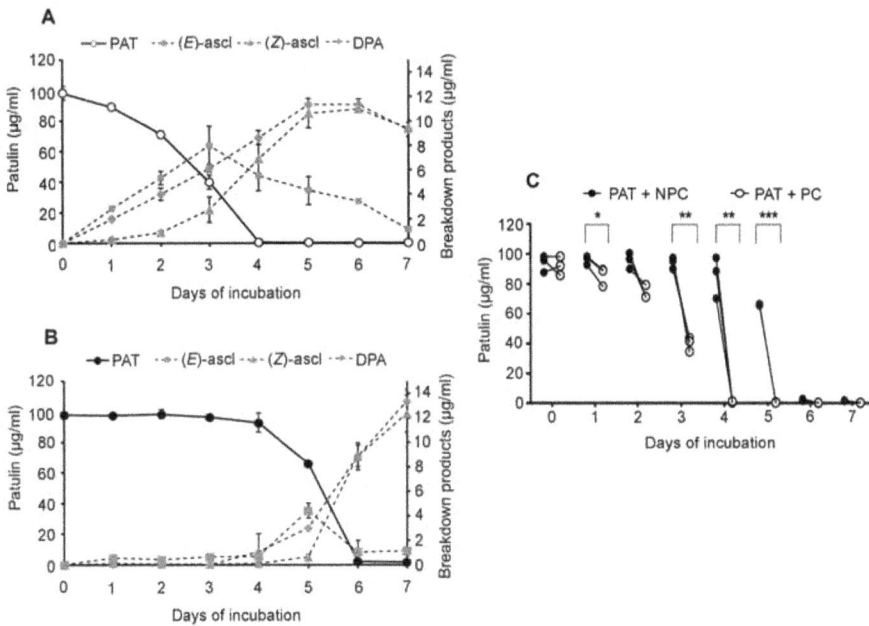

Figure 4. Time course of PAT decrease (full line) and onset of breakdown products (dotted lines) of *Sporobolomyces* PC (**A**) and NPC (**B**) grown in the presence of 100 µg/mL of PAT. PAT is patulin, (E)-ascl and (Z)-ascl are (E)-ascladiol and (Z)-ascladiol, respectively. Values are expressed as µg/mL and represent the means ± standard deviation (*n* = 9, three measurements in three experiments); (**C**) Fate of 100 µg/mL of PAT in the presence of *Sporobolomyces* PC and NPC. Symbols indicate significant differences for $p < 0.05$ (*), $p < 0.01$ (**), and $p < 0.001$ (***) when analyzed by Student's *t* test.

2.2. Intracellular Extracts from Sporobolomyces sp. Cells Preincubated with Patulin Degrade the Mycotoxin More Rapidly Than Extracts from Non-Preincubated Cells

Intracellular extracts from PC (EPC, induced extracts) were obtained from *Sporobolomyces* sp. cells in their active phase of PAT degradation, which was monitored by TLC analysis as described in Section 5.5; intracellular extracts from non-preincubated cells (ENPC) were obtained from *Sporobolomyces* sp. cells grown in LiBa medium without PAT when the growth reached the same

OD_{595} as the culture with PAT. Using this rationale, cells for intracellular extracts were expected to be at the same metabolic stage, with the only difference being the presence or the absence of the mycotoxin.

In Figure 5A,B, the fate of 30 and 50 µg/mL of PAT in the presence of EPC and ENPC of *Sporobolomyces* sp, respectively, is reported. With 30 µg/mL of PAT, even after 30 min of incubation, there was a significant difference between EPC and ENPC, with a PAT recovery of 16.09 ± 0.13 µg/mL and 27.68 ± 1.30 µg/mL, respectively; from 60 min to 360 min (6 h) of incubation; the differences in PAT recovery between EPC and ENPC were still significant, and in both cases only traces of mycotoxin were found after 1440 min (24 h) of incubation (Figure 5A). When 50 µg/mL of mycotoxin was used, significant differences between the PAT recovery in the presence of EPC and ENPC were found after 60 and 90 min of incubation, and since then they were progressively lower until only traces of mycotoxin were found after 1440 min of incubation (Figure 5B). At both concentrations used (30 µg/mL and 50 µg/mL), no PAT decrease was observed when incubated in the presence of boiled ENPC and EPC.

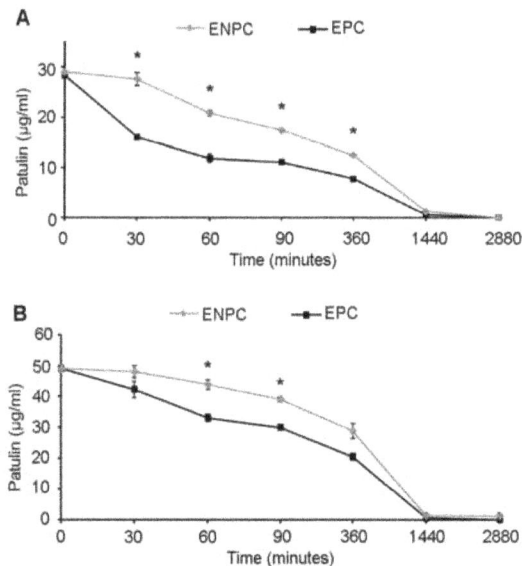

Figure 5. Fate of 30 µg/mL (**A**) and 50 µg/mL (**B**) of PAT in the presence of induced intracellular extracts (EPC, black lines) and non-induced intracellular extracts (ENPC, grey lines) of *Sporobolomyces*. Values are expressed as µg/mL and represent the means ± standard deviation ($n = 9$, three measurements in three experiments). Symbols indicate significant differences for $p < 0.05$ (*) when analyzed by Student's *t* test.

3. Discussion

PAT is a worldwide mycotoxin that represents an economic challenge to pome fruit food industries and a health hazard to consumers. The reduction of PAT contamination through the use of beneficial microbes has been known for long time and it is still an active field of research [14].

Pucciniomycotina red yeasts have emerged amongst the microbes that exhibit an elevated tolerance to PAT toxicity and have a constitutive arsenal to breakdown the PAT molecule. Previous research showed that the red yeast *Sporobolomyces* sp. IAM 13481 used in the present study is able to degrade PAT in vitro by forming desoxypatulinic acid (DPA) and (Z)-ascladiol as the final products of degradation, whereas (E)-ascladiol is only formed as a transient product [15]. The red yeasts

Rhodotorula kratochvilovae LS11 and *R. paludigenum* (originally classified as *Rhodosporidium* but renamed following recent taxonomy analysis [20]) are also able to degrade the mycotoxin PAT, but only DPA was found as a final product of degradation [17,18]; at least for *R. kratochvilovae* LS11, both ascladiol isomers were detected as transient products (our unpublished data). We also found that several other red yeast species degrade PAT in a similar manner to *Sporobolomyces* or *Rhodotorula* (our unpublished data).

Therefore, it seems likely that *Pucciniomycotina* red yeasts degrade PAT through two different pathways, and their regulation differs between the species under examination. The hypothesis of the two distinct pathways is supported by the chemical structures of DPA and ascladiol, as they derive from the breakage of bonds that are located in separate ends of the PAT molecule (Figure 1). Interestingly, the production of DPA has been described only in red yeasts, whereas ascadiol was found following PAT degradation both by yeasts (i.e., *Saccharomyces cerevisiae*) and bacteria (*Gluconobacter oxydans* and *Lactobacillus plantarum* [21–23]). This suggests that while the ascladiol-forming degradation pathway is likely to be conserved across fungi and bacteria, the DPA-forming pathway has independently evolved in the red yeast group within the *Pucciniomycotina*. The importance of these mechanisms of degradation is supported by the low toxicity of the PAT breakdown products: DPA is less toxic to different microorganisms, to human lymphocytes and hepatocytes, as well as to *Arabidopsis thaliana* [17,18,24,25]; ascladiol was originally reported as retaining a quarter of the toxicity displayed by PAT [26], and a more recent investigation demonstrated that it is not toxic to the porcine intestinal tissue [27].

Noteworthy, the sum of the breakdown products was found to be less than the initial concentration of PAT. Although there is evidence of adsorption of PAT on the yeast cell wall [28], this phenomenon does not occur in *Sporobolomyces* sp. [15], *R. paludigenum* [17] and *R. kratochvilovae* [18], indicating that beside the ascladiol- and DPA- forming degradation pathways, in red yeasts part of the mycotoxin is probably transformed in other compounds that can enter other metabolic pathways for energy or other cellular functions, or might be internalized in the cell and/or in organelles.

In the present work, we further examined the response of *Sporobolomyces* to PAT and demonstrated that the resistance to and/or degradation of the mycotoxin by *Sporobolomyces* sp. IAM 13481 are inducible processes. In particular, *Sporobolomyces* cells that were pretreated (PC) with a low concentration of PAT showed a faster growth (Figure 2) and a faster kinetics of PAT degradation compared to *Sporobolomyces* cells that were not preincubated with PAT (NPC) (Figures 3 and 4), with differences being more evident when high concentrations of PAT (100 or 200 µg/mL versus 50 µg/mL) were used after the preincubation. Counts of viable *Sporobolomyces* cells after 1 day of incubation revealed that with 50 µg/mL of PAT, the growth of PC and NPC increased in a similar manner (5.3×10^7 CFU/mL and 4.0×10^7 CFU/mL, respectively) compared to the initial cellular concentration (i.e., 1×10^7 CFU/mL). Conversely, in the presence of 100 µg/mL and 200 µg/mL of PAT, the growth of NPC dropped respectively to 1.7×10^5 and 4.3×10^3 CFU/mL, with that of PC being instead 1.2×10^7 and 6.4×10^5 CFU/mL, respectively. Taken together, these data indicate that *Sporobolomyces* PC have the highest tolerance to the initial stresses caused by PAT toxicity and this resulted in a faster growth and degradation of the mycotoxin. This is also demonstrated by the similar growth (Figure 2) and kinetics of PAT degradation of *Sporobolomyces* NPC incubated with 50 µg/mL of mycotoxin (Figure 3B) and *Sporobolomyces* PC incubated with 100 µg/mL of mycotoxin (Figure 4A). Similar results were also obtained for *R. kratochvilovae* LS11 (our unpublished data) and *S. cerevisiae* [29]. Based on our recent RNAseq study performed with *Sporobolomyces* sp. IAM 13481 incubated in the presence or absence of 5 µg/mL and 50 µg/mL of PAT [19], it is likely that the pre-incubation step with the mycotoxin induces the earlier activation of (i) genes encoding proteins involved in response to oxidative and other stresses related to PAT cytotoxicity; (ii) antioxidants systems (glutathione and thioredoxin) essential for restoring the cellular redox homeostasis; (iii) export and detoxification proteins that are predicted to be involved in PAT efflux and PAT degradation. Therefore, *Sporobolomyces* PC cells are already in an active/alert stage that enables them to better counteract the stronger toxicity impact due to the use of higher PAT concentrations. Other data that corroborate this interpretation of the obtained results come from our screening of T-DNA insertional mutants of *Sporobolomyces* sp. IAM

13481; we found that the inactivation of genes involved in the resistance to oxidative, genotoxic and other stresses caused by PAT resulted in an increased sensitivity of the mutants to the mycotoxin as compared to the wild type strain [15].

A caveat of using intact cells of *Sporobolomyces* sp. IAM 13481 consisted in the difficulty in evaluating the induction of mechanisms involved in the direct degradation of the PAT, which might be hidden by the multiplicity of resistance mechanisms activated in response to the mycotoxin. Therefore, with the rationale to focus only on the degradation mechanisms and excluding those of resistance, the fate of PAT was also tested in the presence of intracellular proteins extracted from *Sporobolomyces* sp. IAM 13481 cells grown in the presence (EPC) and in the absence of PAT (ENPC). Results presented in Figure 5A,B showed that EPC degrades PAT more promptly as compared to ENPC, although in all cases there were no differences in the length of the degradation period as only traces of PAT were detected after 24 h of incubation (1440 min). Corroborating data obtained for *R. kratochvilovae* LS11 (our unpublished data), *R. paludigenum* [17] and *S. cerevisiae* [29], our results confirm that PAT degradation is an enzyme-mediated mechanism and that the synthesis of the relevant enzyme(s) is likely induced by PAT treatment, further suggesting that PAT metabolization is itself a mechanism of resistance that contributes to overcoming PAT toxicity together with the activation of defense mechanisms described above. Several genes encoding for enzymes specifically involved in PAT degradation have been predicted [19], and they include a glucose–methanol–choline (GMC) oxidoreductase, a protein subunit of aromatic ring-opening dioxygenase, the vacuolar proteins Env9 and Ycf1, and several short and medium chain dehydrogenases. Their target mutagenesis is in progress to elucidate their role in the hydrolysis of the mycotoxin, and to allow the identification of the cellular and molecular components that control their expression. Last but not least, the evaluation of the biocontrol activity of these mutants against *P. expansum* will help to understand whether PAT degradation is also an additional mechanism of action used by the biocontrol agents to counteract postharvest pathogens, since some recent reports point to PAT as a factor involved in the virulence/aggressiveness of *P. expansum*, at least on some cultivars of apples [13].

4. Conclusions

In conclusion, in the present study, we used the model organism *Sporobolomyces* sp. IAM 13481 to evaluate the induction of mechanisms of PAT resistance and degradation; the induction step was performed by pretreating *Sporobolomyces* cells with a low concentration of PAT. Our results suggest that intact yeast cells were appropriate for the evaluation of the mechanisms of PAT resistance and degradation together, while the use of intracellular extracts allowed to evaluate more specifically the mechanisms of PAT degradation. We found that the mechanisms that control both PAT resistance and degradation can be induced by pretreatment with the mycotoxin, and that both processes are related to each other and overall, they contribute both to reduce PAT concentration and mitigate its toxicity. These results contribute to the achievements of long-term goals, namely the production of the enzymes responsible for PAT detoxification that could be the base for detoxification processes of products and juices derived from pome fruits.

5. Materials and Methods

5.1. Strain Used in This Study

The red yeast *Sporobolomyces* sp. strain IAM 13481 was used in this study. This strain was routinely cultivated on yeast peptone dextrose (YPD: Yeast extract 10 g/L, Peptone 20 g/L, Glucose 20 g/L, Agar 20 g/L).

5.2. Patulin

Commercial standards of PAT were purchased from Sigma-Aldrich (Milan, Italy) and from A.G. Scientific, Inc. (San Diego, CA, USA). PAT was dissolved in ethyl acetate and stored at −20 °C.

For experiments, the appropriate aliquots of PAT were withdrawn and the ethyl acetate evaporated under nitrogen stream. The dry PAT was subsequently resuspended at the desired concentrations in working solutions that were LiBa medium (LiBa, 10.0 g of D-glucose, 2.0 g of L-asparagine, 1.0 g of KH_2PO_4, 0.5 g of $MgSO_4·7H_2O$, 0.01 mg of $FeSO_4·7H_2O$, 8.7 mg of $ZnSO_4·7H_2O$, 3.0 mg of $MnSO_4·H_2O$, 0.1 mg of Biotin, and 0.1 mg of Thiamine, per Liter) [30] for assays with intact cells, or in buffer potassium phosphate 0.1 M pH 6 [31] for assays with intracellular extracts. The PAT working solutions were filter-sterilized (0.2 µm) prior to use.

5.3. Patulin Degradation Assays with PC and NPC of Sporobolomyces sp. IAM 13481

Analysis of PAT persistence in the presence of PC and NPC of *Sporobolomyces* sp. was performed as previously reported [15] with minor modifications. Cells of *Sporobolomyces* sp. were grown overnight in LiBa medium in the absence and in the presence of 15 µg/mL of PAT. Cultures were centrifuged for 5 min at 4000 rpm, the cells were resuspended in LiBa, and their concentration was adjusted to 1×10^7 CFU/mL. Three millilitres of this suspension was incubated at 24 °C in sterile 25 mL-flasks in the presence of 50 and 100 µg/mL of PAT. Yeasts growth was monitored on a daily basis by reading OD at 595 nm in a microplate Reader (Bio-Rad Laboratories, Hercules, CA, USA) [32], and at the same time points, the time-course of PAT degradation was assessed by means of HPLC analysis in the same samples as described in Section 5.5. Data from the experiments were pooled since they were similar in the three repetitions.

5.4. Patulin Degradation Assays Using Intracellular Protein Extracts Obtained from PC and NPC of Sporobolomyces sp. IAM 13481

In order to obtain induced intracellular extracts of *Sporobolomyces* sp. IAM 13481, an overnight culture of yeast cells was diluted to 1×10^7 CFU/mL, inoculated in LiBa medium with 15 µg/mL of PAT and incubated on a rotary shaker at 24 °C. The yeast growth was monitored by reading the OD at 595 nm, and the degradation of PAT and the onset of breakdown products were monitored by TLC analysis as previously described [15,16]. When the intensity decrease of the PAT spot paralleled the onset of the degradation products, achieved at OD_{595} of ~0.25, intracellular proteins were extracted as described below. Non-pretreated intracellular samples were extracted from *Sporobolomyces* sp. cells grown in LiBa (same initial concentration as in PC) without PAT until the OD_{595} value reached the same value as the culture grown in the presence of the mycotoxin.

Cells were collected by centrifugation for 10 min at 4000 rpm and then lyophilized. Five hundred milligrams of lyophilized pellets was resuspended with lysis buffer [buffer potassium phosphate 50 mM pH 6 + phenylmethylsulfonyl fluoride (PMSF) 1 mM], 300 mg of acid-washed glass beads was added and the samples vortexed for 1 min. Samples were kept in liquid nitrogen for 1 additional min, then thawed and subsequently centrifuged at 10,000 rpm for 5 min at 4 °C. Supernatants were kept on ice. This lysis step was repeated five times, at the end of which all the supernatants of the same sample were pooled. In order to eliminate residual pellet, the supernatants were filtered with 0.45 µm and 0.25 µm filters (Barloworld Scientific, Stone, Staffordshire, UK). Sodium azide (0.02% w/v) was added and the samples were dialyzed (dialysis tubes with cutoff 12,000 Da) in water (pH 6) for 24 h with three changes. Samples were lyophilized, suspended in buffer phosphate 50 mM pH 6 (without PMSF) and the intracellular proteins concentration was quantified using the Bradford assay (Bio-Rad) according to the manufacturer's instructions, with the BSA used as reference.

PAT degradation assays were carried out in 50 mL falcon tubes each containing 3 mL of incubation mixture. Tubes were kept on a rotary shaker at 160 rpm and 24 °C. Amounts of 30 and 50 µg/mL of PAT were added to phosphate buffer 50 mM pH 6 containing 0.4 mg/mL of intracellular extracts. Controls were PAT in buffer phosphate, and PAT incubated in the presence of boiled intracellular extracts. PAT fate was assessed by HPLC analysis according to the same conditions as those described above after 30 min, 60 min, 90 min, 6 h (360 min), 24 h (1440 min) and 48 h (2880 min). At each time point, 100 μL of each suspension was withdrawn, extracted twice with ethyl acetate, dried and resuspended with buffer phosphate 50 mM pH 6 for HPLC analysis.

5.5. TLC and HPLC Analyses

TLC qualitative analysis and HPLC quantitative analyses were performed as previously reported [15,16,18].

For TLC analysis, samples were pooled, centrifuged at 14,000 rpm for 5 min for pelleting the cells, and the supernatant was extracted twice with ethyl acetate adjusted to pH 2. Extracted samples were dried under a nitrogen stream, resuspended in 10 μL of ethyl acetate and loaded on aluminium-backed silica gel 60 F254 plates (EMD Chemicals, Gibbstown, NJ, USA). Chromatography was performed at room temperature in glass tanks by using toluene/ethyl acetate/formic acid 5:4:1 ($v/v/v$) as the solvent system. After development, plates were dried, observed and photographed under UV light (λ = 254 nm).

Quantitative analysis of PAT and breakdown products formed by *Sporobolomyces* sp. IAM13481 was performed by HPLC based on MacDonald et al. [33] with slight modifications. Three samples per time point were centrifuged and filter-sterilized to remove cells, then injected for analysis. The HPLC apparatus was a Dionex (Sunnyvale, CA, USA) analytical system consisting of a P680 solvent delivery system and a 20 μL injector loop (Rheodyne, Cotati). The UVD170 detector (Dionex, Sunnyvale, CA, USA) set at 276 nm was connected to a data integration system (Dionex Chromeleon Version 6.6). Data from the experiments were pooled, since they were similar in the three repetitions, and expressed as μg/mL of patulin, desoxypatulinic acid and ascladiols ± standard deviation (n = 9).

5.6. Statistical Analysis

Patulin persistance in the biodegradation assays carried out using intact *Sporobolomyces* PC and NPC and their intracellular extracts (EPC and ENPC) was analyzed by Student's t test using the software GraphPad Prism 7.00 for Mac (La Jolla, CA, USA), and differences were considered statistically significant when p-value was lower than 0.05.

Acknowledgments: This research was supported by PRIN 2008—prot. 2008JKH2MM—"Exploitation of genes and proteins from different biological sources for limiting patulin contamination caused by the pathogenic fungus *Penicillium expansum*" (to Raffaello Castoria) and by a grant from the Dottorato di Ricerca in Difesa e Qualità delle Produzioni Agro-alimentari e Forestali (to Giuseppe Ianiri). This research was also partly supported by a grant from MINECO (AGL2015-65684-C2-1-R) (to Cristina Pinedo).

Author Contributions: G.I. and R.C. conceived and designed the experiments; G.I. and C.P. performed the experiments; G.I., C.P. and A.F. analyzed the data; R.C. and G.P. contributed reagents/materials/analysis tools; G.I. and R.C. wrote the paper. All the authors read and approved the final version of the manuscript.

Conflicts of Interest: The authors declare no conflict of interest. The founding sponsors had no role in the design of the study; in the collection, analyses, or interpretation of data; in the writing of the manuscript, and in the decision to publish the results.

References

1. Moake, M.M.; Padilla-Zakour, O.I.; Worobo, R.W. Comprehensive review of patulin control methods in foods. *Compr. Rev. Food Sci. Food Saf.* **2005**, *1*, 8–21. [CrossRef]
2. IARC. Some naturally occurring and synthetic food components, furocoumarins and ultraviolet radiation. In *IARC Monographs on the Evaluation of Carcinogenic Risk of Chemicals to Humans*; IARC: Lyon, France, 1986; Volume 40, pp. 83–98.
3. WHO. *World Health Organization, 44th Report of the Joint FAO/WHO Expert Committee on Food Additives*; Technical Report Series; WHO: Geneva, Switzerland, 1995; Volume 859.
4. Harris, K.L.; Bobe, G.; Bourquin, L.D. Patulin surveillance in apple cider and juice marketed in michigan. *J. Food Prot.* **2009**, *72*, 1255–1261. [CrossRef] [PubMed]
5. Ritieni, A. Patulin in italian commercial apple products. *J. Agric. Food Chem.* **2003**, *51*, 6086–6090. [CrossRef] [PubMed]

6. Morales, H.; Marín, S.; Ramos, A.J.; Sanchis, V. Influence of post-harvest technologies applied during cold storage of apples in *Penicillium expansum* growth and patulin accumulation: A review. *Food Control* **2010**, *21*, 953–962. [CrossRef]

7. Puel, O.; Galtier, P.; Oswald, I.P. Biosynthesis and toxicological effects of patulin. *Toxins* **2010**, *2*, 613–631. [CrossRef] [PubMed]

8. Glaser, N.; Stopper, H. Patulin: Mechanism of genotoxicity. *Food Chem. Toxicol.* **2012**, *50*, 1796–1801. [CrossRef] [PubMed]

9. Tannous, J.; El Khoury, R.; Snini, S.P.; Lippi, Y.; El Khoury, A.; Atoui, A.; Lteif, R.; Oswald, I.P.; Puel, O. Sequencing, physical organization and kinetic expression of the patulin biosynthetic gene cluster from *Penicillium expansum*. *Int. J. Food Microbiol.* **2014**, *189*, 51–60. [CrossRef] [PubMed]

10. Sanzani, S.M.; Reverberi, M.; Punelli, M.; Ippolito, A.; Fanelli, C. Study on the role of patulin on pathogenicity and virulence of *Penicillium expansum*. *Int. J. Food Microbiol.* **2012**, *153*, 323–331. [CrossRef] [PubMed]

11. Ballester, A.R.; Marcet-Houben, M.; Levin, E.; Sela, N.; Selma-Lázaro, C.; Carmona, L.; Wisniewski, M.; Droby, S.; González-Candelas, L.; Gabaldón, T. Genome, transcriptome, and functional analyses of *Penicillium expansum* provide new insights into secondary metabolism and pathogenicity. *Mol. Plant Microbe Interact.* **2015**, *28*, 232–248. [CrossRef] [PubMed]

12. Li, B.; Zong, Y.; Du, Z.; Chen, Y.; Zhang, Z.; Qin, G.; Zhao, W.; Tian, S. Genomic characterization reveals insights into patulin biosynthesis and pathogenicity in *Penicillium* species. *Mol. Plant Microbe Interact.* **2015**, *28*, 635–647. [CrossRef] [PubMed]

13. Snini, S.P.; Tannous, J.; Heuillard, P.; Bailly, S.; Lippi, Y.; Zehraoui, E.; Barreau, C.; Oswald, I.P.; Puel, O. Patulin is a cultivar-dependent aggressiveness factor favouring the colonization of apples by *Penicillium expansum*. *Mol. Plant Pathol.* **2016**, *17*, 920–930. [CrossRef] [PubMed]

14. Mahunu, G.K.; Zhang, H.; Yang, Q.; Li, C.; Zheng, X. Biological control of patulin by antagonistic yeast: A case study and possible model. *Crit. Rev. Microbiol.* **2016**, *42*, 643–655. [CrossRef] [PubMed]

15. Ianiri, G.; Idnurm, A.; Wright, S.A.; Durán-Patrón, R.; Mannina, L.; Ferracane, R.; Ritieni, A.; Castoria, R. Searching for genes responsible for patulin degradation in a biocontrol yeast provides insight into the basis for resistance to this mycotoxin. *Appl. Environ. Microbiol.* **2013**, *79*, 3101–3115. [CrossRef] [PubMed]

16. Castoria, R.; Morena, V.; Caputo, L.; Panfili, G.; De Curtis, F.; De Cicco, V. Effect of the biocontrol yeast *Rhodotorula glutinis* strain LS11 on patulin accumulation in stored apples. *Phytopathology* **2005**, *95*, 1271–1278. [CrossRef] [PubMed]

17. Zhu, R.; Feussner, K.; Wu, T.; Yan, F.; Karlovsky, P.; Zheng, X. Detoxification of mycotoxin patulin by the yeast *Rhodosporidium paludigenum*. *Food Chem.* **2015**, *179*, 1–5. [CrossRef] [PubMed]

18. Castoria, R.; Mannina, L.; Durán-Patrón, R.; Maffei, F.; Sobolev, A.P.; De Felice, D.V.; Pinedo-Rivilla, C.; Ritieni, A.; Ferracane, R.; Wright, S.A. Conversion of the mycotoxin patulin to the less toxic desoxypatulinic acid by the biocontrol yeast *Rhodosporidium kratochvilovae* strain LS11. *J. Agric. Food Chem.* **2011**, *59*, 11571–11578. [CrossRef] [PubMed]

19. Ianiri, G.; Idnurm, A.; Castoria, R. Transcriptomic responses of the basidiomycete yeast *Sporobolomyces* sp. to the mycotoxin patulin. *BMC Genom.* **2016**, *17*, 210. [CrossRef] [PubMed]

20. Wang, Q.M.; Yurkov, A.M.; Göker, M.; Lumbsch, H.T.; Leavitt, S.D.; Groenewald, M.; Theelen, B.; Liu, X.Z.; Boekhout, T.; Bai, F.Y. Phylogenetic classification of yeasts and related taxa within Pucciniomycotina. *Stud. Mycol.* **2015**, *81*, 149–189. [CrossRef] [PubMed]

21. Moss, M.O.; Long, M.T. Fate of patulin in the presence of the yeast *Saccharomyces cerevisiae*. *Food Addit. Contam.* **2002**, *19*, 387–399. [CrossRef] [PubMed]

22. Ricelli, A.; Baruzzi, F.; Solfrizzo, M.; Morea, M.; Fanizzi, F.P. Biotransformation of patulin by *Gluconobacter oxydans*. *Appl. Environ. Microbiol.* **2007**, *73*, 785–792. [CrossRef] [PubMed]

23. Hawar, S.; Vevers, W.; Karieb, S.; Ali, B.K.; Billington, R.; Beal, J. Biotransformation of patulin to hydroascladiol by *Lactobacillus plantarum*. *Food Control* **2013**, *34*, 502–508. [CrossRef]

24. Scott, P.M.; Kennedy, B.; Walbeek, W.V. Desoxypatulinic acid from a patulin-producing strain of *Penicillium patulum*. *Experientia* **1972**, *28*, 1252. [CrossRef] [PubMed]

25. Wright, S.A.; De Felice, D.V.; Ianiri, G.; Pinedo-Rivilla, C.; De Curtis, F.; Castoria, R. Two rapid assays for screening of patulin biodegradation. *Int. J. Environ. Sci. Technol.* **2014**, *11*, 1387–1398. [CrossRef]

26. Suzuki, T.; Takeda, M.; Tanabe, H. A new mycotoxin produced by *Aspergillus clavatus*. *Chem. Pharm. Bull.* **1971**, *19*, 1786–1788. [CrossRef] [PubMed]

27. Maidana, L.; Gerez, J.R.; El Khoury, R.; Pinho, F.; Puel, O.; Oswald, I.P.; Bracarense, A.P. Effects of patulin and ascladiol on porcine intestinal mucosa: An ex vivo approach. *Food Chem. Toxicol.* **2016**, *98*, 189–194. [CrossRef] [PubMed]
28. Guo, C.; Yue, T.; Hatab, S.; Yuan, Y. Ability of inactivated yeast powder to adsorb patulin from apple juice. *J. Food Prot.* **2012**, *75*, 585–590. [CrossRef] [PubMed]
29. Sumbu, Z.L.; Thonart, P.; Bechet, J. Action of patulin on a yeast. *Appl. Environ. Microbiol.* **1983**, *45*, 110–115. [PubMed]
30. Lilly, V.G.; Barnett, H.L. *Physiology of the Fungi*; McGraw-Hill: New York, NY, USA, 1951.
31. Sambrook, J.; Fritsch, E.F.; Maniatis, T. *Molecular Cloning: A Laboratory Manual*; Cold Spring Harbor Laboratory: Cold Spring Harbor, NY, USA, 1989.
32. López-García, B.; González-Candelas, L.; Pérez-Payá, E.; Marcos, J.F. Identification and characterization of a hexapeptide with activity against phytopathogenic fungi that cause postharvest decay in fruits. *Mol. Plant Microbe Interact.* **2000**, *13*, 837–846. [CrossRef] [PubMed]
33. MacDonald, S.; Long, M.; Gilbert, J.; Felgueiras, I. Liquid chromatographic method for determination of patulin in clear and cloudy apple juices and apple puree: Collaborative study. *J. AOAC Int.* **2000**, *83*, 1387–1394. [PubMed]

toxins

MDPI

Article

Detoxification of Deoxynivalenol via Glycosylation Represents Novel Insights on Antagonistic Activities of *Trichoderma* when Confronted with *Fusarium graminearum*

Ye Tian [1], Yanglan Tan [1], Na Liu [1], Zheng Yan [1], Yucai Liao [2], Jie Chen [3], Sarah de Saeger [4], Hua Yang [5], Qiaoyan Zhang [5] and Aibo Wu [1,*]

[1] SIBS-UGENT-SJTU Joint Laboratory of Mycotoxin Research, Key Laboratory of Food Safety Research, Institute for Nutritional Sciences, Shanghai Institutes for Biological Sciences, Chinese Academy of Sciences, University of Chinese Academy of Sciences, 294 Taiyuan Road, Shanghai 200031, China; tianye@sibs.ac.cn (Y.T.); yltan@sibs.ac.cn (Y.T.); liuna@sibs.ac.cn (N.L.); zyan@sibs.ac.cn (Z.Y.)

[2] College of Plant Science and Technology, Huazhong Agricultural University, Wuhan 430070, China; yucailiao@mail.hzau.edu.cn

[3] Department of Resources and Environment Sciences, School of Agriculture and Biology, Shanghai Jiaotong University, 800 Dongchuan Road, Shanghai 200240, China; jiechen59@sjtu.edu.cn

[4] Laboratory of Food Analysis, Department of Bioanalysis, Faculty of Pharmaceutical Sciences, Ghent University, Ottergemsesteenweg 460, B-9000 Ghent, Belgium; Sarah.DeSaeger@UGent.be

[5] State Key Laboratory Breeding Base for Zhejiang Sustainable Pest and Disease Control, Institute of Quality and Standard for Agro-Products, Zhejiang Academy of Agricultural Sciences, Hangzhou 310021, China; yanghua806@hotmail.com (H.Y.); yanyan0014@163.com (Q.Z.)

* Correspondence: abwu@sibs.ac.cn; Tel.: +86-21-5492-0716

Academic Editor: Ting Zhou
Received: 22 September 2016; Accepted: 10 November 2016; Published: 15 November 2016

Abstract: Deoxynivalenol (DON) is a mycotoxin mainly produced by the *Fusarium graminearum* complex, which are important phytopathogens that can infect crops and lead to a serious disease called *Fusarium* head blight (FHB). As the most common B type trichothecene mycotoxin, DON has toxic effects on animals and humans, which poses a risk to food security. Thus, efforts have been devoted to control DON contamination in different ways. Management of DON production by *Trichoderma* strains as a biological control-based strategy has drawn great attention recently. In our study, eight selected *Trichoderma* strains were evaluated for their antagonistic activities on *F. graminearum* by dual culture on potato dextrose agar (PDA) medium. As potential antagonists, *Trichoderma* strains showed prominent inhibitory effects on mycelial growth and mycotoxin production of *F. graminearum*. In addition, the modified mycotoxin deoxynivalenol-3-glucoside (D3G), which was once regarded as a detoxification product of DON in plant defense, was detected when *Trichoderma* were confronted with *F. graminearum*. The occurrence of D3G in *F. graminearum* and *Trichoderma* interaction was reported for the first time, and these findings provide evidence that *Trichoderma* strains possess a self-protection mechanism as plants to detoxify DON into D3G when competing with *F. graminearum*.

Keywords: mycotoxin; toxigenic *Fusarium*; biological control; *Trichoderma*; modified mycotoxin

---

1. Introduction

Mycotoxins are toxic secondary metabolites that are produced by toxigenic molds, and may contaminate different cereal grains [1]. Deoxynivalenol (DON), also known as vomitoxin, is mainly produced by the *Fusarium graminearum* species that can infect crops and cause devastating diseases called *Fusarium* head blight (FHB) or scab [2]. DON is an inhibitor of protein, DNA, and RNA synthesis at the molecular level, and exerts toxic potential to plants, animals, and humans [3]. The functions of DON in interactions of *F. graminearum* and other organisms have been studied before. As a crucial secondary metabolite in development of *F. graminearum*, DON is beneficial for *Fusarium* to deal with a complex environment and compete with other organisms [4,5]. *Fusarium* may utilize DON to disrupt plant defense system in the infection process. DON is regarded as a virulence factor when infecting plants, and it can facilitate disease spread in infected plant tissues. Genetic studies have proved that DON non-producing *Fusarium* mutants showed less virulence on crops than wild-type strains [6]. On the other hand, plants usually have the self-defense mechanisms to cope with mycotoxins, such as conjugating them with endogenous metabolites to less toxic products. An important detoxification process reducing the toxicity of DON in plants is glycosylation catalyzed by a special kind of glycosyltransferases [7], and a common detoxification product of DON in this reaction is deoxynivalenol-3-glucoside (D3G) (Figure 1). This conjugated mycotoxin was originally termed as a masked mycotoxin, because its structure has been changed and it may escape routine detection by conventional analytical methods [8].

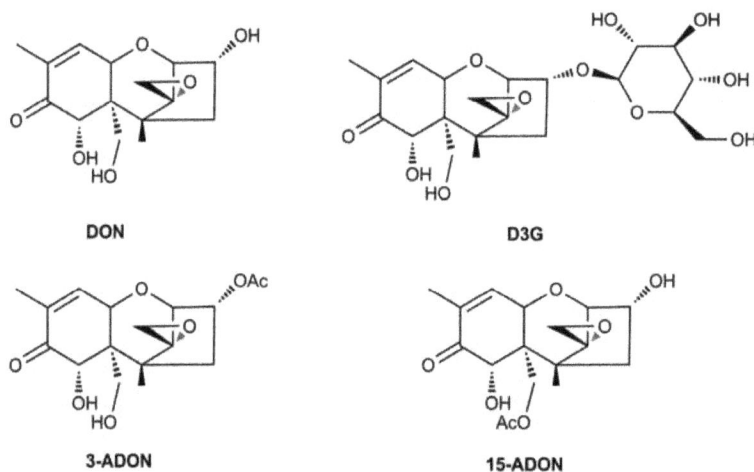

Figure 1. Chemical structures of deoxynivalenol (DON), deoxynivalenol-3-glucoside (D3G), and acetylated derivatives of DON: 3-acetyl-deoxynivalenol (3-ADON) and 15-acetyl-deoxynivalenol (15-ADON).

The pathogens causing FHB are also capable of producing 3-acetyl-deoxynivalenol (3-ADON) or 15-acetyl-deoxynivalenol (15-ADON), which are commonly-detected acetylated derivatives of DON in contaminated grains or food commodities [2]. Both DON and ADONs belong to B-type trichothecene mycotoxins (Figure 1) characterized by a keto-group at position C-8 in their molecule structures [2]. The epoxide group is considered as a key factor determining the toxicity of trichothecene mycotoxins [9], and ADONs also possess toxic effects [10,11]. In order to protect consumers, DON and its acetylated types have all been included in the group provisional maximum tolerable daily intake (PMTDI) according to the maximum tolerated levels of DON in food enacted by the Joint FAO/WHO Expert Committee on Food Additives (JECFA) [10].

To reduce the risks caused by DON contamination in the food chain, plenty of measures, including selection of resistant cultivars, application of fungicide, and biological control agents (BCAs), have been studied and used to control DON-producing pathogens [12,13]. Among these mentioned strategies, biological control emerging as a green approach has captured significant attention recently.

Natural fungicides, antagonistic microorganisms, and detoxification enzymes are the most-concerned functional BCAs to control DON contamination [14]. Antagonistic microorganisms could inhibit the development and mycotoxin production of toxigenic pathogens. *Trichoderma* strains, *Bacillus* strains, *Clonostachys rosea*, and *Cladosporium cladosporioides* are promising antagonistic microorganisms to manage DON contamination [15,16]. Among them, *Trichoderma* strains have been widely investigated and applied as beneficial BCAs in agriculture. They can protect crops against plant pathogens environment-friendly, and biological control activity of *Trichoderma* is mainly due to antibiotic production to inhibit the growth of pathogens, substrate competition to restrict colony areas of pathogens, and the ability to activate plant defense responses against pathogens [17,18].

The aim of our work was to study the potential antagonistic activities of selected *Trichoderma* strains for management of *F. graminearum* by dual culture assay. We assessed the anti-growth activity of the different *Trichoderma* strains on *F. graminearum*. Meanwhile, DON and its acetylated or glycosylated forms were monitored to evaluate the anti-toxigenic activity of antagonists. Interestingly, the modified mycotoxin D3G, which was previously regarded as a detoxification product in plant defense against DON-producing *Fusarium* [7], was detected when *F. graminearum* was co-cultured with *Trichoderma* strains. To our knowledge, this is the first time that it has been reported that the D3G is a detoxification product of *Trichoderma* when in competition with *F. graminearum*, which may provide new understandings on the interaction of antagonistic microbes and toxigenic *Fusarium*.

2. Results

2.1. Impact of Trichoderma Strains on Growth of F. graminearum 5035 in Dual Culture

Trichoderma is recognized as a non-pathogenic genus majorly found in soil and plants, and has been widely studied due to its biological control-related characteristics [19]. In our antagonistic assay, the growth rates of selected *Trichoderma* strains were much faster than that of *F. graminearum* 5035 (Figure 2A). The mycelia of tested fungi began to contact two days after incubation. The faster growth rate made *Trichoderma* quickly occupy the living space and surround the colony of *Fusarium*, and then *Trichoderma* effectively suppressed the mycelial growth of *Fusarium*. The inhibition rate of *Trichoderma* on mycelial growth of *Fusarium* ranged from 73% to 77% (Figure 2B). For the pathogen, it was hard to overgrow antagonists for further extension on PDA medium.

Moreover, we observed the colony morphology of *Fusarium* when alone or in the presence of antagonists after incubation (Figure 3), finding that four *Trichoderma* strains including *T. longibranchiatum* GIM 3.534, *T. harzianum* Q710613, *T. atroviride* Q710251 and *T. asperellum* Q710682 were able to overgrow and sporulate on the colony of *F. graminearum* 5035, and the mycelia of *Fusarium* in these combinations were relatively flat, while the mycelia of *Fusarium* in other combinations were much denser. These signs demonstrated that the four *Trichoderma* strains capable of overgrowing the pathogen were more effective in inhibiting the mycelia spread of *Fusarium* on PDA medium.

Trichoderma strains have a stronger capacity to occupy living space and take up nutrients than other microbes [20]. It seems that growth suppression is the major inhibition pattern of *Trichoderma* when in competition with pathogens.

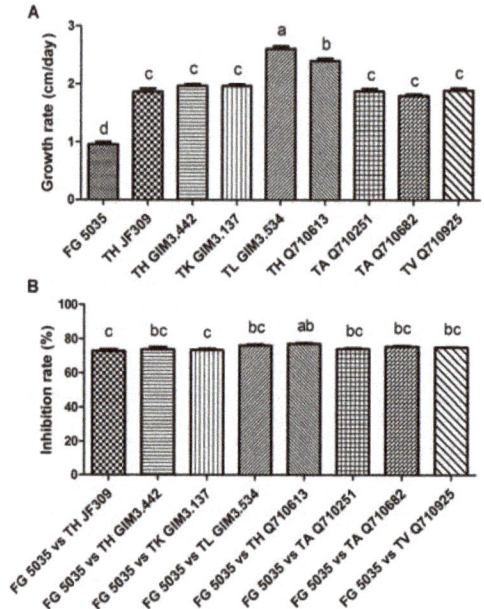

Figure 2. Growth rate (cm/day) of *F. graminearum* 5035 and *Trichoderma* strains (**A**). From left to right: the growth rate of *F. graminearum* 5035, *T. harzianum* JF309, *T. harzianum* GIM3.442, *T. koningii* GIM3.137, *T. longibranchiatum* GIM3.534, *T. harzianum* Q710613, *T. atroviride* Q710251, *T. asperellum* Q710682, and *T. virens* Q710925, respectively. Inhibition rate (%) of *Trichoderma* strains on mycelial growth of *F. graminearum* 5035 in dual culture (**B**). From left to right: *F. graminearum* 5035 grew against *T. harzianum* JF309, *T. harzianum* GIM3.442, *T. koningii* GIM3.137, *T. longibranchiatum* GIM3.534, *T. harzianum* Q710613, *T. atroviride* Q710251, *T. asperellum* Q710682, and *T. virens* Q710925, respectively. Data were from two independent experiments and shown as the mean ± SEM. Bars with different letters were significantly different according Tukey's Test ($\alpha < 0.05$) following one-way ANOVA (analysis of variance) analysis.

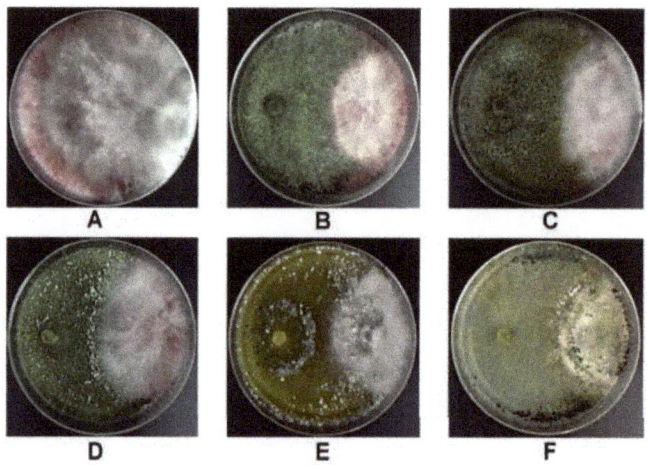

Figure 3. *Cont.*

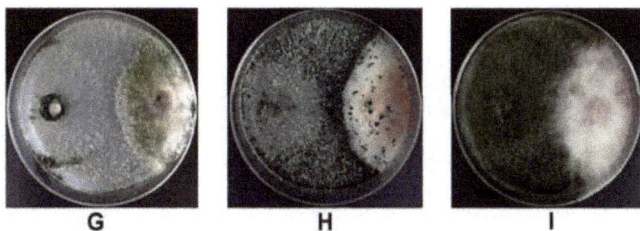

Figure 3. Colony morphology of *F. graminearum* 5035 in dual culture tests after incubation on the potato dextrose agar (PDA) medium. *F. graminearum* 5035 grew alone (**A**); *F. graminearum* 5035 grew against *T. harzianum* JF309 (**B**); *T. harzianum* GIM3.442 (**C**); *T. koningii* GIM3.137 (**D**); *T. longibranchiatum* GIM3.534 (**E**); *T. harzianum* Q710613 (**F**); *T. atroviride* Q710251 (**G**); *T. asperellum* Q710682 (**H**); and *T. virens* Q710925 (**I**).

2.2. Effect of Trichoderma Strains on Mycotoxin Production of F. graminearum 5035 in Dual Culture

DON-producing *Fusarium* could infect crops in field and subsequently cause DON accumulation in cereal grains. Prevention before harvest is an important strategy to manage DON contamination, and antagonistic *Trichoderma* strains are promising BCAs to achieve this goal [21]. Matarese et al. has revealed that a *Trichoderma gamsii* strain 6085 could strongly reduce DON production of *Fusarium* on rice medium [17]. In order to accurately reveal the effect of *Trichoderma* strains on mycotoxin production of *F. graminearum*, DON and ADONs were all monitored by a liquid chromatography tandem mass spectrometry (LC-MS/MS) method in our work.

The data showed that *F. graminearum* 5035 was able to produce 57 µg/g DON and 20 µg/g 15-ADON on PDA medium when alone, indicating that the tested *F. graminearum* was a 15-ADON producer mainly producing DON and 15-ADON [2]. When *F. graminearum* 5035 was co-cultured with *Trichoderma*, the amount of DON produced by *F. graminearum* reduced significantly due to the strong antagonistic action of *Trichoderma*. When in the presence of *T. harzianum* Q710613, *T. atroviride* Q710251 and *T. asperellum* Q710682, the amount of DON produced by *F. graminearum* 5035 decreased more than 90% compared with the control. In the presence of other *Trichoderma* strains, the inhibition rate of DON production ranged from 70% to 88% (Figure 4A). With regard to 15-ADON, the antagonists seemed to exhibit stronger inhibition capacity: the inhibition rate of 15-ADON production ranged from 86% to 98% (Figure 4B). In consideration of the non-ignorable toxicity of ADONs [22], DON and its acetylated forms should be all monitored when assessing the anti-toxigenic activity of antagonistic microbes.

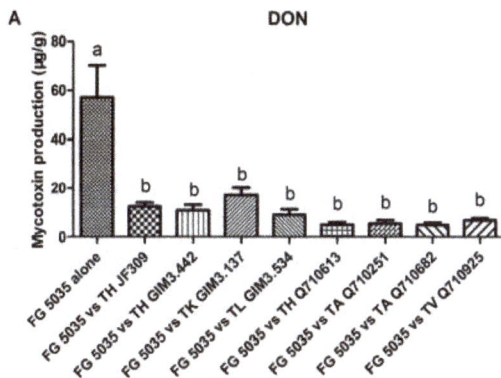

Figure 4. *Cont.*

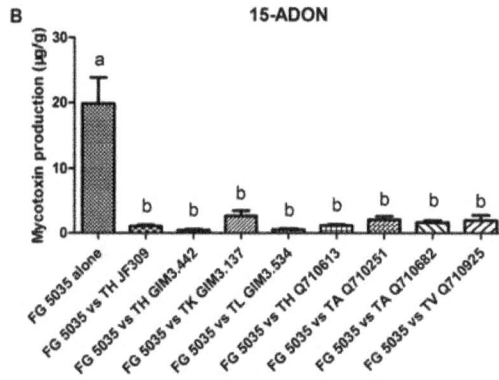

Figure 4. Mycotoxin production (DON and 15-ADON) of *F. graminearum* 5035 when alone or against different *Trichoderma* strains. From left to right: *F. graminearum* 5035 grew alone, and grew against *T. harzianum* JF309, *T. harzianum* GIM3.442, *T. koningii* GIM3.137, *T. longibranchiatum* GIM3.534, *T. harzianum* Q710613, *T. atroviride* Q710251, *T. asperellum* Q710682 and *T. virens* Q710925 (**A,B**). Data were from two independent experiments and shown as the mean ± SEM. Bars with different letters were significantly different according to Tukey's Test ($\alpha < 0.05$) following one-way ANOVA analysis.

2.3. Occurrence of Modified Mycotoxin D3G in Dual Culture of Trichoderma and F. graminearum

Modified mycotoxin D3G is the most prevalent detoxification product of DON in self-defense of plants, and detoxification via glycosylation is very general in plants. However, this biotransformation process is unusual in microbes [23]. A previous study has emphasized that DON production of *F. graminearum* could not reduce the biomass of *Trichoderma* when in competition [24]. Therefore, we speculated that *Trichoderma* strains were very likely to possess some underlying detoxification mechanisms against DON-producing *F. graminearum*. In order to verify whether *Trichoderma* strains could detoxify DON via glycosylation when confronted with *F. graminearum*, D3G was monitored in our work.

Based on our LC-MS/MS method, the glycosylation form of DON (D3G), was detected when *F. graminearum* 5035 was co-cultured with *Trichoderma* strains, while there was no D3G detected in the control group (Figure 5). Glycosylation of DON catalyzed by UDP-glucosyltransferases (UGTs) is a well-known detoxification process in plants, and the first UGT capable of transforming DON to D3G was identified from *Arabidopsis thaliana* [7]. However, no studies have reported the occurrence of modified mycotoxin D3G in interactions between microbes so far. In order to confirm the presence of D3G in the co-culture of *Trichoderma* and *F. graminearum*, prepared samples of dual culture assay were analyzed by liquid chromatography high resolution mass spectrometry (LC-HRMS). Shown as in Figure S1, the MS/MS spectra of the precursor ion (m/z 517.1927, $[M + Ac]^-$) in samples of the dual culture test were in good agreement with the MS/MS spectra of D3G standard. The key fragments m/z 427.16, 457.17, 409.15, 277.11, 247.10 for $[D3G + Ac]^-$ were also observed in the previous study [25]. These results provided evidences for the existence of D3G in dual cultures of *Trichoderma* and DON-producing *Fusarium*. The occurrence of D3G in *Trichoderma* and *Fusarium* interaction indicated that D3G was a detoxification product in self-defense of *Trichoderma* against DON-producing *Fusarium*.

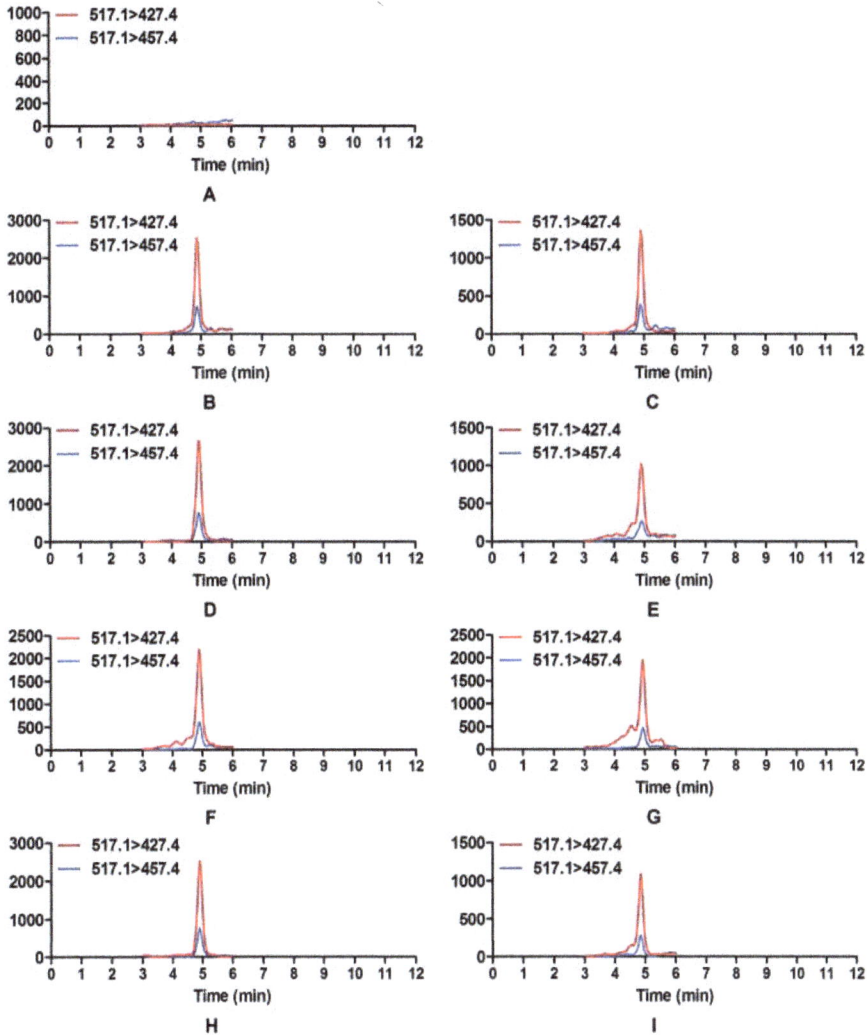

Figure 5. Selected reaction monitoring (SRM) chromatograms of D3G in samples of dual culture test: *F. graminearum* 5035 grew against *T. harzianum* JF309 (**B**); *T. harzianum* GIM3.442 (**C**); *T. koningii* GIM3.137 (**D**); *T. longibranchiatum* GIM3.534 (**E**); *T. harzianum* Q710613 (**F**); *T. atroviride* Q710251 (**G**); *T. asperellum* Q710682 (**H**); and *T. virens* Q710925 (**I**). Red line (m/z 517.1 > m/z 427.4) and blue line (m/z 517.1 > m/z 457.4) show the SRM traces of D3G. There was no D3G detected when *F. graminearum* 5035 grew alone on PDA medium (**A**).

Our preliminary results provide evidence that *Trichoderma* strains not only possess the strong inhibitory action on *F. graminearum*, but also possess the detoxification capability to glycosylate DON when competing with *F. graminearum*. All of these advantageous features make *Trichoderma* strains promising BCAs against plant disease in agriculture.

2.4. A Significant Difference in D3G/DON Ratios of Trichoderma Strains when Confronted with F. graminearum 5035

Previous studies regarding pathogen and plant interactions have illustrated that D3G/DON ratios of various wheat lines can be regarded as their resistance indicators against FHB [8,26,27]. Similarly, D3G/DON ratios of different *Trichoderma* strains used in this study may also be related to their resistance to *Fusarium*. In our work, we assessed the D3G/DON ratios of tested *Trichoderma* strains when dual cultured with DON-producing *F. graminearum* 5035. Interestingly, the D3G/DON ratios of different *Trichoderma* strains that might reflect their resistance to *Fusarium* were significantly different (Figure 6). *T. harzianum* Q710613, *T. atroviride* Q710251 and *T. asperellum* Q710682 showed relatively higher ratios that exceeded 20%, while D3G/DON ratios of other *Trichoderma* strains were relatively lower, which were in the range of 4% to 12%. It seems that DON-resistant antagonists have more potential in further investigation and application, because recent studies have showed that over-expression of UGT genes capable of detoxifying DON into D3G in crops could improve their resistance against *Fusarium* [28,29], and the reported functional UGT genes were all derived from plants. Thus, our data revealed that *Trichoderma* might be a potential source to seek efficient UGT genes, which can be expressed to detoxify DON directly or be genetically-engineered in crops to improve their resistance against FHB causal agents [30].

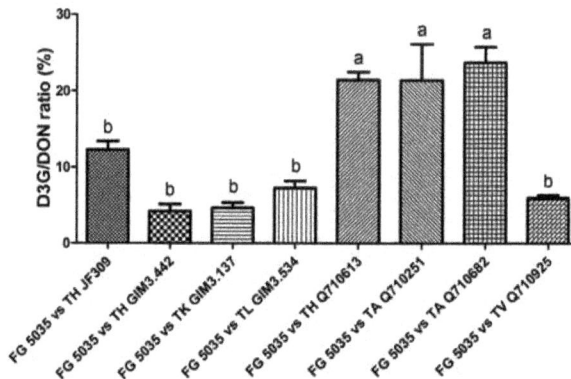

Figure 6. D3G/DON ratios of different *Trichoderma* strains when confronted with *F. graminearum* 5035 in dual culture. *F. graminearum* 5035 grew against *T. harzianum* JF309, *T. harzianum* GIM3.442, *T. koningii* GIM3.137, *T. longibranchiatum* GIM3.534, *T. harzianum* Q710613, *T. atroviride* Q710251, *T. asperellum* Q710682; and *T. virens* Q710925, respectively. Data were from two independent experiments and shown as the mean ± SEM. Bars with different letters were significantly different according Tukey's Test ($\alpha < 0.05$) following one-way ANOVA analysis.

Understanding the mechanisms of the antagonistic action of BCAs on pathogens is essential for searching potential *Trichoderma* strains to manage plant disease. Lutz et al. suggested that the role of DON in competitiveness of *F. graminearum* was a negative signal, which down-regulated one chitinase gene (*nag1*) involved in biological control activity of an antagonistic *T. atroviride* strain [31]. However, our study provided evidence for a new role of DON in the competition of *Trichoderma* and *Fusarium*: DON not only modulated biological-control related genes of *Trichoderma*, but also induced *Trichoderma* self-defense related genes to detoxify mycotoxin DON. The responses of *Trichoderma* to DON-producing *F. graminearum* 5035 were different according to the D3G/DON ratios, which may represent their resistance against DON-producing pathogens, and this hypothesis needs to be validated in future work. Generally, the metabolic fate of DON in *Trichoderma* needs more detailed study to present new antagonistic mechanisms of *Trichoderma*.

3. Discussion

FHB caused by *F. graminearum* species could reduce the yield and quality of cereal grains around the world, and DON contamination in the food chain leads to deleterious effects on the health of animals and humans [12]. Biological control of *F. graminearum* and mycotoxin DON contamination would be an important part in integrated FHB management, which may avoid negative effects and minimize environmental impacts caused by conventional approaches [15]. Application of *Trichoderma* strains on crops or crop residuals to inhibit the development and DON production of *Fusarium* is an available biological-based strategy. As potential BCAs, *Trichoderma* strains are capable of producing a large number of enzymes and secondary metabolites, which are effective components against pathogens and play an important role in the antagonistic action of *Trichoderma* [32]. Hydrolytic enzymes (such as chitinase and proteases) and antibiotics (such as polyketides and terpenes) derived from *Trichoderma* act as "weapons" against pathogens when in competition [32]. In addition, the potential UGTs inactivating DON in *Trichoderma* strains, act as "shields" to protect themselves against DON-producing pathogens. For *Trichoderma*, "weapons" and "shields" are all significant agents for control of *Fusarium*. For plants, glycosylation of DON catalyzed by the UGTs is an effective defense strategy to resist pathogens [27], and our present study indicates that *Trichoderma* also have a defense capacity to detoxify DON as plants when against DON-producing *Fusarium*. Our findings are supported by the detection of modified mycotoxin D3G in a dual culture of *Trichoderma* and *Fusarium*. Suzuki et al. have already elucidated that D3G exhibited extremely low toxicity to yeast and algae, indicating that detoxification via glycosylation may not be exclusive to plants [33]. Moreover, our data confirmed that glycosylation of DON was a self-defense process for *Trichoderma* as well, which is in accordance with a previous study of Suzuki et al. A novel understanding on the role of DON in the competitiveness of *Fusarium* would promote deeper investigation of the interaction between antagonists and toxigenic *Fusarium* species.

The metabolism of mycotoxins in plants has been studied previously. Plants have the detoxification metabolic systems to counter phytotoxic compounds. Mycotoxins can be conjugated to endogenous metabolites by plants, generating modified mycotoxins with decreased toxicity [27]. In plants, detoxification of mycotoxins usually includes three phases. In phase I, mycotoxins are oxidized or hydrolyzed to generate some active sites for the next reaction. In phase II, low-toxic or non-toxic metabolites are formed by conjugating some hydrophilic molecules, such as sugars or amino acids, and D3G is the major metabolite of DON in this phase. Finally, the metabolites from phase II are irreversibly transported to a vacuole or accumulated in the cell wall in phase III [8]. Glycosylation catalyzed by UGTs is a major process of phase II in detoxification reactions of plants [33]. A recent study has clearly validated that over-expressing such UGT genes able to detoxify DON in *Brachypodium* could confer increased resistance to *Fusarium* [34]. Glycosyltransferases are a series of enzymes that can catalyze the formation of a glycosidic bond by transferring sugars to substrates, such as plant secondary metabolites and mycotoxins. Identification of the UGTs inactivating DON into D3G is rather complex because UGTs are encoded by a large family: more than 100 putative gene members in plants [35]. To date, some induced functional UGT genes during the DON-producing *Fusarium* infection process have been studied and identified in different plants [28,34–38]. However, little is known about such UGTs possessing DON-detoxification ability in *Trichoderma* strains. According to previous studies, *Trichoderma* can produce non-phytotoxic trichothecenes, such as trichodermin and harzianum [18,39], but the mechanism of DON metabolism in *Trichoderma* strains is unclear. Taken together, the studies on identifying these specific UGT genes from *Trichoderma* strains by transcriptome analysis of *Trichoderma* and DON-producing *Fusarium* interaction should be conducted in the future. The details about the endogenous self-protection mechanism against DON in *Trichoderma* strains needs further investigation and interpretation after the detoxification UGT genes from *Trichoderma* are identified. On the other hand, other detoxification products of DON have been identified in plant detoxification processes, such as DON-sulfates, DON-15-glucoside, and DON-glutathione [25,40,41].

Seeking and identifying other modified mycotoxins generated from *Trichoderma* and *Fusarium* are also essential to reveal the mycotoxin detoxification mechanism of *Trichoderma*.

4. Conclusions

In the present study, eight *Trichoderma* strains were selected to evaluate their antagonistic activities against *F. graminearum* by dual culture assay. Results suggested that *T. harzianum* Q710613, *T. atroviride* Q710251, and *T. asperellum* Q710682 were more effective antagonists to control the pathogen, not only because of the efficacy in inhibiting mycelia spread and mycotoxin production, but also because of the relatively higher D3G/DON ratios that may reflect their resistance to *Fusarium*. In addition, the occurrence of modified mycotoxin D3G in a dual culture of *Fusarium* and *Trichoderma* provides new insights into antagonistic activities of *Trichoderma*.

5. Materials and Methods

5.1. Standards and Chemicals

The mycotoxin standards of DON, D3G, 3-ADON, and 15-ADON were purchased from Sigma-Aldrich (St. Louis, MO, USA). Methanol and acetonitrile (ACN) were all high performance liquid chromatography(HPLC) grade, and obtained from Merck (Darmstadt, Germany). Other solvents and chemicals of HPLC or analytical grade were provided by Aladdin (Shanghai, China). Ultrapure water (18.2 MΩ·cm) was obtained from Millipore (Bedford, MA, USA), and used throughout the experiments.

5.2. Fungal Strains and Culture Medium

A total of eight *Trichoderma* strains and one *F. graminearum* strain were used in the present study (Table 1). *F. graminearum* 5035 was from the Huazhong Agricultural University [42]. *T. harzianum* JF309 was isolated from *Lentinus edodes* in our lab. *T. harzianum* GIM3.442, *T. koningii* GIM3.137, and *T. longibranchiatum* GIM3.534 were purchased from the Microbial Culture Collection Center of Guangdong Institute of Microbiology (GIMCC). *T. harzianum* Q710613, *T. atroviride* Q710251, *T. asperellum* Q710682, and *T. virens* Q710925 were from the Center for Culture Collection of *Trichoderma* (CCCT), Shanghai Jiaotong University (SJTU). All of the fungal strains were maintained as spore suspensions in 30% glycerol at −80 °C. Before the start of antagonistic assay, the fungal strains were grown on PDA medium at 25 °C for five days for activation.

Table 1. The information of the fungal strains used in this study.

Strain Code	Strain	Source
5035	*Fusarium graminearum*	Huazhong Agricultural University, China
JF309	*Trichoderma harzianum*	Isolated from *Lentinus edodes* in our lab
GIM3.442	*Trichoderma harzianum*	GIMCC, China
GIM3.137	*Trichoderma koningii*	GIMCC, China
GIM3.534	*Trichoderma longibranchiatum*	GIMCC, China
Q710613	*Trichoderma harzianum*	CCCT, SJTU, China
Q710251	*Trichoderma atroviride*	CCCT, SJTU, China
Q710682	*Trichoderma asperellum*	CCCT, SJTU, China
Q710925	*Trichoderma virens*	CCCT, SJTU, China

GIMCC: Microbial Culture Collection Center of Guangdong Institute of Microbiology; CCCT, SJTU: Center for Culture Collection of *Trichoderma*, Shanghai Jiaotong University.

5.3. Antagonistic Activities of Trichoderma on Growth and Mycotoxin Production of F. graminearum

The antagonistic activities of the chosen *Trichoderma* strains on *F. graminearum* were assessed by dual culture assay according to Matarese et al. [17] with minor modifications. In short, 6 mm-diameter mycelial discs of tested *Trichoderma* and *Fusarium* strains were removed from the edge of actively-growing colonies, and then the mycelial discs (*Fusarium* and *Trichoderma* combinations) were placed at a distance of 45 mm on a 90 mm-diameter Petri dish containing 20 mL PDA medium. Meanwhile, a mycelial disc of *F. graminearum* 5035 was placed with an agar disc on the dish as the control. The pathogen-antagonist combinations and the controls were all set up in three biological replicates, and incubated at 25 °C for 14 days. The mycelial growth rates (cm/day) of *Trichoderma* strains and *F. graminearum* were recorded before their mycelia began to interact. After incubation, the radii of *Fusarium* colonies facing the antagonists were measured and used for calculation of the inhibition rate [32]. The inhibition rate of *Trichoderma* on mycelial growth of tested *F. graminearum* was calculated based on the following formula: $I = (C - T)/C \times 100$, where I is the inhibition rate (%), C is the growth radius of pathogen when alone, and T is the growth radius of the pathogen when co-cultured with antagonists [32].

5.4. Fusarium Mycotoxin Extraction and Preparation

After incubation, the PDA medium containing mycelia of tested fungi strains in Petri dish was dried until a constant weight, and then the sample was ground into a homogenized powder for preparation as described previously [43].

The homogenized powder was weighed and put into a 50 mL centrifuge tube, and then 10 mL of ACN/water (84/16, v/v) solution was added. The mixture was shaken for 5 min, and then subjected to ultrasonication for 30 min. Next the mixture was centrifuged at 3200× g for 30 min. Then 2 mL of supernatant was transferred into a centrifuge tube, followed by adding 150 mg of $MgSO_4$. The mixture was vortexed for 30 s, and centrifuged at 3200× g for 30 min. Thereafter, the remaining supernatant was moved into a new tube, and evaporated to dryness by a stream of nitrogen gas at 40 °C. At last, the residue was redissolved with 1 mL of water containing 5 mM ammonium acetate. After redissolution, the mixture was passed through a 0.22-μm filter, and then injected into LC-MS/MS for analysis.

5.5. Mycotoxin Determination by LC-MS/MS

Mycotoxin concentrations were determined according to a published method with minor modifications [43]. Briefly, chromatographic separation was performed on an Agilent Extend—C18 column (100 mm × 4.6 mm, 3.5 μm) at 30 °C using a Thermo Scientific Accela 1250 UPLC system (Thermo Fisher Scientific, San Jose, CA, USA). The mobile phase consisted of water containing 5 mM ammonium acetate (A) and methanol (B), and a gradient elution program was as follows: 0 min 15% B, 1 min 15% B, 6.5 min 90% B, 8.5 min 90% B, 9 min 15% B, 12 min 15% B. The injection volume was 10 μL and the flow rate was 0.35 mL/min.

MS/MS analysis was carried out using a Thermo Scientific TSQ VantageTM (Thermo Fisher Scientific, San Jose, CA, USA) triple stage quadrupole mass spectrometer in both positive and negative electrospray ionization (ESI+/ESI−) modes. The ionization source parameters were set as follows: interface voltage of 3.5 kV (ESI+) or 3.0 kV (ESI−); desolvation line (DL) temperature of 250 °C; nebulizing gas (N_2) and drying gas (N_2) pressure of 30 psi (207 kPa) and 20 psi (138 kPa), respectively; heat block temperature of 350 °C. A series of standard solutions of analytes (10–1000 ng/mL) were used for calibration to absolutely quantify the targeted mycotoxins in the samples, and the quantitation and identification of target mycotoxins were performed in selected reaction monitoring (SRM) mode. The MS/MS parameters of detected mycotoxins were summarized in Table 2. All data acquisition and processing were achieved with Xcalibur™ Software (2.2, Thermo Fisher Scientific, San Jose, CA, USA, 2011).

Table 2. The tandem mass spectrometry (MS/MS) parameters for target mycotoxins in selected reaction monitoring (SRM) mode.

Mycotoxin	Precursor Ion (*m/z*)	Retention Time (min)	Primary Product Ion (*m/z*)	Collision Energy (ev)	Secondary Product Ion (*m/z*)	Collision Energy (ev)
DON	355.1 [M + Ac]⁻	5.04	265.0 [a]	17	247.2 [b]	18
D3G	517.1 [M + Ac]⁻	4.89	427.4 [a]	22	457.4 [b]	14
3-ADON	397.1 [M + Ac]⁻	6.54	307.1 [a]	16	173.1 [b]	15
15-ADON	356.1 [M + NH₄]⁺	6.52	321.0 [a]	15	137.0 [b]	12

[a] Ion for quantification; [b] Ion for identification.

5.6. LC-HRMS Analysis of D3G

LC–HRMS analysis of the fragmentation pattern of D3G was conducted on a UHPLC system (1290 series, Agilent Technologies, Santa Clara, CA, USA) coupled to a quadruple time-of-flight (Q-TOF) mass spectrometer (Agilent 6550 iFunnel Q-TOF, Agilent Technologies, Santa Clara, CA, USA). Chromatographic separation was performed on an Agilent Extend—C18 column (100 mm × 4.6 mm, 3.5 µm) at 30 °C. The mobile phase consisted of water containing 5 mM ammonium acetate (A) and methanol (B), and a gradient elution program was as follows: 0 min 15% B, 1 min 15% B, 6.5 min 90% B, 8.5 min 90% B, 9 min 15% B, and 12 min 15% B. The flow rate was 0.35 mL/min and the sample injection volume was 5 µL.

To confirm the presence of D3G in samples of dual culture test, the negative precursor ion *m/z* 517.1927 (theoretical mass of [D3G + Ac]⁻ based on molecular formula) was ionized in an electrospray ionization (ESI) source and mass isolated by quadruple mass filter, following by collision-induced dissociation (CID) with the collision energy of −20 eV and, finally, the fragments and the remaining precursor were detected in the TOF tube. The parameters of high resolution mass spectrometer were set as follows: sheath gas (N₂) temperature, 300 °C; dry gas (N₂) temperature, 170 °C; sheath gas flow, 12 L/min; dry gas flow, 16 L/min; capillary voltage, 3.0 kV in negative mode; nozzle voltage, 0 V; fragmentor, 175 V; and nebulizer pressure, 40 psi (276 kPa). Data analysis was achieved with Agilent software MassHunter B.06.00 (Agilent Technologies, Santa Clara, CA, USA, 2012).

5.7. Statistical Analysis

Data were subjected to one-way ANOVA (analysis of variance) analysis followed by Tukey's multiple comparisons test using Graphpad Prism 5.0 (GraphPad Software, San Diego, CA, USA, 2007). All data were shown as the mean ± standard error of the mean (SEM), and $p < 0.05$ was considered to be significantly different.

Supplementary Materials: The following are available online at www.mdpi.com/2072-6651/8/11/335/s1, Figure S1: Tandem mass spectra of the precursor ion (*m/z* 517.1927, [M + Ac]⁻) in negative mode. (A) Product ion spectra of the D3G standard (1000 ng/mL); (B–I) product ion spectra of the precursor ion (*m/z* 517.1927) with the same chromatographic retention time as the D3G standard in samples of dual culture test: *F. graminearum* 5035 grew against *T. harzianum* JF309 (B); *T. harzianum* GIM3.442 (C); *T. koningii* GIM3.137 (D); *T. longibranchiatum* GIM3.534 (E); *T. harzianum* Q710613 (F); *T. atroviride* Q710251 (G); *T. asperellum* Q710682 (H); and *T. virens* Q710925 (I).

Acknowledgments: This work was supported by the National Basic Research Program of China (Grant 2013CB127801), National Natural Science Foundation of China (31471661), Shanghai Technical Standards Project (15DZ0503800), China Agriculture Research System (CARS-02), and Zhejiang Province Major Program (2015C02G4010111). We thank Zheng-Jiang Zhu and Fangfang Chen in Interdisciplinary Research Center on Biology and Chemistry, and Yuanhong Shan in the Core Facility Centre of the institute of Plant Physiology and Ecology for mass spectrometry assistance.

Author Contributions: Aibo Wu and Ye Tian conceived and designed the experiments; Ye Tian, Yanglan Tan, Na Liu and Zheng Yan performed the experiments; Aibo Wu and Ye Tian. analyzed the data and wrote the paper; Yucai Liao, Jie Chen, Sarah De Saeger, Hua Yang and Qiaoyan Zhang contributed materials and amended the manuscript; Aibo Wu supervised the whole experiments and revised the paper before submission.

Conflicts of Interest: The authors declare no conflict of interest.

References

1. Goswami, R.S.; Kistler, H.C. Heading for disaster: *Fusarium graminearum* on cereal crops. *Mol. Plant Pathol.* **2004**, *5*, 515–525. [CrossRef] [PubMed]

2. Alexander, N.J.; Mccormick, S.P.; Waalwijk, C.; Lee, T.V.D.; Proctor, R.H. The genetic basis for 3-ADON and 15-ADON trichothecene chemotypes in *Fusarium*. *Fungal Genet. Biol.* **2011**, *48*, 485–495. [CrossRef] [PubMed]

3. Pestka, J.J. Deoxynivalenol: Mechanisms of action, human exposure, and toxicological relevance. *Arch. Toxicol.* **2010**, *84*, 663–679. [CrossRef] [PubMed]

4. Audenaert, K.; Vanheule, A.; Höfte, M.; Haesaert, G. Deoxynivalenol: A major player in the multifaceted response of *Fusarium* to its environment. *Toxins* **2013**, *6*, 1–19. [CrossRef] [PubMed]

5. Ponts, N. Mycotoxins are a component of *Fusarium graminearum* stress-response system. *Front. Microbiol.* **2015**, *6*, 1234. [CrossRef] [PubMed]

6. Bai, G.-H.; Desjardins, A.; Plattner, R. Deoxynivalenol-nonproducing *Fusarium graminearum* causes initial infection, but does not cause disease spread in wheat spikes. *Mycopathologia* **2002**, *153*, 91–98. [CrossRef] [PubMed]

7. Poppenberger, B.; Berthiller, F.; Lucyshyn, D.; Sieberer, T.; Schuhmacher, R.; Krska, R.; Kuchler, K.; Glössl, J.; Luschnig, C.; Adam, G. Detoxification of the *Fusarium* mycotoxin deoxynivalenol by a UDP-glucosyltransferase from *Arabidopsis thaliana*. *J. Biol. Chem.* **2003**, *278*, 47905–47914. [CrossRef] [PubMed]

8. Berthiller, F.; Lemmens, M.; Werner, U.; Krska, R.; Hauser, M.; Adam, G.; Schuhmacher, R. Short review: Metabolism of the *Fusarium* mycotoxins deoxynivalenol and zearalenone in plants. *Mycotoxin Res.* **2007**, *23*, 68–72. [CrossRef] [PubMed]

9. Sundstol Eriksen, G.; Pettersson, H.; Lundh, T. Comparative cytotoxicity of deoxynivalenol, nivalenol, their acetylated derivatives and de-epoxy metabolites. *Food Chem. Toxicol.* **2004**, *42*, 619–624. [CrossRef] [PubMed]

10. Evaluations of the Joint FAO/WHO Expert Committee on Food Additives (JECFA). Available online: http://apps.who.int/food-additives-contaminants-jecfa-database/chemical.aspx?chemID=2947 (accessed on 25 November 2015).

11. Broekaert, N.; Devreese, M.; Demeyere, K.; Berthiller, F.; Michlmayr, H.; Varga, E.; Adam, G.; Meyer, E.; Croubels, S. Comparative in vitro cytotoxicity of modified deoxynivalenol on porcine intestinal epithelial cells. *Food Chem. Toxicol.* **2016**, *95*, 103–109. [CrossRef] [PubMed]

12. Wegulo, S.N.; Baenziger, P.S.; Hernandez Nopsa, J.; Bockus, W.W.; Hallen-Adams, H. Management of *Fusarium* head blight of wheat and barley. *Crop Prot.* **2015**, *73*, 100–107. [CrossRef]

13. Yuen, G.Y.; Schoneweis, S.D. Strategies for managing *Fusarium* head blight and deoxynivalenol accumulation in wheat. *Int. J. Food Microbiol.* **2007**, *119*, 126–130. [CrossRef] [PubMed]

14. Tian, Y.; Tan, Y.; Liu, N.; Liao, Y.; Sun, C.; Wang, S.; Wu, A. Functional agents to biologically control deoxynivalenol contamination in cereal grains. *Front. Microbiol.* **2016**, *7*, 395. [CrossRef] [PubMed]

15. Schoneberg, A.; Musa, T.; Voegele, R.T.; Vogelgsang, S. The potential of antagonistic fungi for control of *Fusarium graminearum* and *Fusarium crookwellense* varies depending on the experimental approach. *J. Appl. Microbiol.* **2015**, *118*, 1165–1179. [CrossRef] [PubMed]

16. Zhao, Y.; Selvaraj, J.N.; Xing, F.; Zhou, L.; Wang, Y.; Song, H.; Tan, X.; Sun, L.; Sangare, L.; Folly, Y.M.; et al. Antagonistic action of *Bacillus subtilis* strain SG6 on *Fusarium graminearum*. *PLoS ONE* **2014**, *9*, e92486. [CrossRef] [PubMed]

17. Matarese, F.; Sarrocco, S.; Gruber, S.; Seidl-Seiboth, V.; Vannacci, G. Biocontrol of *Fusarium* head blight: Interactions between *Trichoderma* and mycotoxigenic *Fusarium*. *Microbiology* **2012**, *158*, 98–106. [CrossRef] [PubMed]

18. Malmierca, M.G.; McCormick, S.P.; Cardoza, R.E.; Alexander, N.J.; Monte, E.; Gutierrez, S. Production of trichodiene by *Trichoderma harzianum* alters the perception of this biocontrol strain by plants and antagonized fungi. *Environ. Microbiol.* **2015**, *17*, 2628–2646. [CrossRef] [PubMed]

19. Mukherjee, P.K.; Horwitz, B.A.; Herrera-Estrella, A.; Schmoll, M.; Kenerley, C.M. *Trichoderma* research in the genome era. *Ann. Rev. Phytopathol.* **2013**, *51*, 105–129. [CrossRef] [PubMed]

20. Benitez, T.; Rincon, A.M.; Limon, M.C.; Codon, A.C. Biocontrol mechanisms of *Trichoderma* strains. *Int. Microbiol.* **2004**, *7*, 249–260. [PubMed]

21. Jard, G.; Liboz, T.; Mathieu, F.; Guyonvarc'h, A.; Lebrihi, A. Review of mycotoxin reduction in food and feed: From prevention in the field to detoxification by adsorption or transformation. *Food Addit. Contam. Part A Chem. Anal. Control Expo. Risk Assess.* **2011**, *28*, 1590–1609. [CrossRef] [PubMed]

22. Pinton, P.; Tsybulskyy, D.; Lucioli, J.; Laffitte, J.; Callu, P.; Lyazhri, F.; Grosjean, F.; Bracarense, A.-P.; Martine, K.-C.; Oswald, I.P. Toxicity of deoxynivalenol and its acetylated derivatives on the intestine: Differential effects on morphology, barrier function, tight junctions proteins and mitogen-activated protein kinases. *Toxicol. Sci.* **2012**, *130*, 180–190. [CrossRef] [PubMed]

23. McCormick, S.P.; Price, N.P.J.; Kurtzman, C.P. Glucosylation and other biotransformations of T-2 toxin by yeasts of the *Trichomonascus* clade. *Appl. Environ. Microb.* **2012**, *78*, 8694–8702. [CrossRef] [PubMed]

24. Naef, A.; Senatore, M.; Défago, G. A microsatellite based method for quantification of fungi in decomposing plant material elucidates the role of *Fusarium graminearum* DON production in the saprophytic competition with *Trichoderma atroviride* in maize tissue microcosms. *Fems Microbiol. Ecol.* **2006**, *55*, 211–220. [CrossRef] [PubMed]

25. Schmeitzl, C.; Warth, B.; Fruhmann, P.; Michlmayr, H.; Malachova, A.; Berthiller, F.; Schuhmacher, R.; Krska, R.; Adam, G. The metabolic fate of deoxynivalenol and its acetylated derivatives in a wheat suspension culture: Identification and detection of DON-15-o-glucoside, 15-acetyl-DON-3-o-glucoside and 15-acetyl-DON-3-sulfate. *Toxins* **2015**, *7*, 3112–3126. [CrossRef] [PubMed]

26. Martina, C.; Silvia, G.; Andrea, D.E.; Pietro, L.; Gianluca, F.; Andrea, M.; Gianni, G.; Chiara, D.A. Durum wheat (*Triticum durum* Desf.) lines show different abilities to form masked mycotoxins under greenhouse conditions. *Toxins* **2014**, *6*, 81–95.

27. Berthiller, F.; Crews, C.; Dall'Asta, C.; Saeger, S.D.; Haesaert, G.; Karlovsky, P.; Oswald, I.P.; Seefelder, W.; Speijers, G.; Stroka, J. Masked mycotoxins: A review. *Mol. Nutr. Food Res.* **2013**, *57*, 165–186. [CrossRef] [PubMed]

28. Li, X.; Shin, S.; Heinen, S.; Dill-Macky, R.; Berthiller, F.; Nersesian, N.; Clemente, T.; Mccormick, S.; Muehlbauer, G.J. Transgenic wheat expressing a barley UDP-glucosyltransferase detoxifies deoxynivalenol and provides high levels of resistance to *Fusarium graminearum*. *Mol. Plant-Microbe Interact.* **2015**, *28*, 1237–1246. [PubMed]

29. Shin, S.; Torres-Acosta, J.A.; Heinen, S.J.; McCormick, S.; Lemmens, M.; Paris, M.P.K.; Berthiller, F.; Adam, G.; Muehlbauer, G.J. Transgenic *Arabidopsis thaliana* expressing a barley UDP-glucosyltransferase exhibit resistance to the mycotoxin deoxynivalenol. *J. Exp. Bot.* **2012**, *63*, 4731–4740. [CrossRef] [PubMed]

30. He, J.; Zhou, T.; Young, J.C.; Boland, G.J.; Scott, P.M. Chemical and biological transformations for detoxification of trichothecene mycotoxins in human and animal food chains: A review. *Trends Food Sci. Technol.* **2010**, *21*, 67–76. [CrossRef]

31. Lutz, M.P.; Feichtinger, G.; Defago, G.; Duffy, B. Mycotoxigenic *Fusarium* and deoxynivalenol production repress chitinase gene expression in the biocontrol agent *Trichoderma atroviride* P1. *Appl. Environ. Microbiol.* **2003**, *69*, 3077–3084. [CrossRef] [PubMed]

32. Li, Y.; Sun, R.; Yu, J.; Saravanakumar, K.; Chen, J. Antagonistic and biocontrol potential of *Trichoderma asperellum* zjsx5003 against the maize stalk rot pathogen *Fusarium graminearum*. *Indian J. Microbiol.* **2016**, *56*, 318–327. [CrossRef] [PubMed]

33. Suzuki, T.; Iwahashi, Y. Low toxicity of deoxynivalenol-3-glucoside in microbial cells. *Toxins (Basel)* **2015**, *7*, 187–200. [CrossRef] [PubMed]

34. Pasquet, J.-C.; Changenet, V.; Macadré, C.; Boex-Fontvieille, E.; Soulhat, C.; Bouchabké-Coussa, O.; Dalmais, M.; Atanasova-Pénichon, V.; Bendahmane, A.; Saindrenan, P.; et al. A *Brachypodium* UDP-glycosyltransferase confers root tolerance to deoxynivalenol and resistance to *Fusarium* infection. *Plant Physiol.* **2016**, *172*, 559–574. [CrossRef] [PubMed]

35. Schweiger, W.; Pasquet, J.C.; Nussbaumer, T.; Paris, M.P.K.; Wiesenberger, G.; Macadre, C.; Ametz, C.; Berthiller, F.; Lemmens, M.; Saindrenan, P.; et al. Functional characterization of two clusters of *Brachypodium distachyon* UDP-glycosyltransferases encoding putative deoxynivalenol detoxification genes. *Mol. Plant-Microbe Interact.* **2013**, *26*, 781–792. [CrossRef] [PubMed]

36. Schweiger, W.; Boddu, J.; Shin, S.; Poppenberger, B.; Berthiller, F.; Lemmens, M.; Muehlbauer, G.J.; Adam, G. Validation of a candidate deoxynivalenol-inactivating UDP-glucosyltransferase from barley by heterologous expression in yeast. *Mol. Plant-Microbe Interact.* **2010**, *23*, 977–986. [CrossRef] [PubMed]

37. Lulin, M.; Yi, S.; Aizhong, C.; Zengjun, Q.; Liping, X.; Peidu, C.; Dajun, L.; Xiu-E, W. Molecular cloning and characterization of an up-regulated UDP-glucosyltransferase gene induced by DON from *Triticum aestivum* L. cv. Wangshuibai. *Mol. Biol. Rep.* **2010**, *37*, 785–795. [CrossRef] [PubMed]
38. Michlmayr, H.; Malachova, A.; Varga, E.; Kleinova, J.; Lemmens, M.; Newmister, S.; Rayment, I.; Berthiller, F.; Adam, G. Biochemical characterization of a recombinant UDP-glucosyltransferase from rice and enzymatic production of deoxynivalenol-3-O-β-D-glucoside. *Toxins* **2015**, *7*, 2685–2700. [CrossRef] [PubMed]
39. Tijerino, A.; Elena Cardoza, R.; Moraga, J.; Malmierca, M.G.; Vicente, F.; Aleu, J.; Collado, I.G.; Gutiérrez, S.; Monte, E.; Hermosa, R. Overexpression of the trichodiene synthase gene tri5 increases trichodermin production and antimicrobial activity in *Trichoderma brevicompactum*. *Fungal Genet. Biol.* **2011**, *48*, 285–296. [CrossRef] [PubMed]
40. Kluger, B.; Bueschl, C.; Lemmens, M.; Michlmayr, H.; Malachova, A.; Koutnik, A.; Maloku, I.; Berthiller, F.; Adam, G.; Krska, R.; et al. Biotransformation of the mycotoxin deoxynivalenol in *Fusarium* resistant and susceptible near isogenic wheat lines. *PLoS ONE* **2015**, *10*, e0119656. [CrossRef] [PubMed]
41. Warth, B.; Fruhmann, P.; Wiesenberger, G.; Kluger, B.; Sarkanj, B.; Lemmens, M.; Hametner, C.; Fröhlich, J.; Adam, G.; Krska, R. Deoxynivalenol-sulfates: Identification and quantification of novel conjugated (masked) mycotoxins in wheat. *Anal. Bioanal. Chem.* **2015**, *407*, 1033–1039. [CrossRef] [PubMed]
42. Gong, A.D.; Li, H.P.; Yuan, Q.S.; Song, X.S.; Yao, W.; He, W.J.; Zhang, J.B.; Liao, Y.C. Antagonistic mechanism of iturin a and plipastatin a from *Bacillus amyloliquefaciens* s76-3 from wheat spikes against *Fusarium graminearum*. *PLoS ONE* **2015**, *10*, e0116871. [CrossRef] [PubMed]
43. Zhao, Z.; Rao, Q.; Song, S.; Na, L.; Zheng, H.; Hou, J.; Wu, A. Simultaneous determination of major type B trichothecenes and deoxynivalenol-3-glucoside in animal feed and raw materials using improved DSPE combined with LC-MS/MS. *J. Chromatogr. B Anal. Technol. Biomed. Life Sci.* **2014**, *963*, 75–82. [CrossRef] [PubMed]

toxins

Review

Strategies and Methodologies for Developing Microbial Detoxification Systems to Mitigate Mycotoxins

Yan Zhu, Yousef I. Hassan, Dion Lepp, Suqin Shao and Ting Zhou *

Guelph Research and Development Centre, Agriculture and Agri-Food Canada, Guelph, ON N1G5C9, Canada; yan.zhu@agr.gc.ca (Y.Z.); yousef.hassan@agr.gc.ca (Y.I.H.); dion.lepp@agr.gc.ca (D.L.); Suqin.Shao@agr.gc.ca (S.S.)
* Correspondence: ting.zhou@agr.gc.ca; Tel.: +1-226-217-8084; Fax: +1-519-829-2600

Academic Editor: Massimo Reverberi
Received: 13 February 2017; Accepted: 4 April 2017; Published: 7 April 2017

Abstract: Mycotoxins, the secondary metabolites of mycotoxigenic fungi, have been found in almost all agricultural commodities worldwide, causing enormous economic losses in livestock production and severe human health problems. Compared to traditional physical adsorption and chemical reactions, interest in biological detoxification methods that are environmentally sound, safe and highly efficient has seen a significant increase in recent years. However, researchers in this field have been facing tremendous unexpected challenges and are eager to find solutions. This review summarizes and assesses the research strategies and methodologies in each phase of the development of microbiological solutions for mycotoxin mitigation. These include screening of functional microbial consortia from natural samples, isolation and identification of single colonies with biotransformation activity, investigation of the physiological characteristics of isolated strains, identification and assessment of the toxicities of biotransformation products, purification of functional enzymes and the application of mycotoxin decontamination to feed/food production. A full understanding and appropriate application of this tool box should be helpful towards the development of novel microbiological solutions on mycotoxin detoxification.

Keywords: mycotoxin; detoxification; biodegradation; biotransformation; enzyme; microorganism identification

---

1. Introduction

Over the past several decades, research interest in the mitigation of mycotoxins, the toxic secondary metabolites produced by specific fungi, has continuously increased due to concerns over human and animal health, economic losses and food safety and security [1]. Important mycotoxins include aflatoxin B_1 (AFB$_1$), aflatoxin G_1 (AFG$_1$), ochratoxin A (OTA), deoxynivalenol (DON), nivalenol (NIV), fumonisin (FUM), zearalenone (ZEA), patulin (PAT) and citrinin (CIT), which are mainly produced by the fungal genera *Aspergillus*, *Fusarium* and *Penicillium*. It has been estimated that the economic costs of crop losses from major mycotoxins (aflatoxins, fumonisins and deoxynivalenol) in the United States are as great as $932 million per year, in addition to mitigation costs of $466 million and livestock costs of $6 million [2]. In Europe, although no data regarding economic losses caused by mycotoxins are available, the direct and indirect losses due to a wheat epidemic in 1998 in Hungary were estimated at 100 million euros [3]. Moreover, severe human health effects can result from the exposure of mycotoxins through either ingestion, absorption or inhalation routes [4]. In 2015, the Rapid Alert System for Food and Feed (RASFF) reported 475 notifications in Europe on mycotoxin exposure in food, most related to the presence of aflatoxins [5].

Pre-harvest control of mycotoxin production and post-harvest mitigation of contamination are the main strategies to limit mycotoxins in food and feed. Good agricultural practices (GAPs), including crop rotation, soil management, choice of varieties and correct fungicide use, have been recommended by the European Commission (EC) to prevent the contamination of *Fusarium* toxins in cereals [6]. Strategies for post-harvest mitigation can be categorized as chemical, physical and biological. Chemical strategies use acids, bases, oxidizing agents, aldehydes or bisulfite gases to change the structure of mycotoxins, which has led to increased public concerns over the chemical residues in food and feed. Furthermore, negative effects on the nutrition and palatability of food and feed may result from chemical treatments [7–9]. For physical strategies, the application of adsorption agents has become popular since the European Union (EU) allowed substrates that suppress or reduce the absorption, promote the excretion of mycotoxins or modify their mode of action to be used as feed additives [10]. However, the efficacy of adsorption agents in reducing mycotoxin contamination is variable, and most of the commercial binding agents presented no sufficient effect against DON [11]. Compared with chemical and physical approaches, biological detoxification methods, which biotransform mycotoxins into less toxic metabolites, are generally more specific, efficient and environmentally friendly.

As more and more researchers enter the field of biodetoxification, guidance is needed on the various approaches and methodologies that may be employed. In this paper, we aim to review the entire process of the discovery and development of biological mitigation systems, focusing on the strategies and methodologies at each research stage from initial microorganism screening to final application. Although the methods in the steps of enrichment, isolation, strain identification, chemical analysis, mycotoxin biotransformation, toxicity evaluation and enzyme extraction were developed based on specific mycotoxins and detoxifying microorganisms, the approaches used should be generally applied to studies seeking microbial-based solutions for mycotoxin mitigation.

2. Workflow

An integrated research program on the microbial detoxification of mycotoxins is expected to proceed through up to five stages. First, mycotoxin-biotransforming microorganisms (MBMs) or consortia should be obtained and identified from either environmental sources or be screened from a group of previously-identified candidates. Secondly, the efficacy and influencing factors on the detoxification activities of the selected microorganisms, as well as the biotransformation products should be investigated and identified. In the third stage, safety assessments of both functional strains and biotransformation products should be performed. Biotransformation may not necessarily result in a less toxic secondary product, so the reduced toxicity of the biotransformation products must also be confirmed. In the fourth, and potentially most difficult, stage, the enzymes responsible for the biotransformation are isolated, identified and/or cloned and expressed. Finally, the feasibility of mycotoxin-detoxifying applications in food and feed should be validated. The complete workflow is outlined in Figure 1. It should be noted that it is not necessary to endure all five stages in an individual research project. For example, studies on the detoxification activities of known strains would not require the first screening stage. Furthermore, the isolation of the enzymes may be optional if the functional microorganisms are to be used in the application.

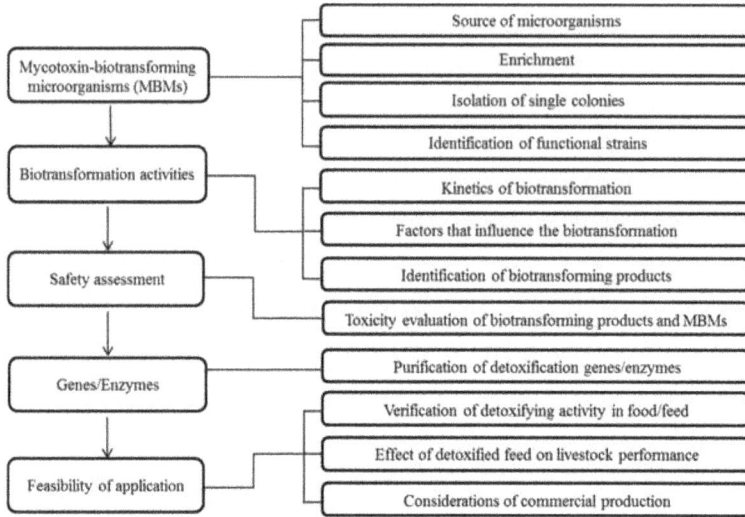

Figure 1. Workflow of research on microbial detoxifications of mycotoxins.

3. Strategies and Methodologies

3.1. Sources of Microorganisms

Undoubtedly, the successful isolation of MBMs is one of the critical steps of the whole research project. A careful selection of environmental sources more likely to harbor MBMs should increase the probability of finding functional microorganisms. Previous studies have isolated microorganisms from a range of environmental [12–29], plant [30–38] and animal [39–48] sources.

The existence of MBMs was initially hypothesized based on the fact that chemically-stable mycotoxins do not accumulate in agricultural soil [25,30,46]. Soil is therefore a likely reservoir for functional MBMs, in particular that on which crops susceptible to mycotoxigenic fungi have been grown previously. In addition, nearby aqueous environments to agricultural soil have also been used for the successful isolation of DON-degrading bacteria [15].

Rumen fluid and intestinal contents are hosts to a diverse microbiota, which not only contribute to host metabolism, but have also been shown in some cases to have biotransformation activities. This is partly demonstrated by the reduced sensitivity of ruminants to mycotoxins and in particular to trichothecenes [40,41]. Some early studies reported that the rumen microbiota has the capability to transform a wide range of mycotoxins, including OTA, ZEA, DON, T-2 toxin and diacetoxyscirpenol [40,42]. In addition, chicken large intestinal contents have been shown to harbor functional bacteria with DON deepoxidation activity [39,43].

From a practical standpoint, it is advantageous to isolate MBMs from an environment similar to that to which they will be applied, as they will then be more likely to grow and express detoxification activities during application. For example, *Bacillus subtilis* strain UTBSP1, isolated from mature pistachio nut fruits in Iran, showed AFB_1 biotransformation activity in this agriculture crop [38]. Similarly, the patulin in apple juice was degraded into less toxic E- and Z-ascladiol by *Gluconobacter oxydans*, a bacterium isolated from rotten apple puree [37].

In contrast to screening microbes in a particular environment, an alternative strategy is to focus on only a certain taxonomic group [19,35,45,49] or already available strains [50–53]. These strains may possess specific properties to facilitate isolation or provide additional benefits to commercial applications. For instance, *Bacillus* spp. have been targeted due to their tolerance of adverse environmental conditions, application as probiotics, antimicrobial activities and the production of insect toxins and extracellular enzymes [19,35,45,47]. Petchkongkaew et al. (2008) [35] screened for *Bacillus* spp. from Thai fermented soybean using Gram-staining and the API 50CH system, and they isolated three strains with highly efficient AFB$_1$ and OTA biotransformation capabilities. Lactic acid bacteria (LAB) are also good candidates for mycotoxin biotransforming bacteria since their additional probiotic properties promote their application in food and feed. *Pediococcus parvulus* UTAD 473 was selected from a collection of 19 LAB strains and has the ability to biotransform 90% of OTA to the less toxic OTα [53]. An isolated CIT biotransforming bacterium, *Moraxella* sp. MB1, has the particular benefit of being able to perform detoxification in solvent environment, as it belongs to the class of organic solvent-tolerant microorganisms (OSTMs) [50]. Rodriguez et al. (2011) [51] considered strains belonging to *Rhodococcus*, *Pseudomonas* and *Brevibacterium*, which have the capability to biotransform aromatic compounds, to have a higher potential to degrade mycotoxins. The authors successfully obtained *Brevibacterium casei* RM101 and *Brevibacterium linens* DSM 20425^T, which can completely biotransform OTA at concentrations up to 40 μg/mL [51].

3.2. Enrichment

Due to the complexity of microbial communities obtained from environmental or animal sources, screening may fall into a "trial-and-error" phase, leading to high costs of labour and consumables. Enrichment of particular microorganism taxa, by either enhancing the growth of potentially functional strains or supressing unwanted ones, is therefore used to reduce the diversity of the microbial consortium.

Enrichment strategies vary based on sources of microbiota, the mycotoxins to be targeted and possible biotransformation pathways (Table 1). Media used for enrichment may be classified into three categories: nutrient medium, minimal medium and sole carbon source medium. The latter normally contains the salt medium with or without vitamin mixtures as the medium base and supplemented with the specific mycotoxin or related compounds as a sole carbon source.

The application of selective pressures encountered in the environments of plant materials contaminated with mycotoxigenic fungi is an effective strategy to screen for functional microbes. In this way, the potential functional strains may become predominant, which aids in further isolation. He et al. (2016) [12] hypothesized that DON-biotransforming microorganisms would have the ability to tolerate and grow in the presence of DON-producing *Fusarium graminearum* and using this approach eventually isolated *Devosia mutans* 17-2-E-8, a new bacterial species that detoxifies DON into 3-*epi*-DON with high efficiency. An in situ plant enrichment (isPE) strategy has been developed to screen functional strains on wheat heads that were artificially contaminated with DON and grown in situ for one month. This approach yielded 17 colonies with DON-biotransformation activities among 60 colonies examined [32]. In order to screen for biotransforming bacteria in the chicken intestinal tract, in vivo enrichment was performed by feeding the chickens moldy wheat contaminated with DON (10 mg/kg), which was shown to improve the activities of the digesta contents in biotransforming DON [43].

Another well-known and effective enrichment strategy that has been widely employed involves the use of a specific mycotoxin as a sole carbon source in a basal mineral medium. It has been reported that DON, ZEA, PAT and CIT can be utilized as a sole carbon source and ultimately biotransformed by a number of microbial strains [13,15,18,23,24,26,27,29,32,41,54]. Due to the high cost and toxicity of many pure mycotoxins, chemically-related compounds may be substituted as the sole carbon source when screening for potential MBMs [25,46,55,56]. One successful application of this strategy involved the use of coumarin as a surrogate for AFB$_1$, which shares the basic structure of this extremely toxic mycotoxin, but is a much safer and less expensive alternative [25,46]. Guan et al. (2008) [46] first developed this strategy to isolate 25 single colonies grown in medium supplemented with coumarin as the sole carbon source, which were all shown to have AFB$_1$ biotransformation activities (between 9.18% and 82.50%), as determined by HPLC, indicating the highly selective and accurate nature of this method.

Table 1. Strategies and methodologies in the enrichment and isolation of mycotoxin biotransforming microorganisms. DGGE, denaturing gradient gel electrophoresis; T-RFLP, terminal restriction fragment length polymorphism.

Mycotoxin	Enrichment		Isolation		Biotransforming Strains	Reference
	Medium	Strategy/Methodology	Medium	Strategy/Methodology		
DON	Corn meal broth	Soil samples enriched from the corn contaminated by the DON-producing fungi	Corn meal agar	Single colonies screening; extended incubation time for slow-growth strains	*Devosia mutans* 17-2-E-8	[12]
	Anaerobic incubation medium with 10% chicken cecal digesta extract	In vivo enrichment with moldy wheat; antibiotics treatment; guiding the enrichment by PCR-DGGE	L10 agar	Single colonies screening	*Bacillus* sp. LS-100	[43]
	M10 medium + DON (100 µg/mL)	Treatment by antibiotics and hemin	M1 medium	Single colonies screening	*Eubacterium* sp. BBSH 797	[55,56]
	Mineral salts with peptone medium + DON (50 µg/mL)	Antibiotics and heat treatment; guiding the enrichment by T-RFLP	Mineral salts with peptone agar	Single colonies screening; extended incubation time for slow-growth strains	Microbial consortium with at least 6 bacterial genera	[14]
	Mineral medium + DON (100 µg/mL)	DON as a sole carbon source	1/100 nutrient agar	Single colonies screening	*Nocardioides* sp. WSN05-2	[13]
	Mineral salt medium + DON (100 µg/mL)	In situ plant enrichment in contaminated wheat head by spraying DON	MRDG medium	Single colonies screening; using gellan gum rather than agar	*Marmoricola* sp. MIM116	[32]
	Mineral medium + DON (100 µg/mL)	DON as a sole carbon source	Reasoner's 2A (R2A) agar, 1/100 nutrient agar	Single colonies screening	9 *Nocardioides* spp. and 4 *Devosia* spp.	[15]
	BYE medium + DON (200 µg/mL)	Repeated sub-culturing in fresh medium with high level of DON (200 µg/mL)	1/10 nutrient agar	Single colonies screening	E3-39 (belonging to *Agrobacterium* or *Rhizobium*)	[16]
	Inorganic salt culture medium + DON (4 µg/mL)	Enrichment with minimal nutrients	Czapek's agar, LB agar	Single colonies screening	*Aspergillus tubingensis* NJA-1	[57]
	Mineral salts with peptone medium + DON (50 µg/mL)	In situ soil enrichment by spraying DON; guiding the enrichment by PCR-DGGE	-	-	Microbial consortium	[17]
ZEA	Minimal salt medium + ZEA (2 µg/mL)	ZEA as a sole carbon source	LB agar	Single colonies screening	*Pseudomonas alcaliphila* TH-C1, *Pseudomonas plecoglossicida* TH-L1	[18]
	LB broth	Selective screening *Bacillus* strains by heat treatment	LB agar	Single colonies screening	*Bacillus subtilis* ANSB01G	[19]
	Minimal salt medium + ZEA (2 µg/mL)	ZEA as a sole carbon source	LB agar	Single colonies screening	*Pseudomonas otitidis* TH-N1	[41]
	M1 + ZEA (25 µg/mL) + nystatin (15 µg/mL), M2 + ZEA (500 µg/mL)	ZEA as a sole carbon source	Nutrient agar	Single colonies screening	*Acinetobacter* sp. SM04	[23]
	M9 medium + ZEA (50 µg/mL)	ZEA as a sole carbon source	LB agar	Single colonies screening	Microbial consortium	[24]
AFB$_1$	Coumarin medium (with 1% coumarin)	Coumarin, a basic molecular structure of aflatoxins, as a sole carbon source	Coumarin medium	Single colonies screening; coumarin as a sole carbon source	*Stenotrophomonas maltophilia* 35-3	[46]

Table 1. Cont.

Mycotoxin	Enrichment Medium	Enrichment Strategy/Methodology	Isolation Medium	Isolation Strategy/Methodology	Biotransforming Strains	Reference
AFB1	Coumarin medium (with 1% coumarin)	Coumarin, a basic molecular structure of aflatoxins, as a sole carbon source	Coumarin medium	Single colonies screening; coumarin as a sole carbon source	Pseudomonas aeruginosa N17-1	[25]
		-	Modified Hornisch medium (with 0.1% coumarin)	Coumarin as a sole carbon source	Stenotrophomonas sp. NMO-3	[58]
	Nutrient broth	Non-selective enrichment	Coumarin medium (with 0.1% coumarin)	Single colonies screening; coumarin as a sole carbon source; K-B disk diffusion	Aspergillus niger ND-1	[34]
	Minimal salt/vitamin medium + fluoranthene (10 mg/mL)	Fluoranthene as a sole carbon source	R2A agar	Single colonies screening	Mycobacterium fluoranthenivorans sp. nov.	[59]
	Minimal salt medium + AFB1 (10 µg/mL)	AFB1 as a sole carbon source	Minimal salt agar + AFB1 (10 µg/mL)	Single colonies screening	Bacillus sp. TUBF1	[31]
		-	Nutrient agar	Single colonies screening	Bacillus licheniformis CM21, Bacillus subtilis MHS 13	[35]
AFB1, AFM1, AFG1	LB broth	Selective screening Bacillus strains by heat treatment	LB agar	Single colonies screening	Bacillus subtilis ANSB060	[45]
PAT	Mineral salt medium + increased concentration (300–600 µg/mL) of PAT	PAT as a sole carbon source	Mineral salt agar + PAT (600 µg/mL)	Single colonies screening	Byssochlamys nivea FF1-2	[54]
		-	YEPD medium + PAT (10 µg/mL)	Screening in liquid medium	Kodameae ohmeri HYJM34	[60]
CIT	Mineral broth + (1–4 µg/mL) of CIT	CIT as a sole carbon source	Mineral salt agar + CIT (10 µg/mL)	Single colonies screening	Klebsiella pneumoniae NPUST-B11	[26]
	Mineral broth + CIT (1 µg/mL)	CIT as a sole carbon source	Mineral salt agar + CIT (1–5 µg/mL)	Single colonies screening	Rhizobium borbori PS45	[27]
			Nutrient agar	Screening strains by disc plate diffusion assay (50 µg/disk of CIT)	Moraxella sp. MB1	[50]
FUB1	BYE medium + FUB1 (500 µg/mL)	Increasing population of FUB1-transforming microbes; antibiotics treatment	Nutrient agar (NA), NA + sucrose, NA + skim milk, PYEI agar, BYE agar	Single colonies screening	NCB 1492 (belonging to Delftia or Comamonas)	[28]
OTA		-	YES medium + OTA (2 µg/mL)	Screening in liquid medium	Aspergillus niger CBS 120.49	[61]
		-	Czapek-Dox medium + OTA (40 µg/plate)	Screening point-pated colonies by observing the loss of fluorescence	Acinetobacter calcoaceticus NRRL B-551	[62]
		-	LB agar + OTA (3 µg/mL); medium with isocoumarin as the sole carbon source	Screening microbes using isocoumarin as a sole carbon source	Bacillus subtilis CW 14	[47]

Individual or combinations of antibiotics can also be used as a selective factor to inhibit the growth of unwanted microorganisms according to the different antimicrobial spectra [14,28,43]. Yu et al. (2010) [43] investigated the growth and DON-biotransformation activity of a bacterial culture treated with different combinations of 10 antibiotics, at various concentrations, demonstrating that one combination of virginiamycin (20 μg/mL), lincomycin (60 μg/mL) and salinomycin (5 μg/mL) significantly reduced microbial growth, without a loss of biotransforming activity.

Heat treatment is another effective method to enrich for heat-resistant strains, such as *Bacillus* spp. [14,19]. In order to enrich for desirable *Bacillus* strains with mycotoxin biotransforming activities, samples were treated at 80 °C for 15 min to inactive non-thermophiles [19,45].

The changes that occur in a microbial community following enrichment, such as the anticipated dominance of functional microbes and the decrease in the population diversity, are often difficult to track using traditional microbiological approaches. Fortunately, a number of molecular techniques, such as polymerase chain reaction with denaturing gradient gel electrophoresis (PCR-DGGE), terminal restriction fragment length polymorphism (T-RFLP) or 16S rRNA gene sequencing, can be applied to assess the microbial diversity and guide the screening process [14,17,43]. By monitoring the number of bands in PCR-DGGE bacterial profiles, Yu et al. (2010) [43] were able to show a reduction in the diversity of a microbial community with DON-biotransformation activity following several rounds of antibiotics and medium-based selection. Guided by PCR-DGGE, 10 positive isolates were finally obtained from 196 single colonies, which is more efficient than traditional blind screenings.

3.3. Isolation of Single Colonies

Once an enriched microflora containing MBMs is obtained, further screening to isolate single active colonies can be performed. Traditional plating methods have been widely used for this purpose. In order to selectively support the growth of potential MBMs, mycotoxins or mycotoxin-like compounds have also been introduced as sole carbon sources [26,27,31,34,46,54,58,62] or used to apply selective pressure in media containing another carbon source [47,60], as mentioned above. A summary of the various media, isolation strategies and isolated functional strains is given in Table 1.

Due to the physiological characteristics of the potential MBMs, they may be difficult to culture under the conditions used in the laboratory [63]. Efforts such as extending the incubation time, choosing media with inorganic nitrogen sources and changing the solidifying agent in the media, have been adopted to improve culturability [13,14,32,57]. The use of longer incubation times (i.e., >5 days) could help in the isolation of slow-growing bacteria, which are prone to suppression by more predominant bacteria [13,14]. In regard to the media-solidifying agent, it was reported that gellan gum, rather than agar, better supported the growth of soil-based bacteria [64]. Ito et al. (2012) [32] isolated a DON-biotransforming strain, *Marmoricola* sp. MIM116, from 1/3 R2A-gellan gum media and also observed better growth of the strain in media containing gellan gum as the solidification agent as compared with agar.

The mycotoxin biotransformation activities of isolated microbes are typically detected by chemical analysis methods, such as TLC, HPLC and LC-MS, which will be discussed further below. Some rapid detection methods have also been developed [34,50,62]. The basic concepts for these methods were originally based on the Kirby–Bauer disc plate diffusion assay. Devi et al. (2006) [50] examined six marine bacteria belonging to *Moraxella* spp. using the disc plate diffusion assay with 50 μg/disk of CIT. A strain showing high tolerance to mycotoxin was selected and examined for biotransformation activity. In another study screening for MBMs, a sterile filter paper disk inoculated with a culture was incubated on a nutrient agar plate coated with 0.4 mL of AFB_1 (10 μg/mL). Another successful application involved visualizing single colonies able to biotransform AFB_1 through the disappearance in fluorescence surrounding the colony, which is contributed by the coumarin ring structure in AFB_1, indicating the degradation of AFB_1 [34]. Similarly, this method has been applied to screening single strains with OTA biotransformation activity, which indicated that none of the strains had the ability to break the isocoumarin ring of OTA [62].

3.4. Identification of Functional Microorganisms

Microbial isolates capable of biotransforming mycotoxins may be identified and further characterized based on genome sequence information, which can range from the analysis of a single marker gene to entire genomes. The most commonly-used marker gene for bacterial identification and phylogenetic analysis is the 16S small ribosomal RNA subunit gene (rDNA) [65], which can be used to characterize both individual isolates and complex microbiota, though other conserved genes, such as *rpoA*, may also be used [66]. Several characteristics make the 16S gene well suited: it is present among all bacteria; it contains regions of a conserved sequence that allow for the design of universal PCR primers, but is also interspersed with regions of variability sufficient to discern phylogenetic relationships often to the genus and, occasionally, the species, level [67–69]. Furthermore, several databases of high-quality full-length 16S rDNA sequences and taxonomical assignments have been developed, such as SILVA [70], the Ribosomal Database Project (RDP) [71] and GreenGenes [72], and many bioinformatics tools are available for processing the sequence data. Many universal 16S primers are available in the literature, and several systematic comparisons have been carried out to evaluate their performance in terms of taxonomic coverage and discriminatory power [73–75].

The general workflow for taxonomic classification of a single isolate involves amplification of the 16S gene region, cloning of the amplicon and subsequent sequencing by the Sanger method. The taxonomy is then assigned by comparison of the sequence against one of the established reference databases (e.g., SILVA, RDP or GreenGenes), in which the taxonomy has already been established.

Next-generation DNA sequencing (NGS) technologies now allow for the taxonomic classification of complex microbiota by sequencing of the rDNA gene en masse, with the added benefit that the rDNA amplicons can be sequenced directly without the need for a cloning step. This approach may be useful during different stages of the isolation process to examine mixed cultures capable of biotransformation and to identify candidate bacteria that might possess activity. Illumina technology is currently the highest throughput and most cost-effective method for microbiota sequencing and can sequence rDNA amplicons up to ~600 bp in length, which is generally adequate for classification to the genus level. Several open-source bioinformatic pipelines, including QIIME [76] and mothur [77], are freely available, which will cluster the rDNA reads into operational taxonomic units (OTUs), assign taxonomy and perform various diversity analyses. It is important to consider several technical factors that could lead to systematic biases in the final community composition, including the choice of universal primers, as mentioned above, as well as the DNA extraction method, which should extract DNA equally from all members of the population [78,79].

A drawback of rDNA-based phylogenetic analysis is that, due to the relatively high conservation of this gene, its resolution is limited and cannot be used to distinguish between different strains. A wide range of other molecular typing techniques is available for strain-level typing, which typically infer sequence variability based on DNA fragment length polymorphisms. These methods generate electrophoretic DNA banding patterns from genomic DNA either through restriction enzyme digestion, (e.g., restriction fragment polymorphism (RFLP), pulsed field gel electrophoresis (PFGE)), PCR amplification (e.g., rapid amplification of polymorphic DNA (RAPD), repetitive sequencing-based PCR (REP-PCR), Multiple-locus variable number tandem repeat analysis (MLVA), Denaturing gradient gel electrophoresis (DGGE)) or a combination of both (e.g., amplified fragment length polymorphism (AFLP)) [80]. Additionally, strain-level sequence-based typing methods, such as multi-locus sequence typing (MLST), use several conserved marker genes to phylogenetically classify members of the same species; however, for these methods, a complete genome sequence is first required to develop the assay. The most definitive phylogenetic resolution, of course, is achieved by whole genome sequencing of the isolate, which is now becoming routine due to the reduced cost made possible by NGS.

Although phylogenetic analysis is a useful tool for the identification of functional strains, traditional morphological and biochemical characterization is still necessary to confirm the physiological characteristics of the isolates. These may include light [18,54], phase contrast, bright field [59] and electron microscopy [12,16,57] to determine cell size, shape and Gram-stain type. Biochemical tests, such as carbon source utilization, enzyme activity and chemical sensitivities [18,19,27,33,38,45,46,50,58,59,81], as well as commercial identification systems, such as BIOLOG or API® strips, may also be used for the identification of MBMs [12,21,35,60]. In addition, fatty acid and ribosomal protein profiles may be determined using gas chromatographic analysis of fatty acid methyl esters (GC-FAME) and matrix assisted laser desorption/ionization-time of flight (MALDI-TOF) techniques, respectively [12,59].

A safety assessment must be performed to exclude potential pathogens, which may produce extracellular toxins that could contaminate food and feed. *Bacillus* spp. with mycotoxin biotransformation activities have been more sought after, due to their perceived probiotic properties. However, some species, such as *Bacillus cereus*, are known toxin producers, and even species outside the *Bacillus cereus* group have been reported to produce related enterotoxins [82]. Yi et al. (2011) [21], using two commercial immunoassay kits, evaluated the enterotoxin-producing ability of *Bacillus licheniformis* CK1 by assaying for the Hbl and Nhe enterotoxins, which are produced by *Bacillus cereus*.

It is recommended to check the functional strains against the U.S. Food and Drug Administration (FDA) Generally Recognized as Safe (GRAS) list and European Food Safety Authority (EFSA) Qualified Presumption of Safety (QPS) list. Biological agents belonging to GRAS are exempt from premarket review and FDA approval. Similarly, biological agents included in the QPS list usually undergo a simplified assessment by the EFSA [83,84].

The identified MBMs are suggested to be registered and deposited in a microorganism collection organization for further research or patent procedures. Those governmental or private non-profit organizations include the Agriculture Research Service Culture Collection (NRRL), the International Depositary Authority of Canada (IDAC), Deutsche Sammlung von Mikroorganismen und Zellkulturen (DSMZ) and the American Type Culture Collection (ATCC).

3.5. Physiological Characterization of Mycotoxin Biotransformation Activity

In this stage, the isolated microorganisms should be investigated to verify the biotransformation activity by determining the efficiency correlation with cell growth, influencing factors and the position of active substrates. A time course showing a decrease in mycotoxin levels combined with an increase in biotransformation products provides clear evidence of the biotransformation activity of the isolated strain [12,14,30,37,53,61,85–87]. The efficiency of biotransformation can be measured based on the relationship between mycotoxin decrease and cell growth. The mycotoxin biotransformation rate (Equation (1)) may be calculated as follows [15,32]:

$$\text{Mycotoxin Biotransformation rate} = \frac{\text{amount of transformation (μg)}}{\text{amount of dry cell (mg)} \times \text{time (hour)}} \quad (1)$$

Knowledge of intrinsic and extrinsic factors that influence the biotransformation is beneficial to understanding the mechanisms of enzymatic biotransformation and for developing and optimizing the detoxification conditions in the complex food/feed matrix. A number of studies have evaluated intrinsic factors, including carbon source [26,34], nitrogen source [26,34,88], vitamins [26], metals ions [12,34,46,89–93], enzyme inhibitors and promoters [52,89–94], initial concentration of mycotoxins [93,95], initial concentration of cells [53,95] and initial pH value [12,14,17,26,31,34,41,46,54,60,88,89,95,96], as well as extrinsic factors, including temperature [12,14,17,25,26,31,34,38,41,46,52,54,60,88,89,95,96], aeration (shaking rate) [26], oxygen preference [14], as well as pre-incubation [26] and incubation time [34,97]. The experimental designs and optimized biotransformation conditions are summarized in Table 2. In addition to the investigation of single factors, researchers have also looked for optimized conditions for multiple variables using mathematic models. One example is the optimization of AFB_1 biotransformation based on six parameters (temperature, pH, liquid volume, inoculum size, agitation speed and incubation time) using a Plackett–Burman design followed by a response surface methodology (RSM) based on a central composite design (CCD). The degradation efficiency reached 95.8% based on the predicted parameters from the mathematic model [97]. Other methods, such as the orthogonal method and one-factor-at-a-time method, were also useful for screening of optimized biotransformation conditions [26,34].

Table 2. Intrinsic and extrinsic factors to influence the biotransformation rates.

Factor	Mycotoxin	Biotransforming Product(s)	Optimal Condition/Reverse Effect
Carbon source	AFB_1	U.I. [a]	Starch (4.0%) [34]
	CIT	U.I.	Glucose (1.2%) [26]
Nitrogen source	AFB_1	U.I.	Yeast extract (0.5%) [88]; tryptone (0.5%) [34]
	CIT	U.I.	Peptone (0.3%) [26]
Vitamins	CIT	U.I.	Vitamin C (100 µg/mL) [26]
Metals ions	DON	3-*epi*-DON	Minerals added in the corn steep liquor and peptone [12]
	ZEA	U.I.	Zn^{2+}, Mn^{2+}, Ca^{2+}, Mg^{2+} (10 mmol/L) [89]
	AFB_1	U.I.	Mg^{2+}, Cu^{2+} (10 mmol/L) [46]; Ca^{2+}, Mg^{2+} (10 mmol/L) [90–92]; Mn^{2+}, Cu^{2+} (10 mmol/L) [25]; Mg^{2+}, Zn^{2+}, Cu^{2+}, Mn^{2+} (10 mmol/L) [93]
Enzyme inhibitor/enhancer	ZEA	U.I.	Reverse effect: chelating agents of EDTA, OPT (10 mmol/L) [89]
	AFB_1	U.I.	Reverse effect: chelating agents of EDTA, OPT (10 mmol/L) [90–92]
	OTA	OTα	Reverse effect: chelating agents of EDTA (10 mmol/L), OPT (1 mmol/L) [52]
	AFB_1	U.I.	Tween 80, Triton X-100 (0.05%) [93]
	AFB_1	U.I.	NADPH (0.2 mmol/L), $NaIO_4$ (3 mmol/L) [94]
Concentration of mycotoxins	AFB_1	U.I.	0.5 µg/mL [93]
	OTA	U.I. [a]	0.1 µg/mL [95]
Concentration of cells	OTA	U.I.	10^8 CFU/mL [95]
	OTA	OTα	10^9 CFU/mL [53]
Initial pH	DON	3-*epi*-DON	pH = 7 [12]
	DON	DOM-1	pH = 6.5–7 [14]; pH = 5–10 [17]
	ZEA	U.I.	pH = 7–8 [89]; pH = 4.5 [41]
	AFB_1	U.I.	pH = 5–6 [96]; pH = 6–7 [31,34,88]; pH = 8 [46]
	PAT	E- and Z-ascladiol	pH = 3–6 [60]
	PAT	U.I.	pH = 3–5 [54]
	CIT	U.I.	pH = 7 [26]
	OTA	U.I.	pH = 4 [95]
Temperature	DON	3-*epi*-DON	20–35 °C [12]
	DON	DOM-1	20–35 °C [14]; 20–37 °C [17]
	ZEA	U.I.	30–37 °C [41]; 42 °C [89]
	AFB_1	U.I.	30–37 °C [31,34,46,88,96]
	PAT	E- and Z-ascladiol	35 °C [60]
	PAT	U.I.	37 °C [54]
	CIT	U.I.	37 °C [26]
	OTA	OTα	25–35 °C [52]
	OTA	U.I.	28 °C [95]
Shaking rate	CIT	U.I.	200 RPM [26]
Oxygen preference	DON	DOM-1	Aerobic condition [14]
Concentration of mycotoxins	OTA	U.I.	0.1 µg/mL [95]
Pre-incubation time	CIT	U.I.	36–48 h [26]

[a] U.I. means unidentified.

As an observed decrease in the amount of mycotoxin may result from absorption rather than biotransformation, identification of the daughter compound(s) is necessary to confirm biotransformation activities. The most direct evidence to support biotransformation is the observation of new product(s) in parallel with a decrease in the amount of the original mycotoxin. In some cases, however, the biotransformation product is not detectable through chemical analysis methods, and indirect methods must therefore be adopted to distinguish between adsorption and biotransformation. The involvement of an enzymatic reaction, rather than simple binding, is evidenced by the loss of biotransformation activity following heat or acid-inactivation of cells or in an extracellular extract treated with enzyme inactivating agents, such as proteinase K, SDS or EDTA [12,15,19,98,99]. Furthermore, the subcellular location of active components responsible for biotransformation can be identified by comparing activities in the whole cell culture, extracellular extract (by centrifugation) and cytoplasmic extract (by cell lysis) [25,27,34,45,50,59,88,99].

3.6. Identification of Mycotoxins and Their Biotransformation Products

To determine the mycotoxin content in a sample, chromatographic techniques are most commonly adopted. Based on the properties of the sample matrices and the mycotoxin itself, a solvent can be chosen for extraction followed by a range of different clean-up procedures. For example, DON was extracted from mouldy corn with 84% acetonitrile (acetonitrile/water 84:16, *v*/*v*) on a rocking platform at 60 RPM for two hours at room temperature [100] before HPLC analysis. ZEA was usually recovered by a mixture of water and methanol (50:50) [18,41] before HPLC analysis. Solvent extraction can be performed under regular shaking conditions or more efficiently with the aid of an ultrasonicator [18]. Table 3 summarizes the solvents used during the extraction of various mycotoxins and can be used as a baseline to choose appropriate extraction solvents for the analysis of mycotoxins.

Table 3. Analytical methods for the detection and identification of mycotoxins and their biotransforming products.

Mycotoxins/Biotransforming Products	Extraction Solvents	Analytical Method
DON	50% methanol [12,43,48]; 84% acetonitrile [100,101]; ethyl acetate [30]	HPLC [12,13,15,100,101]; LC-MS [48,101,102]; ELISA [32]; HSCCC [101]; NMR [102]
3-*epi*-DON	50% methanol [12]; ethyl acetate [13]	HPLC [12,13,15]; LC-MS [102]; NMR [13,102]
DOM-1 (deepoxy DON)	84% acetonitrile [100]; 50% methanol [48]	HPLC [100]; LC-MS [14,48]; GC-MS [17]; MS [43]
3-*keto*-DON	Ethyl acetate [16,30]	MS [16,30]; NMR [16,30]
ZEA	50% methanol [18,23,41]; 90% acetonitrile [19]; 84% acetonitrile [21]; 50% acetonitrile [98]	HPLC [18,19,21,24,41]; LC-MS [23,89,98]; ELISA [24]
1-(3,5-dihydroxy-phenyl)-10′-hydroxy-1′*E*-undecene-6′-one	Chloroform [22]	TLC, MS, NMR [22]
ZEA-sulfate	60% methanol [36]	LC-MS [36]
ZEA-4-*O*-β-glucoside		TLC, MS, NMR, IR [103]
α-ZAL, β-ZAL, α-ZOL, β-ZOL, ZAN, 8′(*S*)-hydroxyzearalenone, ZEA-2,4-bis(methyl ether), ZEA-2-methyl ether	50% chloroform [104]	TLC, MS, NMR, IR [104]
ZEN-4-*O*-sulfate	33% chloroform:methanol (9:1) [105]	MS, NMR, infrared [105]
ZOM-1 ((5*S*)-5-({2,4-dihydroxy-6-[(1*E*)-5-hydroxypent-1-en-1-yl]benzoyl}oxy)hexanoic acid)	Ethyl acetate [106]	HPLC, LC-MS, NMR [106]
α-ZOL, α-ZOL-S, ZEA-14-sulfate, ZEA-16-sulfate, ZEA-14-Glc, ZEA-16-Glc	50% acetonitrile [107]	LC-MS [107]

Table 3. *Cont.*

Mycotoxins/Biotransforming Products	Extraction Solvents	Analytical Method
AFB$_1$	60% methanol [108]; 50% methanol [45]; chloroform [38,46,88,109]; dichloromethane [110]	HPLC [25,38,45,46,88,96,108–110]; TLC [38,96,109]; ESMS [109]; LC-MS [38,96,109]; HR-FTMS [96]; ELISA [34]
AFB$_2$	Chloroform [25]	HPLC [25]
AFG$_1$	60% methanol [108]; 50% methanol [45]	HPLC [45,108]
AFM$_1$	Chloroform [25]; 50% methanol [45]	HPLC [25,45]
AFD$_1$, AFD$_2$, AFD$_3$	Chloroform [111]	TLC, HPLC, GC-MS, FT-IR [111]
PAT	Ethyl acetate [54,60,99]	TLC [85]; HPLC([37,54,60,85]; LC-MS [99]; NMR [85]
DPA (desoxypatulinic acid)	Ethyl acetate [99]	HPLC [112]; LC-MS [99]; NMR [112]
E- and *Z*-ascladiol	Ethyl acetate [60,85]	TLC [85]; HPLC [37,60,85]; LC-MS [37,60]; NMR [37,85]
CIT	Acetone:ethyl acetate (1:1) [27]; ethyl acetate [50]	TLC(C04); HPLC [26,27]
Decarboxycitrinin	Ethyl acetate [50]	MS, NMR [50]
FUB$_1$		TLC [28]; HPLC [28]; GC-MS[28]; LC-MS [113,114]
Heptadecanone, isononadecene, octadecenal, eicosane		GC-MS [28]
Hydrolyzed FUB$_1$		LC-MS [113,114]
2-keto-hydrolyzed FUB$_1$		LC-MS, NMR [115]
OTA	Methanol [51]; dichloromethane [33,52,61]; ethyl acetate [53]	TLC [52,61]; HPLC [33,51–53,61]
OTα	Methanol [51]; dichloromethane [52];	HPLC[51,52]; LC-MS [51]
L-β-phenylalanine	Methanol [51]	HPLC [51]

The strategy for the identification of unknown mycotoxin biotransformation products will be very different from the determination of known mycotoxins in a sample due to the unknown properties of the products. Thus, the sample needs to be subjected to both polar and non-polar solvent extraction and the formation of new compounds identified by comparing the chromatographs of samples before and after biotransformation. Initial identification of the biotransformation products is usually performed using LC/MS, while final identification requires the peaks to be purified at milligram levels and subjected to NMR analysis. For example, the DON biotransformation product 3-*epi*-DON by *D. mutans* 17-2-E-8 was purified from bacterial culture following centrifugation, filtration, freeze-drying and high-speed countercurrent chromatography. The purified 3-*epi*-DON was then subjected to 1D and 2D NMR, and the structure was then identified [102]. Metabolomic strategies have started gaining recognition for the identification of new products resulting from microbial biotransformation. With the progress in metabolomics and its related tools and software, the identification of novel biotransformation mycotoxin products will become easier and faster.

3.7. Evaluation of Toxicity of Biotransforming Product

While many studies report novel structural/chemical modifications that render mycotoxins less toxic [116–119], the final use of any biological detoxification method is dependent on providing empirical evidence for this claim. Furthermore, some country/state regulatory agencies have set up rigid parameters for the assessment of the safety and efficacy of any developed detoxification product, which need to be addressed accordingly before the registration of that product. A very clear example is the EU regulations regarding the additives for use in animal nutrition and the establishment of a new functional group of feed additives (No. 1831/2003 and No. 386/2009).

Different in vitro and in vivo [120–122] approaches were suggested in the past to test for changes in toxicity (Table 4), each with a unique set of advantages and challenges [123,124]. The adopted system should take into consideration: (a) the nature of the tested mycotoxin as screening carcinogens for example with long-term exposure studies is different from screening estrogenic toxins; (b) host suitability and selection since mono-gastric animals for example show higher sensitivity towards certain toxins (such as DON) in comparison to other livestock; (c) the solubility and stability of final biotransformation products (the reduced solubility and stability of 3-*keto*-DON for example necessitates using organic solvents for storing and diluting this metabolite that might interfere with assay outcomes in certain cases.); and finally, (d) the route of exposure; even though the oral route is the main contamination means, symptoms might also develop in animals due to inhalation of grain dust (as the case of aflatoxins for example) [125,126]. Generally speaking, a host highly ranked within the evolutionary tree is more sensitive toward mycotoxins; hence, bacterial hosts are the least preferred for evaluating the bio-toxicity of mycotoxins.

Table 4. Models and methodologies of toxicity evaluation of mycotoxins and their biotransforming products.

Mycotoxins	Biotransforming Products	Model	Methodologies	Reference
DON	3-*epi*-DON	Caco-2 cells	Evaluation of metabolic activity by an MTT cell proliferation assay	[127]
	3-*epi*-DON	3T3 cells	Evaluation of DNA synthesis activity by a cell proliferation ELISA employing BrdU incorporation	[127]
	3-*epi*-DON	Female B6C3F$_1$ mice	Evaluation of effects on body weight gain, relative organ weights, food consumption, hematology and clinical chemistry	[127]
	DOM-1	Swine kidney cells	Evaluation of metabolic activity by an MTT cell proliferation assay	[44]
	DOM-1	Chicken lymphocytes	Evaluation of DNA synthesis activity by a cell proliferation ELISA employing BrdU incorporation	[56]
	DOM-1	Starter pigs	Evaluation of effects on growth performance and serum metabolites	[100]
	3-*keto*-DON	Mouse spleen lymphocytes	Evaluation of immunosuppressive activity by a cell proliferation assay	[16]
ZEA	1-(3,5-dihydroxy-phenyl)-10′-hydroxy-1′E-undecene-6′-one	MCF-7 cells	Evaluation of estrogenic activity by a WST cell proliferation assay	[22]
	ZEA-sulfate	MCF-7 cells	Evaluation of estrogenic activity by an MTS cell proliferation assay	[36]
	α-ZAL, β-ZAL, α-ZOL, β-ZOL	MCF-7 and MDA-MB-231 cells	Evaluation of estrogenic activity by an MTT cell proliferation assay	[128]
	α-ZAL, β-ZAL, α-ZOL, β-ZOL, ZAN, 8′(S)-Hydroxyzearalenone, ZEA-2,4-bis(methyl ether), ZEA-2-methyl ether	Rat uteri	Evaluation of relative binding affinity by an estrogen receptor binding assay	[104]
	ZOM-1	Yeast YZRM7	Evaluation of estrogenic activity by a sensitive yeast assay	[106]
	ZOM-1	Human estrogen receptor-α	Evaluation of estrogenic activity by a HitHunter EFC estrogen chemiluminescence assay	[106]

Table 4. *Cont.*

Mycotoxins	Biotransforming Products	Model	Methodologies	Reference
ZEA	U.I. [a]	Pre-pubertal female gilts	Evaluation of effects on growth performance, genital organs, serum hormones and histopathological changes	[129]
	U.I.	Yeast BLYES	Evaluation of estrogenic activity by a sensitive yeast assay	[130]
	U.I.	Pre-pubertal female rats	Evaluation of estrogenic activity by an immature uterotrophic assay	[130]
AFB$_1$	AFD$_1$, AFD$_2$, AFD$_3$	Hela cells	Evaluation of cytotoxicity by an MTT cell proliferation assay	[111]
	U.I.	L929 cells	Evaluation of cytotoxicity by an MTT cell proliferation assay	[108]
	U.I.	*Salmonella typhimurium* TA100	Evaluation of mutagenicity by an Ames assay	[109]
	U.I.	*Artemia salina*	Evaluation of toxicity by an insect larvae survival assay	[31]
PAT	DPA	*Escherichia coli*	Evaluation of microbial toxicity	[99]
	DPA	Seeds of *Arabidopsis thaliana*	Evaluation of phytotoxicity	[99]
	DPA	Human hepatocytes LO2	Evaluation of cytotoxicity by an MTT cell proliferation assay	[99]
	DPA	Human lymphocytes	Evaluation of cytotoxicity by a trypan blue cell proliferation assay	[112]
CIT	U.I.	*Bacillus subtilis* TISTR	Evaluation of microbial toxicity	[27]
OTA	U.I.	HepG2 cells	Evaluation of cytotoxicity by an MTT cell proliferation assay	[95]
	OTα	Zebrafish (*Danio rerio*) embryo	Evaluation of teratogenicity	[131]

[a] U.I. means unidentified.

The high costs of raising and keeping large animals and strict guidelines for using animals in research [132,133] are other factors to consider when performing preliminary toxicity evaluations. In some cases, the use of primary and secondary cell lines derived from different hosts (mainly porcine/canine) [134–138] may be warranted to replace actual animals. Specific cell lines may be more or less suitable for testing certain toxins. For example, He et al. (2015) [127] used heterogeneous human epithelial colorectal adenocarcinoma (Caco-2) cells and mouse embryonic fibroblast (3T3) cells to assess cell viability and DNA synthesis, respectively, and thereby to illustrate the diminished toxicity of 3-*epi*-DON. Recently, mouse models have been gaining more attention and becoming feasible routes for bio-potency testing. He et al. (2015) [127] followed the above cell culture assays by exposing B6C3F1 mice to DON and 3-*epi*-DON treatments for 14 consecutive days, demonstrating that toxin-induced lesions were only observed in the adrenal glands, thymus, stomach, spleen and colon of the DON-group, but not 3-*epi*-DON. Some alternative approaches for using large animals in toxicity testing have been devised recently. For example, explants of specific and more sensitive tissues were developed by the Oswald group using pig jejunal explants to investigate the toxicity of deepoxy-DON and 3-*epi*-DON [139,140].

Another primary challenge for toxicity testing is the purification of enough starting materials/metabolites with which to proceed. The optimum range to start with can span milligram to gram quantities, based on the selected host, assay and testing approach. In some cases, the development of a large-scale preparative purification of the final biotransformation metabolite(s) is unavoidable in order to proceed with the toxicity testing. He et al. (2015) [102] applied a refined high-speed counter-current chromatography protocol to scale up the purification of 3-*epi*-DON from *D. mutans* 17-2-E-8 bacterial cultures before toxicity assessment.

The stability of the final biotransformation products used in toxicity bio-assays is another factor that should be considered. Conjugation-based modifications (acetylation, glycosylation, phosphorylation) are usually more difficult to address as the attached groups can be hydrolyzed or cleaved under the implemented experimental conditions. Compounds with chemically-active groups, such as ketones [141,142] (3-*keto*-deoxynivalenol for example), are another example of unstable modifications that can give misleading results, as such reactive groups seek the simultaneous reduction to more thermodynamically-stable alcohols (DON or 3-*epi*-DON) in aqueous solutions with the possibility of reverting back to the original toxins (DON in the above example).

Beyond showing a reduction in the toxicity of the final biotransformation product(s), the mechanism of this reduction must also be determined [143,144]. For example, the reduction in DON toxicity due to C3 carbon epimerization was attributed to attenuated binding of 3-*epi*-DON within the ribosomal peptidyl transferase center [139,140], in addition to increased polarity of 3-*epi*-DON and decreased molecular interactions with different cellular targets, such as *Fusarium graminearum* Tri101 acetyltransferase [145]. Similar results were also obtained for deepoxy-DON when transformed anaerobically by strain BBSH 797 [139,140]. This mechanistic information is pivotal for predicting how the final biotransformation product(s) will behave under different scenarios and usage conditions, such as under unfavorable pH conditions, or in the presence of metabolizing bacteria in the animal intestinal tract. Some of the modifications that were promising in early stages (conjugation-based modifications such as acetylation/glycosylation) were subsequently found to be incompatible with practical agricultural and industrial applications as they give rise to so-called masked mycotoxins [146,147], where the modifying chemical groups are easily cleaved or hydrolyzed by intestinal bacteria [148–152].

3.8. Genes and Enzymes

One of the ultimate goals for mycotoxin mitigation strategies is to establish the conditions needed for efficient practical usage [9]. In many agricultural and industrial applications, the use of whole bacterial cells, for instance through incorporation into fermentation matrixes (especially if they belong to the lactic acid group), might be the most feasible approach to mitigation [153,154]. This also reduces the costs associated with enzyme(s) purification and the need to incorporate any co-factors/co-substrates into the final enzymatic reactions. The above scenario is not always possible due to either processing conditions that do not favor the use of microorganisms (human food chain applications) or the presence of natural consortium that outcompete the active strain(s). In such cases, the use of pure enzyme preparations is unavoidable.

The identification of functional genes and the purification of detoxification enzyme(s) is a laborious and time-consuming process that involves working expertise in the fields of microbiology, analytical and synthetic chemistry, molecular biology, enzymology, genomics and bioinformatics. For detailed information regarding such a process and the involved procedures, the readers are referred to other recently published in-depth reviews [9,155–157].

After the responsible enzyme(s) are identified, a second stage of troubleshooting and optimization is initiated to deliver workable enzymatic mixtures that are suitable for the chosen application. Many native enzymes are not stable under certain processing conditions (pH, temperature, presence of inhibitors) or require specific expensive co-factors, such as NADP(H), to function. Such properties can sometimes be manipulated to make these enzymes more industry friendly through targeted protein-engineering [158,159]. The enantioselective epoxidation of styrene by two-component monooxygenases offers a good example of efficient enzyme engineering that was preformed recently. A fusion protein was generated by joining the C-terminus of *StyA* epoxidase with the N-terminus of the *StyB* reductase. Furthermore, the above fusion protein was cloned into a single-vector expression system to couple the epoxidation function to NADH oxidation, thereby enhancing the overall catalytic function of the system. The observed positive changes in the catalytic mechanism were attributed in part to an increased flavin-binding affinity of the *StyB* reductase associated with its N-terminal extension [160]. In a similar optimization, the NADP(H) dependence of hyperthermophilic 6-phosphogluconate dehydrogenase was engineered to favor the less expensive NAD(H) cofactor for a promising industrial application in bio-batteries [161].

3.9. Feasibility of Commercial Application

From the application point of view, there are many additional challenges to overcome in order to achieve success. These obstacles include the validation of detoxification activities in food/feed materials, the industrial-scale production and marketing authorization based on the extensive assessment. Unlike ideal in vitro conditions, many factors such as poor nutrient availability, lower pH, natural temperature, interactions with the microbiota and complicated structures of food/feed materials may become hurdles for successful application. For example, the ensiling procedure for production of silage, an important fodder for ruminants, generates an anaerobic environment with acidic pH, high carbon/nitrogen ratio and low nutrient availability, which inhibits most MBMs that grow well and present detoxification activities in media [162]. During the large-scale production of MBMs, the difficulties to maintain the activity may appear as the media and fermentation conditions vary from those in the laboratory. From an economical point of view, the media used in industrial-scale fermentation processes are usually produced from less expensive and more readily available materials, as opposed to the costly synthetic or semi-synthetic media used in laboratory-scale fermentation. Moreover, the downstream processing steps, such as freeze-drying, spray-drying or preparing a direct-fed microbial (DFM) product, may also influence the detoxification activities. Thus, the optimization and validation of mycotoxin detoxification activities should eventually be performed under conditions of commercial-scale production. In addition, any negative or uncertain effects, such as nutrient/palatability loss and safety of biotransformation agents, must also be considered. Finally, the manufacturer should obtain marketing authority before they may claim their products as anti-mycotoxin additives. In Europe, for example, the dossier of anti-mycotoxin additives registration must include the mycotoxin specificity, the species specificity, the efficacy and the safety. The marketing authority in EU member states is issued by the EC based on positive evaluation by the EFSA [163]. For these reasons, the successful commercial application of microbial detoxification has been limited. To our knowledge, Biomin® BBSH 797 is the sole microorganism product that has obtained market authority in the EU. The product has the capacity of biotransforming DON to harmless products and was approved as a pig feed additive in 2013. Recently, the EFSA published a positive scientific opinion on the safety and efficacy of Biomin® BBSH 797 for application in poultry feed. FUMzyme®, a purified enzyme isolated from the fumonisin-degrading soil bacteria *Sphingopyxis* sp. MTA 144, is the only enzyme product for mycotoxin detoxification that has been approved by the EC [164–166].

The feasibility of commercial applications has been evaluated in a number of research papers and granted patents [8,9]. One of the essential considerations at this stage is to validate the detoxification activity of MBMs in the complex food/feed matrices to which they will be applied, such as corn, wheat, barley, DGGS, peanut meal, pistachio nut, rice straw, apple juice, apple puree and grape must [13,20,32,37–39,47,53,54,100,108,110,167]. The typical procedure includes preparation of samples with either spiked or naturally-contaminated mycotoxins, incubation of samples with the MBMs under natural conditions, followed by extraction and chemical analysis of mycotoxins and their metabolites. Ito et al. (2012) verified a decrease in DON in 1000 kernels of wheat and barley grain by applying *Marmoricola* sp. MIM116. The authors also investigated spreading agents and selected 0.01% Tween 80 for its advantages in facilitating cell growth and DON reduction [32]. *Gluconobacter oxydans*, a bacterium isolated from PAT-contaminated apples, has been reported to degrade 96% of PAT (800 µg/mL) to ascladiols in apple juice. However, a full evaluation of the quality attributes of apple juice, the toxicity of ascladiol and safety assessment of the MBM is needed before any industrial applications of such a microorganism [37].

Since the principle detrimental effect of mycotoxin contamination is the decreased growth of livestock, another consideration for the application of MBMs is the effect of detoxified feed on livestock performance. A two by two factorial design is recommended for these studies. Specifically, the experimental samples should be catalogued into four groups including positive mycotoxin-free control, negative mycotoxin-exposed control, mycotoxin-free with detoxifying agents and mycotoxin-exposed with detoxifying agents. The strength of this design is that it not only verifies the detoxification activity, but also ensures that the bio-availability of essential feed ingredients is not impaired by the detoxifying agent [168]. Li et al. (2011) [100] designed an animal trial to evaluate the growth performance of swine, a DON-sensitive livestock, fed a diet detoxified by the isolate *Bacillus* sp. LS100. Compared to pigs fed a diet with *Fusarium*-infected corn with DON, the daily feed consumption, daily weight gain and feed efficiency of pigs fed the LS100-detoxified diet were significantly improved by 45%, 82% and 32%, respectively. These results demonstrate that microbially-detoxified feed can be used in the livestock industry to reduce the effects of DON toxicity, while not conferring significant negative effects on the nutrition and palatability of the feed.

In addition to reducing mycotoxin contamination through biotransformation, some detoxification strategies have introduced specific microorganisms with additional benefits to commercial applications. For example, *Bacillus subtilis* ANSB01G may be delivered as spores, which have the ability to tolerate the gut environment. In addition, this strain produces antimicrobial compounds against common bacterial pathogens such as *Escherichia coli*, *Salmonella typhimurium* and *Staphylococcus aureus*, which could improve the growth of livestock [19]. In another study, a *Bacillus licheniformis* strain was shown to have higher xylanase, CMCase and protease activities, which may enhance the digestibility of nutrients in feed [21]. Another potential advantage of detoxifying microorganisms is that certain such microorganisms may possess antifungal activity, inhibiting the growth of mycotoxin-producing fungi and further reducing mycotoxin contaminations [33,35,47,95].

4. Conclusions and Research Trends

The biological detoxification of mycotoxins is an attractive and environmentally-friendly alternative to the chemical and physical decontamination methods explored extensively over the past three decades [169]. The recently-reported examples, coupled with the emergence of some efficient commercialized biological/enzymatic agents, highlight the promise of this approach to address the safety of animal feed and human food [170,171].

The introduction of state-of-the-art research tools, such as next-generation sequencing, recombinant-enzyme overexpression and robust HPLC-MS/MS systems, combined with our enhanced understanding of the actual mechanisms underlying the diminished toxicity of final biotransformation products [140,146,172], will immensely aid in the identification [12], optimization and usage [170,171] of such naturally-derived alternatives. It is anticipated that these novel approaches will form the basis for sustainable long-term solutions to the ever-growing problem of mycotoxins in the coming years.

Acknowledgments: This project was supported with research fund from Agriculture and Agri-Food Canada.

Conflicts of Interest: The authors declare no conflict of interest.

Abbreviations

AFB$_1$	aflatoxin B$_1$
AFB$_2$	aflatoxin B$_2$
AFG$_1$	aflatoxin G$_1$
AFM$_1$	aflatoxin M$_1$
CIT	citrinin
DON	deoxynivalenol
DOM-1	deepoxy-deoxynivalenol
DPA	desoxypatulinic acid

FUM	fumonisin
FUB_1	fumonisin B_1
NIV	nivalenol
OTA	ochratoxin A
$OT\alpha$	ochratoxin α
PAT	patulin
ZEA	zearalenone
ZAL	zearalanol
ZOL	zearalenol
ZAN	zearalenone
ZOM-1	((5S)-5-({2,4-dihydroxy-6-[(1E)-5-hydroxypent-1-en-1-yl]benzoyl}oxy)hexanoic acid)

References

1. Bryden, W.L. Mycotoxin contamination of the feed supply chain: Implications for animal productivity and feed security. *Anim. Feed Sci. Technol.* **2012**, *173*, 134–158. [CrossRef]

2. Vardon, P.J.; McLaughlin, C.; Nardinelli, C. Potential economic costs of mycotoxins in the United States. In *Mycotoxins: Risks in Plant, Animal, and Human Systems*; Task Force Report No. 139; Council for Agricultural Science and Technology (CAST): Ames, IA, USA, 2003.

3. Faneli, F.; Logrieko, F. Main mycotoxin concerns in Europe: MYCORED and ISM efforts to harmonize strategies for their reduction in food and feed chain. *Proc. Nat. Sci. Matica Srpska Novi Sad* **2012**, *122*, 7–16. [CrossRef]

4. Peraica, M.; Radić, B.; Lucić, A.; Pavlović, M. Toxic effects of mycotoxins in humans. *Bull. World Health Organ.* **1999**, *77*, 754–766. [PubMed]

5. European Commission. *RASFF—The Rapid Alert System for Food and Feed*; 2015 Annual Report; European Commission: Brussels, Belgium, 2016.

6. European Commission (EC). Commission Regulation (EC) No. 583/2006 of 17 August 2006. Commission recommendation on the prevention and reduction of *Fusarium* toxins in cereals and cereal products. *Off. J. Eur. Union* **2006**, *L234*, 35–40.

7. He, J.; Zhou, T.; Young, J.C.; Boland, G.J.; Scott, P.M. Chemical and biological transformations for detoxification of trichothecene mycotoxins in human and animal food chains: A review. *Trends Food Sci. Technol.* **2010**, *21*, 67–76. [CrossRef]

8. He, J.; Zhou, T. Patented techniques for detoxification of mycotoxins in feeds and food matrices. *Recent Pat. Food Nutr. Agric.* **2010**, *2*, 96–104. [CrossRef] [PubMed]

9. Zhu, Y.; Hassan, Y.I.; Watts, C.; Zhou, T. Innovative technologies for the mitigation of mycotoxins in animal feed and ingredients—A review of recent patents. *Anim. Feed Sci. Technol.* **2016**, *216*, 19–29. [CrossRef]

10. European Commission (EC). Commission regulation (EC) No 386/2009 of 12 May 2009. Amending regulation (EC) No. 1831/2003 of the European Parliament and of the Council as regards the establishment of a new functional group of feed additives. *Off. J. Eur. Union* **2009**, *L118*, 166.

11. Hahn, I.; Kunz-Vekiru, E.; Twaruzek, M.; Grajewski, J.; Krska, R.; Berthiller, F. Aerobic and anaerobic in vitro testing of feed additives claiming to detoxify deoxynivalenol and zearalenone. *Food Addit. Contam. Part A* **2015**, *32*, 922–933. [CrossRef] [PubMed]

12. He, J.; Hassan, Y.I.; Perilla, N.; Li, X.; Boland, G.J.; Zhou, T. Bacterial epimerization as a route for deoxynivalenol detoxification: The influence of growth and environmental conditions. *Front. Microbiol.* **2016**, *7*, 572. [CrossRef] [PubMed]

13. Ikunaga, Y.; Sato, I.; Grond, S.; Numaziri, N.; Yoshida, S.; Yamaya, H.; Hiradate, S.; Hasegawa, M.; Toshima, H.; Koitabashi, M.; et al. *Nocardioides* sp. strain WSN05-2, isolated from a wheat field, degrades deoxynivalenol, producing the novel intermediate 3-*epi*-deoxynivalenol. *Appl. Microbiol. Biotechnol.* **2011**, *89*, 419–427. [CrossRef] [PubMed]

14. Islam, R.; Zhou, T.; Young, J.C.; Goodwin, P.H.; Pauls, K.P. Aerobic and anaerobic de-epoxydation of mycotoxin deoxynivalenol by bacteria originating from agricultural soil. *World J. Microbiol. Biotechnol.* **2012**, *28*, 7–13. [CrossRef] [PubMed]

15. Sato, I.; Ito, M.; Ishizaka, M.; Ikunaga, Y.; Sato, Y.; Yoshida, S.; Koitabashi, M.; Tsushima, S. Thirteen novel deoxynivalenol-degrading bacteria are classified within two genera with distinct degradation mechanisms. *FEMS Microbiol. Lett.* **2012**, *327*, 110–117. [CrossRef] [PubMed]

16. Shima, J.; Takase, S.; Takahashi, Y.; Iwai, Y.; Fujimoto, H.; Yamazaki, M.; Ochi, K. Novel detoxification of the trichothecene mycotoxin deoxynivalenol by a soil bacterium isolated by enrichment culture. *Appl. Environ. Microbiol.* **1997**, *63*, 3825–3830. [PubMed]

17. He, W.; Yuan, Q.; Zhang, Y.; Guo, M.; Gong, A.; Zhang, J.; Wu, A.; Huang, T.; Qu, B.; Li, H.; et al. Aerobic de-epoxydation of trichothecene mycotoxins by a soil bacterial consortium isolated using in situ soil enrichment. *Toxins* **2016**, *8*, 277. [CrossRef] [PubMed]

18. Tan, H.; Hu, Y.; He, J.; Wu, L.; Liao, F.; Luo, B.; He, Y.; Zuo, Z.; Ren, Z.; Zhong, Z.; et al. Zearalenone degradation by two *Pseudomonas* strains from soil. *Mycotoxin Res.* **2014**, *30*, 191–196. [CrossRef] [PubMed]

19. Lei, Y.; Zhao, L.; Ma, Q.; Zhang, J.; Zhou, T.; Gao, C.; Ji, C. Degradation of zearalenone in swine feed and feed ingredients by *Bacillus subtilis* ANSB01G. *World Mycotoxin J.* **2014**, *7*, 143–151. [CrossRef]

20. Cho, K.J.; Kang, J.S.; Cho, W.T.; Lee, C.H.; Ha, J.K.; Song, K.B. In vitro degradation of zearalenone by *Bacillus subtilis*. *Biotechnol. Lett.* **2010**, *32*, 1921–1924. [CrossRef] [PubMed]

21. Yi, P.; Pai, C.; Liu, J. Isolation and characterization of a *Bacillus licheniformis* strain capable of degrading zearalenone. *World J. Microbiol. Biotechnol.* **2011**, *27*, 1035–1043. [CrossRef]

22. Kakeya, H.; Takahashi-Ando, N.; Kimura, M.; Onose, R.; Yamaguchi, I.; Osada, H. Biotransformation of the mycotoxin, zearalenone, to a non-estrogenic compound by a fungal strain of *Clonostachys* sp. *Biosci. Biotechnol. Biochem.* **2002**, *66*, 2723–2726. [CrossRef] [PubMed]

23. Yu, Y.; Qiu, L.; Wu, H.; Tang, Y.; Yu, Y.; Li, X.; Liu, D. Degradation of zearalenone by the extracellular extracts of *Acinetobacter* sp. SM04 liquid cultures. *Biodegradation* **2011**, *22*, 613–622. [CrossRef] [PubMed]

24. Megharaj, M.; Garthwaite, I.; Thiele, J.H. Total biodegradation of the oestrogenic mycotoxin zearalenone by a bacterial culture. *Lett. Appl. Microbiol.* **1997**, *24*, 329–333. [CrossRef] [PubMed]

25. Sangare, L.; Zhao, Y.; Folly, Y.M.E.; Chang, J.; Li, J.; Selvaraj, J.N.; Xing, F.; Zhou, L.; Wang, Y.; Liu, Y. Aflatoxin B_1 degradation by a *Pseudomonas* strain. *Toxins* **2014**, *6*, 3028–3040. [CrossRef] [PubMed]

26. Chen, Y.; Sheu, S.; Mau, J.; Hsieh, P. Isolation and characterization of a strain of *Klebsiella pneumoniae* with citrinin-degrading activity. *World J. Microbiol. Biotechnol.* **2011**, *27*, 487–493. [CrossRef]

27. Kanpiengjai, A.; Mahawan, R.; Lumyong, S.; Khanongnuch, C. A soil bacterium *Rhizobium borbori* and its potential for citrinin-degrading application. *Ann. Microbiol.* **2016**, *66*, 807–816. [CrossRef]

28. Benedetti, R.; Nazzi, F.; Locci, R.; Firrao, G. Degradation of fumonisin B_1 by a bacterial strain isolated from soil. *Biodegradation* **2006**, *17*, 31–38. [CrossRef] [PubMed]

29. Ito, M.; Sato, I.; Ishizaka, M.; Yoshida, S.; Koitabash, M.; Yoshida, S.; Tsushima, S. Bacterial cytochrome p450 system catabolizing the *Fusarium* toxin deoxynivalenol. *Appl. Environ. Microbiol.* **2013**, *79*, 1619–1628. [CrossRef] [PubMed]

30. Völkl, A.; Vogle, B.; Schollenberger, M.; Karlovsky, P. Microbial detoxification of mycotoxin deoxynivalenol. *J. Basic Microbiol.* **2004**, *44*, 147–156. [CrossRef] [PubMed]

31. El-Deeb, B.; Altalhi, A.; Khiralla, G.; Hassan, S.; Gherbawy, Y. Isolation and characterization of endophytic *Bacilli* bacterium from maize grains able to detoxify aflatoxin B_1. *Food Biotechnol.* **2013**, *27*, 199–212. [CrossRef]

32. Ito, M.; Sato, I.; Koitabashi, M.; Yoshida, S.; Imai, M.; Tsushima, S. A novel actinomycete derived from wheat heads degrades deoxynivalenol in the grain of wheat and barley affected by *Fusarium* head blight. *Appl. Microbiol. Biotechnol.* **2012**, *96*, 1059–1070. [CrossRef] [PubMed]

33. Chang, X.; Wu, Z.; Wu, S.; Dai, Y.; Sun, C. Degradation of ochratoxin A by *Bacillus amyloliquefaciens* ASAG1. *Food Addit. Contam. Part A* **2015**, *32*, 564–571. [CrossRef] [PubMed]

34. Zhang, W.; Xue, B.; Li, M.; Mu, Y.; Chen, Z.; Li, J.; Shan, A. Screening a strain of *Aspergillus niger* and optimization of fermentation conditions for degradation of aflatoxin B_1. *Toxins* **2014**, *6*, 3157–3172. [CrossRef] [PubMed]

35. Petchkongkaew, A.; Taillandier, P.; Gasaluck, P.; Lebrihi, A. Isolation of *Bacillus* spp. from Thai fermented soybean (Thua-nao): Screening for aflatoxin B_1 and ochratoxin A detoxification. *J. Appl. Microbiol.* **2008**, *104*, 1495–1502. [CrossRef] [PubMed]

36. Jard, G.; Liboz, T.; Mathieu, F.; Guyonvarc'h, A.; André, F.; Delaforge, M.; Lebrihi, A. Transformation of zearalenone to zearalenone-sulfate by *Aspergillus* spp. *World Mycotoxin J.* **2010**, *3*, 183–191. [CrossRef]

37. Ricelli, A.; Baruzzi, F.; Solfrizzo, M.; Morea, M.; Fanizzi, F. Biotransformation of patulin by *Gluconobacter oxydans*. *Appl. Environ. Microbiol.* **2007**, *73*, 785–792. [CrossRef] [PubMed]

38. Farzaneh, M.; Shi, Z.; Ghassempour, A.; Sedaghat, N.; Ahmadzadeh, M.; Mirabolfathy, M.; Javan-Nikkhah, M. Aflatoxin B₁ degradation by *Bacillus subtilis* UTBSP1 isolated from pistachio nuts of Iran. *Food Control* **2012**, *23*, 100–106. [CrossRef]

39. He, P.; Young, L.G.; Forsberg, C. Microbial transformation of deoxynivalenol (vomitoxin). *Appl. Environ. Microbiol.* **1992**, *58*, 3857–3863. [PubMed]

40. Swanson, S.P.; Nicoletti, J.; Rood, H.D.; Buck, W.B.; Cote, L.M.; Yoshizawa, T. Metabolism of three trichothecene mycotoxins, T-2 toxin, diacetoxyscirpenol and deoxynivalenol, by bovine rumen microorganisms. *J. Chromatogr.* **1987**, *414*, 335–342. [CrossRef]

41. Tan, H.; Zhang, Z.; Hu, Y.; Wu, L.; Liao, F.; He, J.; Luo, B.; He, Y.; Zuo, Z.; Ren, Z.; et al. Isolation and characterization of *Pseudomonas otitidis* TH-N1 capable of degrading zearalenone. *Food Control* **2015**, *47*, 285–290. [CrossRef]

42. Kiessling, K.H.; Pettersson, H.; Sandholm, K.; Olsen, M. Metabolism of aflatoxin, ochratoxin, zearalenone, and three trichothecenes by intact rumen fluid, rumen protozoa, and rumen bacteria. *Appl. Environ. Microbiol.* **1984**, *47*, 1070–1073. [PubMed]

43. Yu, H.; Zhou, T.; Gong, J.; Young, C.; Su, X.; Li, X.; Zhu, H.; Tsao, R.; Yang, R. Isolation of deoxynivalenol-transforming bacteria from the chicken intestines using the approach of PCR-DGGE guided microbial selection. *BMC Microbiol.* **2010**, *10*, 182. [CrossRef] [PubMed]

44. Kollarczik, B.; Gareis, M.; Hanelt, M. In vitro transformation of the *Fusarium* mycotoxins deoxynivalenol and zearalenone by the normal gut microflora of pigs. *Nat. Toxins* **1994**, *2*, 105–110. [CrossRef] [PubMed]

45. Gao, X.; Ma, Q.; Zhao, L.; Lei, Y.; Shan, Y.; Ji, C. Isolation of bacillus subtilis: Screening for aflatoxins B₁, M₁, and G₁ detoxification. *Eur. Food Res. Technol.* **2011**, *232*, 957–962. [CrossRef]

46. Guan, S.; Ji, C.; Zhou, T.; Li, J.; Ma, Q.; Niu, T. Aflatoxin B₁ degradation by *Stenotrophomonas maltophilia* and other microbes selected using coumarin medium. *Int. J. Mol. Sci.* **2008**, *9*, 1489–1503. [CrossRef] [PubMed]

47. Shi, L.; Liang, Z.; Li, J.; Hao, J.; Xu, Y.; Huang, K.; Tian, J.; He, X.; Xu, W. Ochratoxin A biocontrol and biodegradation by *Bacillus subtilis* CW 14. *J. Sci. Food Agric.* **2014**, *94*, 1879–1885. [CrossRef] [PubMed]

48. Guan, S.; He, J.; Young, J.C.; Zhu, H.; Li, X.; Ji, C.; Zhou, T. Transformation of trichothecene mycotoxins by microorganisms from fish digesta. *Aquaculture* **2009**, *290*, 290–295. [CrossRef]

49. Abrunhosa, L.; Serra, R.; Venâncio, A. Biodegradation of ochratoxin A by fungi isolated from grapes. *J. Agric. Food Chem.* **2002**, *50*, 7493–7496. [CrossRef] [PubMed]

50. Devi, P.; Naik, C.G.; Rodrigues, C. Biotransformation of citrinin to decarboxycitrinin using an organic solvent-tolerant marine bacterium, *Moraxella* sp. MB₁. *Mar. Biotechnol.* **2006**, *8*, 129–138. [CrossRef] [PubMed]

51. Rodriguez, H.; Reveron, I.; Doria, F.; Costantini, A.; de Las Rivas, B.; Muñoz, R.; Garcia-Moruno, E. Degradation of ochratoxin A by *Brevibacterium* species. *J. Agric. Food Chem.* **2011**, *59*, 10755–10760. [CrossRef] [PubMed]

52. Péteri, Z.; Téren, J.; Vágvölgyi, C.; Varga, J. Ochratoxin degradation and adsorption caused by astaxanthin-producing yeasts. *Food Microbiol.* **2007**, *24*, 205–210. [CrossRef] [PubMed]

53. Abrunhosa, L.; Inês, A.; Rodrigues, A.I.; Guimarães, A.; Pereira, V.L.; Parpot, P.; Mendes-Faia, A.; Venâncio, A. Biodegradation of ochratoxin A by *Pediococcus parvulus* isolated from Douro wines. *Int. J. Food Microbiol.* **2014**, *188*, 45–52. [CrossRef] [PubMed]

54. Zhang, X.; Guo, Y.; Ma, Y.; Chai, Y.; Li, Y. Biodegradation of patulin by a *Byssochlamys nivea* strain. *Food Control* **2016**, *64*, 142–150. [CrossRef]

55. Binder, J.; Horvath, E.M.; Schatzmayr, G.; Ellend, N.; Danner, H.; Krska, R.; Braun, R. Screening for deoxynivalenol-detoxifying anaerobic rumen microorganisms. *Cereal Res. Commun.* **1997**, *25*, 343–346.

56. Schatzmayr, G.; Zehner, F.; Täubel, M.; Schatzmayr, D.; Klimitsch, A.; Loibner, A.P.; Binder, E.M. Microbiologicals for deactivating mycotoxins. *Mol. Nutr. Food Res.* **2006**, *50*, 543–551. [CrossRef] [PubMed]

57. He, C.; Fan, Y.; Liu, G.; Zhang, H. Isolation and identification of a strain of *Aspergillus tubingensis* with deoxynivalenol biotransformation capability. *Int. J. Mol. Sci.* **2008**, *9*, 2366–2375. [CrossRef] [PubMed]

58. Liang, Z.; Li, J.; He, Y.; Guan, S.; Wang, N.; Ji, C.; Niu, T. AFB1 bio-degradation by a new strain—*Stenotrophomonas*. sp. *Agric. Sci. China* **2008**, *7*, 1433–1437. [CrossRef]

59. Hormisch, D.; Brost, I.; Kohring, G.W.; Giffhorn, F.; Kroppenstedt, R.M.; Stackebrandt, E.; Färber, P.; Holzapfel, W.H. *Mycobacterium fluoranthenivorans* sp. nov., a fluoranthene and aflatoxin B₁ degrading bacterium from contaminated soil of a former coal gas plant. *Syst. Appl. Microbiol.* **2004**, *27*, 653–660. [CrossRef] [PubMed]

60. Dong, X.; Jiang, W.; Li, C.; Ma, N.; Xu, Y.; Meng, X. Patulin biodegradation by marine yeast *Kodameae ohmeri*. *Food Addit. Contam. Part A* **2015**, *32*, 352–360.

61. Varga, J.; Rigó, K.; Téren, J. Degradation of ochratoxin A by *Aspergillus* species. *Int. J. Food Microbiol.* **2000**, *59*, 1–7. [CrossRef]

62. Hwanga, C.; Draughon, F.A. Degradation of ochratoxin A by *Acinetobacter calcoaceticus*. *J. Food Prot.* **1994**, *57*, 410–414. [CrossRef]

63. Barer, M.R.; Harwood, C.R. Bacterial viability and culturability. *Adv. Microb. Physiol.* **1999**, *41*, 93–137. [PubMed]

64. Janssen, P.H.; Yates, P.S.; Grinton, B.E.; Taylor, P.M.; Sait, M. Improved culturability of soil bacteria and isolation in pure culture of novel members of the divisions *Acidobacteria*, *Actinobacteria*, *Proteobacteria*, and *Verrucomicrobia*. *Appl. Environ. Microbiol.* **2002**, *68*, 2391–2396. [CrossRef] [PubMed]

65. Lane, D.J.; Pace, B.; Olsen, G.J.; Stahl, D.A.; Sogin, M.L.; Pace, N.R. Rapid determination of 16S ribosomal RNA sequences for phylogenetic analyses. *Proc. Natl. Acad. Sci. USA* **1985**, *82*, 6955–6959. [CrossRef] [PubMed]

66. Vos, M.; Quince, C.; Pijl, A.S.; Hollander, M.D.; Kowalchuk, G.A. A comparison of rpoB and 16S rRNA as markers in pyrosequencing studies of bacterial diversity. *PLoS ONE* **2012**, *7*, e30600. [CrossRef] [PubMed]

67. Conlan, S.; Kong, H.H.; Segre, J.A. Species-level analysis of DNA sequence data from the NIH Human Microbiome Project. *PLoS ONE* **2012**, *7*, e47075. [CrossRef] [PubMed]

68. Fettweis, J.M.; Serrano, M.G.; Sheth, N.U.; Mayer, C.M.; Glascock, A.L.; Brooks, J.P.; Jefferson, K.K.; Buck, G.A. Species-level classification of the vaginal microbiome. *BMC Genom.* **2012**, *13* (Suppl. S8), S17.

69. Srinivasan, R.; Karaoz, U.; Volegova, M.; MacKichan, J.; Kato-Maeda, M.; Miller, S.; Nadarajan, R.; Brodie, E.L.; Lynch, S.V. Use of 16S rRNA gene for identification of a broad range of clinically relevant bacterial pathogens. *PLoS ONE* **2015**, *10*, e0117617. [CrossRef] [PubMed]

70. Quast, C.; Pruesse, E.; Yilmaz, P.; Gerken, J.; Schweer, T.; Yarza, P.; Peplies, J.; Glöckner, F.O. The SILVA ribosomal RNA gene database project: Improved data processing and web-based tools. *Nucleic Acids Res.* **2013**, *41*, D590–D596. [CrossRef] [PubMed]

71. Cole, J.R.; Wang, Q.; Fish, J.A.; Chai, B.; McGarrell, D.M.; Sun, Y.; Brown, C.T.; Porras-Alfaro, A.; Kuske, C.R.; Tiedje, J.M. Ribosomal Database Project: Data and tools for high throughput rRNA analysis. *Nucleic Acids Res.* **2014**, *42*, D633–D642. [CrossRef] [PubMed]

72. DeSantis, T.Z.; Hugenholtz, P.; Larsen, N.; Rojas, M.; Brodie, E.L.; Keller, K.; Huber, T.; Dalevi, D.; Hu, P.; Andersen, G.L. Greengenes, a chimera-checked 16S rRNA gene database and workbench compatible with ARB. *Appl. Environ. Microbiol.* **2006**, *72*, 5069–5072. [CrossRef] [PubMed]

73. Klindworth, A.; Pruesse, E.; Schweer, T.; Peplies, J.; Quast, C.; Horn, M.; Glöckner, F.O. Evaluation of general 16S ribosomal RNA gene PCR primers for classical and next-generation sequencing-based diversity studies. *Nucleic Acids Res.* **2013**, *41*, e1. [CrossRef] [PubMed]

74. Wang, Y.; Qian, P.-Y. Conservative fragments in bacterial 16S rRNA genes and primer design for 16S ribosomal DNA amplicons in metagenomic studies. *PLoS ONE* **2009**, *4*, e7401. [CrossRef] [PubMed]

75. Liu, Z.; DeSantis, T.Z.; Andersen, G.L.; Knight, R. Accurate taxonomy assignments from 16S rRNA sequences produced by highly parallel pyrosequencers. *Nucleic Acids Res.* **2008**, *36*, e120. [CrossRef] [PubMed]

76. Navas-Molina, J.A.; Peralta-Sánchez, J.M.; González, A.; McMurdie, P.J.; Vázquez-Baeza, Y.; Xu, Z.; Ursell, L.K.; Lauber, C.; Zhou, H.; Song, S.J.; et al. Advancing our understanding of the human microbiome using QIIME. *Methods Enzymol.* **2013**, *531*, 371–444. [PubMed]

77. Schloss, P.D.; Westcott, S.L.; Ryabin, T.; Hall, J.R.; Hartmann, M.; Hollister, E.B.; Lesniewski, R.A.; Oakley, B.B.; Parks, D.H.; Robinson, C.J.; et al. Introducing mothur: Open-source, platform-independent, community-supported software for describing and comparing microbial communities. *Appl. Environ. Microbiol.* **2009**, *75*, 7537–7541. [CrossRef] [PubMed]

78. D'Amore, R.; Ijaz, U.Z.; Schirmer, M.; Kenny, J.G.; Gregory, R.; Darby, A.C.; Shakya, M.; Podar, M.; Quince, C.; Hall, N. A comprehensive benchmarking study of protocols and sequencing platforms for 16S rRNA community profiling. *BMC Genom.* **2016**, *17*, 55. [CrossRef] [PubMed]

79. Fouhy, F.; Clooney, A.G.; Stanton, C.; Claesson, M.J.; Cotter, P.D. 16S rRNA gene sequencing of mock microbial populations—Impact of DNA extraction method, primer choice and sequencing platform. *BMC Microbiol.* **2016**, *16*, 123. [CrossRef] [PubMed]

80. Li, W.; Raoult, D.; Fournier, P.E. Bacterial strain typing in the genomic era. *FEMS Microbiol. Rev.* **2009**, *33*, 892–916. [CrossRef] [PubMed]

81. Molnar, O.; Schatzmayr, G.; Fuchs, E.; Prillinger, H. *Trichosporon mycotoxinivorans* sp. nov., a new yeast species useful in biological detoxification of various mycotoxins. *Syst. Appl. Microbiol.* **2004**, *27*, 661–671. [CrossRef] [PubMed]

82. From, C.; Pukall, R.; Schumann, P.; Hormazábal, V.; Granum, P.E. Toxin-producing ability among *Bacillus* spp. outside the *Bacillus cereus* group. *Appl. Environ. Microbiol.* **2005**, *71*, 1178–1183. [CrossRef] [PubMed]

83. Generally Recognized As Safe (GRAS). Available online: https://www.Fda.Gov/food/ingredientspackaginglabeling/gras/ (accessed on 10 March 2017).

84. Qualified Presumption of Safety (QPS). Available online: https://www.Efsa.Europa.Eu/en/topics/topic/qualified-presumption-safety-qps (accessed on 10 March 2017).

85. Moss, M.O.; Long, M.T. Fate of patulin in the presence of the yeast *Saccharomyces cerevisiae*. *Food Addit. Contam.* **2002**, *19*, 387–399. [CrossRef] [PubMed]

86. Abrunhosa, L.; Santos, L.; Venâncio, A. Degradation of ochratoxin A by proteases and by a crude enzyme of *Aspergillus niger*. *Food Biotechnol.* **2006**, *20*, 231–242. [CrossRef]

87. Bejaoui, H.; Mathieu, F.; Taillandier, P.; Lebrihi, A. Biodegradation of ochratoxin A by *Aspergillus* section *nigri* species isolated from French grapes: A potential means of ochratoxin A decontamination in grape juices and musts. *FEMS Microbiol. Lett.* **2006**, *255*, 203–208. [CrossRef] [PubMed]

88. Guan, S.; Zhao, L.; Ma, Q.; Zhou, T.; Wang, N.; Hu, X.; Ji, C. In vitro efficacy of *Myxococcus fulvus* ANSM068 to biotransform aflatoxin B_1. *Int. J. Mol. Sci.* **2010**, *11*, 4063–4079. [CrossRef] [PubMed]

89. Tinyiro, S.E.; Wokadala, C.; Xu, D.; Yao, W. Adsorption and degradation of zearalenone by *Bacillus* strains. *Folia Microbiol.* **2011**, *56*, 321–327. [CrossRef] [PubMed]

90. D'Souza, D.H.; Brackeet, R.E. The role of trace metal ions in aflatoxin B_1 degradation by *Flavobacterium aurantiacum*. *J. Food Prot.* **1998**, *61*, 1666–1669. [CrossRef] [PubMed]

91. D'Souza, D.H.; Brackett, R.E. The influence of divalent cations and chelators on aflatoxin B_1 degradation by *Flavobacterium aurantiacum*. *J. Food Prot.* **2000**, *63*, 102–105. [CrossRef] [PubMed]

92. D'Souza, D.H.; Brackett, R.E. Aflatoxin B_1 degradation by *Flavobacterium aurantiacum* in the presence of reducing conditions and seryl and sulfhydryl group inhibitors. *J. Food Prot.* **2001**, *64*, 268–271. [CrossRef] [PubMed]

93. Das, A.; Bhattacharya, S.; Palaniswamy, M.; Angayarkanni, J. Biodegradation of aflatoxin B_1 in contaminated rice straw by *Pleurotus ostreatus* MTCC 142 and *Pleurotus ostreatus* GHBBF10 in the presence of metal salts and surfactants. *World J. Microbiol. Biotechnol.* **2014**, *30*, 2315–2324. [CrossRef] [PubMed]

94. Hamid, A.B.; Smith, J.E. Degradation of aflatoxin by *Aspergillus flavus*. *J. Gen. Microbiol.* **1987**, *133*, 2023–2029. [CrossRef] [PubMed]

95. Yang, Q.; Wang, J.; Zhang, H.; Li, C.; Zhang, X. Ochratoxin A is degraded by *Yarrowia lipolytica* and generates non-toxic degradation products. *World Mycotoxin J.* **2016**, *9*, 269–278. [CrossRef]

96. Eshelli, M.; Harvey, L.; Edrada-Ebel, R.; McNeil, B. Metabolomics of the bio-degradation process of aflatoxin B1 by actinomycetes at an initial pH of 6.0. *Toxins* **2015**, *7*, 439–456. [CrossRef] [PubMed]

97. Kong, Q.; Zhai, C.; Guan, B.; Li, C.; Shan, S.; Yu, J. Mathematic modeling for optimum conditions on aflatoxin B_1 degradation by the aerobic bacterium *Rhodococcus erythropolis*. *Toxins* **2012**, *4*, 1181–1195. [CrossRef] [PubMed]

98. Sun, X.; He, X.; Xue, K.; Li, Y.; Xu, D.; Qian, H. Biological detoxification of zearalenone by *Aspergillus niger* strain FS10. *Food Chem. Toxicol.* **2014**, *72*, 76–82. [CrossRef] [PubMed]

99. Zhu, R.; Feussner, K.; Wu, T.; Yan, F.; Karlovsky, P.; Zheng, X. Detoxification of mycotoxin patulin by the yeast *Rhodosporidium paludigenum*. *Food Chem.* **2015**, *179*, 1–5. [CrossRef] [PubMed]

100. Li, X.; Zhu, C.; Lange, C.F.M.d.; Zhou, T.; He, J.; Yu, H.; Gong, J.; Young, J.C. Efficacy of detoxification of deoxynivalenol-contaminated corn by *Bacillus* sp. LS100 in reducing the adverse effects of the mycotoxin on swine growth performance. *Food Addit. Contam. Part A* **2011**, *28*, 894–901. [CrossRef] [PubMed]

101. He, J.; Yang, R.; Zhou, T.; Tsao, R.; Young, J.C.; Zhu, H.; Li, X.-Z.; Boland, G.J. Purification of deoxynivalenol from *Fusarium graminearum* rice culture and mouldy corn by high-speed counter-current chromatography. *J. Chromatogr. A* **2007**, *1151*, 187–192. [CrossRef] [PubMed]

102. He, J.; Yang, R.; Zhou, T.; Boland, G.J.; Scott, P.M.; Bondy, G.S. An epimer of deoxynivalenol: Purification and structure identification of 3-*epi*-deoxynivalenol. *Food Addit. Contam. Part A* **2015**, *32*, 1523–1530. [CrossRef] [PubMed]

103. El-Sharkawy, S.; Abul-Hajj, Y. Microbial transformation of zearalenone, I. Formation of zearalenone-4-O-β-glucoside. *J. Nat. Prod.* **1987**, *50*, 520–521. [CrossRef]

104. El-Sharkawy, S.H.; Abul-Hajj, Y.J. Microbial transformation of zearalenone. 2. Reduction, hydroxylation, and methylation products. *J. Org. Chem.* **1988**, *53*, 515–519. [CrossRef]

105. El-Sharkawy, S.H.; Selim, M.I.; Afifi, M.S.; Halaweish, F.T. Microbial transformation of zearalenone to a zearalenone sulfate. *Appl. Environ. Microbiol.* **1991**, *57*, 549–552.

106. Vekiru, E.; Hametner, C.; Mitterbauer, R.; Rechthaler, J.; Adam, G.; Schatzmayr, G.; Krska, R.; Schuhmacher, R. Cleavage of zearalenone by *Trichosporon mycotoxinivorans* to a novel nonestrogenic metabolite. *Appl. Environ. Microbiol.* **2010**, *76*, 2353–2359. [CrossRef] [PubMed]

107. Brodehl, A.; Möller, A.; Kunte, H.J.; Koch, M.; Maul, R. Biotransformation of the mycotoxin zearalenone by fungi of the genera *Rhizopus* and *Aspergillus*. *FEMS Microbiol. Lett.* **2014**, *359*, 124–130. [CrossRef] [PubMed]

108. Chen, Y.; Kong, Q.; Chi, C.; Shan, S.; Guan, B. Biotransformation of aflatoxin B_1 and aflatoxin G_1 in peanut meal by anaerobic solid fermentation of *Streptococcus thermophilus* and *Lactobacillus delbrueckii* subsp. bulgaricus. *Int. J. Food Microbiol.* **2015**, *211*, 1–5. [CrossRef] [PubMed]

109. Alberts, J.F.; Engelbrecht, Y.; Steyn, P.S.; Holzapfel, W.H.; van Zyl, W.H. Biological degradation of aflatoxin B_1 by *Rhodococcus erythropolis* cultures. *Int. J. Food Microbiol.* **2006**, *109*, 121–126. [CrossRef] [PubMed]

110. Das, A.; Bhattacharya, S.; Palaniswamy, M.; Angayarkanni, J. Aflatoxin B_1 degradation during co-cultivation of *Aspergillus flavus* and *Pleurotus ostreatus* strains on rice straw. *3 Biotech* **2015**, *5*, 279–284. [CrossRef] [PubMed]

111. Samuel, M.S.; Sivaramakrishna, A.; Mehta, A. Degradation and detoxification of aflatoxin B_1 by *Pseudomonas putida*. *Int. Biodeterior. Biodegrad.* **2014**, *86*, 202–209. [CrossRef]

112. Castoria, R.; Mannina, L.; Durán-Patrón, R.; Maffei, F.; Sobolev, A.P.; de Felice, D.V.; Pinedo-Rivilla, C.; Ritieni, A.; Ferracane, R.; Wright, S.A. Conversion of the mycotoxin patulin to the less toxic desoxypatulinic acid by the biocontrol yeast *Rhodosporidium kratochvilovae* strain LS11. *J. Agric. Food Chem.* **2011**, *59*, 11571–11578. [CrossRef] [PubMed]

113. Heinl, S.; Hartinger, D.; Thamhesl, M.; Vekiru, E.; Krska, R.; Schatzmayr, G.; Moll, W.D.; Grabherr, R. Degradation of fumonisin B_1 by the consecutive action of two bacterial enzymes. *J. Biotechnol.* **2010**, *145*, 120–129. [CrossRef] [PubMed]

114. Heinl, S.; Hartinger, D.; Thamhesl, M.; Schatzmayr, G.; Moll, W.D.; Grabherr, R. An aminotransferase from bacterium ATCC 55552 deaminates hydrolyzed fumonisin B_1. *Biodegradation* **2011**, *22*, 25–30. [CrossRef] [PubMed]

115. Hartinger, D.; Moll, W.D. Fumonisin elimination and prospects for detoxification by enzymatic transformation. *World Mycotoxin J.* **2011**, *4*, 271–283. [CrossRef]

116. De Zutter, N.; Audenaert, K.; Arroyo-Manzanares, N.; de Boevre, M.; van Poucke, C.; de Saeger, S.; Haesaert, G.; Smagghe, G. Aphids transform and detoxify the mycotoxin deoxynivalenol via a type II biotransformation mechanism yet unknown in animals. *Sci. Rep.* **2016**, *6*, 38640. [CrossRef] [PubMed]

117. Mannaa, M.; Kim, K.D. Microbe-mediated control of mycotoxigenic grain fungi in stored rice with focus on aflatoxin biodegradation and biosynthesis inhibition. *Mycobiology* **2016**, *44*, 67–78. [CrossRef] [PubMed]

118. Vanhoutte, I.; Audenaert, K.; de Gelder, L. Biodegradation of mycotoxins: Tales from known and unexplored worlds. *Front. Microbiol.* **2016**, *7*, 561. [CrossRef] [PubMed]

119. Tian, Y.; Tan, Y.; Liu, N.; Liao, Y.; Sun, C.; Wang, S.; Wu, A. Functional agents to biologically control deoxynivalenol contamination in cereal grains. *Front. Microbiol.* **2016**, *7*, 395. [CrossRef] [PubMed]

120. Fernandez-Blanco, C.; Frizzell, C.; Shannon, M.; Ruiz, M.J.; Connolly, L. An in vitro investigation on the cytotoxic and nuclear receptor transcriptional activity of the mycotoxins fumonisin B_1 and beauvericin. *Toxicol. Lett.* **2016**, *257*, 1–10. [CrossRef] [PubMed]

121. Reisinger, N.; Dohnal, I.; Nagl, V.; Schaumberger, S.; Schatzmayr, G.; Mayer, E. Fumonisin B$_1$ (FB$_1$) induces lamellar separation and alters sphingolipid metabolism of in vitro cultured hoof explants. *Toxins* **2016**, *8*, 89. [CrossRef] [PubMed]

122. Wang, X.; Wu, Q.; Wan, D.; Liu, Q.; Chen, D.; Liu, Z.; Martinez-Larranaga, M.R.; Martinez, M.A.; Anadon, A.; Yuan, Z. Fumonisins: Oxidative stress-mediated toxicity and metabolism in vivo and in vitro. *Arch. Toxicol.* **2016**, *90*, 81–101. [CrossRef] [PubMed]

123. Peng, Z.; Chen, L.; Nussler, A.K.; Liu, L.; Yang, W. Current sights for mechanisms of deoxynivalenol-induced hepatotoxicity and prospective views for future scientific research: A mini review. *J. Appl. Toxicol.* **2016**. [CrossRef] [PubMed]

124. Boevre, M.D.; Graniczkowska, K.; Saeger, S.D. Metabolism of modified mycotoxins studied through in vitro and in vivo models: An overview. *Toxicol. Lett.* **2015**, *233*, 24–28. [CrossRef] [PubMed]

125. Chen, X.; Murdoch, R.; Shafer, D.J.; Ajuwon, K.M.; Applegate, T.J. Cytotoxicity of various chemicals and mycotoxins in fresh primary duck embryonic fibroblasts: A comparison to HepG2 cells. *J. Appl. Toxicol.* **2016**, *36*, 1437–1445. [CrossRef] [PubMed]

126. Jakimiuk, E.; Gajecka, M.; Jana, B.; Brzuzan, P.; Zielonka, L.; Skorska-Wyszynska, E.; Gajecki, M. Factors determining sensitivity of prepubertal gilts to hormonal influence of zearalenone. *Pol. J. Vet. Sci.* **2009**, *12*, 149–158. [PubMed]

127. He, J.; Bondy, G.S.; Zhou, T.; Caldwell, D.; Boland, G.J.; Scott, P.M. Toxicology of 3-*epi*-deoxynivalenol, a deoxynivalenol-transformation product by *Devosia mutans* 17-2-E-8. *Food Chem. Toxicol.* **2015**, *84*, 250–259. [CrossRef] [PubMed]

128. Minervini, F.; Giannoccaro, A.; Cavallini, A.; Visconti, A. Investigations on cellular proliferation induced by zearalenone and its derivatives in relation to the estrogenic parameters. *Toxicol. Lett.* **2005**, *159*, 272–283. [CrossRef] [PubMed]

129. Zhao, L.; Lei, Y.; Bao, Y.; Jia, R.; Ma, Q.; Zhang, J.; Chen, J.; Ji, C. Ameliorative effects of *Bacillus subtilis* ANSB01G on zearalenone toxicosis in pre-pubertal female gilts. *Food Addit. Contam. Part A* **2015**, *32*, 617–625. [CrossRef] [PubMed]

130. Kriszt, R.; Krifaton, C.; Szoboszlay, S.; Cserháti, M.; Kriszt, B.; Kukolya, J.; Czéh, Á.; Fehér-Tóth, S.; Török, L.; Szőke, Z.; et al. A new zearalenone biodegradation strategy using non-pathogenic *Rhodococcus pyridinivorans* K408 strain. *PLoS ONE* **2012**, *7*, e43608. [CrossRef] [PubMed]

131. Haq, M.; Gonzalez, N.; Mintz, K.; Jaja-Chimedza, A.; Jesus, C.L.D.; Lydon, C.; Welch, A.Z.; Berry, J.P. Teratogenicity of ochratoxin A and the degradation product, ochratoxin α, in the zebrafish (*Danio rerio*) embryo model of vertebrate development. *Toxins* **2016**, *8*, 40. [CrossRef] [PubMed]

132. Dearfield, K.L.; Gollapudi, B.B.; Bemis, J.C.; Benz, R.D.; Douglas, G.R.; Elespuru, R.K.; Johnson, G.E.; Kirkland, D.J.; LeBaron, M.J.; Li, A.P.; et al. Next generation testing strategy for assessment of genomic damage: A conceptual framework and considerations. *Environ. Mol. Mutagen.* **2016**. [CrossRef] [PubMed]

133. Lean, I.J.; Lucy, M.C.; McNamara, J.P.; Bradford, B.J.; Block, E.; Thomson, J.M.; Morton, J.M.; Celi, P.; Rabiee, A.R.; Santos, J.E.; et al. Invited review: Recommendations for reporting intervention studies on reproductive performance in dairy cattle: Improving design, analysis, and interpretation of research on reproduction. *J. Dairy Sci.* **2016**, *99*, 1–17. [CrossRef] [PubMed]

134. Park, S.H.; Kim, J.; Kim, D.; Moon, Y. Mycotoxin detoxifiers attenuate deoxynivalenol-induced pro-inflammatory barrier insult in porcine enterocytes as an in vitro evaluation model of feed mycotoxin reduction. *Toxicol. In Vitro* **2017**, *38*, 108–116. [CrossRef] [PubMed]

135. Juan-Garcia, A.; Manyes, L.; Ruiz, M.J.; Font, G. Applications of flow cytometry to toxicological mycotoxin effects in cultured mammalian cells: A review. *Food Chem. Toxicol.* **2013**, *56*, 40–59. [CrossRef] [PubMed]

136. Guillouzo, A.; Corlu, A.; Aninat, C.; Glaise, D.; Morel, F.; Guguen-Guillouzo, C. The human hepatoma HepaRG cells: A highly differentiated model for studies of liver metabolism and toxicity of xenobiotics. *Chem. Biol. Interact.* **2007**, *168*, 66–73. [CrossRef] [PubMed]

137. Berger, V.; Gabriel, A.F.; Sergent, T.; Trouet, A.; Larondelle, Y.; Schneider, Y.J. Interaction of ochratoxin A with human intestinal Caco-2 cells: Possible implication of a multidrug resistance-associated protein (MRP2). *Toxicol. Lett.* **2003**, *140–141*, 465–476. [CrossRef]

138. Mace, K.; Offord, E.A.; Harris, C.C.; Pfeifer, A.M. Development of in vitro models for cellular and molecular studies in toxicology and chemoprevention. *Arch. Toxicol. Suppl.* **1998**, *20*, 227–236. [PubMed]

139. Pierron, A.; Mimoun, S.; Murate, L.S.; Loiseau, N.; Lippi, Y.; Bracarense, A.P.; Schatzmayr, G.; He, J.; Zhou, T.; Moll, W.D.; et al. Microbial biotransformation of DON: Molecular basis for reduced toxicity. *Sci. Rep.* **2016**, *6*, 29105. [CrossRef] [PubMed]

140. Payros, D.; Alassane-Kpembi, I.; Pierron, A.; Loiseau, N.; Pinton, P.; Oswald, I.P. Toxicology of deoxynivalenol and its acetylated and modified forms. *Arch. Toxicol.* **2016**, *90*, 2931–2957. [CrossRef] [PubMed]

141. Harbeson, S.L.; Abelleira, S.M.; Akiyama, A.; Barrett, R., 3rd; Carroll, R.M.; Straub, J.A.; Tkacz, J.N.; Wu, C.; Musso, G.F. Stereospecific synthesis of peptidyl alpha-keto amides as inhibitors of calpain. *J. Med. Chem.* **1994**, *37*, 2918–2929. [CrossRef] [PubMed]

142. Wu, Y.; Levons, J.; Narang, A.S.; Raghavan, K.; Rao, V.M. Reactive impurities in excipients: Profiling, identification and mitigation of drug-excipient incompatibility. *AAPS PharmSciTech* **2011**, *12*, 1248–1263. [CrossRef] [PubMed]

143. Burgess, K.M.; Renaud, J.B.; McDowell, T.; Sumarah, M.W. Mechanistic insight into the biosynthesis and detoxification of fumonisin mycotoxins. *ACS Chem. Biol.* **2016**, *11*, 2618–2625. [CrossRef] [PubMed]

144. Wetterhorn, K.M.; Newmister, S.A.; Caniza, R.K.; Busman, M.; McCormick, S.P.; Berthiller, F.; Adam, G.; Rayment, I. Crystal structure of Os79 (Os04g0206600) from Oryza sativa: A UDP-glucosyltransferase involved in the detoxification of deoxynivalenol. *Biochemistry* **2016**, *55*, 6175–6186. [CrossRef] [PubMed]

145. Hassan, Y.I.; Zhu, H.L.; Zhu, Y.; Zhou, T. Beyond ribosomal binding: The increased polarity and aberrant molecular interactions of 3-*epi*-deoxynivalenol. *Toxins* **2016**, *8*, E261. [CrossRef] [PubMed]

146. Kovalsky, P.; Kos, G.; Nahrer, K.; Schwab, C.; Jenkins, T.; Schatzmayr, G.; Sulyok, M.; Krska, R. Co-occurrence of regulated, masked and emerging mycotoxins and secondary metabolites in finished feed and maize—An extensive survey. *Toxins* **2016**, *8*, E363. [CrossRef] [PubMed]

147. Berthiller, F.; Crews, C.; Dall'Asta, C.; Saeger, S.D.; Haesaert, G.; Karlovsky, P.; Oswald, I.P.; Seefelder, W.; Speijers, G.; Stroka, J. Masked mycotoxins: A review. *Mol. Nutr. Food Res.* **2013**, *57*, 165–186. [CrossRef] [PubMed]

148. Gratz, S.W.; Dinesh, R.; Yoshinari, T.; Holtrop, G.; Richardson, A.J.; Duncan, G.; MacDonald, S.; Lloyd, A.; Tarbin, J. Masked trichothecene and zearalenone mycotoxins withstand digestion and absorption in the upper GI tract but are efficiently hydrolyzed by human gut microbiota in vitro. *Mol. Nutr. Food Res.* **2016**. [CrossRef] [PubMed]

149. Cirlini, M.; Barilli, A.; Galaverna, G.; Michlmayr, H.; Adam, G.; Berthiller, F.; Dall'Asta, C. Study on the uptake and deglycosylation of the masked forms of zearalenone in human intestinal Caco-2 cells. *Food Chem. Toxicol.* **2016**, *98*, 232–239. [CrossRef] [PubMed]

150. Dall'Erta, A.; Cirlini, M.; Dall'Asta, M.; Del Rio, D.; Galaverna, G.; Dall'Asta, C. Masked mycotoxins are efficiently hydrolyzed by human colonic microbiota releasing their aglycones. *Chem. Res. Toxicol.* **2013**, *26*, 305–312. [CrossRef] [PubMed]

151. Dellafiora, L.; Galaverna, G.; Righi, F.; Cozzini, P.; Dall'Asta, C. Assessing the hydrolytic fate of the masked mycotoxin zearalenone-14-glucoside—A warning light for the need to look at the "maskedome". *Food Chem. Toxicol.* **2017**, *99*, 9–16. [CrossRef] [PubMed]

152. Stoev, S.D. Foodborne mycotoxicoses, risk assessment and underestimated hazard of masked mycotoxins and joint mycotoxin effects or interaction. *Environ. Toxicol. Pharmacol.* **2015**, *39*, 794–809. [CrossRef] [PubMed]

153. Bullerman, L.B.; Giesova, M.; Hassan, Y.; Deibert, D.; Ryu, D. Antifungal activity of sourdough bread cultures. *Adv. Exp. Med. Biol.* **2006**, *571*, 307–316. [PubMed]

154. Hassan, Y.I.; Bullerman, L.B. Cell-surface binding of deoxynivalenol to *lactobacillus paracasei* subsp. *tolerans isolated from sourdough starter culture*. *J. Microbiol. Biotechnol. Food Sci.* **2013**, *2*, 2323–2325.

155. Abe, K.; Ichikawa, H. Gene overexpression resources in cereals for functional genomics and discovery of useful genes. *Front. Plant Sci.* **2016**, *7*, 1359. [CrossRef] [PubMed]

156. Adesioye, F.A.; Makhalanyane, T.P.; Biely, P.; Cowan, D.A. Phylogeny, classification and metagenomic bioprospecting of microbial acetyl xylan esterases. *Enzym. Microb. Technol.* **2016**, *93*, 79–91. [CrossRef] [PubMed]

157. Chu, Q.; Ma, J.; Saghatelian, A. Identification and characterization of sORF-encoded polypeptides. *Crit. Rev. Biochem. Mol. Biol.* **2015**, *50*, 134–141. [CrossRef] [PubMed]

158. Wrenbeck, E.E.; Faber, M.S.; Whitehead, T.A. Deep sequencing methods for protein engineering and design. *Curr. Opin. Struct. Biol.* **2016**, *45*, 36–44. [CrossRef] [PubMed]

159. Mehta, D.; Satyanarayana, T. Bacterial and archaeal alpha-amylases: Diversity and amelioration of the desirable characteristics for industrial applications. *Front. Microbiol.* **2016**, *7*, 1129. [CrossRef] [PubMed]

160. Heine, T.; Tucker, K.; Okonkwo, N.; Assefa, B.; Conrad, C.; Scholtissek, A.; Schlomann, M.; Gassner, G.; Tischler, D. Engineering styrene monooxygenase for biocatalysis: Reductase-epoxidase fusion proteins. *Appl. Biochem. Biotechnol.* **2016**, 1–21. [CrossRef] [PubMed]

161. Chen, H.; Zhu, Z.; Huang, R.; Zhang, Y.P. Coenzyme engineering of a hyperthermophilic 6-phosphogluconate dehydrogenase from NADP+ to NAD+ with its application to biobatteries. *Sci. Rep.* **2016**, *6*, 36311. [CrossRef] [PubMed]

162. Wambacq, E.; Vanhoutte, I.; Audenaert, K.; Gelder, L.D.; Haesaert, G. Occurrence, prevention and remediation of toxigenic fungi and mycotoxins in silage: A review. *J. Sci. Food Agric.* **2016**, *96*, 2284–2302. [CrossRef] [PubMed]

163. European Food Safety Authority (EFSA). Guidance for the preparation of dossiers for technological additives—EFSA panel on additives and products or substances used in animal feed (FEEDAP). *EFSA J.* **2012**, *10*, 2528.

164. EU Authorizations for Mycofix® Secure and Biomin® BBSH 797. First-Ever Products with Official Anti-Mycotoxin Claim. Available online: http://www.Biomin.Net/en/press-releases/eu-authorizations-for-mycofixr-secure-and-biominr-bbsh-797-first-ever-products-with-official-anti-mycotoxin-claim/ (accessed on 10 March 2017).

165. Positive EFSA Opinion for Mycofix® Ingredient Biomin® BBSH 797 for All Avian Species. Available online: http://www.Biomin.Net/en/press-releases/positive-efsa-opinion-for-mycofixr-ingredient-biominr-bbsh-797-for-all-avian-species/ (accessed on 10 March 2017).

166. EU Authorization for FUMzyme® Proves Fumonisin Biotransformation. Available online: http://www.Biomin.Net/en/press-releases/eu-authorization-for-fumzymer-proves-fumonisin-biotransformation/ (accessed on 10 March 2017).

167. Varga, J.; Péteri, Z.; Tábori, K.; Téren, J.; Vágvölgyi, C. Degradation of ochratoxin A and other mycotoxins by rhizopus isolates. *Int. J. Food Microbiol.* **2005**, *99*, 321–328. [CrossRef] [PubMed]

168. Döll, S.; Dänicke, S. In vivo detoxification of *Fusarium* toxins. *Arch. Anim. Nutr.* **2004**, *58*, 419–441. [CrossRef] [PubMed]

169. Kong, C.; Shin, S.Y.; Kim, B.G. Evaluation of mycotoxin sequestering agents for aflatoxin and deoxynivalenol: An in vitro approach. *SpringerPlus* **2014**, *3*, 346. [CrossRef] [PubMed]

170. Masching, S.; Naehrer, K.; Schwartz-Zimmermann, H.E.; Sarandan, M.; Schaumberger, S.; Dohnal, I.; Nagl, V.; Schatzmayr, D. Gastrointestinal degradation of fumonisin B_1 by carboxylesterase fumD prevents fumonisin induced alteration of sphingolipid metabolism in turkey and swine. *Toxins* **2016**, *8*, 84. [CrossRef] [PubMed]

171. Lee, J.T.; Jessen, K.A.; Beltran, R.; Starkl, V.; Schatzmayr, G.; Borutova, R.; Caldwell, D.J. Effects of mycotoxin-contaminated diets and deactivating compound in laying hens: 2. Effects on white shell egg quality and characteristics. *Poult. Sci.* **2012**, *91*, 2096–2104. [CrossRef] [PubMed]

172. Pinton, P.; Tsybulskyy, D.; Lucioli, J.; Laffitte, J.; Callu, P.; Lyazhri, F.; Grosjean, F.; Bracarense, A.P.; Kolf-Clauw, M.; Oswald, I.P. Toxicity of deoxynivalenol and its acetylated derivatives on the intestine: Differential effects on morphology, barrier function, tight junction proteins, and mitogen-activated protein kinases. *Toxicol. Sci.* **2012**, *130*, 180–190. [CrossRef] [PubMed]

toxins

Review

Mycotoxin Biotransformation by Native and Commercial Enzymes: Present and Future Perspectives

Martina Loi [1,2,†], Francesca Fanelli [1,†,*], Vania C. Liuzzi [1], Antonio F. Logrieco [1] and Giuseppina Mulè [1]

1   Institute of Sciences of Food Production, National Research Council, via Amendola 122/O, Bari 70126, Italy; martina.loi@ispa.cnr.it (M.L.); vania.liuzzi@ispa.cnr.it (V.C.L.); antonio.logrieco@ispa.cnr.it (A.F.L.); giuseppina.mule@ispa.cnr.it (G.M.)
2   Department of Economics, University of Foggia, via Napoli 25, Foggia 71122, Italy
*   Correspondence: francesca.fanelli@ispa.cnr.it; Tel.: +39-080-592-9317
†   These authors contribute equally to this work.

Academic Editor: Ting Zhou
Received: 31 January 2017; Accepted: 18 March 2017; Published: 24 March 2017

Abstract: Worldwide mycotoxins contamination has a significant impact on animal and human health, and leads to economic losses accounted for billions of dollars annually. Since the application of pre- and post- harvest strategies, including chemical or physical removal, are not sufficiently effective, biological transformation is considered the most promising yet challenging approach to reduce mycotoxins accumulation. Although several microorganisms were reported to degrade mycotoxins, only a few enzymes have been identified, purified and characterized for this activity. This review focuses on the biotransformation of mycotoxins performed with purified enzymes isolated from bacteria, fungi and plants, whose activity was validated in in vitro and in vivo assays, including patented ones and commercial preparations. Furthermore, we will present some applications for detoxifying enzymes in food, feed, biogas and biofuel industries, describing their limitation and potentialities.

Keywords: mycotoxins; biotransformation; degradation; enzymes; application

1. Introduction

Mycotoxins are secondary toxic metabolites produced by filamentous fungi mainly belonging to *Fusarium*, *Aspergillus* and *Penicillium* genera. They infect cereals, seeds and fruits both in the field and during storage, and can be found as common contaminants in food and feed supply chains [1,2].

Mycotoxins poisoning of staple food commodities has a significant impact on worldwide health, especially in developing countries [3,4]. Both animals and humans may develop acute and chronic mycotoxicosis, depending on several factors, which include the type of mycotoxin, the amount and the duration of the exposure, etc. Mycotoxins can be classified as hepatotoxins, nephrotoxins, neurotoxins, immunotoxins and so forth, based on the organ they affect [1]. Main target tissues include gastrointestinal and breathing apparatus, endocrine, exocrine, reproductive, nervous and immune system [5,6]. Aflatoxin B_1 (AFB_1) is the most toxic mycotoxin and has been classified in group 1, carcinogenic to humans, by the International Agency of Research on Cancer (IARC) in 2002 [7]. Other mycotoxins, such as aflatoxin M_1 (AFM_1), ochratoxin A (OTA) and fumonisin B_1 (FB_1), are classified within group 2B, thus possibly carcinogenic to humans, due to the limited availability of their toxicological data [7–9].

Synergistic and additive effects of different mycotoxins have been documented; however, the worsening and toxic outcome of multiple exposure cannot be predicted by summing up the individual toxicities. Furthermore, the mechanisms of interactions among toxins are still poorly understood [10]. In rats FB$_1$ synergistically promotes liver lesions, hepatocyte dysplasia and, by long term exposure, tumors initiated by AFB$_1$ [11]. AFB$_1$ also acts synergistically with zearalenone (ZEN) in decreasing egg production, feed intake, feed conversion ratio and eggshell strength in birds [12]. In vitro cytotoxicity of AFM$_1$ on Caco-2 cells is greatly enhanced by the presence of OTA, ZEN and/or α-zearalenol (α-ZEL), which often co-contaminate milk and infant formulas [13]. OTA and citrinin have often been reported to act in a synergistic mode in relation to their cytotoxic [14–17] and genotoxic effects [18].

Global trade significantly contributes to mycotoxin spread. The economic losses associated with mycotoxin contamination in commodities account for billions of dollars annually [19]. They can be categorized into direct and indirect losses: direct losses are related to lowered crop yields, reduction of animal performance and costs derived from diseases for livestock producers; indirect losses are very challenging to quantify, and are linked to the increased use of fungicide, the reduction of the marketable value of the commodities, the management, health-care, veterinary-care costs, and investments in the development of reducing strategies and in research programs.

Mycotoxins levels are regulated in many countries worldwide. The European Union has implemented the most extensive and detailed food regulation for mycotoxins within the Commission Regulation (EC) No. 1881/2006 [20], which set mycotoxins maximum levels discriminating among food products and consumers (e.g., adults or infants). Indicative levels in cereals and cereals products [21] have been established for the trichothecenes T-2 and HT-2 toxins (T-2, HT-2). A scientific opinion on their toxicity was released in 2001 by the European Commission Scientific Committee on Food (SCF) [22], highlighting the need for further studies on the occurrence, daily intake, analytical methods development and induced hematotoxicity and immunotoxicity. However, for other so-called emerging mycotoxins (e.g., enniatins, beauvericin, fusaric acid and moniliformin), whose increasing occurrence has been clearly evidenced, maximum levels have not been yet established. This delay exists since certified analytical methods for their determination, complete surveys on their occurrence and defined scientific opinions on their toxic effect and health associated risks are still being developed [21,23].

Mycotoxin contamination can be prevented in the field through the application of good agricultural practices (GAPs), such as the choice of resistant varieties, harvesting at the right time, crop rotation, and the use of fungicides [20,24]. Nevertheless, pre-harvest strategies are not completely effective, and fungal contamination of raw materials can lead to mycotoxin accumulation during storage. Furthermore, mycotoxins are extremely stable and resistant to the commonly used physical and chemical treatments of food and feed processing.

Since the application of GAPs, proper storage and risk management procedures might only mitigate mycotoxin occurrence, the development of alternative strategies to reduce mycotoxins contamination is considered a relevant, innovative, urgent, yet challenging research topic.

In this review, we will briefly present the current methods to reduce mycotoxins contaminations in food and feed, and then we will focus on the biotransformation of mycotoxins performed with purified enzymes isolated from bacteria, fungi and plants, whose activity was validated in in vitro and in vivo assays, including patented ones and commercial preparations. We will also describe enzyme potential applications and limitations in food, feed and bioenergy in compliance with the European Regulation.

Methods for Mycotoxins Reduction

Current methods to reduce mycotoxin contamination in food and feed can be classified into physical, chemical and biological. It must be underlined that EU regulation does not authorize any detoxification methods for those commodities, intended for food production, exceeding mycotoxins limit levels [20].

(1) Physical methods comprise the mechanical removal of highly contaminated fractions from raw materials (by sorting, cleaning, milling, dehulling), the application of heat, the irradiation and the use of adsorbents, which is still limited to feed production [25]. The latter approach is considered promising,

although some possible negative drawbacks related to unspecific binding of essential nutrients and antibiotics exist. In addition, these adsorbents have diversified efficacy towards different classes of mycotoxins, with trichothecenes being the most difficult to target. Furthermore, since co-occurrence is much more common than individual contamination, complementary reduction strategies should be implemented.

Recently, the application of unconventional strategies such as cold plasma [26–28], photoirradiation [29,30] and microwave treatments [31] have been proposed and/or introduced in food processing, as sterilization or degrading methods. Nevertheless, the partial knowledge about degradation products (DPs) and the nutritional/organoleptic changes induced by these treatments still confine their application.

(2) Chemical methods such as ammoniation [32], acid treatments [33], alkaline hydrolysis [34], peroxidation [35], ozonation [36], and the use of bisulphites [37] have been tested, but their application in food and feed is limited due to their potential toxicity, poor efficacy, high costs and negative effects on the quality of raw materials.

(3) Biological methods consist of the use of microorganisms or enzymes, which are able to metabolize, destroy or deactivate toxins into stable, less toxic, up to harmless compounds [38]. Biological agents and their enzymes allow a specific, most likely irreversible, environmental friendly and effective approach, with minor impact on food sensory and nutritional quality with respect to chemical ones.

2. Mycotoxin Biotransformation by Enzymes

Mycotoxin biotransformation is defined as "the degradation of mycotoxins into non-toxic metabolites by using bacteria/fungi or enzymes" [39]. The possibility to use living microorganisms as whole cell biocatalysts for mycotoxins degradation has cost advantages. This represents a valid strategy, especially if multi step reactions are required, or if the microorganism is already implemented within industrial processes [40,41]. On the other hand, in case of high levels of mycotoxins contamination, the growth and physiology of such microorganisms might be altered or inhibited, thus requiring longer time for adaptation before achieving satisfactory decontamination levels.

The majority of the research papers which describes microbial degrading activity, rarely discriminates between physical adsorption and enzymatic degradation. This greatly complicates the identification of DPs and the evaluation of their toxicity. This knowledge is relevant in the evaluation of biotrasforming enzymes, especially since not every reaction leads to a real detoxification. Indeed, the metabolized mycotoxin can acquire greater toxic properties than the parent compound. This is the case, for example, of ZEN biotransformation performed in vivo by yeasts, which reduce the toxin to α-ZEL, actually more estrogenic than ZEN [42].

The identification and characterization of a degrading enzyme (DE) can be challenging and time consuming, but it is a necessary step to understand the mechanism of degradation, towards its optimization and the development of mycotoxins reducing methods. Enzymes guarantee reproducible and homogeneous performances, with ease-of-handling, no risks of contamination and no safety concerns for operators compared to the use of living microorganisms.

Screening microbial population from different mycotoxin-contaminated environment is an efficient, fast and promising strategy to discover degrading microbes and activities [43–45], which could be enhanced by coupling advanced approaches, such as metagenomics or functional metagenomics.

In addition, genetic engineering enables to clone and express heterologous enzymes in bacterial, yeast, fungal and plant cells for massive, less expensive and less laborious productions. If no structural-functional data are available, enzyme efficiency, stability and tolerance to organic solvents can also be improved by computational approaches, aiming at targeted or combinatorial semi-rational mutagenesis.

Many enzymes have been reported to remove or reduce mycotoxin contamination both in vitro and in real matrices. Nonetheless, their application in feed is very limited, due to the lack of information about the potential toxic effects of generated products and their influence on nutritional quality of feed. These data are mandatory to be authorized as possible biotransforming agent in Europe [39].

Very few commercial biotransforming feed additives are available: Mycofix®, FUMzyme®, Biomin® BBSH 797 and Biomin® MTV (Biomin Holding GmbH, Getzersdorf, Austria) are some examples, but only FUMzyme® exploits a purified enzyme, an esterase, to perform fumonisin degradation. Below we will discuss in details the biotransformation of the main mycotoxins by native and commercial enzymes.

2.1. Aflatoxins

The term aflatoxins (AFs) includes more than 20 fungal secondary metabolites produced by fungi belonging to *Aspergillus* genus [1]. They are classified into two main groups according to their chemical structure. The difurocoumarocyclopentenone group includes AFB_1, aflatoxin B_2, aflatoxin B2a, AFM_1, aflatoxin M_2, aflatoxin Q_1 (AFQ_1), and aflatoxicol (AFL), while the difurocoumarolactone group comprises aflatoxin G_1, aflatoxin G_2 and aflatoxin G2a (Figure 1).

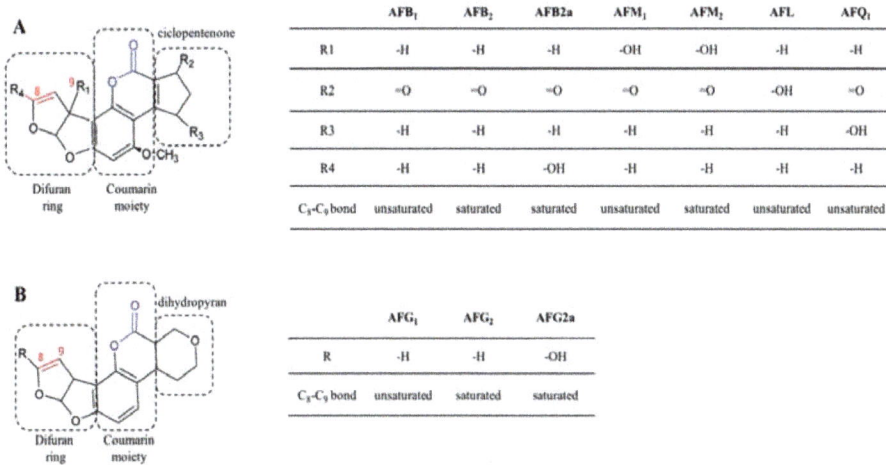

	AFB_1	AFB_2	AFB2a	AFM_1	AFM_2	AFL	AFQ_1
R1	-H	-H	-H	-OH	-OH	-H	-H
R2	=O	=O	=O	=O	=O	-OH	=O
R3	-H	-H	-H	-H	-H	-H	-OH
R4	-H	-H	-OH	-H	-H	-H	-H
C_8-C_9 bond	unsaturated	saturated	saturated	unsaturated	saturated	unsaturated	unsaturated

	AFG_1	AFG_2	AFG2a
R	-H	-H	-OH
C_8-C_9 bond	unsaturated	saturated	saturated

Figure 1. Chemical structure and features of ciclopentenone (**A**) and difurocoumarolactone (**B**) aflatoxin (AF) series. Coloured bonds indicate reactive groups involved in AFs toxicity. The double bond leading to 8,9-epoxide upon metabolic activation is indicated in red, while the lactone bond is indicated in blue. Tables show substituent groups and saturation of the C_8-C_9 bond in different AF analogues.

While AFs of the B and G series co-occur in cereals and their derived products, fruits, oilseeds, nuts, tobacco and spices, AFM_1 AFM_2, AFL and AFQ_1 are detected in food as carry-over products of AFB_1 contaminated feeds. In vivo AFB_1 is readily metabolized through hydroxylation (to AFM_1 or, to a lesser extent, to AFQ_1) or reduction (to AFL).

AFs are difuranocoumarin derivatives composed by two furan rings, linked together to a coumarin moiety. Furofuran and cumarin rings are arranged in a planar configuration which is responsible for conjugation leading to the typical AFs fluorescence.

The furofuran ring has been recognized as responsible for the toxic and carcinogenic activity upon metabolic activation of the C_8-C_9 double bond to 8–9 epoxide [9] (Figure 1).

The epoxidation is a crucial reaction for AFs carcinogenicity, since it allows the binding to N7-guanine and the subsequent G to T transversions in the DNA molecule [46]. Activated AFs are also able to form schiff bases with cellular and microsomal proteins (via methionine, histidine and lysine), thus leading to acute toxicity [47]. The lactone ring also plays a role in AFs toxicity and carcinogenicity: upon ammoniation it is hydrolyzed, forming aflatoxin D_1 (AFD_1) which still retains the 8,9-dihydrofuran double bond; AFD_1 lacks the strong in vivo DNA binding activity of AFB_1, demonstrating that DNA alkylation depends upon both difuranocumarin and lactone moieties [48].

Several authors addressed lactone hydrolysis, reduction or addition reactions as possible mechanisms of degradation, since the putative hydrolyzed products showed greatly reduced mutagenic activity in vitro [49,50].

Table 1 summarizes the purified enzymes identified as capable of degrading AFs, and their main features. Although direct comparison is not possible, since the enzymes and the experimental conditions used by the authors are not equivalent, reaction parameters and AFs concentrations are indicated.

Table 1. Aflatoxins (AFs) degrading enzymes.

Enzyme	Accession/EC	Producing Organism	AF Target	Toxin Concentration	In Vitro/In Matrix Degrading Conditions	In Vitro/In Matrix Degradation	Toxicity/Mutagenicity Test	Reference
aflatoxin oxidase enzyme (AFO)	EC 1.1	Armillariella tabescens	AFB$_1$	0.05 µg/mL	• PBS buffer 0.02 M pH 6; • incubation at 28 °C for 30 min	• 100% with 0.2 mg of enriched preparation; • NQ* with pure enzyme	• reduced liver toxicity in rats; • reduced mutagenicity in Salmonella typhimurium TA 98; • reduced genotoxicity on chicken embryos	[51]
peroxidase	EC 1.11.1.7	horseradish (Armoracia rusticana)	AFB$_1$	312 µg/mL	• phosphate buffer 50 mM pH 6; • incubation at 20 °C for 60 min; • 0.2U/mL of enzyme	42.2%		[52]
				440 µg/mL	• 100 g of defatted groundnut kernels; • 2-16U of enzyme; • 50 mM phosphate buffer pH 6 and 20 mM hydrogen peroxide; • incubation at room temperature up to 24 h	41.1%	reduced toxicity on Bacillus megaterium	
laccase	EC 1.10.3.2	Trametes versicolor (commercial enzyme from Sigma-Aldrich, Missouri, U.S.)	AFB$_1$	1.40 µg/mL	• phosphate buffer 0.2 M pH 6.5; • 1 U/mL of enzyme; • incubation at 30 °C for 72 h	87%	reduced mutagenicity on Salmonella typhimurium TA 100	[53]
laccase	EC 1.10.3.2	Streptomyces coelicor	AFB$_1$	9.36 µg/mL	• sodium acetate buffer 100 mM pH 7; • 0.1 mg/mL of enzyme protein and 0.2 mM mediator; • incubation at 37 °C for 24 h	100%	n.p.	[54]
F420H2-dependent reductases	E.C. 1.5.8	Mycobacterium smegmatis	AFB$_1$ AFB$_2$ AFG$_1$ AFG$_2$	n.p.	n.p.	n.p.	n.p.	[55]
Mn peroxidase	EC 1.11.1.7	Pleurotus ostreatus	AFB$_1$	0.31 µg/mL	• sodium lactate buffer 50 mM pH 4.5; • 1.5 U/mL of enzyme; • incubation at 30 °C for 48 h	90%	n.p.	[56]
aflatoxin degradation enzyme	n.p.	Pleurotus ostreatus	AFB$_1$	5 µg/mL	• sodium acetate buffer 0.1 M pH 5; • incubation at 25 °C for 1 h	n.q.	n.p.	[50]

Table 1. Cont.

Enzyme	Accession/EC	Producing Organism	AF Target	Toxin Concentration	In Vitro/In Matrix Degrading Conditions	In Vitro/In Matrix Degradation	Toxicity/Mutagenicity Test	Reference
myxobacteria aflatoxin degrading enzyme (MADE)	n.p.	Myxococcus fulvus ANSM068	AFB_1 / AFG_1 / AFM_1	0.1 µg/mL	• citrate phosphate buffer 0.1 M pH 6; • 100 U/mL of enzyme • incubation at 30 °C for 48 h	72% with culture filtrates / 97% / 96%	n.p.	[57]
laccase (lac2)	EC 1.10.3.2	Pleurotus pulmonarius (ITEM 17144)	AFB_1 / AFM_1	1 µg/mL / 0.05 µg/mL	• sodium acetate buffer 1mM pH 5; • 5 U/mL enzyme and redox mediator; • incubation at 25 °C for 72 h	90% / 100%	n.p.	[58]
Ery4	CAO79915.1/EC 1.10.3.2	Pleurotus eryngii (PS419)	AFM_1	0.05 µg/mL	• sodium acetate buffer 1 mM pH 5; • 5 U/mL enzyme and redox mediator; • incubation at 25 °C for 72 h • artificially spiked skim UHT milk; • 5 U/mL enzyme and redox mediator; • incubation at 25 °C for 72 h	100%	n.p.	[59]

n.q. = not quantitative; n.p. = not provided.

Most of these enzymes are comprised in the oxidoreductase (EC 1) group. The first identified was the AF oxidase enzyme (AFO) isolated from the edible fungus *Armillariella tabescens* by Liu and colleagues in 1998 [51]. AFO is an oxygen dependent reductase, releasing H_2O_2 as by product [51,60]. One mg/mL of enriched preparation and 30 min of incubation at 28 °C were needed to completely degrade 150 ng of AFB_1. AFO did not affect AFB_1 fluorescence properties, indicating that the conjugation system, between the coumarin moiety and the lactone ring, was not disrupted by the degrading reaction. The mechanism proposed by the authors consisted of the enzymatic cleavage of the bisfuran ring. This hypothesis, although not yet experimentally verified, is in agreement with the reduced mutagenicity, toxicity and genotoxicity of the DPs and with their differential pulse voltammetry response similar to furofuran analogs [61,62]. Products of AFO treatment were found to exert neither liver toxicity in rats, nor mutagenicity activity on *Salmonella typhimurium* TA98, and not to be genotoxic to chicken embryos [61].

Three different peroxidases (EC 1.11.1.7) were also reported to possess AFB_1 degrading capabilities. They were extracted and purified from horseradish (*Armoracia rusticana*) [52], *Phanerochaete sordida* YK-624 [63] and *Pleurotus ostreatus* [56]. The horseradish peroxidase was tested in vitro towards AFB_1, which was reduced by 42% after 1h with 0.2 U/mL of enzyme. Similar results were also obtained in real matrix: AFB_1 was reduced by 41% in artificially spiked groundnut [64]. A lowered toxicity of the DPs was also registered by inhibition growth assay on *Bacillus megaterium* [51]. The two manganese peroxidases (MnPs) from the two different white rot fungi were studied [56,63] in in vitro assays with similar reaction conditions, allowing a direct comparison. They were tested towards 0.3 µg/mL of AFB_1, achieving 86% reduction after 24 h of incubation, and up to 90% after 48 h by using the MnP from *P. sordida* YK-624 and the one from *P. ostreatus* respectively. Wang and colleagues [63] also registered 69.2% reduction of mutagenicity of DPs compared to AFB_1, based on the umu test performed on *Salmonella typhimurium* TA1535 and on S9 liver homogenate. According to Proton Nuclear Magnetic Resonance (^{1}H-NMR) and high resolution electrospray ionization mass spectrometry (HR-ESI-MS) analysis, the authors hypothesized that MnP treatment converted AFB_1 to AFB_1-8,9-dihydrodiol.

Laccases (EC 1.10.3.2) (LCs) have been intensively used in bioremediation [65] and were recently proposed for ZEN and AFs biotransformation [59,66,67].

Alberts and colleagues [53] made as firsts a decisive step towards the correlation between AFs degradation and LCs activity. The authors used different LCs, including native LCs from representatives of *Peniophora* genus and *P. ostreatus*, one commercial LC from *Trametes versicolor*, and one recombinant LC from *Aspergillus niger*. By treatment with 1 U/mL of the commercial preparation AFB_1 (1.4 µg/mL) was reduced by 87.34% after 3 days of incubation at 30 °C. DPs were not identified, but proved to be less mutagenic than the parent compound. However, none of the preparations used by the authors was purified to homogeneity. The commercial product from *T. versicolor* was indeed an enriched preparation, which included additional proteins and different laccase isoforms [68]; thus, an unambiguous assignment of the degrading activity to a specific LCs would be improper.

Recently the effectiveness of purified LCs towards AFs and ZEN was acknowledged [58,67,68]. Pure LCs from *P. pulmonarius* and *P. eryngii* were used by Loi and colleagues [58,59] towards both AFB_1 and AFM_1. The authors performed in vitro and *in matrix* tests with 1 µg/mL of AFB_1 and 0.05 µg/mL for AFM_1. Interestingly, they reported that, while LC alone is poorly able to degrade these toxins, the addition of a redox mediator at 10 mM concentration increased the degrading percentages from 23% up to 90% for AFB_1, and up to 100% for AFM_1 after 72 h [58,59].

The laccase-mediator approach patented by Novozymes in 2008 [54], consisted of the use of LCs, preferably from *Streptomyces coelicolor*, and methylsyringate as mediator. In this case, AFB_1 was completely removed after 24 h of incubation with 0.1 mg/mL of enzyme and 0.2 mM of mediator at 37 °C.

Members of the reductase family have also been studied for AFs degrading capabilities. Among these, the F420H2-dependent reductases (EC 1.5.8.), isolated from *Mycobacterium smegmatis*, were tested towards AFB$_1$, AFB$_2$, AFG$_1$ and AFG$_2$; no details were provided in relation to efficacy and assay conditions [55].

Besides oxidases above described, Guan et al. [69] identified the myxobacteria aflatoxin DE (MADE), isolated from *Myxococcus fulvus*, which was further characterized in 2011 [57]. Pure MADE (100 U/mL) was tested towards 0.1 µg/mL each of AFG$_1$ and AFM$_1$, achieving 98 and 97% of degradation after 48 h of incubation. No further details about the nature of the enzyme, nor about the DPs were given.

2.2. Ochratoxin A

OTA is a phenylalanine-dihydroisocoumarine derivative, composed of a 7-carboxy-5-chloro-8-hydroxy-3,4-dihydro-3-R-methylisocoumarin (ochratoxin α—OTα) moiety and a L-β-phenylalanine molecule (Phe), which are linked at the 7-carboxy group by an amide bond (Figure 2).

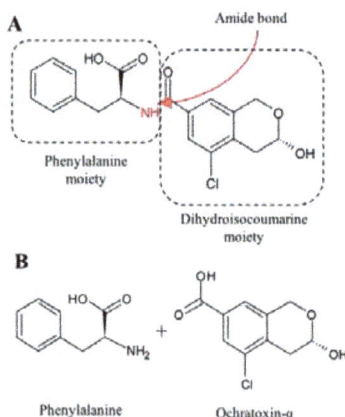

Figure 2. Chemical structures of (**A**) ochratoxin A and (**B**) its degradation products, ochratoxin-α and phenylalanine. The amide bond hydrolyzed by the main degrading pathway is indicated in red.

Due to its structural analogy to the amino acid Phe, the toxin can competitively inhibit tRNA phenylalanine synthetases and, consequently, block protein synthesis [70]. Furthermore, OTA causes the formation of DNA adducts, indirect oxidative DNA damage and activates a network of interacting epigenetic mechanisms [71,72].

The main OTA detoxification pathway consists in the hydrolysis of the amide bond between the isocoumarin residue and phenylalanine, resulting in the formation of Phe and OTα (Figure 2). The former is considered to be a non-toxic compound, with a 10-times shorter elimination half-life than OTA [73].

Numerous enzymes were hypothesized to hydrolyze the OTA amide bond, but only few of them were isolated and characterized [74].

Two classes of carboxypeptidases (EC 3.4) have been associated with OTA degradation: carboxypeptidase A (CPA) and carboxypeptidase Y (CPY) class. CPA uses one zinc ion within the protein for hydrolysis (EC 3.4.24), while CPY is a serine-type carboxypeptidase (EC 3.4.16) and does not contain any zinc ion in its active site. The first peptidase reported as able to hydrolyze OTA was a CPA isolated from bovine pancreas, which resulted able to perform the degrading reaction with a Km value of 1.5×10^{-4} M at 25 °C [75]. A CPY isolated from *Saccharomyces cerevisiae* was demonstrated to hydrolyze OTA with optimum at pH 5.6 and 37 °C; its specific activity was very low considering that only 52% of OTA was converted into OTα after five days of incubation [74]. The same enzyme was efficiently immobilized on electroactive surfaces in order to develop a biosensor system for the direct detection of OTA in olive oil, with promising results [76]. Many carboxypeptidases have high optimal reaction temperatures (30 °C or higher); this might not hamper detoxifying applications for food and feed [25].

Other enzymes are also able to perform OTA hydrolysis, such as lipases (EC 3.1), amidases (EC 3.5) and several commercial proteases (EC 3.4) [77–79].

By screening different commercial hydrolases, a lipase preparation from *Aspergillus niger* (Amano A) was shown to hydrolyze OTA into OTα and Phe. Single-step purification, by anion exchange chromatography, allowed the isolation of the pure protein. The lipase nature of the enzyme was confirmed by assaying the cleavage of p-nitrophenyl palmitate. The purified enzyme resulted able to completely hydrolyze 50 μg of OTA into OTα after 120 min of incubation, in 1 ml of reaction mixture [77].

Several commercial proteases were also reported to hydrolyze OTA to OTα, such as Protease A from *A. niger* and pancreatin from porcine pancreas. These enzymes showed a significant hydrolytic activity at pH 7.5, which resulted in the cleavage of 87.3% and 43.4% of 1 μg of OTA respectively, after 25 h of incubation in 1 ml reaction mixture [78].

Finally, an amidase and the feed or food additive comprising it, capable of degrading OTA, were patented by Dalsgaard et al. in 2010 [79]. The protein, named amidase 2, is encoded by an open reading frame of *A. niger*. The degrading assay was performed in 300 μL of reaction mixture, containing 160 ng/mL amidase 2 and 50 μg/mL of OTA. This enzyme was able to reduce OTA concentration by 83%. Amidase 2 was tested also in some food preparations. The patented additive decreased OTA content, from 47 ppb to undetectable level (<2 ppb), after 2.5 h-incubation in contaminated milk. Similarly, OTA concentration in corn flour was reduced from 38 ppm to less than 2 ppb after 20 h of incubation.

2.3. Fumonisins

Fumonisins are a group of mycotoxins associated with several mycotoxicoses, including equine leukoencephalomalacia, porcine pulmonary edema and experimental kidney and liver cancer in rats [80]. Chemically, fumonisins are diesters of propane-1,2,3-tricarboxylic acid and similar long-chain aminopolyol backbones. Structurally they resemble the sphingoid bases sphinganine (SA) and sphingosine (SO), with tricarboxylic acid groups added at the C_{14} and C_{15} positions (Figure 3). This structural similarity is responsible for the fumonisin mechanism of action. It was described to act by disturbing the sphingolipids metabolism, by inhibiting the enzyme ceramide synthase, and leading to accumulation of sphinganine in cells and tissues [81–83].

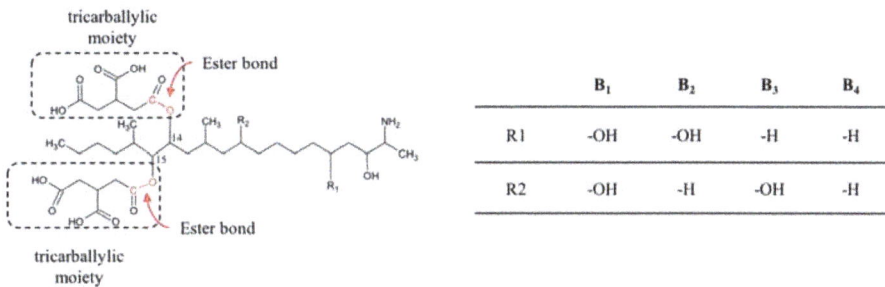

	B_1	B_2	B_3	B_4
R1	-OH	-OH	-H	-H
R2	-OH	-H	-OH	-H

Figure 3. Type B fumonisins chemical structure. The ester bonds hydrolyzed by the main degrading pathways, leading to the formation of HFB_1 and the two tricarballylic acid moieties, are indicated in red. The table shows substituent groups of different fumonisin analogues.

There are at least 28 different forms of fumonisins, designated as A-series, B-series, C-series, and P-series [84]. The B-series (FB) is the most abundant and important with respect to toxicity.

Pre-harvest strategies to reduce fumonisins contamination are based on the bio-control of the spread of fumonisin-producing fungi. Post-harvest methods include the application of natural clay adsorbents during food processing. While they do not lead to a real detoxification of fumonisin [85,86], different microorganisms were reported to transform these class of mycotoxins. Among them, only few enzymes have been identified, molecularly and biochemically characterized, and patented.

FB$_1$ was reported to be degraded by the consecutive action of a carboxylesterase (EC 3.1.1) [87] and an aminotransferase (EC 2.6.1) [88,89] (Table 2).

By the deesterification action of the carboxylesterase, the two tricarballylic acid moieties are released, resulting in hydrolyzed FB$_1$ (HFB$_1$), also known as aminopentol 1 (AP1). After the oxidative deamination, HFB$_1$ is converted to *N*-acetyl HFB$_1$ and 2-oxo-12,16-dimethyl-3,5,10,14,15-icosanepentol hemiketal [88]. The action of these enzymes, isolated from *Sphingomonas* sp. ATCC 55552 [90], was worldwide patented in 1994 [89].

Table 2. Fumonisin B$_1$ degrading enzymes.

Enzyme	Producing Organism	Accession/EC	Toxin Concentration	Degrading Conditions	Degradation	Reference
carboxylesterase and aminotransferase	*Sphingomonas* sp. ATCC55552	E.C. 3.1.1, E.C. 2.6.1	1000 μg/mL	• citrate buffer 0.1 M pH 3 or citrate-phosphate buffer 0.1 M pH 4; • overnight incubation at 37 °C	100%	[89]
carboxylesterase B and aminotransferase	*Sphingopyxis* sp. MTA144	E.C. 3.1.1, E.C. 2.6.1/ FJ426269.1	3.6 μg/mL	• Tris-HCl 20 mM pH 8, 0.1 mg/mL BSA; • incubation at 30 °C for 2 h	100%	[80]
fumonisin esterase	*Sphingopyxis* sp. MTA144	E.C. 3.1.1.87	60 μg/mL	• unspecified buffer pH 8; • 18 U/L of enzyme • incubation at 30 °C for 15 min	100% conversion to HFB$_1$	[91]

The same degrading activity was shown by *Sphingopyxis* sp. MTA144, isolated from composted earth [92]. In this strain the two key DEs were identified within a gene cluster: *fumD*, encoding a type B carboxylesterase, and *fumI*, encoding an aminotransferase [80,93,94]. The authors also demonstrated the degrading activities of the two recombinant enzymes by in vitro assays. The first enzyme, encoded by *fumD*, was able to catalyze the complete deesterification of FB_1 (3.34 µg/mL) to HFB_1 within 15 min. The oxygen independent aminotransferase, encoded by *fumI*, was shown to deaminate HFB_1 (8.9 µg/mL), in presence of pyruvate and pyridoxal phosphate, within 44 h. Furthermore, the authors described the additional genes located in this putative degrading cluster identified by genome walking (NCBI Accession n. FJ426269). The cluster includes one transporter, one permease, dehydrogenases, and transcriptional regulators and was U.S. patented by Moll et al. within the "Method for the production of an additive for the enzymatic decomposition of mycotoxins, additive and use thereof" [95].

The esterase (EC 3.1.1.87) from *Sphingopyxis* sp. MTA144, produced by a genetically modified strain of *Komagataella pastoris* (formerly *Pichia pastoris*), has been included in a patented formulation, FUMzyme® (Biomin Holding GmbH, Getzersdorf, Austria). This enzyme-based feed additive, whose safety and efficacy has been recently evaluated by EFSA [91], is intended to degrade fumonisins found as contaminants in feeds for growing pigs. The esterase partially degrades FB_1 and related fumonisins by cleavage of the diester bonds and release of the tricarballylic acid.

2.4. Trichothecenes

Trichothecenes are a large group of sesquiterpenoid metabolites sharing a common core structure comprised in a rigid tetracyclic ring (Figure 4). To date, more than 190 trichothecenes and derivatives have been described [96,97].

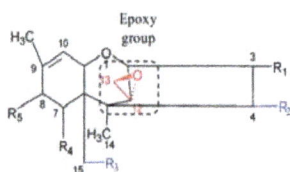

	T-2	HT-2	DON	NIV
R1	-OH	-OH	-OH	-OH
R2	-OAc	-OH	-H	-OH
R3	-OAc	-OAc	-OH	-OH
R4	-H	-H	-OH	-OH
R5	-OCOCH₂CH(CH₃)₂	-OCOCH₂CH(CH₃)₂	=O	=O

Figure 4. Chemical structure of trichothecenes. Groups responsible for trichothecenes toxicity are highlighted in red (epoxide) and blue (substituent groups, see the text for further details). The table shows substituent groups of different trichothecene analogues.

Their synthesis starts with the formation of trichodiene via the cyclization of farnesyl pyrophosphate. Trichodiene then undergoes a series of oxygenation, cyclizations, isomerization and esterification needed for the bioactivation of the molecule [98].

Trichothecenes are classified into 4 groups according to the functional groups associated to the core molecule: type A trichothecenes (e.g., T-2 and HT-2) do not contain a carbonyl group at the C-8 position; type B trichothecenes (mainly represented by deoxynivalenol (DON) and nivalenol (NIV) contain a carbonyl group at the C-8 position; trichothecenes of type C (e.g., crotocin and baccharin) have an additional epoxy ring between C-7 and C-8, or between C-9 and C-10; trichothecenes of type D (e.g., satratoxin and roridin), contain a macrocyclic ring between C-4 and C-15 (Figure 4).

The results of trichothecenes exposure on eukaryotic cells was reviewed by Rocha et al. [99], while T-2 and HT-2 toxicity data were reported to EFSA in 2010 by Schuhmacher-Wolz and colleagues [100]. Trichothecenes are not degraded during normal food processing, they are stable at neutral and acidic pH and, consequently, they are not hydrolyzed in the stomach after ingestion [99].

Type A trichothecenes arise special interest, being more toxic than the other food-borne trichothecenes [101]. Among the effects exerted by T-2 and HT-2 toxins several studies reported the inhibition of DNA, RNA and protein synthesis, and apoptotic effects in mammalian cell cultures in vitro and ex vivo [22,102].

T-2 is rapidly metabolized to HT-2 in the gut. Data indicate that these toxins induce acute toxic effects with similar severity. For these reasons, the toxicity of T-2 in vivo is considered to include that of HT-2 [102]. Type B trichothecene have a relatively low toxicity compared to type A trichothecene, but it varies depending on the species and the cell type. The general toxic mechanism of trichothecenes is induced by the interference with protein synthesis, through 60S ribosome binding, leading to translation inhibition [103].

The 12,13-epoxy ring is the most important structural determinant of trichothecenes toxicity. In addition, the presence of hydroxyl or acetyl groups at appropriate positions on the trichothecene core [104], the presence of substituents on C_{15} and C_4 [105,106], and of the side groups, are implicated in defining the degree of toxicity [107]. Both acetylation and de-acetylation may reduce toxicity of trichothecenes, as well as epimerization, oxidation [108–110].

Trichothecenes are noticeably stable molecules under different treatments, including both thermal and chemical ones. Several microorganisms, isolated from disparate complex environments, such as rumen or soil, have been reported to degrade trichothecenes into de-acetylated and/or de-epoxidated products [25]. However, the majority of the studies focused on this topic never achieved the identification of the enzymes responsible for the biotransformation. Nevertheless, considering the reaction products, acetylase, deacetylase or de-epoxidase activities are surely involved in the process.

In addition, hydroxylation and glycosylation of trichothecenes generate less toxic derivatives [111]. In this case, these compounds might undergo reverse reaction in the digestive tract of humans and animals, limiting the use of these enzymes at least in feed-related applications.

Recently, the bacterial cytochrome P450 system (EC 1.14) from *Sphingomonas* sp. strain KSM1, was reconstructed in vitro and demonstrated to hydroxylate DON, NIV and 3-acetyl DON [112] (Table 3). The system includes the cytP450, encoded by the gene ddnA, and the endogenous redox partners. The DON catabolic product, 16-HDON, is used by the *Sphingomonas* strain as a carbon source, and was shown to exert a reduced phytotoxicity to wheat.

Poppengberger et al. [113] identified a UDP-glycosyltransferase from *Arabidopsis thaliana* (the deoxynivalenol-glucosyl-transferase DOGT1) able to catalyze the transfer of glucose from UDP-glucose to the hydroxyl group at C3 of DON forming 3-*O*-glucopyranosyl-4-DON. The overexpression of this enzyme in *A. thaliana* enhanced DON resistance. The reaction products were identified by in vitro assays, using the enzyme purified from recombinant *Escherichia coli* cells.

Commercial products containing biotransformant agents have already been developed and patented. As an example, Biomin® BBSH 797 includes a pure culture of *Eubacterium* BBSH 797 isolated from bovine rumen fluid; it is capable of converting DON into DOM-1 and with de-epoxydase activity towards NIV, T-2, tetraol, scirpentriol, and HT-2. Nevertheless, purified enzymes with specific trichothecenes degrading activity have not been yet reported.

However, the recent increasing of microorganisms isolated from trichothecenes contaminated environments and capable of transforming this class of toxins led to the assumption that this goal will not be so distant [109,110,114].

Table 3. Trichothecenes degrading enzymes.

Enzyme	Accession/EC	Producing Organism	Trichothecene Target	Toxin Concentration	Degrading Conditions	In Vitro Degradation	Toxicity-Mutagenicity Test	Reference
cytochrome P450 system (Ddna + Kdx + KdR)	E.C. 1.14 AB744215.1 AB744217.1	*Sphingomonas* sp. strain KSM1	DON	99.86 µg/mL	• potassium phosphate buffer 10 mM pH 7.5; • 10% glycerol, 0.2 µM DdnA, 1.2 µM Kdx, 1.2 µM KdR, 1 mM NADH, and 100 mg/ml bovine liver catalase; • overnight incubation at 30 °C	100% after 3 days	reduced phytotoxicity to wheat	[112]
			NIV	105.25 µg/mL		100% after 5 days		
UDP-glycosyltransferase	AC006282	*Arabidopsis thaliana*	DON	n.p.	n.p.	n.p.	increased resistance in transgenic *Arabidopsis*	[113]

n.p. = not provided.

2.5. Zearalenone

ZEN is a resorcylic acid lactone, nonsteroidal yet estrogenic mycotoxin, which binds to mammalian estrogen receptors (ER), although with lower affinity than the natural estrogens 17β-estradiol, estriol and estrone [115]. From a structural point of view, ZEN resembles the 17-βestradiol, and this similarity is responsible for its estrogenic potential and ER binding capacity (Figure 5).

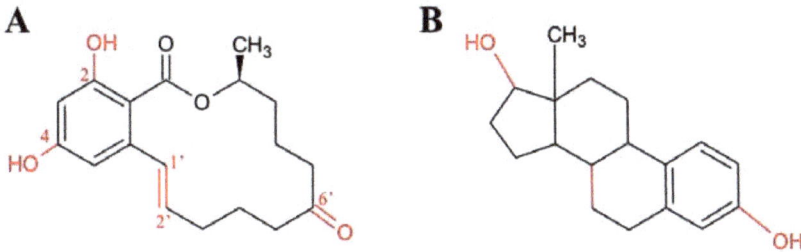

Figure 5. Chemical and structural analogies between zearalenone (**A**) and 17β-estradiol (**B**). The main chemical groups interacting with the estrogen receptors and responsible for zearalenone toxicity are highlighted in red (see the text for further details).

In vivo ZEN undergoes reduction to α-ZEL and β-zearalenol (β-ZEL), with the first being more estrogenic than the parent compound [116]. A further reduction leads to the formation of α- and β-zearalanol (α-ZAL, β-ZAL), which are both less estrogenic than ZEN.

ZEN estrogenicity is greatly enhanced by the reduction of the 6'-ketone group and of the 1'-2'double bond, while it is reduced by the methylation of the hydroxyls in C4 and C2. The OH group in 6'of α and β-ZEL resembles that of C3 in estradiol. This group strongly interacts with ER, while the saturation of the 1'-2'double bond increases the flexibility of the molecule. This allows ZEN and/or ZEL to undergo some slight conformational changes, effective at adapting the molecules to the binding pocket of the receptor. Hydroxyl groups in C4 and C2 also contribute to the binding, with the first being more important than the latter in increasing ZEN estrogenic potential [116] (Figure 5).

ZEN and derivatives detoxification strategies aim at disrupting their estrogenic activity. Among the reported ZEN metabolizing reactions, lactone ring cleavage, catalyzed by esterases, is the prevalent detoxification method described so far. Since the resulting hydroxyketones spontaneously decarboxylate, the reaction is irreversible.

A lactonohydrolase (EC 3.1.1) from the fungus *Clonostachys rosea* has been identified as capable of degrading ZEN. The enzyme was purified to homogeneity, and its gene (namely *zhd101*) was cloned and characterized by heterologous expression in *E. coli* BL21 and *S. cerevisiae* INVSc1 [117,118] (Table 4). The recombinant protein was proved to completely degrade 2 μg/mL of ZEN in in vitro assays, although with minimal yeast mediated conversion of ZEN to β-zearalenol. *Zhd101* gene and transformants carrying lactonohydrolase gene were patented in 2002 [119].

Table 4. Zearalenone degrading enzymes.

Enzyme	Producing Organism	EC	Toxin Concentration	Degrading Conditions	Degradation	Toxicity-Mutagenicity Test	Reference
laccase	*Trametes versicolor* (commercial enzyme from Sigma–Aldrich)	EC 1.10.3.2	6.2×10^{-4} µg/mL	▪ sodium acetate buffer 0.2 M pH 5.2; ▪ incubation at 30 °C for 4 h.	up to 58 %	n.p.	[67]
laccase	*Streptomyces coelicolor*	EC 1.10.3.2	9.36 µg/mL	▪ sodium acetate 0.1 M pH 4.5; ▪ 0.2 mM mediator; ▪ incubation at 37 °C for 24 h	100%	n.p.	[120]
lactono hydrolase	*Clonostachys rosea*	E.C. 3.1.1	2 µg/mL	▪ YPD; ▪ incubation at 28 °C for 4 h or at 37 °C for 2 h	100%	n.p.	[117,118]
2cys-peroxiredoxin	*Acinetobacter* sp. SM04	EC 1.11.1.15	20 µg/mL	▪ Tris-HCl 50 mM pH 9; ▪ $H_2O_2 \geq 20$ mM; ▪ incubation at 30 °C for 4 h	up to 95%	reduced MCF-7 cells proliferation by 75%	[121]
			1 µg/mL	▪ 1 mL of 0.8 M H_2O_2 and 19 mL of purified recombinant Prx solution pH 9; ▪ incubation at 40 °C for 6 h.	up to 90%		

Besides the above-mentioned activity towards AFs, laccases are also able to degrade ZEN. Novozyme patented the laccase-mediator degradation of ZEN in 2007 [120], using *Streptomyces coelicolor* LC and, preferably, phenotiazin-10-propionic acid or methylsyringate as redox mediators. By using this system, ZEN was completely removed by 0.1 mg/mL of LC and 0.2 mM of mediator at 37 °C, within 24 h. Laccase degrading capabilities were also studied by Banu and colleagues [67] using one LC-enriched *Trametes versicolor* preparation. Reactions, containing 6.2×10^{-4} µg/mL of ZEN, were incubated with 0.4 mg/mL of laccase for 4 h at 30 °C, achieving a maximum of 81.7% of degradation.

Another ZEN degrading oxidoreductase is a 2-Cys peroxiredoxin (Prx) (EC 1.11.1.15) extracted from *Acinetobacter* sp. SM04 [121,122]. Prxs are peroxidases containing redox-active cysteine, which can be oxidized to cystine by using H_2O_2 as cofactor. Prx degraded up to 95% of ZEN (20 µg/mL) in in vitro assays, when added with 20mM H_2O_2; nearly 90% of the toxin was instead degraded in contaminated corn sample (ZEN levels of nearly 1000 µg/mL), treated for 6 h at 30 °C with purified recombinant Prx, plus 0.09% (mass fraction, mol/mol) H_2O_2. Toxicity bioassays were also performed. The proliferative effect on MCF-7 cells by ZEN-DPs was reduced by 75%. Altalhi and colleagues [123] identified a 5.5 kb gene fragment from *Pseudomonas putida* pZEA-1, which encodes for DE(s) not yet characterized. *E. coli* DH5 cells, expressing the 5.5 kb gene fragment, were able to reduce 100 µg/mL of ZEN by 85% after 72 h, while no degradation was observed by wild type cells. ZEN DPs were also shown to possess a marked reduced toxicity on *Artemia salina* larvae.

3. Potentialities and Limitations of Mycotoxin Degrading Enzymes in Food, Feed, and Bioenergy

The use of enzymes in the food, feed, biogas and biofuel industries is not new, as biocatalysts have been increasingly used in the last 30 years. They allowed reducing the employment of hazardous chemicals, to use mild working conditions, to increase specificity, to speed up a process or to simply create new products.

Enzyme biotechnological applications are widespread: the dairy industry has a long story of protease use for cheese manufacturing, while bakery employs laccase and transglutaminase to achieve doughs strengthening; phytases are used in non-ruminant animal nutrition for the degradation of phytic acid, which interferes with mineral absorption; pretreatment of lignocellulosic biomass for bioethanol production can be achieved using a cocktail of cellulases (endoglucanase, EC 3.2.1.4, exoglucanase, EC 3.2.1.91 and cellobiohydrolase, EC 3.2.1.21) instead of employing acid or alkali treatments [124,125].

Mandatory requirements for enzyme application in large scale industry are (a) safety; (b) effectiveness; (c) low cost of production and purification for both enzyme and cofactors, if needed; d) stability to wide ranges of temperature, pH and organic solvents, and thus compatibility to productive processes. All those characteristics make the enzyme use advantageous from both technological and economic points of view. Native enzymes usually do not respond to each distinctive requirement of a perfect industrial enzyme, but these features can be achieved via molecular engineering and structure-function modifications by targeted or random mutagenesis.

The most important limitation related to the application of mycotoxin DEs in real matrices is represented by the reduced effectiveness of the process due to matrix effects. The physicochemical properties of food, such as the moisture and fat content, acidity, texture etc., greatly influence the success of the detoxification process. Moreover, inhibitory compounds may be present in raw materials and mycotoxins can occur in masked forms in plant tissues [126]; thus, their bioavailability for the enzyme catalysis may be further reduced. These implications might require pretreatments, additional time and costs, which must be taken into account in the development of industrial applications.

Despite these limitations, the potentialities of enzyme use in food and feed industries remain widespread. Their application is versatile, since they can be used both in free or immobilized form and easily applied to well established industrial processes (fermentations, ripening, brewing or cheese manufacturing, feed and bioenergy production).

Moreover, DEs can be heterologously expressed by industrial microorganisms, such as lactic acid bacteria or yeasts, to perform in situ bioremediation. New smart and edible packaging with enzymes are also under study [127].

3.1. Food

The European Community established the last regulation on food enzymes in 2008 [128]. This regulation covers "enzymes added to food to perform a technological function in the manufacture, processing, preparation, treatment, packaging, transport or storage of such food, including enzymes used as processing aids". All enzymes are submitted to this regulation as processing aids, with the exception of lysozyme and invertase, considered as additives.

Article 6 states that necessary conditions to authorize a food enzyme are that (i) its use does not pose a safety concern for the health of the consumer; (ii) it responds to a technological need that cannot be achieved by other economically and technologically practicable means; and (iii) it does not mislead the consumer.

Only recently, acceptability criteria for detoxification treatments, including biotransformation, have been set for commodities intended for animal nutrition [38].

Some of the enzyme recognized to have mycotoxins degrading capabilities have been proposed and studied for applications in food industry, even if thought for other purposes.

Laccases have been studied for their introduction in bakery and dairy as crosslinking agent, in brewing, wine and fruit juice production as clarifying agent and polyphenol remover [129–131]. Peroxidases have also been studied for their crosslinking activity in bakery and dairy industry, but the use of H_2O_2 as cofactor limits their real application in food.

Amano enzyme Inc. (United Kingdom) requested EFSA to perform a scientific risk assessment on one laccase from *Trametes hirsuta*, currently in progress.

The employment of DEs in the food industry, especially in immobilized forms, is not unrealistic. Still, it must overcome the gap of knowledge related to the effects that these enzymes exert on the nutritional and organoleptic qualities of raw materials, as well as fill a legislative void, which, so far, does not foresee the possibility of detoxifying treatments for food.

3.2. Feed

Enzyme use in animal diets has a long history, starting from the 1920s, when a commercial enzyme preparation, produced by *Aspergillus orizae*, known as Protozyme, was applied in poultry diets [132]. Phytases, proteases, α-galactosidases, glucanases, xylanases, α-amylases, and polygalacturonases are commercially available and increasingly used in animal nutrition, with huge benefits in terms of nutritional value and micronutrients availability [133].

For authorizing their use in feed, EFSA requires solid supporting information about production, safety, efficacy and non-interference with the nutritional and organoleptic quality of the feed [134].

However, to become a commercial reality, an effective and safe DE must be formulated in order to guarantee its stability and efficacy during storage, and to ensure detoxification once released in the animal digestive tract.

There are two ways to apply a DE to feed: (a) as stabilized by a suitable formulation for becoming a feed additive or (b) used in a detoxifying process for contaminated raw materials intended for feed.

(a) During feed production enzymes are usually added before pelleting, in premixes with other additives (such as vitamins), or after pelleting as a liquid application. The most convenient approach is to introduce them as protected in formulations before pelleting: indeed, in this condition enzymes are less exposed to high moisture content and high temperature [135], thus preventing their inactivation.

High temperature protection is usually achieved by coating. Its challenging development has to succeed in both stabilizing the enzyme during feed processing, and in fast dissolving in the animal gut, thus releasing the active enzyme. To this aim new coated and tough (CT) granulated coatings have been developed by different companies (Novozymes®, Bagsvaerd, Denmark; DSM™, The Netherlands).

Within the CT granulates, enzymes are enclosed in a core matrix of minerals and carbohydrates. The outer layer can consist of vegetable oil, kaolin, and calcium carbonate, which are readily degraded upon ingestion [136]. On 5 May 2014 Biomin® (Holding GmbH) received the first-ever EFSA positive opinion, for using a purified enzyme in feed. This is an esterase, embodied in a 10% maltodextrin matrix and spray dried, capable of biotransforming fumonisins.

(b) The development of a detoxifying application at industrial scale should allow the treatment of highly contaminated batches, with significant reduction of mycotoxin under the limits imposed by the current regulation. The achieved detoxification could result worthless in case of residual contamination by toxigenic fungi, if long term storage of the treated material is expected.

Free enzymes can be delivered in the feed productive process both in liquid and in solid forms (some prior wetting in dissolving feeders might be needed), through pumps, and mixed with the raw material, usually through a vertically mounted, shaft-driven impeller. This mode of application well fits a batch type process.

The major limitation of this approach is the difficulty to achieve a homogenous enzyme delivery throughout the entire material. In addition, the enzyme must target hot spots of contamination, typical of the distribution of toxigenic fungi and associated mycotoxins.

Among the environmental parameters that should be well set to preserve enzyme catalytic properties, moisture is by far the most critical. A high water content can increase enzyme activity, but simultaneously trigger fungal spread.

Immobilized enzymes encounter the same limitations, but they present the advantage of allowing the setup of continual processes, enzyme reuse and costs reduction.

3.3. Bioenergy

In biofuel production, enzymes are used to break down complex carbohydrates and to release ready fermentable sugars from cheap raw materials. Cellulose hydrolysis preceding fermentation is usually performed with cellulases, cellobiohydrolases and endoglucanases [137]. Pre-treatment of lignocellulosic biomasses prior to saccharification by laccases is under study, both to degrade lignin and to detoxify phenolic inhibitory compounds in bioethanol production [138,139]. The same biomass breakdown is realized during biogas production, especially in maize-fed biogas plants, to adapt and accommodate raw materials and to enhance their suitability for the activity of methanogenic microorganisms [140,141]. In this case, the mixture may include cellulose, hemicellulose, pectin and starch DEs, but also proteins able to breakdown long chain fatty acids, whose accumulation is responsible for the decrease of the process stability.

Despite these applications, no enzymes have been yet developed to reduce mycotoxin level in maize silage, plants or any other substrate used for bioenergy production. Their content, which mirrors fungal contamination in raw materials, affects the efficiency of the anaerobic digestion and ethanol/methane accumulation: indeed, fungal metabolites, including mycotoxins, act as antimicrobial or inhibitory compounds, or lead to excessive foam formation [142].

Storm et al. [143] recently reviewed fungal and mycotoxin contamination in maize feeding biogas plants. The authors assessed that pre-harvest contamination of grass and maize by *Fusarium* spp., *Aspergillus* spp. and *Alternaria* spp. can lead to DON, ZEN, FBs and AFs accumulation, but with concentration usually lower than regulatory limits. Salati et al. [144] reported that the addition of AFB_1 in a lab scale anaerobic digestion trials did not affect biogas production. In a recent conference abstract presented by De Gelder et al. [145] the authors evaluated the biodegradability of mycotoxins during anaerobic digestion. They performed lab scale degradation tests with digestate spiked with different mycotoxins. After 30 days of mesophilic and thermophilic digestion, each tested mycotoxin resulted absent in the final products, suggesting the activation of some adsorption/detoxification mechanism during the fermentation. The adsorption mechanism appears the most probable, since (1) as discussed in previous paragraphs mycotoxins are greatly stable and (2) they strongly bind to soil clay minerals and organic matter present in the reactors. However, no experiment has been performed in real biogas plants under conventional operating conditions.

Recent studies demonstrated that the addition of enzymes directly into biogas reactors has no significant effect on biogas process [146], since the enzymes are quickly degraded [147]. This obviously might require a continuous addition, and thus increases costs in comparison to raw materials pre-treatment. The development of enzymatic formulates able to couple biomass breakdown and mycotoxin degrading activity could allow the simultaneous increase of biogas yield and the safe utilization of byproducts, for which mycotoxins levels are not yet regulated.

The fate of mycotoxins during biofuel production has been only recently investigated. Dzuman et al. [148], performed a study on five different batches of the ethanol–distiller's grains and soluble (DDGS) production process. They reported a significant increase of deoxynivalenol and its glycosylated form, DON-3-glucoside, during the first part of fermentation, when hydrolytic enzymes were added. After yeast addition, total DON content rapidly decreased. They observed an opposite trend for FB_1, since yeast addition contributed to its increase. The same results were discussed by Wu and Munkvold [149], who estimated that mycotoxins are concentrated up to three times in DDGS compared to the feeding grain.

Despite these data, the second generation of bioethanol industry is not greatly concerned by mycotoxin contamination since: (a) less susceptible commodities can be used as raw material; (b) mycotoxins occurrence in bioethanol byproducts is not yet regulated neither in Europe or in U.S. and (c) the growth and productivity of highly tolerant fermenting yeasts are only slightly influenced by mycotoxin contamination [150,151].

The re-use of contaminated biomasses for renewable energy production is raising great interest both for economic and environmental reasons. However, more detailed studies should evaluate side effects not sufficiently considered, since by-products use for animal nutrition in Europe, Canada and U.S. [152–155] is increasing.

Sosa et al. [156] performed a preliminary economic analysis evaluating the feasibility of using fumonisin highly contaminated corns as feedstock for ethanol production. The authors calculated that the advantages deriving from the lower price of contaminated raw materials is balanced by the increase of the operating costs and the decrease of the final ethanol production by the fermenter. Thus, the performance of the process must be optimized, using process alternatives (e.g., both microfiltration and pervaporation membranes) coupled by the introduction of ethanol tolerant yeast strains, in order to effectively increase the economic value of the process.

DDGS are considered valuable by-products, thanks to the low carbon-nitrogen ratio and their high proteins and micronutrients content [157]. However, their utilization should be responsibly addressed, since ecological outcomes of mycotoxin contaminated fertilizers and health effects are still poorly investigated.

As for biogas production, the implementation of DEs for biofuel application has not been developed. Still, it benefits of less limitations than those encountered by the food and feed industries. Cloning DEs in industrial hosts could ideally allow to simultaneously perform detoxification and fermentation of contaminated biomasses.

4. Conclusions and Future Perspectives

Within this review we gave a global view of the identified mycotoxins DEs, their mechanisms, their current and possible application in food, feed, biogas and biofuel industries. We also presented patented commercial preparations used by feed industries.

Although this paper is focused on the main mycotoxins, for which a high level of contamination and health implications have driven the majority of economic and research efforts, the discovery of novel DEs is becoming an interesting and stimulating topic also in relation to other mycotoxins.

Several patents have been deposited, mostly in U.S. and China, describing microorganisms and methods to degrade minor mycotoxins, such as patulin [158], beauvericin [159] and moniliformin [160] in plants, maize or grains. These patents are mainly related to microorganisms or crude extracts preparation, and no enzymes have been yet characterized as responsible for the degrading activity.

It should not be excluded that enzymes efficient towards minor mycotoxins could be effective towards structurally similar main ones.

In addition to the previously described ones, another challenge related to the application of DEs, is represented by mycotoxins co-occurrence, which should require the simultaneous use, and thus optimization, of different preparations and protocols. To overcome this barrier, the combination of different strategies, such as adsorbents mixture, binding microorganisms and enzymes, suitable for the inclusion within the same procedure, should be considered.

In the optimization process, great potentialities are ascribed to structural modelling and design of experiments (DOE) technologies. These approaches are able to identify structural determinants responsible for the degradation mechanism and to improve substrate-enzyme affinity by setting the conditions, which maximize enzyme efficiency. In combination with targeted mutagenesis methods, they strongly reduce lab-optimization processes and accelerate the industrial scale up.

Transcriptional analysis of degrading microorganisms [161] has recently led to the identification of a DE from the novel *Acinetobacter* sp. *neg-1* species, described by Fanelli et al. [162] as capable of degrading OTA. This high-throughput approach, as other advanced Next Generation Sequencing technologies, such as functional metagenomic, could give a huge contribution in the identification of similar classes of DEs from complex and unexplored environments, such as contaminated soils, water or rumen. The advantage of these approaches is represented by the possibility to investigate non-culturable organisms, as well as those living in complex environmental niches, hardly analyzable under laboratory conditions.

The application of DEs could be simplified by the biotechnological advances in the field of immobilization and encapsulation techniques. Finally, an interesting perspective is represented by the development of transgenic crops able to counteract mycotoxins formation in the field. EU does not authorize the genetic modification of plants as a breeding technique. In contrast, the U.S. have 20 years of history of genetically engineered crops: herbicide-tolerant and insect-resistant crops were introduced in 1996 and today more than 90% of the corn cultivated in the U.S. is genetically modified [163].

Syngenta patented trichothecene-resistant transgenic plants bearing the *Fusarium graminearum* Tri101 gene [164], which encodes for a 3-O-acetyltransferase [165,166], which catalyzes the transfer of an acetate to the C3 position of trichothecenes. The company performed field trials, in Canada, the U.S., Argentina, and three European countries [108], proving that the development of self-defending crops is already achievable.

Acknowledgments: This work was financially supported by H2020-E.U.3.2-678781-MycoKey-Integrated and innovative key actions for mycotoxin management in the food and feed chain.

Author Contributions: M.L. and A.F.L. conceived the review. M.L., V.C.L. and F.F. performed the bibliographic research and wrote the paper. F.F. and G.M. coordinated the contributions. All authors reviewed the manuscript. F.F. was responsible for the submission.

Conflicts of Interest: The authors declare no conflict of interest.

Abbreviations

^{1}H-NMR	Proton Nuclear Magnetic Resonance
AFB_1	aflatoxin B_1
AFD_1	aflatoxin D_1
AFL	aflatoxicol
AFM_1	aflatoxin M_1
AFO	aflatoxin oxidase enzyme
AFQ_1	aflatoxin Q_1
AFs	aflatoxins
AP1	aminopentol 1

CPA	carboxypeptidase A
CPY	carboxypeptidase Y
CT	coated and tough
DDGS	distiller's grains and soluble
DE	degrading enzyme
DOE	Design of experiments
DON	deoxynivalenol
DPs	degradation products
EDC	endocrine disrupting chemical
EFSA	European Food Safety Authority
ER	estrogen receptors
FB_1	fumonisin B_1
GAPs	Good Agricultural Practices
HFB	hydrolyzed FB_1
HR-ESI-MS	High Resolution electrospray ionization mass spectrometry
HT-2	HT-2 toxin
IARC	International Agency of Research on Cancer
LCs	laccases
MADE	Myxobacteria Aflatoxin Degrading Enzyme
MnPs	Manganese Peroxidases
NIV	nivalenol
OTA	ochratoxin A
OTα	ochratoxin α
Phe	phenylalanine molecule
Prx	peroxiredoxin
T-2	T-2 toxin
ZEN	zearalenone
α-ZAL	α-zearalanol
α-ZEL	α-zearalenol
β-ZAL	β-zearalanol
β-ZEL	β-zearalenol

References and Notes

1. Bennett, J.W.; Klich, M. Mycotoxins. *Clin. Microbiol. Rev.* **2003**, *16*, 497–516. [CrossRef] [PubMed]
2. Logrieco, A.F.; Mulé, G.; Moretti, A.; Bottalico, A. Toxigenic Fusarium species and mycotoxins associated with maize ear rot in Europe. *Eur. J. Plant. Pathol.* **2002**, *108*, 597–609.
3. Bryden, W.L. Mycotoxins in the food chain: Human health implications. *Asia Pac. J. Clin. Nutr.* **2007**, *16*, 95–101. [PubMed]
4. Shephard, G.S. Impact of mycotoxins on human health in developing countries. *Food Addit. Contam.* **2008**, *25*, 146–151.
5. Sharma, R.P. Immunotoxicity of mycotoxins. *J. Dairy Sci.* **1993**, *76*, 892–897. [CrossRef]
6. Zain, M.E. Impact of mycotoxins on humans and animals. *J. Saudi Chem. Soc.* **2011**, *15*, 129–144. [CrossRef]
7. The International Agency for Research on Cancer (IARC). Some traditional herbal medicines, some mycotoxins, naphthalene and styrene. In *IARC Monographs on the Evaluation of Carcinogenic Risks to Humans*; World Health Organization: Lyon, France, 2002; Volume 82, pp. 1–556.
8. The International Agency for Research on Cancer (IARC). Some naturally occurring substances: Food items and constituents, heterocyclic aromatic amines and mycotoxins. In *IARC Monographs on the Evaluation of Carcinogenic Risks to Humans*; World Health Organization: Lyon, France, 1993; Volume 56, pp. 489–521.
9. The International Agency for Research on Cancer (IARC). Chemical agents and related occupations. In *IARC Monographs on the Evaluation of Carcinogenic Risks to Humans*; World Health Organization: Lyon, France, 2012; Volume 100F, pp. 1–599.

10. Alassane-Kpembi, I.; Schatzmayr, G.; Taranu, I.; Marin, D.; Puel, O.; Oswald, I.P. Mycotoxins co-contamination: Methodological aspects and biological relevance of combined toxicity studies. *Crit. Rev. Microbiol.* **2016**. [CrossRef] [PubMed]

11. Gelderblom, W.C. Interaction of fumonisin B(1) and aflatoxin B(1) in a short-term. *Toxicology* **2002**, *171*, 161–173. [CrossRef]

12. Ji, C.; Fan, Y.; Zhao, L. Review on biological degradation of mycotoxins. *Anim. Nutr.* **2016**, *2*, 127–133.

13. Gao, Y.N.; Wang, J.Q.; Li, S.L.; Zhang, Y.D.; Zheng, N. Aflatoxin M1 cytotoxicity against human intestinal Caco-2 cells is enhanced in the presence of other mycotoxins. *Food Chem. Toxicol.* **2016**, *96*, 79–89. [CrossRef] [PubMed]

14. Roth, A.; Creppy, E.E.; Kane, A.; Bacha, H.; Steyn, P.S.; Roschenthaler, R.; Dirheimer, G. Influence of ochratoxin B on the ochratoxin A inhibition of phenylalanyl-tRNA formation In vitro and protein synthesis in hepatoma tissue culture cells. *Toxicol. Lett.* **1989**, *45*, 307–313. [PubMed]

15. Bouslimi, A.; Bouaziz, C.; Ayed-Boussema, I.; Hassen, W.; Bacha, H. Individual and combined effects of ochratoxin A and citrinin on viability and DNA fragmentation in cultured Vero cells and on chromosome aberrations in mice bone marrow cells. *Toxicology* **2008**, *251*, 1–7. [CrossRef] [PubMed]

16. Bouslimi, A.; Ouannes, Z.; El Golli, E.; Bouaziz, C.; Hassen, W.; Bacha, H. Cytotoxicity and oxidative damage in kidney cells exposed to the mycotoxins ochratoxin A and citrinin: Individual and combined effects. *Toxicol. Mech. Method* **2008**, *18*, 341–349.

17. Klaric, M.S.; Zeljezic, D.; Rumora, L.; Peraica, M.; Pepeljnjak, S.; Domijan, A.M. A potential role of calcium in apoptosis and aberrant chromatin forms in porcine kidney PK15 cells induced by individual and combined ochratoxin A and citrinin. *Arch. Toxicol.* **2012**, *86*, 97–107. [CrossRef] [PubMed]

18. Knasmuller, S.; Cavin, C.; Chakraborty, A.; Darroudi, F.; Majer, B.J.; Huber, W.W.; Ehrlich, V.A. Structurally related mycotoxins ochratoxin A, ochratoxin B, and citrinin differ in their genotoxic activities and in their mode of action in human-derived liver (HepG2) cells: Implications for risk assessment. *Nutr. Cancer* **2004**, *50*, 190–197. [PubMed]

19. Wu, F. Measuring the economic impacts of Fusarium toxins in animal feed. *Anim. Feed Sci. Technol.* **2007**, *137*, 363–374.

20. Commission Regulation 2006/1881/EC of 19 December 2006 Setting Maximum Levels for Certain Contaminants in Food stuffs. Available online: http://eur-lex.europa.eu/legal-content/EN/TXT/PDF/?uri=CELEX:32006R1881&from=en (accessed on 03 November 2016).

21. Commission Recommendation 2013/165/EU of 27 March 2013 on the Presence of T-2 and HT-2 Toxin in Cereals and Cereal Products Text with EEA Relevance. Available online: http://eur-lex.europa.eu/legal-content/EN/TXT/PDF/?uri=CELEX:32013H0165&from=EN (accessed on 03 November 2016).

22. Opinion of the Scientific Committee on Food (SCF) on Fusarium Toxins, Part 5: T-2 toxin and HT-2 Toxin. 2001. Available online: https://ec.europa.eu/food/sites/food/files/safety/docs/cs_contaminants_catalogue_out88_en.pdf (accessed on 03 November 2016).

23. Commission Recommendation 2006/576/EC of 17 August 2006 on the Presence of Deoxynivalenol, Zearalenone, Ochratoxin A, T-2 and HT-2 and Fumonisins in Products intended for Animal Feeding. Available online: http://eur-lex.europa.eu/legal-content/EN/TXT/PDF/?uri=CELEX:32006H0576&from=EN (accessed on 03 November 2016).

24. Commission Recommendation 2006/583/EC of 17 August 2006 on the Prevention and Reduction of Fusarium Toxins in Cereals and Cereal Products. Available online: http://eur-lex.europa.eu/legal-content/EN/TXT/PDF/?uri=CELEX:32006H0583&from=EN (accessed on 03 November 2016).

25. Vanhoutte, I.; Audenaert, K.; De Gelder, L. Biodegradation of Mycotoxins: Tales from Known and Unexplored Worlds. *Front. Microbiol.* **2016**, *7*, 561. [CrossRef] [PubMed]

26. Schlüter, O.; Ehlbeck, J.; Hertel, C.; Habermeyer, M.; Roth, A.; Engel, K.H.; Holzhauser, T.; Knorr, D.; Eisenbrand, G. Opinion on the use of plasma processes for treatment of foods. *Mol. Nutr. Food Res.* **2013**, *57*, 920–927. [PubMed]

27. Bong, J.P.; Kosuke, T.; Yoshiko, S.K.; Ik-Hwi, K.; Mi-Hee, L.; Dong-Wook, H.; Kie-Hyung, C.; Soon, O.H.; Jong-Chul, P. Degradation of mycotoxins using microwave-induced argon plasma at atmospheric pressure. *Surf. Coat. Technol.* **2007**, *201*, 5733–5737.

28. Kriz, P.; Petr, B.; Zbynek, H.; Jaromir, K.; Pavel, O.; Petr, S.; Miroslav, D. Influence of plasma treatment in open air on mycotoxin content and grain nutriments. *Plasma Med.* **2015**, *5*, 145–158. [CrossRef]

29. Herzallah, S.; Al Shawabkeh, K.; Al Fataftah, A. Aflatoxin decontamination of artificially contaminated feeds by sunlight, g-radiation, and microwave heating. *J. Appl. Poult. Res.* **2008**, *17*, 515–521.

30. Fanelli, F.; Geisen, R.; Schmidt-Heydt, M.; Logrieco, A.F.; Mulè, G. Light regulation of mycotoxin biosynthesis: New perspectives for food safety. *World Mycotoxin J.* **2016**, *9*, 129–146. [CrossRef]

31. Bretz, M.; Beyer, M.; Cramer, B.; Knecht, A.; Humpf, H.-U. Thermal degradation of the *Fusarium* mycotoxin deoxynivalenol. *J. Agric. Food Chem.* **2006**, *54*, 6445–6451. [CrossRef]

32. Park, D.L.; Lee, L.S.; Price, R.L.; Pohland, A.E. Review of the decontamination of aflatoxins by ammoniation: Current status and regulation. *J. Assoc. Off. Anal. Chem.* **1988**, *71*, 685–703. [PubMed]

33. Aiko, V.; Edamana, P.; Mehta, A. Decomposition and detoxification of aflatoxin B1 by lactic acid. *J. Sci. Food Agric.* **2016**, *96*, 1959–1966. [PubMed]

34. Müller, H.M. Entgiftung von Mykotoxinen: II. Chemische Verfahren und Reaktion mit Inhaltsstoffen von Futtermitteln. *Übers Tierernährg.* **1983**, *11*, 47–80.

35. Fouler, S.G.; Trivedi, A.B.; Kitabatake, N. Detoxification of citrinin and ochratoxin a by hydrogen peroxide. *J. Assoc. Off. Agric. Chem.* **1994**, *77*, 631–637.

36. Maeba, H.; Takamoto, Y.; Kamimura, M.I.; Miura, T.O. Destruction and detoxification of aflatoxins with ozone. *J. Food Sci.* **1988**, *53*, 667–668. [CrossRef]

37. Altug, T.; Youssef, A.E.; Marth, E.H. Degradation of aflatoxin B1 in dried figs by sodium bisulfite with or without heat, ultraviolet energy or hydrogen peroxide. *J. Food Prot.* **1990**, *53*, 581–628.

38. Commission Regulation 2015/786/EU Defining Acceptability Criteria for Detoxification Processes Applied to Products Intended for Animal Feed as Provided for in Directive 2002/32/EC of the European Parliament and of the Council. Available online: http://extwprlegs1.fao.org/docs/pdf/eur144560.pdf (accessed on 3 November 2016).

39. Boudergue, C.; Burel, C.; Dragacci, S.; Favrot, M.; Fremy, J.; Massimi, C.; Prigent, P.; Debongnie, P.; Pussemier, L.; Boudra, H.; et al. Review of mycotoxin-detoxifying agents used as feed additives: mode of action, efficacy and feed/food safety. *EFSA Support. Publ.* **2009**, *6*. [CrossRef]

40. Hassan, Y.I.; Zhou, T.; Bullerman, L.B. Sourdough lactic acid bacteria as antifungal and mycotoxin-controlling agents. *Food Sci. Technol. Int.* **2015**, *22*, 79–90.

41. Hassan, Y.I.; Bullerman, L.B. Cell-surface binding of deoxynivalenol to *Lactobacillus paracasei* subsp. *tolerans* isolated from sourdough starter culture. *JMBFS* **2013**, *2*, 2323–2325.

42. Böswald, C.; Engelhardt, G.; Vogel, H.; Wallnöfer, P.R. Metabolism of *Fusarium* mycotoxins zearalenone and deoxynivalenol by yeast strains of technological relevance. *Nat. Toxins* **1995**, *3*, 138–144.

43. De Bellis, P.; Tristezza, M.; Haidukowski, M.; Fanelli, F.; Sisto, A.; Mulè, G.; Grieco, F. Biodegradation of ochratoxin A by bacterial strains isolated from vineyard soils. *Toxins* **2015**, *7*, 5079–5093. [CrossRef]

44. Sato, I.; Ito, M.; Ishizaka, M.; Ikunaga, Y.; Sato, Y.; Yoshida, S.; Koitabashi, M.; Tsushima, S. Thirteen novel deoxynivalenol-degrading bacteria are classified within two genera with distinct degradation mechanisms. *FEMS Microbiol. Lett.* **2012**, *327*, 110–117. [PubMed]

45. He, W.J.; Yuan, Q.S.; Zhang, Y.B.; Guo, M.W.; Gong, A.D.; Zhang, J.B.; Wu, A.B.; Huang, T.; Qu, B.; Li, H.P.; et al. Aerobic de-epoxydation of trichothecene mycotoxins by a soil bacterial consortium isolated using in situ soil enrichment. *Toxins* **2016**, *8*, 277.

46. Essigmann, J.M.; Croy, R.G.; Nadzan, A.M.; Busby, W.F., Jr.; Reinhold, V.N.; Büchi, G.; Wogan, G.N. Structural identification of the major DNA adduct formed by aflatoxin B1 In vitro. *Proc. Nat. Acad. Sci. USA* **1977**, *74*, 1870–1874. [PubMed]

47. Eaton, D.L.; Gallagher, E.P. Mechanisms of aflatoxin carcinogenesis. *Annu. Rev. Pharmacol. Toxicol.* **1994**, *34*, 135–172. [PubMed]

48. Schroeder, T.U.; Zweifel, P.; Sagelsdorff, U.; Friederich, J.; Luthy, C.; Schlatter, J. Ammoniation of aflatoxin-containing corn: Distribution, *in vivo* covalent deoxyribonucleic acid binding, and mutagenicity of reaction products. *Agric. Food Chem.* **1985**, *33*, 311–316. [CrossRef]

49. Méndez-Albores, A.; Arámbula-Villa, G.; Loarca-Piña, M.G.; Castaño-Tostado, E.; Moreno-Martínez, E. Safety and efficacy evaluation of aqueous citric acid to degrade B-aflatoxins in maize. *Food Chem. Toxicol.* **2005**, *43*, 233–238. [PubMed]

50. Motomura, M.; Toyomasu, T.; Mizuno, K.; Shinozawa, T. Purification and characterization of an aflatoxin degradation enzyme from *Pleurotus ostreatus*. *Microbiol. Res.* **2003**, *158*, 237–242. [PubMed]

51. Liu, D.L.; Yao, D.S.; Liang, R.; Ma, L.; Cheng, W.Q.; Gu, L.Q. Detoxifcation of aflatoxin B1 by enzymes isolated from *Armillariella tabescens*. *Food Chem. Toxicol.* **1998**, *36*, 563–574.

52. Chitrangada, D.; Mishra, H.N. In vitro degradation of aflatoxin B1 by horseradish peroxidase. *Food Chem.* **2000**, *68*, 309–313.

53. Alberts, J.F.; Gelderblom, W.C.; Botha, A.; van Zyl, W.H. Degradation of aflatoxin B(1) by fungal laccase enzymes. *Int. J. Food Microbiol.* **2009**, *30*, 47–52.

54. Novozymes A/S. Detoxification of Aflatoxin in Feed Products. World Patent 2009109607, 5 March 2009.

55. Taylor, M.C.; Jackson, C.J.; Tattersall, D.B.; French, N.; Peat, T.S.; Newman, J.; Briggs, L.J.; Lapalikar, G.V.; Campbell, P.M.; Scott, C.; et al. Identification and characterization of two families of F420H2-dependent reductases from *Mycobacteria* that catalyse aflatoxin degradation. *Mol. Microbiol.* **2010**, *78*, 561–575. [PubMed]

56. Yehia, R.S. Aflatoxin detoxification by manganese peroxidase purified from *Pleurotus ostreatus*. *Braz. J. Microbiol.* **2014**, *45*, 127–133. [PubMed]

57. Zhao, L.H.; Guan, S.; Gao, X.; Ma, Q.G.; Lei, Y.P.; Bai, X.M.; Ji, C. Preparation, purification and characteristics of an aflatoxin degradation enzyme from *Myxococcus fulvus* ANSM068. *J. Appl. Microbiol.* **2011**, *110*, 147–155. [CrossRef]

58. Loi, M.; Fanelli, F.; Zucca, P.; Liuzzi, V.C.; Quintieri, L.; Cimmarusti, M.T.; Monaci, L.; Haidukovski, M.; Logrieco, A.F.; Sanjust, E.; et al. Aflatoxin B_1 and M_1 degradation by Lac2 from *Pleurotus pulmonarius* and redox mediators. *Toxins* **2016**, *8*, 245. [CrossRef] [PubMed]

59. Loi, M.; Quintieri, L.; Liuzzi, V.C.; Haidukovski, M.; Logrieco, A.F.; Sanjust, E.; Fanelli, F.; Mulè, G. Aflatoxin M_1 removal and biotechnological application of a laccase from *Pleurotus eryngii* for milk safety. *Scienza e Tecnica Lattiero Casearia.* In press.

60. Cao, H.; Liu, D.; Mo, X.; Xie, C.; Yao, D. A fungal enzyme with the ability of aflatoxin B1 conversion: Purification and ESI-MS/MS identification. *Microbiol. Res.* **2011**, *166*, 475–483. [CrossRef] [PubMed]

61. Liu, D.L.; Ma, L.; Gu, L.Q.; Liang, R.; Yao, D.S.; Chen, W.Q. *Armillariella tabescens* enzymatic detoxification of aflatoxin B1. Part III. Immobilized enzymatic detoxification. *Ann. N. Y. Acad. Sci.* **1998**, *864*, 592–599. [PubMed]

62. Wu, Y.Z.; Lu, F.P.; Jiang, H.L.; Tan, C.P.; Yao, D.S.; Xie, C.F.; Liu, D.L. The furofuran-ring selectivity, hydrogen peroxide-production and low K_m value are the three elements for highly effective detoxification of aflatoxin oxidase. *Food Chem. Toxicol.* **2015**, *76*, 125–131.

63. Wang, J.; Ogata, M.; Hirai, H.; Kawagishi, H. Detoxification of aflatoxin B1 by manganese peroxidase from the white-rot fungus *Phanerochaete sordida* YK-624. *FEMS Microbiol. Lett.* **2011**, *314*, 164–169. [CrossRef] [PubMed]

64. Chitrangada, D.; Mishra, H.N. In vitro degradation of aflatoxin B1 in groundnut (*Arachis hypogea*) meal by horseradish peroxidase. *Lebensm. Wiss. Technol.* **2000**, *33*, 308–312.

65. Viswanath, B.; Rajesh, B.; Janardhan, A.; Kumar, A.P.; Narasimha, G. Fungal laccases and their applications in bioremediation. *Enzyme Res.* **2014**, *2014*, 1–21.

66. Banu, I.; Lupu, A.; Aprodu, I. Degradation of zearalenone by laccase enzyme. *Sci. Study Res.* **2013**, *14*, 79–84.

67. Novozymes A/S. Detoxification of Aflatoxin in Feed Products. European Patent 2252163, 5 March 2009.

68. Margot, J.; Bennati-Granier, C.; Maillard, J.; Blánquez, P.; Barry, D.A.; Holliger, C. Bacterial versus fungal laccase: Potential for micropollutant degradation. *AMB Express* **2013**, *3*, 63–77. [CrossRef]

69. Guan, S.; Zhao, L.; Ma, Q.; Zhou, T.; Wang, N.; Hu, X.; Ji, C. In vitro efficacy of *Myxococcus fulvus* ANSM068 to biotransform aflatoxin B_1. *Int. J. Mol. Sci.* **2010**, *11*, 4063–4079.

70. Dirheimer, G.; Creppy, E.E. Mechanism of action of ochratoxin A. *IARC Sci. Publ.* **1991**, *115*, 171–186.

71. Pfohl-Leszkowicz, A.; Manderville, R.A. An update on direct genotoxicity as molecular mechanism of ochratoxin A carcinogenicity. *Chem. Res. Toxicol.* **2012**, *25*, 252–262. [PubMed]

72. Vettorazzi, A.; van Delft, J.; López de Cerain, A. A review on ochratoxin A transcriptomic studies. *Food Chem. Toxicol.* **2013**, *59*, 766–783. [PubMed]

73. Kőszegi, T.; Poór, M. Ochratoxin A: Molecular interactions, mechanisms of toxicity and prevention at the molecular Level. *Toxins* **2016**, *8*, 111. [CrossRef] [PubMed]

74. Abrunhosa, L.; Paterson, R.R.; Venâncio, A. Biodegradation of Ochratoxin A for Food and Feed Decontamination. *Toxins* **2010**, *2*, 1078–1099. [CrossRef]

75. Pitout, M.J. The hydrolysis of Ochratoxin A by some proteolytic enzymes. *Biochem. Pharmacol.* **1969**, *18*, 485–491. [PubMed]

76. Dridi, F.; Marrakchi, M.; Gargouri, M.; Saulnier, J.; Jaffrezic-Renault, N.; Lagarde, F. Comparison of carboxypeptidase Y and thermolysin for ochratoxin A electrochemical biosensing. *Anal. Methods* **2015**, *7*, 8954–8960.

77. Stander, M.A.; Bornscheuer, U.T.; Henke, E.; Steyn, P.S. Screening of commercial hydrolases for the degradation of ochratoxin A. *J. Agric. Food Chem.* **2000**, *48*, 5736–5739. [PubMed]

78. Abrunhosa, L.; Santos, L.; Venâncio, A. Degradation of ochratoxin A by proteases and by a crude enzyme of *Aspergillus niger*. *Food Biotechnol.* **2006**, *20*, 231–242.

79. Danisco A/S. Food Additive Comprising an Amidase for Detoxifying Ochratoxin. World Patent 2012032472, 6 September 2011.

80. Heinl, S.; Hartinger, D.; Moll, W.D.; Schatzmayr, G.; Grabherr, R. Identification of a fumonisin B1 degrading gene cluster in *Sphingomonas* sp. MTA144. *New Biotechnol.* **2009**, *25*, S61–S62. [CrossRef]

81. Wang, E.; Norred, W.P.; Bacon, C.W.; Riley, R.T.; Merrill, A.H., Jr. Inhibition of sphingolipid biosynthesis by fumonisins: Implications for diseases associated with *Fusarium moniliforme*. *J. Biol. Chem.* **1991**, *266*, 14486–14490. [PubMed]

82. Merrill, A.H.; Wang, E.; Vales, T.R.; Smith, E.R.; Schroeder, J.J.; Menaldino, D.S.; Alexander, C.; Crane, H.M.; Xia, J.; Liotta, D.C.; et al. Fumonisin toxicity and sphingolipid biosynthesis. *Adv. Exp. Med. Biol.* **1996**, *392*, 297–306. [PubMed]

83. Soriano, J.M.; González, L.; Catala, A.I. Mechanism of action of sphingolipids and their metabolites in the toxicity of fumonisin B1. *Progr. Lipid Res.* **2005**, *44*, 345–356. [CrossRef]

84. Rheeder, J.P.; Marasas, W.F.O.; Vismer, H.F. Production of fumonisin analogs by *Fusarium* species. *Appl. Environ. Microbiol.* **2002**, *68*, 2101–2105. [PubMed]

85. Aly, S.E.; Abdel-Galil, M.M.; Abdel-Wahhab, M.A. Application of adsorbent agents technology in the removal of aflatoxin B1 and fumonisin B1 from malt extract. *Food Chem. Toxicol.* **2004**, *42*, 1825–1831.

86. Robinson, A.; Johnson, N.M.; Strey, A.; Taylor, J.F.; Marroquin-Cardona, A.; Mitchell, N.J.; Afriyie-Gyawu, E.; Ankrah, N.A.; Williams, J.H.; Wang, J.S.; et al. Calcium montmorillonite clay reduces urinary biomarkers of fumonisin B1 exposure in rats and humans. *Food Addit. Contam.* **2012**, *29*, 809–818.

87. Pioneer Hi-Bred International, Inc. Fumonisin detoxification compositions and methods. U.S. Patent 5792931, 7 June 1995.

88. Duvick, J. Prospects for reducing fumonisin contamination of maize through genetic modification. *Environ. Health Persp.* **2001**, *109*, 337–342. [CrossRef]

89. Pioneer Hi-Bred International, Inc. Fumonisin-Detoxifying Enzymes. World Patent 1996006175, 11 August 1995.

90. Duvick, J.; Rood, T.; Maddox, J.; Gilliam, J. Detoxification of mycotoxins in planta as a strategy for improving grain quality and disease resistance: Identification of fumonisin-degrading microbes from maize. In *Molecular Genetics of Host-Specific Toxins in Plant Disease*; Kohmoto, K., Yoder, O., Eds.; Kluwer Academic Publishers: Dordrecht, The Netherlands, 1998; pp. 369–381.

91. Scientific Opinion on the safety and efficacy of fumonisin esterase (FUMzyme®) as a technological feed additive for pigs. *EFSA J.* **2014**, *12*, 3667.

92. Täubel, M. Isolierung und Charakterisierung von Mikroorganismen zur Biologischen Inaktivierung von Fumonisinen. Ph.D. Thesis, University of Natural Resources and Applied Life Sciences, Vienna, Austria, 2005.

93. Hartinger, D.; Schwartz, H.; Hametner, C.; Schatzmayr, G.; Haltrich, D.; Moll, W.D. Enzyme characteristics of aminotransferase FumI of *Sphingopyxis* sp. MTA144 for deamination of hydrolyzed fumonisin B1. *Appl. Microbiol. Biotechnol.* **2011**, *91*, 757–768. [CrossRef] [PubMed]

94. Heinl, S.; Hartingerb, D.; Thamhesl, M.; Schatzmayrb, G.; Moll, W.D.; Grabherra, R. An aminotransferase from bacterium AtrichotheceneC 55552 deaminates hydrolyzed fumonisin B1. *Biodegradation* **2011**, *22*, 25–30. [CrossRef] [PubMed]

95. Method for the Production of an Additive for the Enzymatic Decomposition of Mycotoxins, Additive, and Use Thereof. U.S. Patent 8,703,460, 2008.

96. Ueno, Y. Trichothecenes. In *Chemical, Biological and Toxicological Aspects*; Elsevier Scientific Publishers: Amsterdam, The Netherlands, 1983.

97. Zöllner, P.; Mayer-Helm, B. Trace mycotoxin analysis in complex biological and food matrices by liquid chromatography-atmospheric pressure ionisation mass spectrometry. *J. Chromatogr. A* **2006**, *1136*, 123–169. [CrossRef] [PubMed]

98. Grove, J.F. The trichothecenes and their biosynthesis. *Prog. Chem. Org. Nat. Prod.* **2007**, *88*, 63–130.
99. Rocha, O.; Ansari, K.; Doohan, F.M. Effects of trichothecene mycotoxins on eukaryotic cells: A review. *Food Addit. Contam.* **2005**, *22*, 369–378. [CrossRef]
100. Schuhmacher-Wolz, U.; Heine, K.; Schneider, K. Report on Toxicity Data on Trichothecene Mycotoxins HT-2 and T-2 Toxins. CT/EFSA/CONTAM/2010/03. Available online: http://www.efsa.europa.eu/en/scdocs/doc/65e.pdf (accessed on 10 November 2016).
101. Miller, J.D. Aspects of the ecology of *Fusarium* toxins in cereals. *Adv. Exp. Med. Biol.* **2002**, *504*, 19–27.
102. World Health Organization. WHO Food Additives Series: 47. Safety Evaluation of Certain Mycotoxins in Food. Prepared by the Fifty-Sixth Meeting of the Joint FAO/WHO Expert Committee on Food Additives (JECFA). 2001. Available online: http://www.inchem.org/documents/jecfa/jecmono/v47je01.htm (accessed on 10 November 2016).
103. Pestka, J.J. Deoxynivalenol: Mechanisms of action, human exposure, and toxicological relevance. *Arch. Toxicol.* **2010**, *84*, 663–679. [CrossRef] [PubMed]
104. Thompson, W.L.; Wannemacher, R.W., Jr. Structure–function relationship of 12,13-epoxytrichothecene mycotoxins in cell culture: Comparison of whole animal lethality. *Toxicon* **1986**, *24*, 985–994. [CrossRef]
105. Cundliffe, E.; Cannon, M.; Davies, J. Mechanism of inhibition of eukaryotic protein synthesis by trichothecene fungal toxins. *Proc. Natl. Acad. Sci. USA* **1974**, *71*, 30–34. [PubMed]
106. Cundliffe, E.; Davies, J.E. Inhibition of initiation, elongation, and termination of eukaryotic protein synthesis by trichothecene fungal toxins. *Antimicrob. Agents Chemother.* **1977**, *11*, 491–499. [CrossRef] [PubMed]
107. Sudakin, D.L. Trichothecenes in the environment: Relevance to human health. *Toxicol. Lett.* **2003**, *20*, 97–107.
108. Karlovsky, P. Biological detoxification of the mycotoxin deoxynivalenol and its use in genetically engineered crops and feed additives. *Appl. Microbiol. Biotechnol.* **2011**, *91*, 491–504. [PubMed]
109. Hassan, Y.I.; Zhu, H.; Zhu, Y.; Zhou, T. Beyond ribosomal binding-the increased polarity and aberrant molecular interactions of 3-epi-deoxynivalenol. *Toxins* **2016**, *8*, 261.
110. He, J.W.; Hassan, Y.I.; Perilla, N.; Li, X.Z.; Boland, G.J.; Zhou, T. Bacterial epimerization as a route for deoxynivalenol detoxification: The influence of growth and environmental conditions. *Front. Microbiol.* **2016**, *7*, 572. [CrossRef] [PubMed]
111. He, J.; Zhou, T.; Young, J.C.; Boland, G.J.; Scott, P.M. Chemical and biological transformations for detoxification of trichothecene mycotoxins in human and animal food chains: A review. *Trend Food Sci. Technol.* **2010**, *21*, 67–76. [CrossRef]
112. Ito, M.; Sato, I.; Ishizaka, M.; Yoshida, S.; Koitabashi, M.; Yoshida, S.; Tsushima, S. Bacterial cytochrome P450 system catabolizing the *Fusarium* toxin deoxynivalenol. *Appl. Environ. Microbiol.* **2013**, *79*, 1619–1628. [CrossRef] [PubMed]
113. Poppenberger, B.; Berthiller, F.; Lucyshyn, D.; Sieberer, T.; Schuhmacher, R.; Krska, R.; Kuchler, K.; Glössl, J.; Luschnig, C.; Adam, G. Detoxification of the *Fusarium* mycotoxin deoxynivalenol by a UDP-glucosyltransferase from *Arabidopsis thaliana*. *J. Biol. Chem.* **2003**, *278*, 47905–47914. [CrossRef] [PubMed]
114. Pierron, A.; Mimoun, S.; Murate, L.S.; Loiseau, N.; Lippi, Y.; Bracarense, A.P.; Schatzmayr, G.; He, J.W.; Zhou, T.; Moll, W.D.; et al. Microbial biotransformation of DON: Molecular basis for reduced toxicity. *Sci. Rep.* **2016**, *6*, 29105.
115. Zinedine, A.; Soriano, J.M.; Molto, J.C.; Manes, J. Review on the toxicity, occurrence, metabolism, detoxification, regulations and intake of zearalenone: An oestrogenic mycotoxin. *Food Chem. Toxicol.* **2007**, *45*, 1–18. [CrossRef] [PubMed]
116. Shier, W.T.; Shier, A.C.; Xie, W.; Mirocha, C.J. Structure-activity relationships for human estrogenic activity in zearalenone mycotoxins. *Toxicon* **2001**, *39*, 1435–1438. [CrossRef]
117. Takahashi-Ando, N.; Kimura, M.; Kakeya, H.; Osada, H.; Yamaguchi, I. A novel lactonohydrolase responsible for the detoxification of zearalenone: Enzyme purification and gene cloning. *Biochem. J.* **2002**, *365*, 1–6.
118. Takahashi-Ando, N.; Tokai, T.; Hamamoto, H.; Yamaguchi, I.; Kimura, M. Efficient decontamination of zearalenone, the mycotoxin of cereal pathogen, by transgenic yeasts through the expression of a synthetic lactonohydrolase gene. *Appl. Microbiol. Biotechnol.* **2005**, *67*, 838–844. [PubMed]
119. Riken. Zearalenone-Detoxifying Enzyme Gene and Transformant Having the Gene Transferred Thereinto. World Patent 2003080842, 25 March 2003.
120. Novozymes A/S. Detoxification of Feed Products. World Patent 2009109607, 5 March 2009.

121. Yu, Y.; Wu, H.; Tang, Y.; Qiu, L. Cloning, expression of a peroxiredoxin gene from *Acinetobacter* sp. SM04 and characterization of its recombinant protein for zearalenone detoxification. *Microbiol. Res.* **2012**, *167*, 121–126. [PubMed]

122. Tang, Y.; Xiao, J.; Chen, Y.; Yu, Y.; Xiao, X.; Yu, Y.; Wu, H. Secretory expression and characterization of a novel peroxiredoxin for zearalenone detoxification in *Saccharomyces cerevisiae*. *Microbiol. Res.* **2013**, *168*, 6–11.

123. Altalhi, A.D.; El-Deeb, B. Localization of zearalenone detoxification gene(s) in pZEA-1 plasmid of *Pseudomonas putida* ZEA-1 and expressed in *Escherichia coli*. *J. Hazard. Mater.* **2009**, *161*, 1166–1172. [CrossRef]

124. Patel, A.K.; Singhania, R.R.; Pandey, A. Novel enzymatic processes applied to the food industry. *Curr. Opin. Food Sci.* **2016**, *7*, 64–72. [CrossRef]

125. Sun, Y.; Cheng, J. Hydrolysis of lignocellulosic materials for ethanol production: A review. *Biores. Technol.* **2002**, *83*, 1–11.

126. Berthiller, F.; Crews, C.; Dall'Asta, C.; Saeger, S.D.; Haesaert, G.; Karlovsky, P.; Oswald, I.P.; Seefelder, W.; Speijers, G.; Stroka, J. Masked mycotoxins: A review. *Mol. Nutr. Food Res.* **2013**, *57*, 165–186. [CrossRef]

127. Baldino, L.; Cardea, S.; Reverchon, E. Supercritical assisted enzymatic membranes preparation, for active packaging applications. *J. Memb. Sci.* **2014**, *453*, 409–418.

128. Regulation (EC) No 1332/2008 of the European Parliament and of the Council of 16 December 2008 on food enzymes and amending Council Directive 83/417/EEC, Council Regulation (EC) No 1493/1999, Directive 2000/13/EC, Council Directive 2001/112/EC and Regulation (EC) No 258/97.

129. Osma, J.F.; Toca-Herrera, J.L.; Rodrıguez-Couto, S. Uses of Laccases in the Food Industry. *Enzym. Res.* **2010**, *2010*, 918761. [CrossRef]

130. Gassara-Chatti, F.; Brar, S.K.; Ajila, C.M.; Verma, M.; Tyagi, R.D.; Valero, J.R. Encapsulation of ligninolytic enzymes and its application in clarification of juice. *Food Chem.* **2013**, *137*, 18–24.

131. Lettera, V.; Pezzella, C.; Cicatiello, P.; Piscitelli, A.; Giacobelli, V.G.; Galano, E.; Amoresano, A.; Sannia, G. Efficient immobilization of a fungal laccase and its exploitation in fruit juice clarification. *Food Chem.* **2016**, *196*, 1272–1278.

132. Clickner, F.H.; Follwell, E.H. Application of 'Protozyme' (*Aspergillus orizae*) to poultry feeding. *Poult. Sci.* **1925**, *5*, 241–247. [CrossRef]

133. Menezes-Blackburn, D.; Greiner, R. Enzymes used in animal feed: Leading technologies and forthcoming developments. In *Functional Polymers in Food Science: From Technology to Biology, Volume 2: Food Processing*; Cirillo, G., Spizzirri, U.G., Iemma, F., Eds.; Wiley: Hoboken, NJ, USA, 2015.

134. Regulation (EC) No 1831/2003 of the European Parliament and of the Council of 22 September 2003 on Additives for Use in Animal Nutrition. Available online: http://eur-lex.europa.eu/legal-content/EN/TXT/PDF/?uri=CELEX:32003R1831&from=EN (accessed on 17 November 2016).

135. Beaman, K.R.; Lilly, K.G.S.; Gehring, C.K.; Turk, P.J.; Moritz, J.S. Influence of pelleting on the efficacy of an exogenous enzyme cocktail using broiler performance and metabolism. *J. Appl. Poult. Res.* **2012**, *21*, 744–775. [CrossRef]

136. Novo Nordisk A/S. Enzyme-Containing Granules and Process for the Production Thereof. World Patent 1997039116, 14 April 1997.

137. Xiros, C.; Topakas, E.; Christakopoulos, P. Hydrolysis and fermentation for cellulosic ethanol production. *WIREs Energy Environ.* **2013**, *2*, 633–654. [CrossRef]

138. Kudanga, T.; Le Roes-Hill, M. Laccase applications in biofuels production: Current status and future prospects. *Appl. Microbiol. Biotechnol.* **2014**, *98*, 6525–6542.

139. Jurado, M.; Prieto, A.; Martínez-Alcalá, A.; Martínez, A.T.; Martínez, M.J. Laccase detoxification of steam-exploded wheat straw for second generation bioethanol. *Bioresour. Technol.* **2009**, *100*, 6378–6384. [PubMed]

140. Christy, P.M.; Divya, D.; Gopinath, L.R. A review on anaerobic decomposition and enhancement of biogas production through enzymes and microorganisms. *Renew. Sustain. Energy Rev.* **2014**, *34*, 167–173. [CrossRef]

141. Parawira, W. Enzyme research and applications in biotechnological intensification of biogas production. *Crit. Rev. Biotechnol.* **2012**, *32*, 172–186.

142. Effenberger, M.; Lebuhn, M.; Gronauer, A. Fermentermanagement-Stabiler Prozess bei NawaRo-Anlagen. *Biogas Wandel* **2007**, *16*, 99–105.

143. Storm, I.M.L.D.; Sørensen, J.L.; Rasmussen, R.R.; Nielsen, K.F.; Thrane, U. Mycotoxins in silage. *Stewart Posthar. Rev.* **2008**, 1–12. [CrossRef]

144. Salati, S.; D'Imporzano, G.; Panseri, S.; Pasquale, E.; Adani, F. Degradation of aflatoxin b1 during anaerobic digestion and its effect on process stability. *Int. Biodeter. Biodegrad.* **2014**, *94*, 19–23.

145. De Gelder, L.; Audenaert, K.; Willems, B.; Schelfhout, K.; De Saeger, S.; De Boevre, M. Biodegradability of Mycotoxins during Anaerobic Digestion, *Abstract in* De Saeger, S.; Audenaert, K.; Croubels, S. Report from the 5th International Symposium on Mycotoxins and Toxigenic Moulds: Challenges and Perspectives (MYTOX) Held in Ghent, Belgium, May 2016. *Toxins* **2016**, *8*, 146.

146. Rintala, J.; Ahring, B.K. A two-stage thermophilic anaerobic process for the treatment of source sorted household solid waste. *Biotechnol. Lett.* **1994**, *16*, 1097–1102. [CrossRef]

147. Binner, R.; Menath, V.; Huber, H.; Thomm, M.; Bischof, F.; Schmack, D.; Reuter, M. Comparative study of stability and half-life of enzymes and enzyme aggregates implemented in anaerobic biogas processes. *Biomass Convers. Biorefin.* **2011**, *1*, 1–8. [CrossRef]

148. Dzuman, Z.; Stranska-Zachariasova, M.; Vaclavikova, M.; Tomaniova, M.; Veprikova, Z.; Slavikova, P.; Hajslova, J. Fate of Free and conjugated mycotoxins within the production of distiller's dried grains with solubles (DDGS). *J. Agric. Food Chem.* **2016**, *64*, 5085–5092. [CrossRef]

149. Wu, F.; Munkvold, G. Mycotoxins in ethanol co-products: Modeling economic impacts on the livestock industry and management strategies. *J. Agric. Food Chem.* **2008**, *56*, 3900–3911.

150. Nathanail, A.V.; Gibson, B.; Han, L.; Peltonen, K.; Ollilainen, V.; Jestoi, M.; Laitila, A. The lager yeast *Saccharomyces pastorianus* removes and transforms *Fusarium* trichothecene mycotoxins during fermentation of brewer's wort. *Food Chem.* **2016**, *203*, 448–455. [CrossRef]

151. Boeira, L.; Bryce, J.; Stewart, G.; Flannigan, B. Inhibitory effect of *Fusarium* mycotoxins on growth of brewing yeasts. 1 zearalenone and fumonisin B1. *J. Inst. Brew.* **1999**, *105*, 366–375. [CrossRef]

152. Commission Regulation (EU) No 68/2013 of 16 January 2013 on the Catalogue of Feed Materials. Available online: http://eur-lex.europa.eu/legal-content/EN/TXT/PDF/?uri=CELEX:32013R0068&from=EN (accessed on 17 November 2016).

153. Feeds Regulations (SOR/83-593), Feeds act. Government of Canada, 1983. Available online: http://laws-lois.justice.gc.ca/PDF/SOR-83-593.pdf (accessed on 17 November 2016).

154. Code of Federal Regulations (CFR), Title 21, Sections 573 Food Additives Permitted in Feed and Drinking Water of Animals. Available online: https://www.accessdata.fda.gov/scripts/cdrh/cfdocs/cfcfr/CFRSearch.cfm?CFRPart=573&showFR=1 (accessed on 17 November 2016).

155. Regulation (EC) No 1069/2009 of the European Parliament and of the Council of 21 October 2009 Laying down Health Rules as Regards Animal by-Products and Derived Products Not Intended for Human Consumption and Repealing Regulation (EC) No 1774/2002 (Animal by-Products Regulation). Available online: http://eur-lex.europa.eu/legal-content/EN/TXT/PDF/?uri=CELEX:32009R1069&from=EN (accessed on 17 November 2016).

156. Sosa, M.A.; Chovau, S.; Van der Bruggen, B.; Espinosa, J. Ethanol production from corn contaminated with fumonisins: A preliminary economic analysis including novel processing alternatives. *Ind. Eng. Chem. Res.* **2013**, *52*, 7504–7513. [CrossRef]

157. Liu, K. Chemical Composition of Distillers Grains, a Review. *J. Agric. Food Chem.* **2011**, *59*, 1508–1526. [CrossRef]

158. Zhejiang University. A Method for Reducing the Content of Patulin in Apple Juice Concentrate. Chinese Patent 103859016, 26 February 2014.

159. Pioneer Hi-Bred International, Inc. Beauvericin Detoxification Method Using Bacteria. U.S. Patent 6126934, 29 January 1998.

160. Pioneer Hi-Bred International, Inc. Moniliformin detoxification compositions and methods. Canadian Patent 2272554, 12 November 1997.

161. Liuzzi, V.C.; Fanelli, F.; Tristezza, M.; Haidukowski, M.; Picardi, E.; Manzari, C.; Lionetti, C.; Grieco, F.; Logrieco, A.F.; Thon, M.R.; et al. Transcriptional analysis of *Acinetobacter* sp. *neg1* capable of degrading ochratoxin A. *Front. Microbiol.* **2017**, *7*, 2162.

162. Fanelli, F.; Chiara, M.; Liuzzi, V.C.; Haidukowski, M.; Tristezza, M.; Manzari, C.; D'Erchia, A.M.; Pesole, G.; Horner, D.S.; Mulè, G. Draft genome sequence of *Acinetobacter* sp. *neg1* capable of degrading ochratoxin A. *FEMS Microbiol. Lett.* **2015**, *362*, 7. [CrossRef]

163. USDA Economic Research Service. Adoption of Genetically Engineered Crops in the U.S. USDA: Washington, DC, USA, 2011. Available online: https://www.ers.usda.gov/data-products/adoption-of-genetically-engineered-crops-in-the-us.aspx (accessed on 17 November 2016).

164. Hohn, T.M.; Peters, C.; Salmeron, J. Trichothecene-Resistant Transgenic Plants. U.S. Patent 20020162136, 12 February 2002.

165. Kimura, M.; Kaneko, I.; Komiyama, M.; Takatsuki, A.; Koshino, H.; Yoneyama, K.; Yamaguchi, I. Trichothecene 3-*O*-acetyltransferase protects both the producing organism and transformed yeast from related mycotoxins. *J. Biol. Chem.* **1998**, *273*, 1654–1661.

166. Kimura, M.; Shingu, Y.; Yoneyama, K.; Yamaguchi, I. Features of Tri101, the trichothecene 3-*O*-acetyltransferase gene, related to the self-defense mechanism in *Fusarium graminearum*. *Biosci. Biotech. Bioch.* **1998**, *62*, 1033–1036. [CrossRef]

toxins

MDPI

Article

Degradation of Aflatoxins by Means of Laccases from *Trametes versicolor*: An In Silico Insight

Luca Dellafiora [1], Gianni Galaverna [1,*], Massimo Reverberi [2] and Chiara Dall'Asta [1]

[1]   Department of Food Science, University of Parma, 43124 Parma, Italy; luca.dellafiora@unipr.it (L.D.); chiara.dallasta@unipr.it (C.D.)

[2]   Department of Environmental Biology, Sapienza University, 00185 Rome, Italy; massimo.reverberi@uniroma1.it

*   Correspondence: gianni.galaverna@unipr.it; Tel.: +39-0521-906-270

Academic Editor: Ting Zhou
Received: 16 November 2016; Accepted: 26 December 2016; Published: 1 January 2017

Abstract: Mycotoxins are secondary metabolites of fungi that contaminate food and feed, and are involved in a series of foodborne illnesses and disorders in humans and animals. The mitigation of mycotoxin content via enzymatic degradation is a strategy to ensure safer food and feed, and to address the forthcoming issues in view of the global trade and sustainability. Nevertheless, the search for active enzymes is still challenging and time-consuming. The in silico analysis may strongly support the research by providing the evidence-based hierarchization of enzymes for a rational design of more effective experimental trials. The present work dealt with the degradation of aflatoxin B_1 and M_1 by laccase enzymes from *Trametes versicolor*. The enzymes–substrate interaction for various enzyme isoforms was investigated through 3D molecular modeling techniques. Structural differences among the isoforms have been pinpointed, which may cause different patterns of interaction between aflatoxin B_1 and M_1. The possible formation of different products of degradation can be argued accordingly. Moreover, the laccase gamma isoform was identified as the most suitable for protein engineering aimed at ameliorating the substrate specificity. Overall, 3D modeling proved to be an effective analytical tool to assess the enzyme–substrate interaction and provided a solid foothold for supporting the search of degrading enzyme at the early stage.

Keywords: aflatoxins; biotransformation; enzymatic detoxification; laccase; mild technologies; food safety; mycotoxins mitigation

1. Introduction

Mycotoxins are low-molecular weight molecules produced as secondary metabolites by several species of fungi. They may enter the feed and food production chains worldwide upon the infection of crops and commodities intended for animal and human consumption. The contamination of food and feed by mycotoxins poses major concerns for the public health and welfare as the dietary exposure may cause disorders, dysfunctions and alterations of physiological states in both humans and animals [1,2].

Many countries have adopted regulations to reduce the possible dietary intake, thereby preserving the health of animals and consumers (for Europe: EC No 1881/2006, EU No 165/2010, EU No 105/2010). However, the allowed levels of contamination are not harmonized among countries, and this may cause trade frictions at the global level. De facto, the management of risks related to foodborne mycotoxins must consider several factors and have to reach controversial socio-economical tradeoffs, being primarily influenced by the availability of a secure food supply. The developing areas are the most damaged in terms of health and international exchanges as the contamination levels commonly found in traded commodities do not often comply with those enforced by industrialized countries, also considering that low-income countries lack regulatory actions and monitoring plans. As an example, the restrictions of the European Regulation cause huge economic losses to Africa, unreasonably exceeding the limits to effectively safeguard the public health in the European countries, as commented by the past Secretary-General of the United Nations, Kofi Annan [3].

On this basis, the mitigation of mycotoxin content in food and feed is a critical foothold to address the forthcoming challenges in view of the sustainability and global trade. On the one side, the reduction of contamination levels in food and feed can effectively ameliorate the health and welfare of both humans and animals. On the other side, the implementation of cost-effective strategies for recovering contaminated food and raw materials after spoilage may concretely allow the weaker markets to reenter the global trade. In this scenario, the development of affordable and straightforward strategies for mitigating mycotoxin content is definitely a major task for the scientific research.

A wide number of strategies for the control and mitigation of mycotoxins content in food and feed are currently under consideration. For instance, the strategies of biocontrol with non-toxigenic fungi, conventional breeding and genetic engineering are promising methodologies aimed at preventing the accumulation of mycotoxins on the field at the pre-harvest level [4]. Instead, food processing, physical methods (e.g., irradiation and adsorption) and microbial/biochemical transformation of mycotoxins to non- or less toxic compounds may act post-harvest on raw material or intermediate products [4]. The main advantage of the latter approach on the former is the possible application to low- or non-compliant food batches to reduce and/or reuse wastes. Among these, the enzymatic transformation seems to be the most promising tool for the mitigation in situ [5].

The high-throughput search and optimization of effective enzymes to be used for mitigation is thus of primary interest, even though highly challenging and time-consuming. Typically, the first steps in the conventional large-scale research process of active enzymes inevitably include the coarse-grained selection of candidate enzymes, which drastically depends on the realistic affordability of proteins in sufficient amounts and/or in the active forms. Commonly, this hardly complies with the rational criteria of exclusion in terms of possible effectiveness. However, the search of putative active enzymes in the early steps can be effectively boosted in a straightforward and cost-effective manner by using screening procedures in silico. Indeed, the upstream use of computational approaches to screen the libraries of candidate enzymes may support a wide-ranging and evidence-based selection of those enzymes to be investigated experimentally. In particular, the use of the 3D modeling by means of the computational estimate of the interaction at the enzymes binding site deepens the structural aspects underlying the enzyme–substrate interaction and succeeds in identifying substrates (e.g., ref. [6,7]). Hence, it can be a reliable tool for providing the rational and evidence-based hierarchization of enzymes on the basis of the computed capability to allow a favorable arrangement of substrates.

In this framework, the present study addressed the enzymatic degradation of aflatoxins (AFs) by laccase enzymes from *Trametes versicolor*. AFs are difuranocoumarin derivatives produced by *Asperigillus sect. Flavi* that can be found as contaminants primarily in cereals, maize, oilseeds and nuts [8]. AFs are mutagenic, genotoxic and carcinogenic compounds that cause both acute and chronic health effects [9]. Among the 20 AFs identified so far, aflatoxin B_1 (AFB_1) is the most widespread and harmful in terms of both acute and chronic toxicity [10], while aflatoxin M_1 (AFM_1)—its major hepatic metabolite in mammals—raises concern as it can be found in milk and dairy products [11]. In recent years, severe outbreaks related to aflatoxin contamination in feed have been reported in Europe, mainly in the Mediterranean and Balkan areas [12]. Besides health concerns, these events caused significant losses in terms of veterinary costs and managing of incompliant feed and milk batches.

Laccases are multi-copper containing enzymes capable of performing one electron oxidation of a broad range of substrates [13]. The laccase enzymes from *T. versicolor* have been identified as a promising route for the low-cost and effective reduction of AFB_1 content [14,15], while the use of these enzymes has been never considered before to degrade AFM_1. Moreover, the strategies commonly rely on the use of mixtures of the various laccase isoforms. The relative activity of the various isoforms and the structural aspects of aflatoxin interaction have not been elucidated yet, making the design of rational strategies based on selected isoforms difficult.

As a proof of concept, the present work aimed at modeling the interaction of AFB_1 and AFM_1 (Figure 1) with three out four laccase isoforms from *T. versicolor* (namely, the beta, delta and gamma isoforms) in order to find out possible differences among the enzymes in terms of pocket–ligand recognition. The complementarity of both AFB_1 and AFM_1 towards the various catalytic sites has been assessed using a previously validated structure-based molecular modeling workflow based on docking simulation and rescoring procedures. The pharmacophoric analysis of catalytic sites and the comparison of structures and sequences of the various isoforms have been done to better understand the basis of the enzymes–substrates interaction at a molecular level.

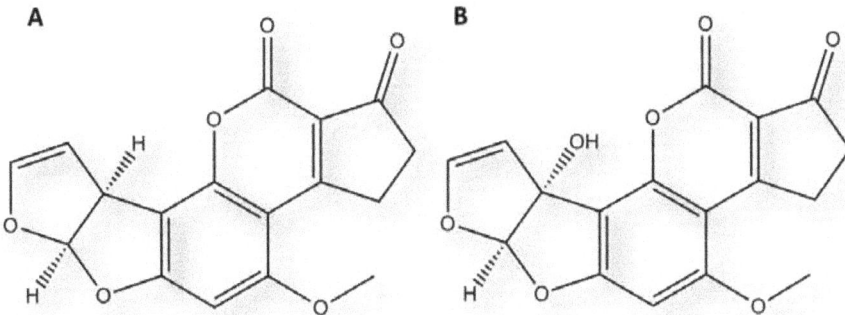

Figure 1. Chemical structure of aflatoxin B_1 (**A**) and aflatoxin M_1 (**B**).

2. Results

2.1. Sequence Analysis and Pocket Anatomy

Laccases from *T. versicolor* show a globular structure with approximate dimensions of $70 \times 50 \times 50$ Å with a topology consisting mainly of antiparallel β-barrels [16]. The overall amino acids sequence alignment of the laccase enzymes (from here on referred to as models) revealed that the delta and gamma isoforms showed, respectively, 71.6% and 71.8% of the identity with the beta isoform, and 77% between themselves (Figure 2A).

Concerning the primary structure, the sequence alignment revealed that the binding site of the beta and delta isoforms appeared fairly comparable to each other (12 residues out of 16 are conserved). Instead, the gamma model showed a higher divergence mainly due to the presence of an extend loop with five and four additional residues in respect to the beta and delta isoforms, respectively (Figure 2A,D).

The pharmacophoric analysis of the various binding site environments revealed that all the pockets appeared prevalently hydrophobic, as the hydrophobic environment turned out to exceed in extension the hydrophilic one, with a limited capability to receive polar groups, wherein H-bond donors groups were found to be more energetically favored than the H-bond acceptor ones (Figure 2B–D). In addition, concerning the distribution of the polarity of the space available for ligands, the binding sites of beta and delta isoforms were found to be more similar to each other, while the site of the gamma isoform was found to be less hydrophobic with a more extended volume energetically able to receive H-bond donor groups.

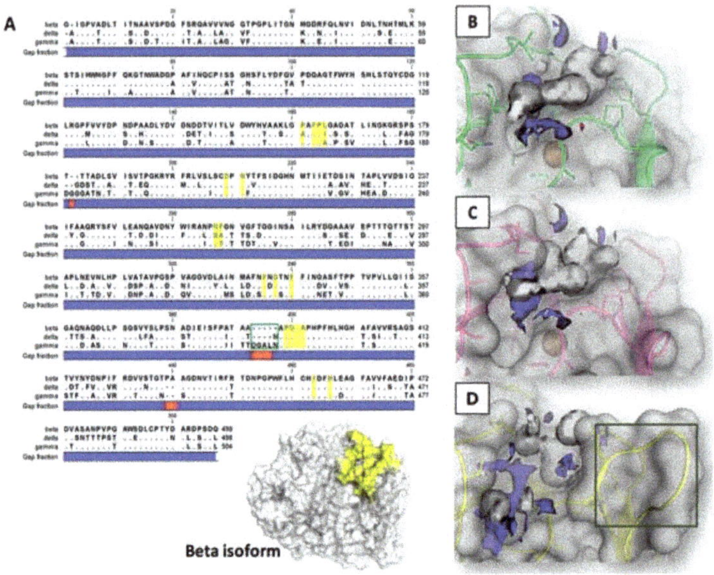

Figure 2. The sequence alignment (**A**) and pharmacophoric analysis of beta (**B**), gamma (**C**) and delta (**D**) laccase isoforms from *T. versicolor*. In the sequence alignment (box **A**), dots represent matching residues while dashes indicate gaps (**red** spots in the gap fraction **blue** bar). Residues of the binding site are highlighted in **yellow** while the **green** box indicates an extended region of the gamma isoform lining the catalytic site. The overall 3D structure of the beta isoform is also reported to provide localization of the binding site (colored in **yellow**). In the pharmacophoric analysis (boxes **B**, **C** and **D**), **white**, **red**, and **blue** contours identify regions sterically and energetically favorable for hydrophobic, H-bond acceptor, and H-bond donor groups, respectively. Spheres indicate Cu ions. The **green** box indicates the extended region of the gamma isoform.

2.2. Assessment of Procedure Reliability

The in silico investigation of AF-laccase interactions relied on the assessment of pocket–substrate complementarity through the coupling of docking simulation to calculate the binding architecture with re-scoring procedures using the HINT (Hydropathic INTeractions) scoring function for the careful estimation of the energetic contributions of the binding event (see Section 5 for further details). Such a procedure already proved to be an effective strategy for investigating the protein–ligand complex formation and the biological activity of chemicals, [17,18], and succeeded in identifying enzymes substrates as well [6]. However, the case-by-case assessment of procedural performances is required to prove the fit-for-purpose reliability among the diverse case studies [19].

Therefore, in the present work, the two benchmark laccase substrates ABTS (2,2'-azino-di-(3-ethylbenzothiazoline)-6-sulfonic acid) and 2,6-dimethoxyphenol [20] have been chosen to assess the case-specific reliability of the computational procedure. Specifically, the procedure has been validated assessing the capability to properly rank the reference compounds in accordance with the experimental data from the literature, and for the capability to reproduce the 3D binding architecture observed in the crystallographic structures available so far. The K_m values indicate the enzymes affinity for substrates. Thus, the K_ms have been chosen among the various biochemical parameters for the comparison with the computed scores since the HINT scoring may correlate the pocket–ligand affinity proportionally [6,21–25].

ABTS showed a clearly defined rank of K_ms among the various isoforms, with the following order: 88 μM in beta < 359 μM in gamma < 2262 μM in delta [26]. As shown in Table 1, the computational ranking of ABTS among the various laccase isoforms was consistent with that observed experimentally.

Table 1. Comparison of experimental affinity and HINT (Hydrophatic INTeractions) scores of ABTS (2,2′-azino-di-(3-ethylbenzothiazoline)-6-sulfonic acid) within the various laccase isoforms.

Laccase Isoform	Experimental Affinity Rank [1]	HINT Score
Beta	1	430
Gamma	2	206
Delta	3	104

[1] As reported by Christensen and co-workers [26].

Moreover, while the beta isoform *wild type* shows a higher affinity for ABTS than for 2,6-dimethoxyphenol, previous studies demonstrated that the D206A mutation at the level of binding site causes the inversion of ranking [20,27]. As reported in Table 2, the computational ranking of ABTS and 2,6-dimethoxyphenol within both the mutated and *wild type* enzyme was consistent with that observed experimentally.

Table 2. Comparison of experimental affinity and HINT scores of ABTS and 2,6-dimethoxyphenol within the *wild type* and mutated form of beta isoform.

Laccase Isoform	ABTS		2,6-dimethoxyphenol	
	Experimental Affinity Rank [1]	HINT Score	Experimental Affinity Rank [1]	HINT Score
Beta *wild type*	2	430	1	500
Beta D206A	1	495	2	203

[1] As reported by Madzak and co-workers [20].

In order to assess the reliability of the procedure in predicting the binding architectures, the computed pose of ABTS within the beta isoform was compared with the crystallographic one within the orthologous laccase enzyme from *Bacillus subtilis*, as no structures of laccases from *T. versicolor* are available so far. They can be consistently compared as orthology allows taking the functional conservation in different species for granted (i.e., the catalytic reaction and the overall organization of substrates in this case) [28]. It should be kept in mind that laccase enzymes catalyze the one-electron oxidation of substrates involving a Cu ion (at the so-called Cu T1 site), coupled to the four-electron reduction of molecular oxygen to water at the tri-nuclear Cu cluster [27]. The catalytic histidine lining the binding site, which may interact with substrates, mediates the electron transfer. For a proper transfer, the distance of the electron–donor regions of substrates from the receiving histidine cannot exceed the 5 Å [26]. The inspection of the crystallographic pose of ABTS revealed that the electron–donor region occupies the deepest region of the binding site, closely arranged to the catalytic histidine, while sulfonic acid groups protrude outside. Therefore, the geometric reliability of the in silico procedure was established by assessing the capability to properly reproduce such binding architecture. As showed in Figure 3, the overall computed organization of ABTS turned out to be consistent with the crystallographic pose and the arrangement of the electron–donor region has been correctly predicted as well, being posed within the 5 Å needed for undergoing the reaction [27].

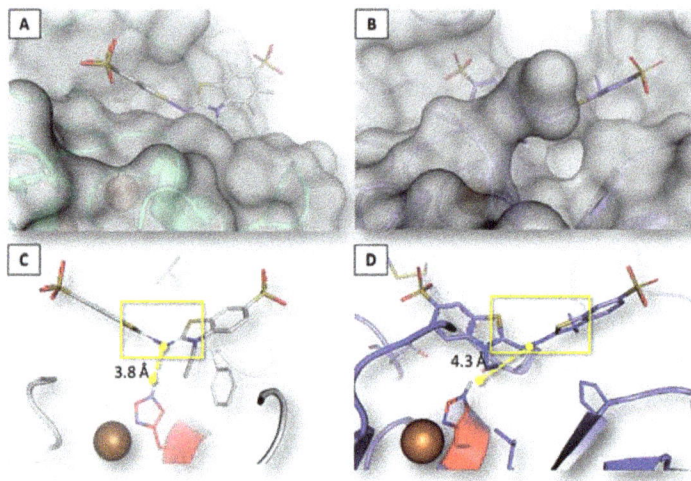

Figure 3. Binding architecture of ABTS. Proteins are represented in cartoon and surface, ABTS and residues of binding sites are represented in sticks. Cu at the T1 site is represented by the **red** sphere and the catalytic histidine is colored in **red**. ABTS electron donor region is highlighted with the **yellow** box, while interatomic distances are indicated by **yellow** dashed lines. (**A**) calculated surface interaction with beta laccase from *T. versicolor*; (**B**) crystallographic surface interaction with CotA laccase from *Bacillus subtilis* [29]; (**C**) detail of binding architecture with beta laccase from *T. versicolor*; and (**D**) detail of binding architecture with CotA laccase from *B. subtilis* [29].

2.3. Interaction of Aflatoxin B_1 and M_1 within Beta, Delta and Gamma Laccase Isoforms

The scores of interactions between AFs and laccase isoforms are reported in Table 3. AFM_1 was found able to interact favorably with all the laccase isoforms herein considered. AFB_1 favorably interacted instead with beta and delta, but not with gamma, as the pharmacophoric requirements were not satisfied (see below).

Table 3. HINT scores of aflatoxin B_1 (AFB_1)and aflatoxin M_1 (AFM_1)within the various laccase isoforms.

Laccase Isoform	HINT Scores	
	AFB_1	AFM_1
Beta	248	373
Gamma	−199	372
Delta	291	339

The close inspection of the computed poses revealed that AFB_1 adopted comparable binding architectures within both the beta and delta laccase isoforms, wherein the methoxy moiety was oriented toward the bottom of the binding sites. Both the interactions were mainly driven by hydrophobic/hydrophobic interaction, in accordance with the marked hydrophobic environment of the pockets. Indeed, a unique polar contact was found, wherein the oxygen of the methoxy group engaged the His458 (according to the amino acids' numeration of the beta isoform) with a hydrogen bond (Figure 4A). AFM_1 engaged the His458 with the oxygen on the difuran ring and used the additional hydroxyl group for engaging Asp206 (according to the amino acids' numeration of the beta isoform), thereby embedding much more into the catalytic sites (Figure 4B) and retracing the mode of interaction proposed for phenolic substrates [27]. In addition, the formation of the additional hydrogen bond was responsible of the higher score recorded for AFM_1 within all of the laccase isoforms herein considered. Conversely, AFB$_1$ was unable to sink into the site to the extent of AFB$_1$ since the hydroxyl-free difuran moiety did not find the energetic favors for being close to the aspartate side chain.

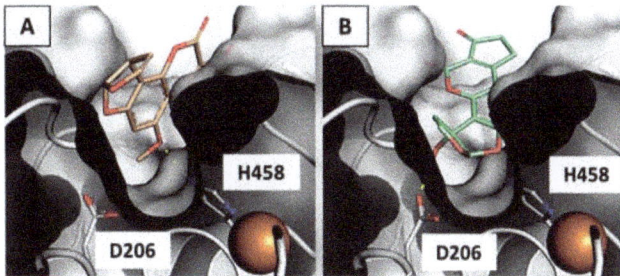

Figure 4. Binding architecture of AFB$_1$ (**A**) and AFM$_1$ (**B**) within the beta isoform. Proteins are represented in with cartoons and cut surfaces, and ligands and amino acids side-chains are represented with sticks. The Cu ions are represented with spheres. **Yellow** dotted lines indicate H-bonds.

The incapability of AFB$_1$ to positively interact with the gamma isoform may be explained by the diverse capability of AFB$_1$ and AFM$_1$ to sink within the catalytic site, where AFM$_1$ was found closer to the bottom of the pocket. Indeed, the additional loop at the entrance of the binding site of the gamma isoform redefined markedly the available space for ligands in the upper portion of the pocket (Figure 5) and prevented the proper accommodation of AFB$_1$.

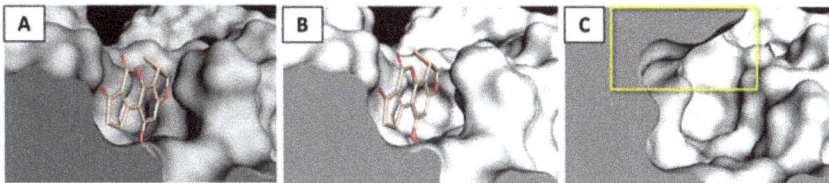

Figure 5. Details of ligand binding site of laccase isoforms. Proteins are represented with cut surfaces and ligands with sticks. The **yellow** box indicates the additional volume in the gamma isoform binding site due to the presence of the extended loop. (**A**) AFB$_1$ within the beta isoform pocket; (**B**) AFB$_1$ within the delta isoform pocket; and (**C**) gamma isoform pocket.

3. Discussion

The mitigation of mycotoxin content in food and feed is undoubtedly a major task for safeguarding health and global trade. Besides the use of good agricultural practices and crop breeding, the biological control of toxin accumulation in the final products by acting at pre- and post-harvest levels and during food processing proved to be an effective strategy [4]. In this framework, the search for enzymes able to convert mycotoxins into non- or low-toxic products has been the object of a growing number of studies. The main advantage of using enzyme-based strategies is the possibility to act at the different stages of food and feed production chains. Indeed, enzymes are currently considered as food/feed additives or agents during the from-field-to-fork pathway, and they are aimed at reducing the carryover and accumulation in the final products [30]. Nonetheless, while the safety and security of genetically modified organisms for health and environment are under a heated scientific debate, the genome engineering by introducing effective enzymes in susceptible hosts or detoxifying microorganisms can be thought as a possible (future) strategy (e.g., ref. [31,32]).

In this scenario, novel strategies for a more effective search of mycotoxin-degrading enzymes should be implemented. The upstream investigation via in silico approaches can be a straightforward choice to significantly extend the explorable space in the early stage providing reliable and informative insights on the enzymes–substrate interaction, also in the view of the evidence-based preliminary hierachization of candidates for the experimental trials.

As a proof of concept, in the present work, we addressed the case study of the interaction between AFs and laccase enzymes from *T. versicolor*. The tertiary structures of gamma and delta isoforms, which are not structurally resolved up to now, were obtained through the homology modeling on the crystallographic structure of the homologous beta isoform (further details are reported in Section 5). On the basis of the fit-for-purpose validation, the 3D receptor modeling turned out to be an effective strategy for estimating the enzyme–substrate interaction for all of the isoforms under investigation. Actually, the reliability of 3D models that were derived from the primary structure of proteins is among the most relevant outcomes pinpointed herein. Indeed, the possibility to use the primary structure for deriving reliable 3D libraries of enzymes may significantly expand the explorable space of research beyond the enzymes that are commercially available or structurally and functionally known so far. In this respect, the advances in structural biology and the whole-genome sequencing data are providing a growing number of high-resolution structures to derive 3D models and a wealth of newly identified sequences of putative proteins for an even more wide-ranging screening.

The degradation by means of laccase enzymes from *T. versicolor* is considered among the most promising and cost-effective enzymatic strategies for the mitigation of AF content [14]. Typically, the development of enzyme-based strategies requires the precise understanding of the specific enzymes–substrate activity, especially in the presence of different isoforms, in order to isolate the most suitable for the purpose. To this end, it is mandatory to gain knowledge on the substrate–enzyme interaction from a structural perspective, thereby understanding in-depth the mechanism of catalysis and deciphering the reasons underlying the formation of degraded products. Taken as a whole, this background of knowledge may reduce the extent of the try-and-error timeframe during the optimization of degradation processes. However, the relative effectiveness of the various laccase isoforms from *T. versicolor* is still unknown, as only unspecified mixtures of the various isoforms have been assessed up to now. Furthermore, the structural organization of AFs within the laccase binding sites has not been elucidated yet. Overall, this scenario eventually makes it hard to refine more effective strategies based on selected enzymes. In this context, the present work investigated the arrangement of AFB_1 and AFM_1 within the binding site of the beta, delta and gamma laccase isoforms with the aim to find out some possible differences in the accommodation of AFs in the catalytic sites. On the basis of our findings, relevant structural differences between gamma and the other isoforms have been pointed out in terms of accessibility of the catalytic site. In fact, AFB_1 were found able to accommodate within beta and delta, but not within gamma due to the extension of a loop lining the binding site. In particular, the re-shaping of the upper portion of the pocket was responsible for the inappropriate accommodation of AFB_1 since the hydrophobic difuran moiety has been unfavorably arranged too close to the hydrophilic space at the bottom of the pocket. Notably, the pathway of entrance toward the catalytic site of enzymes may strongly influence the reaction yield and commonly concur to determine the substrate specificity [33,34]. Specifically, it has been previously reported that mutation at the C-terminus of orthologous fungal laccases may affect the enzyme activity influencing the accessibility to the binding sites [35,36]. Accordingly, this feature can be accounted for gaining more specificity for the laccase enzymes from *T. versicolor* since the low substrate specificity is the major drawback in the application on real food and feed matrices. Indeed, laccases can degrade a wide spectrum of compounds, including a wealth of healthy low molecular weight food and feed constituents (e.g., polyphenols) [37], thus possibly causing an overall pauperization of the treated products. Therefore, the engineering of such a loop might be a straightforward strategy to modulate the substrate specificity of laccases without significantly altering the binding site and preserving as much as possible the catalytic environment. To this end, the gamma isoform can be considered as the most suitable among the various isoforms holding an extended loop at the entrance of the catalytic site, which might be accounted for mutations, while the other isoforms have a direct access.

AFM$_1$ was found able to positively interact within the catalytic sites of all the isoforms herein considered. In this case, the presence of the hydroxyl group on the difuran moiety facilitated a deeper sinking into the pocket that allowed the accommodation also within the gamma isoform. Notably, AFM$_1$ recorded higher scores than AFB$_1$ within all of the catalytic sites due to the formation of an additional hydrogen bond. The gain of enthalpy may be responsible for an increase in substrate–enzyme affinity and, eventually, may cause a higher degrading yield. In this respect, the full degradation of AFM$_1$, by laccases from the edible mushroom *Pleurotus pulmonarius*, was reported by Loi and coworkers [38], thus supporting the strong degradation of hydroxylated forms in accordance with the well-known activity of laccases on poly-phenolic molecules [27].

In the framework of effectively reducing the content of toxicants, the formation of more toxic by-products must be carefully avoided. In this respect, a decreased (geno)toxicity for AFB$_1$ after treatment with laccases has been already reported. The modification at the level of difuran moiety—which is responsible for the toxic action—has been proposed as the possible mechanism [15]. However, neither the exact chemistry of the degrading event nor the structure(s) of reaction product(s) have been elucidated so far. The computed architectures of binding revealed that both AFs structures (including the difuran moieties) were found almost entirely arranged within the proper range of distance from the catalytic hystidine to undergo oxidation. Thus, some other parts of the AFB$_1$ molecules might undergo modifications. Moreover, it is worth mentioning that AFM$_1$ showed some differences in the pattern of interaction with respect to AFB$_1$ as the additional hydroxyl group on the difuran moiety interacted directly with the catalytic core. Accordingly, AFM$_1$ might undergo the same degrading route of phenolic compounds via the electron/proton transfer mechanism [27], which is instead less likely to occur for AFB$_1$ due to the lack of the additional hydroxyl groups. Therefore, differences in terms of sites, type of modification and chemical structures of products between AFB$_1$ and AFM$_1$ cannot be excluded throughout and should be carefully evaluated to rule out the formation of possible toxic byproducts.

4. Conclusions

In conclusion, the in silico simulations proved to be effective analytical tools to investigate the enzyme–substrate interaction, correlating with the affinity of binding. Specifically, the 3D modeling approach provided, for the first time, structural insights on the laccase–aflatoxin interaction, which may be useful for the evidence-based hierarchization of enzymes to be used in further experimental trials.

The modeling of laccase enzyme has been shown to be reliable from previous works (e.g., ref. [39,40]). However, the in silico screening of multiple laccase isoforms in the framework of AF control has been never used before. In more detail, AFM$_1$ was found to be able to arrange positively within all the isoforms herein considered, while AFB$_1$ was able to arrange within beta and delta but not within gamma. Accordingly, the low degradation yield of AFB$_1$ by laccase gamma can be hypothesized. Furthermore, AFB$_1$ and AFM$_1$ showed different binding architectures in arranging within laccase catalytic sites. Therefore, it cannot be excluded that AFB$_1$ and AFM$_1$ undergo different modifications in different regions of the molecule, thus forming degraded products that are chemically different. This might also cause the differential formation of toxic by-products. On the other hand, the effects of the extended loop of the gamma isoform in diversifying the enzyme–substrate recognition can be accounted for developing future strategies to modulate the substrate specificity. Taken together, these results may provide a basic foothold for addressing future studies from a more informed perspective.

Finally, it is worth mentioning the degradation of AFM$_1$, which is the main mammal metabolite. The carryover phenomenon of this metabolite in dairy products poses serious health concerns, and, nowadays, the strategies for reducing the contamination levels in milk and derived products primarily act on reducing the consumption of AFB$_1$-contaminated feed by dairy animals. However, the use of AFM$_1$-degrading enzymes on milk might be an additional strategy to further mitigate the contamination level of AFM$_1$ in milk itself and dairy products.

5. Materials and Methods

5.1. Homology Modeling and Sequence Analysis

The crystallographic structure of the enzyme beta laccase from *Trametes versicolor* (PDB code 1KYA [16]) was the template for the homology modeling of gamma and delta isoforms. The Modeller software, version 9.14 (copyright © 1989–2016 Andrej Sali; maintained by Ben Webb at the Departments of Biopharmaceutical Sciences and Pharmaceutical Chemistry, California Institute for Quantitative Biomedical Research, Mission Bay Byers Hall, University of California San Francisco, San Francisco, CA, USA) was used [41]. The D206A beta isoform model was obtained by manually editing the wild type structure with the software Sybyl, version 8.1 (Certara USA, Princeton, NJ, USA). For sequence analysis, a local pairwise alignment was conducted by using the on-line tool EMBOSS-Water Pairwise Sequence Alignment (EMBL-EBI, Wellcome Genome Campus, Hinxton, Cambridgeshire, UK; http://www.ebi.ac.uk) and the Smith–Waterman algorithm was chosen.

5.2. Molecular Modeling

All protein structures and ligands were processed by using the software Sybyl, version 8.1 (Certara USA, Princeton, NJ, USA). All atoms were checked for atom- and bond-type assignments. Amino- and carboxyl-terminal groups were set as protonated and deprotonated, respectively. Hydrogen atoms were computationally added to the protein and energy-minimized using the Powell algorithm with a coverage gradient of ≤ 0.5 kcal $(\text{mol}\cdot\text{Å})^{-1}$ and a maximum of 1500 cycles.

5.3. Pharmacophore Models

The ligand binding site was defined by using the Flapsite tool of the FLAP (Fingerprint for Ligand And Protein) software version 2.0 (Molecular Discovery Ltd., Borehamwood, Hertfordshire, UK; http://www.moldiscovery.com) [42], while the GRID algorithm [43,44] was used to investigate the corresponding pharmacophoric space. The hydrophobic (DRY) probe was used to describe the potential hydrophobic interactions, while the sp2 carbonyl oxygen (O) and the neutral flat amino (N1) probes were used to describe the hydrogen bond acceptor and donor capacity of the target, respectively. All images were obtained using the software PyMol version 1.7 (Schrödinger, New York, NY, USA; http://www.pymol.org).

5.4. Docking Simulations and Re-Scoring Procedures

The coupling of GOLD (Genetic Optimization for Ligand Docking), as docking software, and HINT [45], as rescoring function, was chosen on the basis of previous studies demonstrating the higher reliability of HINT with respect to other scoring functions in estimating the ligand binding free energies and evaluating the protein–ligand complex formation [19,23,46]. Software setting and rescoring procedures reported by Ehrlich were used [47]. In more detail, HINT score provides the evaluation of thermodynamic benefits of protein–ligand interaction, and relates with the $\Delta G°$ of complex formation [22,24,25]. Specifically, the empirical HINT scoring function implicitly considers enthalpic and entropic aspects of protein–ligand interaction using experimental Log Po/w measurements (partition coefficient for 1-octanol/water) as the basis of its force field. Indeed, from a mechanical point of view, the forces that drive the repartition of molecules between the two solvent phases also underline protein–ligand interaction, as well protein–protein interaction or ligand–ligand interaction. The HINT score is the sum of the all inter-atomic contributions from binding, thereby providing an empirical and quantitative estimate of the favors of the host–guest interaction from an atomic point of view. Thus, the higher the score, the more favored is the arrangement of ligands within the binding site [22,24,25,48]. The HINT equation is the following:

$$\text{HINT score} \ = \ \sum_i \sum_j b_{ij} = \sum_i \sum_j \left(a_i S_i a_j S_j T_{ij} R_{ij} + r_{ij} \right),$$

where b_{ij} is the interaction score between atoms i and j, a is the hydrophobic atomic constant, S represents the solvent accessible surface area, T_{ij} is a logic function assuming +1 or −1 values, depending on the nature of the interatomic interaction, and R_{ij} and r_{ij} are functions of the distance between atoms i and j. Further details on the basic theory of HINT can be found in ref. [49–51].

Acknowledgments: We would like to acknowledge Glen E. Kellogg and Gabriele Cruciani for the courtesy of the HINT scoring function and the Flap software, respectively. We would like to also acknowledge Pietro Cozzini for the kind access to all the facilities of the Molecular Modeling Laboratory (Department of Food Science, University of Parma, Italy) and Elia Vighi for the valuable contribution with doing the analysis.

Author Contributions: L.D., G.G. and C.D. conceived and designed the experiments; L.D. performed the experiments; L.D., C.D., M.R. and G.G. contributed equally to analyzing data; and L.D. and C.D. contributed equally to writing the paper.

Conflicts of Interest: The authors declare no conflict of interest.

References

1. Berthiller, F.; Crews, C.; Dall'Asta, C.; Saeger, S.D.; Haesaert, G.; Karlovsky, P.; Oswald, I.P.; Seefelder, W.; Speijers, G.; Stroka, J. Masked mycotoxins: A review. *Mol. Nutr. Food Res.* **2013**, *57*, 165–168. [CrossRef] [PubMed]

2. Dellafiora, L.; Perotti, A.; Galaverna, G.; Buschini, A.M.; Dall'Asta, C. On the masked mycotoxin zearalenone-14-glucoside. Does the mask truly hide? *Toxicon* **2016**, *111*, 139–142. [CrossRef] [PubMed]

3. Wu, F.; Guclu, H. Aflatoxin regulations in a network of global maize trade. *PLoS ONE* **2012**, *7*, e45151. [CrossRef] [PubMed]

4. Jard, G.; Liboz, T.; Mathieu, F.; Guyonvarc'h, A.; Lebrihi, A. Review of mycotoxin reduction in food and feed: From prevention in the field to detoxification by adsorption or transformation. *Food Addit. Contam. Part A Chem. Anal. Control Expo. Risk Assess.* **2011**, *28*, 1590–1609. [CrossRef] [PubMed]

5. Karlovsky, P. Detoxification strategies for mycotoxins in plant breeding. In *Masked Mycotoxins in Food: Formation, Occurrence and Toxicological Relevance*; Dall'Asta, C., Berthiller, F., Eds.; Royal Society of Chemistry: London, UK, 2016.

6. Dellafiora, L.; Paolella, S.; Dall'Asta, C.; Dossena, A.; Cozzini, P.; Galaverna, G. Hybrid in silico/in vitro approach for the identification of angiotensin I converting enzyme inhibitory peptides from Parma dry-cured ham. *J. Agric. Food Chem.* **2015**, *22*, 6366–6375. [CrossRef] [PubMed]

7. Kalyanaraman, C.; Jacobson, M.P. Studying enzyme-substrate specificity in silico: A case study of the *Escherichia coli* glycolysis pathway. *Biochemistry* **2010**, *49*, 4003–4005. [CrossRef] [PubMed]

8. Wu, F.; Groopman, J.D.; Pestka, J.J. Public health impacts of foodborne mycotoxins. *Annu. Rev. Food Sci. Technol.* **2014**, *5*, 351–372. [CrossRef] [PubMed]

9. European Food Safety Authority (EFSA). Opinion of the scientific panel on contaminants in the food chain [contam] related to the potential increase of consumer health risk by a possible increase of the existing maximum levels for aflatoxins in almonds, hazelnuts and pistachios and derived products. *EFSA J.* **2007**, *446*. [CrossRef]

10. Raiola, A.; Tenore, G.C.; Manyes, L.; Meca, G.; Ritieni, A. Risk analysis of main mycotoxins occurring in food for children: An overview. *Food Chem. Toxicol.* **2015**, *84*, 169–180. [CrossRef] [PubMed]

11. Prandini, A.; Tansini, G.; Sigolo, S.; Filippi, L.; Laporta, M.; Piva, G. On the occurrence of aflatoxin M_1 in milk and dairy products. *Food Chem. Toxicol.* **2009**, *47*, 984–991. [CrossRef] [PubMed]

12. Perrone, G.; Gallo, A.; Logrieco, A.F. Biodiversity of *Aspergillus* section *Flavi* in Europe in relation to the management of aflatoxin risk. *Front. Microbiol.* **2014**, *5*, 377. [CrossRef] [PubMed]

13. Mate, D.M.; Alcalde, M. Laccase: A multi-purpose biocatalyst at the forefront of biotechnology. *Microb. Biotechnol.* **2016**. [CrossRef] [PubMed]

14. Scarpari, M.; Bello, C.; Pietricola, C.; Zaccaria, M.; Bertocchi, L.; Angelucci, A.; Ricciardi, M.R.; Scala, V.; Parroni, A.; Fabbri, A.A.; et al. Aflatoxin control in maize by *Trametes versicolor*. *Toxins (Basel)* **2014**, *6*, 3426–3437. [CrossRef] [PubMed]

15. Zeinvand-Lorestani, H.; Sabzevari, O.; Setayesh, N.; Amini, M.; Nili-Ahmadabadi, A.; Faramarzi, M.A. Comparative study of in vitro prooxidative properties and genotoxicity induced by aflatoxin B_1 and its laccase-mediated detoxification products. *Chemosphere* **2015**, *135*, 1–6. [CrossRef] [PubMed]

16. Bertrand, T.; Jolivalt, C.; Briozzo, P.; Caminade, E.; Joly, N.; Madzak, C.; Mouqin, C. Crystal structure of a four-copper laccase complexed with an arylamine: Insights into substrate recognition and correlation with kinetics. *Biochemistry* **2002**, *41*, 7325–7333. [CrossRef] [PubMed]

17. Dellafiora, L.; Mena, P.; Cozzini, P.; Brighenti, F.; Del Rio, D. Modelling the possible bioactivity of ellagitannin-derived metabolites. In silico tools to evaluate their potential xenoestrogenic behavior. *Food Funct.* **2013**, *4*, 1442–1451. [CrossRef] [PubMed]

18. Dellafiora, L.; Mena, P.; Del Rio, D.; Cozzini, P. Modeling the effect of phase II conjugations on topoisomerase I poisoning: Pilot study with luteolin and quercetin. *J. Agric. Food Chem.* **2014**, *62*, 5881–5886. [CrossRef] [PubMed]

19. Dellafiora, L.; Dall'Asta, C.; Cozzini, P. Ergot alkaloids: From witchcraft till in silico analysis. Multi-receptor analysis of ergotamine metabolites. *Toxicol. Rep.* **2015**, *2*, 535–545. [CrossRef]

20. Madzak, C.; Mimmi, M.C.; Caminade, E.; Brault, A.; Baumberger, S.; Briozzo, P.; Mougin, C.; Jolivalt, C. Shifting the optimal pH of activity for a laccase from the fungus *Trametes versicolor* by structure-based mutagenesis. *Protein Eng. Des. Sel.* **2006**, *19*, 77–84. [CrossRef] [PubMed]

21. Cozzini, P.; Dellafiora, L. In silico approach to evaluate molecular interaction between mycotoxins and the estrogen receptors ligand binding domain: A case study on zearalenone and its metabolites. *Toxicol. Lett.* **2012**, *214*, 81–85. [CrossRef] [PubMed]

22. Cozzini, P.; Fornabaio, M.; Marabotti, A.; Abraham, D.J.; Kellogg, G.E.; Mozzarelli, A. Simple, intuitive calculations of free energy of binding for protein-ligand complexes. 1. Models without explicit constrained water. *J. Med. Chem.* **2002**, *45*, 2469–2483. [CrossRef] [PubMed]

23. Dellafiora, L.; Dall'Asta, C.; Cruciani, G.; Galaverna, G.; Cozzini, P. Molecular modelling approach to evaluate poisoning of topoisomerase I by alternariol derivatives. *Food Chem.* **2015**, *189*, 93–101. [CrossRef] [PubMed]

24. Fornabaio, M.; Cozzini, P.; Mozzarelli, A.; Abraham, D.J.; Kellogg, G.E. Simple, intuitive calculations of free energy of binding for protein-ligand complexes. 2. Computational titration and pH effects in molecular models of neuraminidase-inhibitor complexes. *J. Med. Chem.* **2003**, *46*, 4487–4500. [CrossRef] [PubMed]

25. Fornabaio, M.; Spirakis, F.; Mozzarelli, A.; Cozzini, P.; Abraham, D.J.; Kellogg, G.E. Simple, intuitive calculations of free energy of binding for protein-ligand complexes. 3. The free energy contribution of structural water molecules in HIV-1 protease complexes. *J. Med. Chem.* **2004**, *47*, 4507–4516. [CrossRef] [PubMed]

26. Christensen, N.J.; Kepp, K.P. Setting the stage for electron transfer: Molecular basis of ABTS-binding to four laccases from *Trametes versicolor* at variable pH and protein oxidation state. *J. Mol. Catal. B Enzym.* **2014**, *100*, 68–77. [CrossRef]

27. Galli, C.; Madzak, C.; Vadalà, R.; Jolivalt, C.; Gentili, P. Concerted electron/proton transfer mechanism in the oxidation of phenols by laccase. *ChemBioChem* **2013**, *14*, 2500–2505. [CrossRef] [PubMed]

28. Lee, D.; Redfern, O.; Orengo, C. Predicting protein function from sequence and structure. *Nat. Rev. Mol. Cell Biol.* **2007**, *8*, 995–1005. [CrossRef] [PubMed]

29. Enguita, F.J.; Marçal, D.; Martins, L.O.; Grenha, R.; Henriques, A.O.; Lindley, P.F.; Carrondo, M.A. Substrate and dioxygen binding to the endospore coat laccase from *Bacillus subtilis*. *J. Biol. Chem.* **2004**, *279*, 23472–23476. [CrossRef] [PubMed]

30. Hahn, I.; Thamhesl, M.; Apfelthaler, E.; Klingenbrunner, V.; Hametner, C.; Krska, R.; Schatzmayr, G.; Moll, W.-D.; Berthiller, F.; Schwartz-Zimmermann, H.E. Characterisation and determination of metabolites formed by microbial and enzymatic degradation of ergot alkaloids. *World Mycotoxins J.* **2015**, *8*, 393–404. [CrossRef]

31. Igawa, T.; Takahashi-Ando, N.; Ochiai, N.; Ohsato, S.; Shimizu, T.; Kudo, T.; Yamaguchi, I.; Kimura, M. Reduced contamination by the *Fusarium* mycotoxin zearalenone in maize kernels through genetic modification with a detoxification gene. *Appl. Environ. Microbiol.* **2007**, *73*, 1622–1629. [CrossRef] [PubMed]

32. Munkvold, G.P. Cultural and genetic approaches to managing mycotoxins in maize. *Annu. Rev. Phytopathol.* **2003**, *44*, 99–116. [CrossRef] [PubMed]

33. Gerike, U.; Danson, M.J.; Hough, D.W. Cold-active citrate synthase: Mutagenesis of active-site residues. *Protein Eng.* **2001**, *14*, 655–661. [CrossRef] [PubMed]

34. Leferink, N.G.; Antonyuk, S.V.; Houwman, J.A.; Scrutton, N.S.; Eady, R.R.; Hasnain, S.S. Impact of residues remote from the catalytic centre on enzyme catalysis of copper nitrite reductase. *Nat. Commun.* **2014**, *5*, 4395. [CrossRef] [PubMed]

35. Bleve, G.; Lezzi, C.; Spagnolo, S.; Tasco, G.; Tufariello, M.; Casadio, R.; Mita, G.; Rampino, P.; Grieco, F. Role of the C-terminus of *Pleurotus eryngii* Ery4 laccase in determining enzyme structure, catalytic properties and stability. *Protein Eng. Des. Sel.* **2013**, *26*, 1–13. [CrossRef] [PubMed]

36. Andberg, M.; Hakulinen, N.; Auer, S.; Saloheimo, M.; Koivula, A.; Rouvinen, J.; Kruus, K. Essential role of the C-terminus in *Melanocarpus albomyces* laccase for enzyme production, catalytic properties and structure. *FEBS J.* **2009**, *276*, 6285–6300. [CrossRef] [PubMed]

37. Reiss, R.; Ihssen, J.; Richter, M.; Eichhorn, E.; Schilling, B.; Thöny-Meyer, L. Laccase versus laccase-like multi-copper oxidase: A comparative study of similar enzymes with diverse substrate spectra. *PLoS ONE* **2013**, *8*, e65633. [CrossRef] [PubMed]

38. Loi, M.; Fanelli, F.; Zucca, P.; Liuzzi, V.C.; Quintieri, L.; Cimmarusti, M.T.; Monaci, L.; Haidukowski, M.; Logrieco, A.F.; Sanjust, E.; et al. Aflatoxin B_1 and M_1 degradation by lac2 from *Pleurotus pulmonarius* and redox mediators. *Toxins (Basel)* **2016**, *8*, E245. [CrossRef] [PubMed]

39. Cambria, M.T.; Di Marino, D.; Falconi, M.; Garavaglia, S.; Cambria, A. Docking simulation and competitive experiments validate the interaction between the 2,5-xylidine inhibitor and *Rigidoporus lignosus* laccase. *J. Biomol. Struct. Dyn.* **2010**, *27*, 501–510. [CrossRef] [PubMed]

40. Suresh, P.S.; Kumar, A.; Kumar, R.; Singh, V.P. An in silico [correction of insilico] approach to bioremediation: Laccase as a case study. *J. Mol. Graph. Model.* **2008**, *26*, 845–849. [CrossRef] [PubMed]

41. Sali, A.; Blundell, T.L. Comparative protein modelling by satisfaction of spatial restraints. *J. Mol. Biol.* **1993**, *234*, 779–815. [CrossRef] [PubMed]

42. Baroni, M.; Cruciani, G.; Sciabola, S.; Perruccio, F.; Mason, J.S. A common reference framework for analyzing/comparing proteins and ligands. Fingerprints for ligands and proteins (FLAP): Theory and application. *J. Chem. Inf. Model.* **2007**, *47*, 279–294. [CrossRef] [PubMed]

43. Goodford, P.J. A computational procedure for determining energetically favourable binding sites on biologically important macromolecules. *J. Med. Chem.* **1985**, *28*, 849–857. [CrossRef] [PubMed]

44. Carosati, E.; Sciabola, S.; Cruciani, G. Hydrogen bonding interactions of covalently bonded fluorine atoms: From crystallographic data to a new angular function in the GRID force field. *J. Med. Chem.* **2004**, *47*, 5114–5125. [CrossRef] [PubMed]

45. Kellogg, E.G.; Abraham, D.J. Hydrophobicity: Is $LogP_{o/w}$ more than the sum of its parts? *Eur. J. Med. Chem.* **2000**, *37*, 651–661. [CrossRef]

46. Dellafiora, L.; Galaverna, G.; Dall'Asta, C.; Cozzini, P. Hazard identification of cis/trans-zearalenone through the looking-glass. *Food Chem. Toxicol.* **2015**, *86*, 65–71. [CrossRef] [PubMed]

47. Ehrlich, V.A.; Dellafiora, L.; Mollergues, J.; Dall'Asta, C.; Serrant, P.; Marin-Kuan, M.; Lo Piparo, E.; Schilter, B.; Cozzini, P. Hazard assessment through hybrid in vitro/in silico approach: The case of zearalenone. *ALTEX* **2015**, *32*, 275–286. [CrossRef] [PubMed]

48. Marabotti, A.; Spyrakis, F.; Facchiano, A.; Cozzini, P.; Alberti, S.; Kellogg, G.E.; Mozzarelli, A. Energy-based prediction of amino acid-nucleotide base recognition. *J. Comput. Chem.* **2008**, *29*, 1955–1969. [CrossRef] [PubMed]

49. Kellogg, G.E.; Burnett, J.C.; Abraham, D.J. Very empirical treatment of solvation and entropy: A force field derived from $LogP_{o/w}$. *J. Comput. Aided Mol. Des.* **2001**, *15*, 381–393. [CrossRef] [PubMed]

50. Kellogg, G.E.; Fornabaio, M.; Spyrakis, F.; Lodola, A.; Cozzini, P.; Mozzarelli, A.; Abraham, D.J. Getting it right: Modeling of pH, solvent and "Nearly" Everything else in virtual screening of biological targets. *J. Mol. Graph. Model.* **2004**, *22*, 479–486. [CrossRef] [PubMed]

51. Sarkar, A.; Kellogg, G.E. Hydrophobicity-shake flasks, protein folding and drug discovery. *Curr. Top. Med. Chem.* **2010**, *10*, 67–83. [CrossRef] [PubMed]

toxins

MDPI

Article

Novel Aflatoxin-Degrading Enzyme from *Bacillus shackletonii* L7

Liang Xu [1,2,†], Mohamed Farah Eisa Ahmed [1,†], Lancine Sangare [1], Yueju Zhao [1,2], Jonathan Nimal Selvaraj [1,‡], Fuguo Xing [1,2], Yan Wang [1,2], Hongping Yang [3] and Yang Liu [1,2,*]

[1] Institute of Food Science and Technology, Chinese Academy of Agricultural Sciences, 1 Nongda South Road, Xibeiwang Town, Haidian District, Beijing 100193, China; xuliang19824@163.com (L.X.); elfarahy89@hotmail.com (M.F.E.A.); lancin.sangar@gmail.com (L.S.); zhaoyueju@caas.cn (Y.Z.); sjonnim@gmail.com (J.N.S.); xingfuguo@caas.cn (F.X.); awangyan@126.com (Y.W.)

[2] Key Laboratory of Agro-products Processing, Ministry of Agriculture, 1 Nongda South Road, Xibeiwang Town, Haidian District, Beijing 100193, China

[3] Shenyang Institute of Engineering, No.18 Puchang Road, Shenbei New District, Shenyang 110136, China; yanghp@sie.edu.cn

* Correspondence: liuyang01@caas.cn; Tel./Fax: +86-10-6281-5874

† These authors contributed equally to this work.

‡ Presently address: College of Life Sciences, Hubei University, Wuhan 430062, China.

Academic Editor: Ting Zhou

Received: 22 September 2016; Accepted: 11 January 2017; Published: 14 January 2017

Abstract: Food and feed contamination by aflatoxin (AF)B_1 has adverse economic and health consequences. AFB_1 degradation by microorganisms or microbial enzymes provides a promising preventive measure. To this end, the present study tested 43 bacterial isolates collected from maize, rice, and soil samples for AFB_1-reducing activity. The higher activity was detected in isolate L7, which was identified as *Bacillus shackletonii*. L7 reduced AFB_1, AFB_2, and AFM_1 levels by 92.1%, 84.1%, and 90.4%, respectively, after 72 h at 37 °C. The L7 culture supernatant degraded more AFB_1 than viable cells and cell extracts; and the degradation activity was reduced from 77.9% to 15.3% in the presence of proteinase K and sodium dodecyl sulphate. A thermostable enzyme purified from the boiled supernatant was designated as Bacillus aflatoxin-degrading enzyme (BADE). An overall 9.55-fold purification of BADE with a recovery of 39.92% and an activity of 3.85×10^3 U·mg^{-1} was obtained using chromatography on DEAE-Sepharose. BADE had an estimated molecular mass of 22 kDa and exhibited the highest activity at 70 °C and pH 8.0, which was enhanced by Cu^{2+} and inhibited by Zn^{2+}, Mn^{2+}, Mg^{2+}, and Li^+. BADE is the major protein involved in AFB_1 detoxification. This is the first report of a BADE isolated from *B. shackletonii*, which has potential applications in the detoxification of aflatoxins during food and feed processing.

Keywords: aflatoxin B_1; aflatoxin-degrading enzyme; biodegradation; *Bacillus shackletonii*; purification

1. Introduction

Aflatoxins (AFs) are secondary metabolites produced by *Aspergillus* species, mainly *A. flavus* and *A. parasiticus* [1], that are carcinogenic, teratogenic, and hepatotoxic to both humans and animals, which has grave economic and health consequences [2–4]. AFB_1, the most abundantly produced isotype, is highly toxic [5,6]. Cytochrome P450-associated enzymes in animal liver can metabolize AFB_1 to AFM_1, which is detected in the milk of dairy cows that consume AFB_1-infected feeds [7,8].

Methods for decontamination and reduction of AFs in food sources have been widely investigated. AF inactivation by physical and chemical means is not highly effective or economically feasible [6]; a promising alternative approach is biological detoxification of AF-contaminated food and feed. AFB_1 biodegradation by fungi and bacteria or their secondary metabolites or enzymes has been widely reported using *Nocardia corynebacterioides* (formerly *Lavobacterium aurantiacum*) [9,10], *Armillariella tabescens* [11,12], *Pleurotus ostreatus* [13], *Bacillus licheniformis* [14], *Mycobacterium fluoranthenivorans* [15,16], *Rhodococcus erythropolis* [16–18], *Stenotrophomonas maltophilia* [19], *Myxococcus fulvus* [20], and *Bacillus subtilis* ANSB060 [21]. This approach has the advantage of being highly target-specific, effective, and environmentally safe, as the decontaminated food or feed products can be subsequently used [20].

AF-degrading enzymes have been isolated from a variety of microorganisms. Recently, an aflatoxin oxidase from *A. tabescens* [12] and manganese peroxidase from the white-rot fungus *Phanerochaete sordida* YK-624 [22] were shown to have AFB_1-degrading ability. It was also reported that a recombinant *Trametes versicolor* laccase enzyme expressed in *Aspergillus niger* degrades AFB_1 [23]. Nine *Mycobacterium smegmatis* enzymes belonging to two $F_{420}H_2$-dependent reductase families were found to catalyze the reduction of the α, β-unsaturated ester moiety of AFs by spontaneous hydrolysis [24]. However, most of these enzymes are intracellular and have been isolated from fungi. The process of crushing mycelia to recover enzymes can compromise their activity, preventing their large-scale production. This problem can be circumvented by obtaining AF-degrading enzymes from bacteria.

To this end, a screening method was developed in the present study to isolate AFB_1-degrading microbes from soils and contaminated kernels using coumarin medium. Several new AFB_1-degrading bacterial strains were thus identified; among them, *Bacillus shackletonii* strain L7 showed the strongest activity. We evaluated the degradation efficiency of strain L7 against various AFs and purified and characterized a thermostable enzyme named Bacillus aflatoxin-degrading enzyme (BADE) responsible for AFB_1 degradation activity. We also analyzed the genotoxicity of AFB_1 degradation products treated with proteins from strain L7.

2. Results

2.1. Screening for AFB_1-Degrading Microorganisms

In total, 43 single-colony bacterial isolates were obtained from 247 samples collected from different sources, all of which were able to reduce AFB_1 to varying degrees in nutrient broth (NB) after incubation for 72 h at 37 °C (Table S1). Eight of the isolates (belonging to the genera *Stenotrophomonas*, *Pseudomonas*, *Arthrobacter*, *Bacillus*, and the family *Flavobacteriaceae*) showed more than 70% AFB_1 degradation in the medium, with the highest value (71.7%) observed for isolate L7.

2.2. Identification of Isolate L7

Isolate L7 was a rod-shaped Gram-positive bacterium that could grow under extreme conditions (55 °C and pH 9.0). It utilized simple sugars like glucose, maltose, and sucrose as a sole carbon source, and hydrolyzed casein but not gelatin, amylum, butyrin, or Tween 80 (Table S2). Based on a 16S rRNA gene sequence analysis, L7 was identified as *Bacillus shackletonii* strain LMG 18435 (99% sequence similarity). This is the first report of a bacterium of this genus exhibiting mycotoxin-degrading ability. The partial 16S rRNA gene sequence of L7 was submitted to GenBank (access. no. KX364157), and the strain was deposited at the China General Microbiological Culture Collection Center (CGMCC8868).

2.3. Degradation of AFs by Isolate L7

The degradation activity of isolate L7 towards AFB_1, AFB_2, AFG_1, AFG_2, and AFM_1 was 92.1%, 84.1%, 63.6%, 76.1%, and 90.4%, respectively, when cultured in NB at 37 °C for 72 h (Figure 1).

The culture supernatant of isolate L7 was more effective than viable cells and cell extracts in degrading higher concentrations of AFB_1 after 72 h (77.9% vs. 28.6% and 17.2%, respectively; $p < 0.05$) (Figure 2). The AFB_1-degrading ability of the supernatant declined to 52.6% and 15.3% upon treatment with proteinase K without and with sodium dodecyl sulphate (SDS), respectively (Figure 3). These results suggest that proteins/enzymes secreted by L7 are involved in AFB_1 degradation.

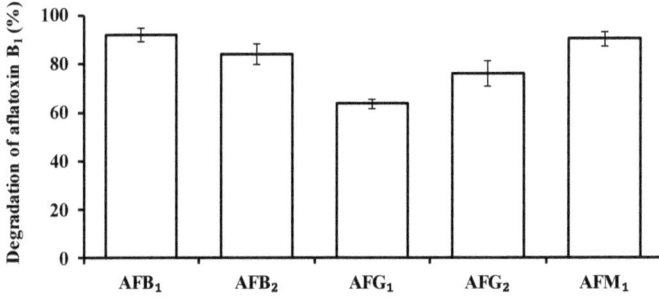

Figure 1. AF degradation by isolate L7. Values represent the means of three replicates and their standard errors.

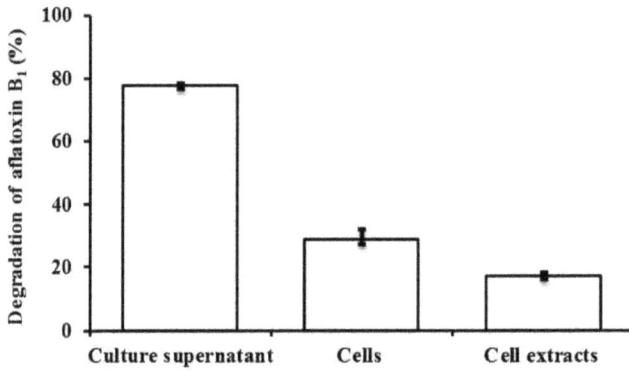

Figure 2. AFB$_1$ degradation by L7 culture supernatant, viable cells, and cell extracts after 72 h of incubation. Values represent the means of three replicates and their standard errors.

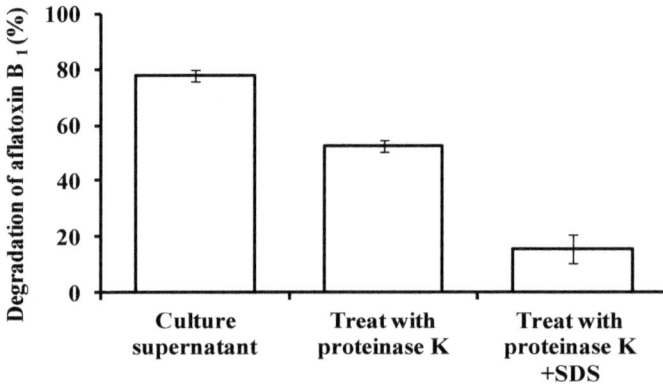

Figure 3. Effect of proteinase K and SDS on AFB$_1$ degradation by L7 culture supernatant. Values represent means of three replicates and their standard errors.

AFB$_1$ degradation by the culture supernatant of isolate L7 proceeded relatively rapidly and continuously, with 40.9, 77.9, and 90.3% reduction observed in the first 12 h and after 72 h and 5 days, respectively (Figure 4).

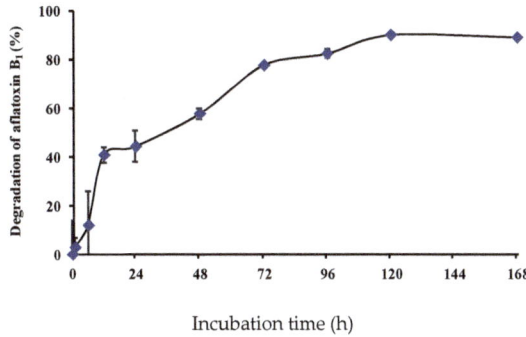

Figure 4. Dynamics of AFB$_1$ degradation by isolate L7 culture supernatant at indicated time points. Values represent the means of three replicates and their standard errors.

2.4. Evaluation of Genotoxicity

The genotoxicity of the samples was evaluated using the SOS Chromotest, with the results expressed as induction factor ± standard deviation (Table 1). The induction factors of degraded samples incubated in the dark at 37 °C for 72 h with the culture supernatant of isolate L7 were in line with the negative control but not the positive control. At higher concentrations (25, 50, and 100%), the differences between the positive control and degraded samples were significant ($p < 0.05$). These results indicate that AFB$_1$ treated with the culture supernatant of isolate L7 had no genotoxicity in the presence of the S9 crude enzyme preparation.

Table 1. Induction factors produced by samples in the SOS Chromotest.

Induction Factor ± Standard Deviation				
Concentration (%)	2-Aminoanthracene [a]	Positive Control [b]	Degraded Sample [c]	Negative Control [d]
100	3.49 ± 0.08	2.53 ± 0.06	1.06 ± 0.00	1.09 ± 0.05
50	3.02 ± 0.06	1.73 ± 0.14	1.02 ± 0.03	1.12 ± 0.10
25	2.29 ± 0.09	1.08 ± 0.06	1.00 ± 0.01	1.14 ± 0.09
12.5	1.78 ± 0.05	1.18 ± 0.09	1.02 ± 0.01	1.13 ± 0.04
6.25	1.42 ± 0.12	1.04 ± 0.05	0.96 ± 0.06	0.98 ± 0.09
3.125	1.45 ± 0.07	1.07 ± 0.08	1.07 ± 0.03	1.01 ± 0.11

[a] Six two-fold dilutions of 2-aminoanthracene (initial concentration = 100 µg/mL) were prepared in 10% dimethyl sulphoxide/saline. A 10-µL volume of diluted 2-aminoanthracene was used as a positive S9 control. [b] A 0.06-mL volume of AFB$_1$ solution (50 mg/L) was added to 1.44 mL of NB to obtain a final concentration of 2 mg/L. After incubation in the dark at 37 °C for 72 h, six two-fold dilutions were prepared, and a 10-µL volume was used as a positive control. [c] A 0.06-mL volume of AFB$_1$ solution (50 mg/L) was added to 1.44 mL of culture supernatant of isolate L7 to obtain a final concentration of 2 mg/L. After incubation in the dark at 37 °C for 72 h (AFB$_1$ degradation ratio = 77.9% ± 2.3%), six two-fold dilutions were prepared, and a 10-µL volume was used as the degraded sample. [d] A 0.06-mL volume of methanol was added to 1.44 mL of culture supernatant of isolate L7. After incubation in the dark at 37 °C for 72 h, six two-fold dilutions were prepared, and a 10-µL volume was used as a negative control.

2.5. Preliminary Characterization of B. shackletonii BADE

The AFB$_1$-degrading ability of the culture supernatant after incubation for 24 h increased from 47.6% to 70.1% upon concentration by 50-fold by ultrafiltration and was positively correlated with protein concentration (tab:toxins-09-00036-t002). When the ultrafiltered culture supernatant was heat-treated (incubated in a boiling water bath for 10 min), there was no decrease in degradation activity (76.7%). These results imply that heat-stable proteins or enzymes in the L7 culture supernatant are responsible for AFB$_1$ degradation.

Table 2. Degradation of AFB$_1$ by L7 culture supernatant after 24 h of incubation.

Supernatant Conditions	Protein Concentration (mg/mL)	Degradation (%)
Culture supernatant	0.13 ± 0.03	47.58 ± 1.09
Culture supernatant with ultrafiltration [a]	0.66 ± 0.04	70.12 ± 0.69
Boiled culture supernatant with ultrafiltration [a]	0.10 ± 0.02	76.67 ± 0.85

[a] The culture supernatant was concentrated by ultrafiltration with a cut-off molecular weight of 3 kDa.

2.6. BADE Purification

To isolate the BADE responsible for AFB$_1$ degradation, L7 culture supernatant was heat-treated (incubation in a boiling water bath for 10 min), concentrated by ultrafiltration with a cut-off molecular weight of 3 kDa, and subjected to diethylaminoethanol (DEAE)-sepharose ion-exchange chromatography. The maximum absorption of BADE was observed between 240 and 300 nm. The protein concentration of each fraction was monitored at 280 nm during enzyme purification. The separation profile of the enzyme showed one flow-through fraction (peak 1) and two eluted protein peaks (peaks 2 and 3), and different fractions were collected and dialyzed for 24 h against 20 mmol·L^{-1} phosphate buffer (pH 7.4). Compared to the control (AFB$_1$ with 20 mmol·L^{-1} phosphate buffer (pH 7.4)), peak 2 (40–50 mL) was associated with the highest AFB$_1$-degrading activity (3.85 × 10^3 U·mg^{-1}) (Figure 5), followed by peak 1 (0–20 mL) (2.63 × 10^3 U·mg^{-1}), peak 2 (50–80 mL) (2.39 × 10^3 U·mg^{-1}), peak 3 (80–95 mL) (1.73 × 10^3 U·mg^{-1}) and peak 3 (95–100 mL) ($\approx$ 0.00 U·mg^{-1}). The specific activity after each step of purification is shown in Table 3. BADE (40–50 mL, peak 2) was purified 9.55-fold with a 39.92% yield, and had a molecular mass of 22 kDa as determined by SDS-polyacrylamide gel electrophoresis (PAGE) (Figure 6).

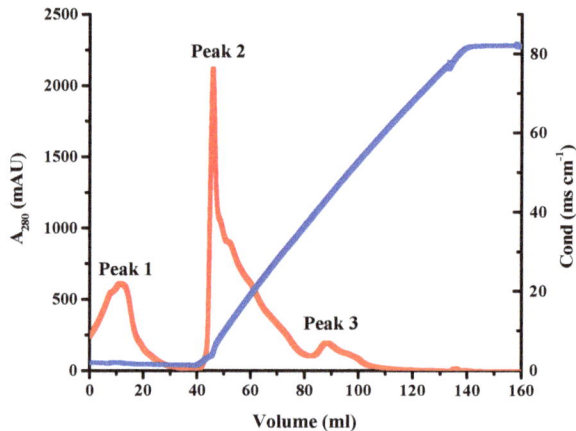

Figure 5. Ion-exchange chromatography using a DEAE-sepharose column. Red and blue lines represent the absorption of the protein at 280 nm and conductance value, respectively. Peak 1 corresponds to the flow-through fraction, and peaks 2 and 3 to the eluted protein.

Table 3. Summary of BADE purification.

Purification Step	Total Protein (mg) × 10^{-3}	Total AFB$_1$ Degradation Activity (U) *	Specific Activity (U·mg^{-1}) × 10^2 *	Purification (Fold)	Yield (%)
Culture filtrate	96.55	38.95	4.03	1.00	100.00
Peak 1 (0–20 mL)	1.35	3.55	26.30	6.53	9.10
Peak 2 (40–50 mL)	4.04	15.55	38.49	9.55	39.92
Peak 2 (50–80 mL)	6.37	15.21	23.88	5.93	39.05
Peak 3 (80–95 mL)	3.26	5.63	17.27	4.29	14.45
Peak 3 (95–100 mL)	0.20	5.5×10^{-5}	0.0028	0.00069	0.00014

* One unit (U) of enzyme activity was defined as the amount of enzyme required to decrease the amount of AFB$_1$ by 1 ng in 72 h at 37 °C.

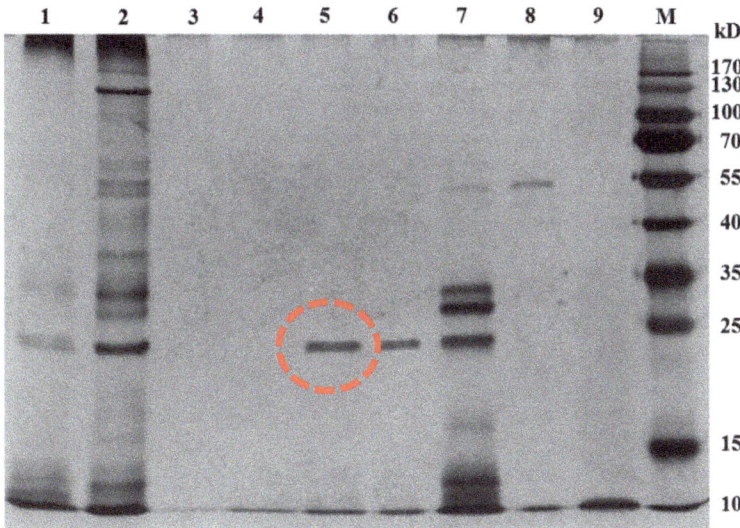

Figure 6. SDS-PAGE gels of proteins obtained during BADE purification. M, protein marker; lane 1, crude protein in culture supernatant of isolate L7 after heat treatment for 10 min; lane 2, crude protein in culture supernatant; lanes 3 and 4, protein in the first flow-through fraction (0–20 mL; peak 1) obtained by ion-exchange chromatography; lanes 5 and 6, protein in the first eluted fraction (40–50 mL, peak 2); lane 7, protein in the peak 2 fraction (50–80 mL); lane 8, protein in the second eluted fraction (80–95 mL, peak 3); lane 9, protein in the peak 3 fraction (95–100 mL).

2.7. Effect of pH on BADE Activity

The extent of AFB$_1$ degradation by BADE (40–50 mL, peak 2) from isolate L7 varied as a function of pH (Figure 7). There was no significant difference in AFB$_1$ degradation among pH 4.0, 5.0, and 7.0 without BADE, and there was little degradation. Compared to the control (AFB$_1$ without BADE), the degradation ratio of AFB$_1$ in the presence of BADE increased significantly at pH 4.0 (10.40%), pH 5.0 (14.30%), and pH 7.0 (30.73%). The degradation ratio of AFB$_1$ without BADE was 13.52% at pH 8.0 (shown in Figure 7), and the degradation ratio of AFB$_1$ in the presence of BADE increased significantly from 13.52% to 44.70% at pH 8.0. The degradation ratio of AFB$_1$ treated with BADE increased following the pH increase, and the degradation rate at pH 8.0 was the highest (31.18%) among different pH values, indicating that pH 8.0 is the optimal pH for BADE activity.

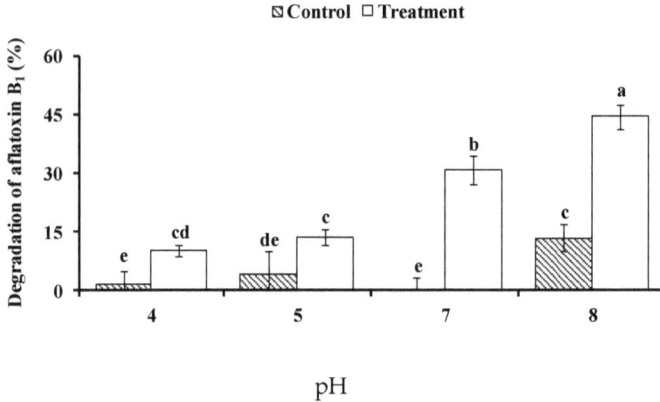

Figure 7. Effect of pH on AFB$_1$ degradation by BADE from isolate L7. Values represent the means of three replicates and their standard errors. Control, AFB$_1$ without BADE; treatment, AFB$_1$ with BADE.

2.8. Effect of Metal Ions on BADE Activity

The effect of various metal ions on BADE (40–50 mL, peak 2) activity was evaluated. There was no significant difference in AFB$_1$ degradation among the no-ion condition, Zn^{2+}, Mn^{2+}, Mg^{2+}, Cu^{2+}, and Li$^+$ in the absence of BADE, and there was little degradation. Compared to the no-ion condition in the presence of BADE, BADE activity was inhibited by Zn^{2+}, Mn^{2+}, Mg^{2+}, and Li$^+$ (10 mmol·L^{-1}) but was enhanced in the presence of Cu^{2+} (Figure 8).

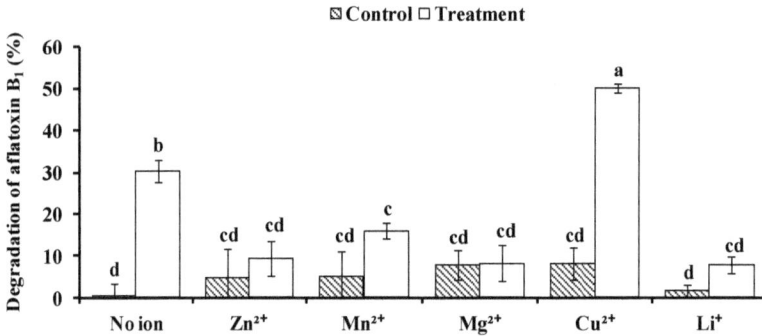

Figure 8. Effect of ions on AFB$_1$ degradation by BADE from isolate L7. Values represent the means of three replicates and their standard errors. Control, AFB$_1$ without BADE; treatment, AFB$_1$ with BADE.

2.9. Effect of Temperature on BADE Activity

The AFB$_1$-degrading activity of BADE (40–50 mL, peak 2) from isolate L7 varied as a function of temperature (Figure 9). There was no significant difference in AFB$_1$ degradation among 16, 28, 37, 55, and 70 °C without BADE, and there was little degradation. Compared to the control (AFB$_1$ without BADE), the degradation ratio of AFB$_1$ in the presence of BADE increased significantly at 16 °C (19.50%), 28 °C (24.40%), 37 °C (34.27%), 55 °C (27.01%), and 70 °C (47.51%). The degradation ratio of AFB$_1$ treated with BADE increased following the increase in temperature, and the degradation rate at 70 °C was the highest (47.51%) among different temperatures, indicating that 70 °C is the optimal temperature for BADE activity.

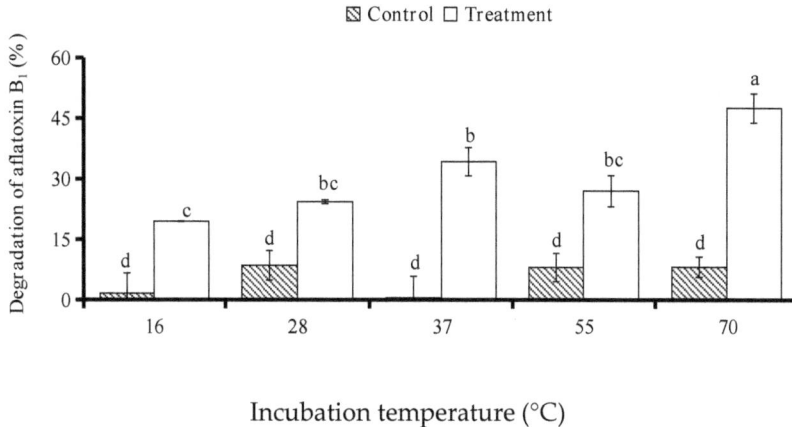

Figure 9. Effect of temperature on AFB$_1$ degradation by BADE from isolate L7. The mixtures were incubated at indicated temperatures for 72 h. Values represent the means of three replicates and their standard errors. Control, AFB$_1$ without BADE; treatment, AFB$_1$ with BADE.

3. Discussion

This is the first study reporting that *B. shackletonii* has AFB$_1$-degrading activity. *B. shackletonii* strain LMG 18435T, a strictly aerobic bacterium isolated from the eastern lava flow of northern Candlemas Island in the South Sandwich archipelago, was reported to be capable of decomposing large molecules into organic acids [25]. In addition, the thermophilic *B. shackletonii* strain K5 isolated from a biotrickling filter of a coal-fired power plant in Guangzhou, China showed a strong capacity for producing polyhydroxybutyrate [26].

We demonstrated that AFB$_1$ in solution could be effectively degraded by the culture supernatant of *B. shackletonii* L7. This was similar to previous findings demonstrating that extracellular extracts from *R. erythropolis* liquid cultures degraded AFB$_1$, with only 33.2% residual toxin after 72 h [17]. Incubation with 5×10^{10} resting cells/mL of *Fusarium aurantiacum* for 4 h completely removed AFM$_1$ (8 µg·mL^{-1}) [27]. Similar to *R. erythropolis* [17] and *S. maltophilia* 35-3 [19], degradation of AFB$_1$ by the culture supernatant produced without pre-exposure to AFB$_1$ showed the constitutive activity of *B. shackletonii* L7. The AFB$_1$-degrading activity of the culture supernatant increased from 47.6% to 70.1% after ultrafiltration, and was decreased from 77.9% to 15.3% upon treatment with proteinase K and SDS, similar to the trends observed for the culture supernatants of *R. erythropolis*, *M. fulvus* ANSM068, and *S. maltophilia* [17,19,20]. These findings imply that the AFB$_1$-degrading enzyme from *B. shackletonii* is an extracellular enzyme, in contrast to intracellular AFB$_1$-degrading enzymes isolated from fungi [11,13].

The increase in AFB$_1$ detoxification over time indicated that the AFB$_1$-degrading enzyme from *B. shackletonii* was stable at 37 °C for one week. Similarly, a 66.8% reduction in AFB$_1$ was observed over 72 h in the presence of *R. erythropolis* culture supernatant [28], whereas the AFB$_1$ level was reduced by 70%–80% within 36 h and AFB$_1$ was completely degraded after 72 h by liquid cultures of *Mycobacterium* strain FA4 [22]. AFB$_1$ was quickly removed from liquid by binding to lactic acid bacteria, although some of the toxin was released over a prolonged period of incubation [10,19]. Binding may not be important in AFB$_1$ reduction, given the continuous increase in detoxification by strain L7 over time. To further verify the role of binding (or adsorption) in the observed reduction in aflatoxin concentration, bacterial pellets of *B. shackletonii* L7 cultured for 24 h were washed twice with and resuspended in phosphate buffer (20 mM, pH 7.0). The resuspended liquid (0.026 g/mL) was sterilized at 121 °C for 20 min and was used for the AFB$_1$ degradation test, using phosphate buffer rather than the resuspended liquid as a control. AFB_1 concentration did not decrease relative to the control upon treatment with the resuspended liquid; the AFB_1 degradation rate was 0% (data not shown). This indicates that binding does not play a role in AFB_1 reduction.

Methods for purifying enzymes from microorganisms should be selected based on the activity and yield of the active component. In this study, crude AFB_1-degrading enzymes were purified from the *B. shackletonii* culture supernatant by ion-exchange chromatography on a DEAE-sepharose column. AFB_1-degrading enzymes have been purified from the fungi *A. tabescens* by ion-exchange chromatography and chromatofocusing chromatography [11], and from *P. ostreatus* by chromatography using DEAE-sepharose and phenyl sepharose columns [13]. In our study, BADE (40–50 mL, peak 2) with an estimated molecular mass of 22 kDa and the highest AFB_1-degrading activity (3.85×10^3 U·mg^{-1}) was purified 9.55-fold from *B. shackletonii* cultures with a 39.92% yield, implying a low processing cost [29] which, along with the reasonably high overall yield [30], makes it suitable for large-scale production.

B. shackletonii BADE was active at a broad range of pH values (pH 4.0–8.0) and temperatures (16 °C–70 °C), which is advantageous for AF degradation in the digestive tracts of animals. Importantly, there was no decrease in enzymatic activity when the culture supernatant of strain L7 was subjected to heat treatment. This was in contrast with a previous study in which AFB_1-degrading activity in the culture supernatant of *S. maltophilia* 35-3 was abolished by heat treatment [19]. These results imply that *B. shackletonii* BADE is heat-stable, similar to the enzyme in thermophilic *B. shackletonii* K5 [26].

The AFB_1-degrading activity of *B. shackletonii* BADE was influenced by the presence of metal ions: the activity was enhanced by Cu^{2+} (10 mM) and inhibited by Zn^{2+}, Mn^{2+}, Mg^{2+}, and Li^+ (10 mM), in agreement with earlier reports [19,31]. We speculate that BADE is an oxidoreductase that utilizes Cu^{2+} as an activator. Cu^{2+} may participate in redox reactions in electron transport, transferring an oxygen atom to the AFB_1 substrate; oxidized AFB_1 would then be hydrolyzed into non-toxic products [32]. In contrast, 10 mM Zn^{2+} inhibited AFB_1 degradation by BADE, which was similar to the one observed in *F. aurantiacum* [33]. Zn^{2+} may cause a conformational change in the enzyme that leads to its inactivation or decreases its affinity for AFB_1 [28].

We evaluated the genotoxicity of AFB_1 degradation products with the SOS Chromotest. The genotoxicity of AFB_1 was significantly reduced by treatment with proteins or enzymes in *B. shackletonii* L7 culture supernatant in the presence of S9. It was previously demonstrated that treatment of AFB_1 with enzymes altered the double bond of the furofuran ring, with consequent changes in the fluorescence and mutagenicity of the molecule [23]. In this study, BADE from strain L7 played a major role in AFB_1 detoxification; we also showed that the chemical properties of AFB_1 degradation products differed from those of the parent molecule, although the detailed mechanism of AFB_1 degradation by *B. shackletonii* BADE has yet to be described.

To date, BADE from *B. shackletonii* is the only AF-degrading enzyme purified from a bacterium. Enzyme purification is the limiting factor in the industrial-scale production of AFs-degrading enzymes. Challenges associated with obtaining BADE by *B. shackletonii* fermentation include contamination of bacterial cultures and low yields. Nonetheless, our results suggest that treatment of food and feed with recombinant *B. shackletonii* BADE can potentially eliminate mycotoxin contamination.

4. Conclusions

We isolated *B. shackletonii* strain L7, which could degrade AFB_1, AFB_2, and AFM_1. A presumed extracellular BADE was purified from strain L7 showed the highest activity at 70 °C and pH 8.0 in the presence of 10 mmol·L^{-1} Cu^{2+}, and was the main protein involved in the detoxification of AFB_1. These findings indicate that BADE from *B. shackletonii* L7 is a promising new agent for AF biodegradation. The cloning and expression of the BADE-encoding gene are currently underway in our laboratory.

5. Materials and Methods

5.1. Chemicals and Media

Standard solutions of AFB_1, AFB_2, AFG_1, AFG_2, and AFM_1 were purchased from Sigma-Aldrich (St. Louis, MO, USA). Stock solutions (500 µg/L) were prepared by diluting the standard solutions with methanol. Coumarin medium (CM) was prepared as previously described [19]. NB was used for bacterial culture.

5.2. Collection of Bacteria

A total of 247 samples were collected from farm soils, maize, and rice in different provinces in China in June 2013. Bacterial strains with the ability to degrade AFB_1 were isolated on CM medium [19]; selected colonies were further tested for AFB_1-degrading activity.

5.3. Analysis of AFs Degradation

Isolates were cultured in NB for 12 h. For inoculation, a 12-h culture broth (1 mL) was transferred to NB (20 mL) and cultured at 37 °C with agitation for 24 h in a shaking incubator. AFB_1 solution (500 µg/L; 0.1 mL) was added to 0.4 mL of microbial culture to obtain a final concentration of 100 µg/L. The detoxification test was performed in the dark at 37 °C without shaking for 72 h; sterile NB containing AFB_1 at a final concentration of 100 µg/L was used as a control.

The reaction mixtures were extracted three times with chloroform according to a standard protocol [34]. The chloroform was evaporated under nitrogen gas at 60 °C and dried extracts were dissolved in 50% methanol in water (1:1, v/v), and the solution was passed through a sterile 0.22-µm pore size filter (Merck Millipore, Darmstadt, Germany) and analyzed by high-performance liquid chromatography using a C18 column (150 × 4.6 mm, 5 µm; Agilent Technologies, Santa Clara, CA, USA) and methanol:water (1:1, v/v) as the mobile phase at a flow rate of 1 mL/min. AFB_1 was derived by a photochemical reactor (Waters, Milford, MA, USA) and detected with a fluorescence detector. Excitation and detection wavelengths were set at 360 and 440 nm, respectively. Percent AFB_1 degradation was calculated using the formula $(1 - AFB_1$ peak area in treatment/AFB_1 peak area in control$) \times 100\%$. AFB_2, AFG_1, AFG_2, and AFM_1 degradations were analyzed in a similar manner.

5.4. Identification of Isolate L7

Biochemical and physiological analyses of isolate L7 were performed following standard protocols [35]. Genomic DNA of strain L7 was extracted using the TIANamp Bacterial DNA kit (Tiangen Biotech, Beijing, China) according to the manufacturer's instructions. The 16S rRNA gene was amplified using a universal primer set consisting of 27F (5'-GAGAGTTTGATCCTGGCTCAG-3') and 1492R (5'-TACCTTGTTACGACTT-3') [36] and then sequenced. Similar sequences were identified by a BLAST search of the NCBI nucleotide database [37]. Strain L7 was preserved at −80 °C until use.

5.5. Degradation of AFB₁ by Cell-Free Supernatant, Bacterial Cells, and Intracellular Cell Extracts of Strain L7

AFB_1 degradation by cell-free supernatant, bacterial cells, and intracellular cell extracts of strain L7 was evaluated as previously described [16,19,38]. Briefly, 5 mL of fresh NB was inoculated with strain L7 at 37 °C with agitation at 180 rpm for 12 h. A 1-mL volume of the culture was transferred to 100 mL of the same medium. After cultivation at 37 °C with shaking at 180 rpm for 72 h, cells were harvested by centrifugation at 8000 rpm for 10 min at 4 °C. The culture supernatant was separated from the cell pellet and passed through a sterile 0.22-µm pore size filter and used for analysis of AFB_1 degradation, as described above. NB was used rather than the supernatant as a control.

Bacterial cell pellets were washed twice with phosphate buffer (50 mM, pH 7.0) and resuspended in the same buffer. AFB_1 degradation was evaluated as described above. Phosphate buffer was used rather than the bacterial cell suspensions as a control.

Cell pellets resuspended in phosphate buffer (50 mM, pH 7.0; 3 mL of buffer per gram of cell mass) were lysed on ice using an ultrasonic cell disintegrator (Ningbo Xinzhi Instruments, Ningbo, China). Cell lysates were centrifuged at $12,000 \times g$ for 10 min at 4 °C, and passed through a sterile 0.22-μm pore size filter. The AFB$_1$ degradation test was performed as described above. Phosphate buffer was used rather than intracellular cell extracts as a control.

5.6. Effects of Incubation Period, Proteinase K + SDS, and Heat Treatment on AF Degradation by Strain L7 Culture Supernatant

The culture supernatant of strain L7 was prepared as described above. A 0.1-mL volume of AFB$_1$ solution (500 μg/L) was combined with 0.4 mL of culture supernatant in a 10-mL tube, and the mixture was incubated in the dark at 37 °C without shaking. For incubation time studies, degradation was measured at 1, 6, 12, 24, 48, 72, 96, 120, and 168 h. The effect of heat treatment was assessed by immersing the culture supernatant in a boiling water bath for 10 min. The effect of proteinase K treatment with and without SDS was evaluated as previously described [19]. To determine the correlation between AFB$_1$ degradation and protein concentration, an Amicon Ultra-15 centrifugal filter unit (Merck Millipore, Tullagreen, Iealand) with a cut-off molecular weight of 3 kDa was used for ultrafiltration of the culture supernatant to obtain different protein concentrations. NB alone processed in a similar manner and treated with AFB$_1$ served as a control.

5.7. Genotoxicity Assay

The genotoxicity of AFB$_1$ degradation products was evaluated using SOS Chromotest Basic S9 Activation Enzymes v.64 (Environmental Bio-detection Products, Mississauga, ON, Canada) according to the manufacturer's instructions. Cell viability was monitored according to alkaline phosphatase activity at 420 nm using a microplate reader [39,40]. Sample concentrations are expressed as the percentage of sample volume in each well of a 96-well microtiter plate. AFB$_1$ solution combined with NB was used as a positive control, AFB$_1$ solution combined with culture supernatant of isolate L7 was used as a degraded sample, and methanol added to the culture supernatant of isolate L7 served as a negative control. After incubation in the dark at 37 °C for 72 h (at which point the AFB$_1$ degradation ratio was 77.9% $\pm$ 2.3%), 10 μL of the diluted solution was dispensed into the appropriate wells, and a range of sample concentrations (100, 50, 25, 12.5, 6.25, and 3.125%) was tested using dimethyl sulphoxide as a diluent and reagent blank and 2-aminoanthracene serving as a positive control for S9.

5.8. Preparation of BADE

Large-scale *B. shackletonii* cultures were prepared in a 1-L working volume of NB (pH 7.4). A 10-mL volume of fresh NB was inoculated with strain L7 at 37 °C, with agitation at 180 rpm for 12 h; 10 mL of the culture was then transferred to 1 L of the same medium. After cultivation at 37 °C with shaking at 180 rpm for 72 h, cells were harvested by centrifugation at 8000 rpm for 10 min at 4 °C. The culture supernatant was separated from the cell pellet and passed through a sterile 0.22-μm pore size filter. The supernatant of the 1-L culture was heated for 10 min and then subjected to ultrafiltration with a cut-off molecular weight of 3 kDa at $5000 \times g$ for 20 min in a refrigerated centrifuge to obtain a final volume of 20 mL, yielding a sample that was concentrated 50-fold. The concentrated sample was dialyzed against 20 mmol·L^{-1} Tris-HCl buffer (pH 8.5) for 24 h at 4 °C.

5.9. Purification of B. shackletonii BADE

The dialyzed concentrated sample was subjected to ion-exchange chromatography on a DEAE-sepharose column (1.6 $\times$ 20 cm; GE Healthcare, Little Chalfont, UK). Proteins were packed in the column pre-equilibrated with 20 mmol·L^{-1} Tris-HCl buffer (pH 8.5) and were eluted with an increasing linear gradient (0%–100%) of 1 mol·L^{-1} NaCl with 20 mmol·L^{-1} Tris-HCl buffer (pH 8.5) at a flow rate of 2 mL·min^{-1} until no protein was detected. The flow-through and eluted enzyme fractions were collected and dialyzed for 24 h against 20 mmol·L^{-1} phosphate buffer (pH 7.4). Each fraction was concentrated by ultrafiltration with a cut-off molecular weight of 3 kDa; the final volumes of each fraction were 20 mL for Peak 1 (0–20 mL), 10 mL for Peak 2 (40–50 mL), 30 mL for Peak 2 (50–80 mL), 15 mL for Peak 3 (80–95 mL), and 5 mL for Peak 3 (95–100 mL). Protein concentration was determined by the Bradford method using Coomassie Brilliant Blue, and AFB_1-degrading activity was evaluated as described above. After each fraction was incubated with AFB_1 for 72 h, residual AFB_1 was detected by high-performance liquid chromatography. To assess the degradation of AFB_1 with the proteins in different peaks, AFB_1 with 20 mmol·L^{-1} phosphate buffer (pH 7.4) was used as a control. One unit (U) of enzyme activity was defined as the amount of enzyme required to decrease the amount of AFB_1 by 1 ng in 72h at 37 °C. This purification step yielded the active fraction, which was tested for AFB_1-degrading activity. The molecular mass of the active fraction was verified by SDS-PAGE.

5.10. Determination of Molecular Mass

The molecular mass of the purified BADE from *B. shackletonii* L7 was determined by SDS-PAGE based on the method of Laemmli [41] on a 12% polyacrylamide gel with molecular weight markers (Thermo Fisher Scientific, Waltham, MA, USA). Protein bands were stained using the MS-Compatible Silver Stain kit (Tiandz, Beijing, China) according to the manufacturer's protocol.

5.11. Characterization of BADE

5.11.1. Determination of Optimum pH and Temperature for Enzymatic Activity

To determine the effects of pH on BADE activity, the initial pH value was adjusted to 4.0 and 5.0 with 0.1 mol·L^{-1} citrate phosphate, and to 7.0 and 8.0 with 0.1 mol·L^{-1} sodium phosphate. The effect of temperature on BADE activity was determined at 16, 28, 37, 55, and 70 °C in 0.1 mol·L^{-1} sodium phosphate buffer (pH 7.0). A 0.1-mL volume of AFB_1 solution (500 µg/L) was added to 0.4 mL of BADE solutions with different pH and at different temperatures, with a solution at pH 7.0 and 37 °C used as a control along with solutions without BADE under the same conditions. The assays were performed in the dark at 37 °C without shaking for 72 h.

5.11.2. Effect of Metal Ions on Enzyme Activity

To determine the effect of different metal ions on BADE activity, Zn^{2+} ($ZnSO_4$), Mn^{2+} ($MnCl_2$), Mg^{2+} ($MgCl_2$), Cu^{2+} ($CuSO_4$), and Li^+ (LiCl) solutions at a final ion concentration of 10 mmol·L^{-1} were added to 0.1 mL of AFB_1 (500 µg/L) combined with 0.4 mL of BADE solution (pH 7.0). BADE solution without any metal ions (no ion) was also used as a treatment. To assess the degradation of AFB_1 in the presence of metal ions, solutions without BADE under the same conditions were used as controls. The assay was performed in the dark at 37 °C without shaking for 72 h.

5.12. Statistical Analysis

Differences between groups were evaluated by analysis of variance with the general linear model procedure in a completely randomized single-factor design using SAS software (SAS Institute, Cary, NC, USA). F tests were performed at the 0.05 level of probability. When a significant F value was detected, significant differences among the means were assessed with Duncan's multiple range test.

Supplementary Materials: Supplementary materials can be accessed at: www.mdpi.com/2072-6651/9/1/36/s1, Table S1: AFB_1 degradation by individual microbial isolates selected using coumarin medium, Table S2: Characteristics of strain L7.

Acknowledgments: This work was supported by the Special Fund for Agro-scientific Research in the Public Interest (201203037), the National Basic Research Program of China (973 program) (2013CB127805), the National Program of China Basic Science and Technology Research (2013FY113400) and China Postdoctoral Science Foundation (2015M570180).

Author Contributions: Liang Xu and Yang Liu conceived and designed the experiments; Liang Xu, Mohamed Farah Eisa Ahmed, and Lancine Sangare performed the experiments. Liang Xu and Yang Liu analyzed the data; Liang Xu, Mohamed Farah Eisa Ahmed, Lancine Sangare, and Hongping Yang contributed reagents/materials/analysis tools; Liang Xu wrote the paper; Yueju Zhao, Jonathan Nimal Selvaraj, Fuguo Xing and Yan Wang contributed suggestions and revision of the paper.

Conflicts of Interest: The authors declare no conflict of interest.

References

1. Shcherbakova, L.; Statsyuk, N.; Mikityuk, O.; Nazarova, T.; Dzhavakhiya, V. Aflatoxin B$_1$ degradation by metabolites of *Phoma glomerata* PG41 isolated from natural substrate colonized by aflatoxigenic *Aspergillus flavus*. *Jundishapur J. Microbiol.* **2015**, *8*. [CrossRef] [PubMed]
2. Guengerich, F.P.; Johnson, W.W.; Ueng, Y.F.; Yamazaki, H.; Shimada, T. Involvement of cytochrome P450, glutathione S-transferase, and epoxide hydrolase in the metabolism of aflatoxin B$_1$ and relevance to risk of human liver cancer. *Environ. Health Perspect.* **1996**, *104*, 557–562. [CrossRef] [PubMed]
3. Jarvis, B.B.; Miller, J.D. Mycotoxins as harmful indoor air contaminants. *Appl. Microbiol. Biotechnol.* **2005**, *66*, 367–372. [CrossRef] [PubMed]
4. Mellon, J.E.; Cotty, P.J.; Dowd, M.K. *Aspergillus flavus* hydrolases: Their roles in pathogenesis and substrate utilization. *Appl. Microbiol. Biotechnol.* **2007**, *77*, 497–504. [CrossRef] [PubMed]
5. Eaton, D.L.; Gallagher, E.P. Mechanisms of aflatoxin carcinogenesis. *Annu. Rev. Pharmacol. Toxicol.* **1994**, *34*, 135–172. [CrossRef] [PubMed]
6. Mishra, H.N.; Das, C. A review on biological control and metabolism of aflatoxin. *Crit. Rev. Food Sci.* **2003**, *43*, 245–264. [CrossRef] [PubMed]
7. Lee, Y.W. Detection of aflatoxin M$_1$ in cow's milk by an enzyme-linked immunosorbent assay. *Korean J. Appl. Microbiol. Biotechnol.* **1996**, *24*, 630–635.
8. Zinedine, A.; González-Osnaya, L.; Soriano, J.M.; Moltó, J.C.; Idrissi, L.; Mañes, J. Presence of aflatoxin M$_1$ in pasteurized milk from Morocco. *Int. J. Food Microbiol.* **2007**, *114*, 25–29. [CrossRef] [PubMed]
9. Line, J.E.; Brackett, R.E.; Wilkinson, R.E. Evidence for degradation of aflatoxin B$_1$ by *Flavobacterium aurantiacum*. *J. Food Prot.* **1994**, *57*, 788–791. [CrossRef]
10. Smiley, R.; Draughon, F. Preliminary evidence that degradation of aflatoxin B$_1$ by *Flavobacterium aurantiacum* is enzymatic. *J. Food Prot.* **2000**, *63*, 415–418. [CrossRef] [PubMed]
11. Liu, D.; Yao, D.; Liang, Y.; Zhou, T.; Song, Y.; Zhao, L.; Ma, L. Production, purification, and characterization of an intracellular aflatoxin-detoxifizyme from *Armillariella tabescens* (E-20). *Food Chem. Toxicol.* **2001**, *39*, 461–466. [CrossRef]
12. Cao, H.; Liu, D.; Mo, X.; Xie, C.; Yao, D. A fungal enzyme with the ability of aflatoxin B$_1$ conversion: Purification and ESI-MS/MS identification. *Microbiol. Res.* **2011**, *166*, 475–483. [CrossRef] [PubMed]
13. Motomura, M.; Toyomasu, T.; Mizuno, K.; Shinozawa, T. Purification and characterization of an aflatoxin degradation enzyme from *Pleurotus ostreatus*. *Microbiol. Res.* **2003**, *158*, 237–242. [CrossRef] [PubMed]
14. Petchkongkaew, A.; Taillandier, P.; Gasaluck, P.; Lebrihi, A. Isolation of *Bacillus* spp. from Thai fermented soybean (Thua-nao): Screening for aflatoxin B$_1$ and ochratoxin A detoxification. *J. Appl. Microbiol.* **2008**, *104*, 1495–1502. [CrossRef] [PubMed]
15. Hormisch, D.; Brost, I.; Kohring, G.W.; Giffhorn, F.; Kroppenstedt, R.M.; Stackebradt, E.; Färber, P.; Holzapfel, W.H. *Mycobacterium fluoranthenivorans* sp. nov., a fluoranthene and aflatoxin B$_1$ degrading bacterium from contaminated soil of a former coal gas plant. *Syst. Appl. Microbiol.* **2004**, *27*, 653–660. [CrossRef] [PubMed]
16. Teniola, O.D.; Addo, P.A.; Brost, I.M.; Färber, P.; Jany, K.D.; Alberts, J.F.; van Zyl, W.H.; Steyn, P.S.; Holzapfel, W.H. Degradation of aflatoxin B$_1$ by cell-free extracts of *Rhodococcus erythropolis* and *Mycobacterium fluoranthenivorans* sp. nov. DSM44556T. *Int. J. Food Microbiol.* **2005**, *105*, 111–117. [CrossRef] [PubMed]
17. Alberts, J.; Engelbrecht, Y.; Steyn, P.; Holzapfel, W.; van Zyl, W. Biological degradation of aflatoxin B$_1$ by *Rhodococcus erythropolis* cultures. *Int. J. Food Microbiol.* **2006**, *109*, 121–126. [CrossRef] [PubMed]
18. Kong, Q.; Zhai, C.; Guan, B.; Li, C.; Shan, S.; Yu, J. Mathematic modeling for optimum conditions on aflatoxin B$_1$ degradation by the aerobic bacterium *Rhodococcus erythropolis*. *Toxins* **2012**, *4*, 1181–1195. [CrossRef] [PubMed]
19. Guan, S.; Ji, C.; Zhou, T.; Li, J.; Ma, Q.; Niu, T. Aflatoxin B$_1$ degradation by *Stenotrophomonas maltophilia* and other microbes selected using coumarin medium. *Int. J. Mol. Sci.* **2008**, *9*, 1489–1503. [CrossRef] [PubMed]

20. Guan, S.; Zhao, L.; Ma, Q.; Zhou, T.; Wang, N.; Hu, X.; Ji, C. In vitro efficacy of *Myxococcus fulvus* ANSM068 to biotransform aflatoxin B_1. *Int. J. Mol. Sci.* **2010**, *11*, 4063–4079. [CrossRef] [PubMed]

21. Gao, X.; Ma, Q.; Zhao, L.; Lei, Y.; Shan, Y.; Ji, C. Isolation of *Bacillus subtilis*: Screening for aflatoxins B_1, M_1, and G_1 detoxification. *Eur. Food Res. Technol.* **2011**, *232*, 957–962. [CrossRef]

22. Wang, J.; Ogata, M.; Hirai, H.; Kawagishi, H. Detoxification of aflatoxin B_1 by manganese peroxidase from the white-rot fungus *Phanerochaete sordida* YK-624. *FEMS Microbiol. Lett.* **2011**, *314*, 164–169. [CrossRef] [PubMed]

23. Alberts, J.F.; Gelderblom, W.C.; Botha, A.; van Zyl, W.H. Degradation of aflatoxin B_1 by fungal laccase enzymes. *Int. J. Food Microbiol.* **2009**, *135*, 47–52. [CrossRef] [PubMed]

24. Taylor, M.C.; Jackson, C.J.; Tattersall, D.B.; French, N.; Peat, T.S.; Newman, J.; Briggs, L.J.; Lapalikar, G.V.; Campbell, P.M.; Scott, C.; et al. Identification and characterization of two families of $F_{420}H_2$-dependent reductases from *Mycobacteria* that catalyse aflatoxin degradation. *Mol. Microbiol.* **2010**, *78*, 561–575. [CrossRef] [PubMed]

25. Logan, N.A.; Lebbe, L.; Verhelst, A.; Goris, J.; Forsyth, G.; Rodríguez-Díaz, M.; Heyndrickx, M.; De Vos, P. *Bacillus shackletonii* sp. Nov., from volcanic soil on Candlemas Island, South Sandwich archipelago. *Int. J. Syst. Evol. Microbiol.* **2004**, *54*, 373–376. [CrossRef] [PubMed]

26. Liu, Y.; Huang, S.; Zhang, Y.; Xu, F. Isolation and characterization of a thermophilic *Bacillus shackletonii* K5 from a biotrickling filter for the production of polyhydroxybutyrate. *J. Environ. Sci.* **2014**, *26*, 1453–1462. [CrossRef] [PubMed]

27. Lillehoj, E.B.; Stubblefield, R.D.; Shannon, G.M.; Shotwell, O.L. Aflatoxin M_1 removal from aqueous solutions by *Flavobacterium aurantiacum*. *Mycopath. Mycol. Appl.* **1971**, *45*, 259–266. [CrossRef]

28. D'Souza, D.H.; Brackett, R.E. The role of trace metal ions in aflatoxin B_1 degradation by *Flavobacterium aurantiacum*. *J. Food Prot.* **1998**, *61*, 1666–1669. [CrossRef] [PubMed]

29. Ibrahim, C.O. Development of applications of industrial enzymes from Malaysian indigenous microbial sources. *Bioresour. Technol.* **2008**, *99*, 4572–4582. [CrossRef] [PubMed]

30. Street, G.; Rodgers, P.J. Large scale industrial enzyme production. *Crit. Rev. Biotechnol.* **1983**, *1*, 59–85. [CrossRef]

31. Sangare, L.; Zhao, Y.; Folly, Y.M.; Chang, J.; Li, J.; Selvaraj, J.N.; Xing, F.; Zhou, L.; Wang, Y.; Liu, Y. Aflatoxin B_1 degradation by a *Pseudomonas* strain. *Toxins* **2014**, *6*, 3028–3040. [CrossRef] [PubMed]

32. Hu, L. Study on Molecular Roots about the Function of Oxidation for Aflatoxin Oxidase. Master's Thesis, Jinan University, Guangzhou, China, 2011.

33. D'Souza, D.H.; Brackett, R.E. Aflatoxin B_1 degradation by *Flavobacterium aurantiacum* in the presence of reducing conditions and seryl and sulfhydryl group inhibitors. *J. Food Prot.* **2001**, *64*, 268–271. [CrossRef] [PubMed]

34. Tosch, D.; Waltking, A.E.; Schlesier, J.F. Comparison of liquid chromatography and high performance thin layer chromatography for determination of aflatoxin in peanut products. *J. Associ. Off. Anal. Chem.* **1984**, *67*, 337–339.

35. Holt, J.G. *Bergey's Manual of Determinative Bacteriology*, 9th ed.; Williams and Wilkins: Baltimore, MD, USA, 1994.

36. Lane, D.J. 16S/23S rRNA sequencing. In *Nucleic Acid Techniques in Bacterial Systematics*; John Wiley and Sons: Chichester, UK, 1991.

37. NCBI nucleotide database. Available online: http://blast.ncbi.nlm.nih.gov/Blast.cgi (accessed on 27 December 2016).

38. El-Nezami, H.; Kankaanpaa, P.; Salminen, S.; Ahokas, J. Ability of dairy strains of lactic acid bacteria to bind a common food carcinogen, aflatoxin B_1. *Food Chem. Toxicol.* **1998**, *36*, 321–326. [CrossRef]

39. Kocak, E.; Yetilmezsoy, K.; Gonullu, M.T.; Petek, M. A statistical evaluation of the potential genotoxic activity in the surface waters of the golden horn estuary. *Mar. Pollut. Bull.* **2010**, *60*, 1708–1717. [CrossRef] [PubMed]

40. Sharma, P.; Mathur, N.; Singh, A.; Sogani, M.; Bhatnagar, P.; Atri, R.; Pareek, S. Monitoring hospital wastewaters for their probable genotoxicity and mutagebicity. *Environ. Monit. Assess.* **2015**, *187*. [CrossRef] [PubMed]

41. Laemmli, U.K. Cleavage of structural proteins during the assembly of the head of bacteriophage T4. *Nature* **1970**, *227*, 680–685. [CrossRef] [PubMed]

toxins

MDPI

Article

Static Hot Air and Infrared Rays Roasting are Efficient Methods for Aflatoxin Decontamination on Hazelnuts

Ilenia Siciliano [1], Barbara Dal Bello [2], Giuseppe Zeppa [2], Davide Spadaro [1,2,]* and Maria Lodovica Gullino [1,2]

[1] Agroinnova—Centre of Competence for the Innovation in the Agro-Environmental Sector,
University of Turin, Largo Paolo Braccini, 2 Grugliasco, Turin 10095, Italy; ilenia.siciliano@unito.it (I.S.);
marialodovica.gullino@unito.it (M.L.G.)
[2] DISAFA—Department of Agricultural, Forest and Food Science, University of Turin, Largo Paolo Braccini,
2 Grugliasco, Turin 10095, Italy; barbara.dalbello@unito.it (B.D.B.); giuseppe.zeppa@unito.it (G.Z.)
* Correspondence: davide.spadaro@unito.it; Tel.: +39-011-6708-942

Academic Editor: Ting Zhou
Received: 16 January 2017; Accepted: 17 February 2017; Published: 21 February 2017

Abstract: Aflatoxins are a group of secondary metabolites produced by members of *Aspergillus* Section *Flavi* that are dangerous to humans and animals. Nuts can be potentially contaminated with aflatoxins, often over the legal threshold. Food processes, including roasting, may have different effects on mycotoxins, and high temperatures have proven to be very effective in the reduction of mycotoxins. In this work, two different roasting methods—traditional static hot air roasting and infra-red rays roasting—were applied and compared for the detoxification of hazelnuts from Italy and Turkey. At the temperature of 140 °C for 40 min of exposure, detoxification was effective for both roasting techniques. Residual aflatoxins after infra-red rays treatments were lower compared to static hot air roasting. On Italian hazelnuts, residual aflatoxins were lower than 5%, while for Turkish hazelnuts they were lower than 15% after 40 min of exposure to an infra-red rays roaster. After roasting, the perisperm was detached from the nuts and analyzed for aflatoxin contents. Residual aflatoxins in the perisperm ranged from 80% up to 100%. After roasting, the lipid profile and the nutritional quality of hazelnuts were not affected. Fatty acid methyl esters analyses showed a similar composition for Italian and Turkish hazelnuts.

Keywords: aflatoxins; *Aspergillus flavus*; *Corylus avellana*; fatty acids; thermal treatment

---

1. Introduction

Aflatoxins (AFs) are a group of secondary metabolites produced by members of *Aspergillus* Section *Flavi*—mainly *A. flavus* and *A. parasiticus* [1]—on a variety of food products, such as nuts, grains, and spices [2]. Since 1960, more than 20 aflatoxins have been identified; only four of them—aflatoxin B_1 (AFB$_1$), aflatoxin B_2 (AFB$_2$), aflatoxin G_1 (AFG$_1$), aflatoxin G_2 (AFG$_2$)—occur naturally [3,4]. AFs are a group of difuranocoumarin derivatives named based on their fluorescence under UV-light. AFB$_1$ and AFB$_2$ have a blue fluorescence due to the difuro-coumaro-cyclopentenone structure, while a six-member lactone ring replacing the cyclopentenone is responsible for the yellow-green fluorescence of AFG$_1$ and AFG$_2$ [5]. AFB$_1$ and AFG$_1$ have also an olefinic double bond at the C_8-C_9 position, whereas AFB$_2$ and AFG$_2$ lack this bond. AFs are toxic, mutagenic, teratogenic, carcinogenic compounds implicated in human hepatic and extrahepatic carcinogenesis [6]. AFB$_1$ is the most widespread among AFs, and has the most potent carcinogenic effect. In fact, AFB$_1$ is the only one classified by International Agency of Research on Cancer (IARC) as a Group 1 carcinogen [7].

Toxicological and epidemiological data, together with occurrence and distribution data, were used to fix limits for food consumption contaminated with mycotoxins. In Europe, thresholds for AFB$_1$ and total AFs in hazelnuts for direct human consumption and for use as ingredient in foodstuffs are 5 and 10 µg/kg, specified by the Commission Regulation (EU) No 165/2010, based on scientific opinion of authoritative bodies, such as JECFA (Joint FAO/WHO Expert Committee on Food Additives), WHO (World Health Organization), FAO (Food and Agriculture Organization of the United Nations), EFSA (European Food Safety Authority). Moreover, the Codex Alimentarius Commission has specified 15 µg/kg as limit for peanuts, almonds, shelled Brazil nuts, hazelnuts, and pistachios [8].

Nuts and dried fruits are known to be possibly contaminated with mycotoxins. Different studies on pistachios, almonds, groundnuts, peanuts, chestnuts, and hazelnuts have demonstrated that contamination with aflatoxins often exceeds the legal threshold in commercial samples [9–11].

Prevention of mycotoxin contamination in the field is the best way to reduce mycotoxins in food and feed, while decontamination is difficult and frequently lowers the quality or quantity of the commercial product. Chemical, physical, and biological approaches for the detoxification of aflatoxins have been reported in previous studies on nuts, and include strategies such as ozone, plasma, UV or gamma irradiation, thermal treatments, or microorganisms [12–15]. However, most of the current methods are not practical, due to time consumption, nutrition losses, or low detoxification efficiency.

Many nuts are predominantly consumed roasted. Thermal treatments such as roasting lead to chemical changes in carbohydrates, proteins, and fats. The flavor, color, texture, and appearance of hazelnuts are significantly enhanced by roasting, and this process is also used to remove the pellicle of kernels, to inactivate enzymes that cause nutrient loss, and to destroy microorganisms, toxins or allergens [16].

Turkey is the main hazelnut producing country in the world, with 660,000 tons in 2012 [17]. The main cultivar produced in Turkey is "Tombul", which occupies 29.8% of all production areas, followed by the cultivars "Çakildak" (15.2%), "Mincane" (14.1%), "İncekara" (12.4%), "Palaz" (11.9%), "Foşa" (7.0%), and "Kalinkara" (2.7%). Italy is the second world hazelnut producer, with about 85,000 tons in 2012 [17], and "Tonda Gentile Trilobata" is the main cultivar in Piedmont, Northern Italy (15,000 tons in 2014). Hazelnuts provide an excellent source of energy due to their high oil content of about 60% and several bioactive nutrients. Hazelnut oil has been reported to be the richest source of vitamin E and fatty acids, in particular oleic acid [18]. Among the fatty acids of hazelnut oil, palmitic acid (2.96%–7.40%) is the main saturated fatty acid (SFA). The highest percentage of fatty acids are instead covered by monounsaturated fatty acid (MUFA) as oleic acid (73.1%–90.7%) and polyunsaturated fatty acid (PUFA) as linoleic acid (4.4%–16%), which confer heart and blood beneficial attributes to the hazelnut oil [14]. Moreover, MUFA and PUFA—and particularly linoleic acid—play an important role not only in terms of health, but also in stability, because their oxidation may cause flavor degradation and rancidity. [19]. Fatty acid composition is closely correlated with the stability and degradation kinetics of hazelnuts. In particular, linoleic acid plays the most important role in the stability, and its oxidation may cause flavor degradation and rancidity.

By considering the possibility of aflatoxin contamination on hazelnuts that is often over the legal threshold and the effect of high temperature on mycotoxin reduction, in this work two different roasting methods—traditional static hot air roasting and infrared rays roasting—were compared in the detoxification of hazelnuts from aflatoxins, and their effect on hazelnut quality.

2. Results and Discussion

2.1. Static Hot Air Treatments

Data of the first trial on hazelnuts roasted with a static hot air roaster are shown in Table 1 for two temperatures (120 °C and 170 °C) at two exposure times (20 and 40 min). Residual aflatoxins after roasting showed a different trend for Italian and Turkish hazelnuts; a decrease was observed at the same temperature with increasing exposure time only for Turkish hazelnuts. With the increase of temperature, differences in reduction were registered, either considering the geographical origin of the hazelnuts or the mycotoxin. In general, the residue was lower in AFB_1 and AFG_1 than AFB_2 and

AFG_2. The influence of temperature was different in both types of hazelnuts used. Turkish hazelnuts showed a decrease in AFs concentration with increasing temperatures and the same exposure time. On the contrary, in Tonga Gentile Trilobata (TGT) hazelnuts, a decrease in the residual concentration was not observed. The decomposition temperature of aflatoxins ranges from 237 °C to 306 °C [20]; in particular, AFB_1 is stable to dry heating at temperature below its thermal decomposition temperature (267 °C), so the temperatures used during the experiments were below this threshold. Normal home cooking conditions failed to totally destroy AFB_1 and AFG_1, and temperatures of approximately 150 °C were necessary to obtain a partial degradation of AFs [21]. In a study on pistachio nuts [22], the reduction of AFs ranged from 17% to 63% and was time and temperature dependent. In coffee beans, a reduction of about 42%–56%—depending on temperature and type of roasting—was achieved [23]. Other parameters seem to influence the rate of AFs destruction: higher moisture content increased AFB_1 destruction on rice. The presence of water could help the lactone ring opening of AFB_1 with the formation of a carboxylic group. On both hazelnut types, fatty acid methyl esters (FAMEs) were analyzed (Supplementary Tables S1 and S2), showing that the nutritional composition was not affected at all tested temperatures and exposure times.

Table 1. Residual (%) aflatoxin B_1 (AFB_1), aflatoxin B_2 (AFB_2), aflatoxin G_1 (AFG_1), and aflatoxin G_2 (AFG_2) after treatment with static hot air in the first set of experiments.

Hazelnut	T (°C)	t (min)	AFB_1	AFB_2	AFG_1	AFG_2
Italian	120	20	17.5 ± 4.3	17.5 ± 3.8	5.62 ± 2.9	9.46 ± 3.8
Turkish			47.2 ± 2.4	81.7 ± 5.1	77.8 ± 10	69.9 ± 6.0
Italian	120	40	11.7 ± 1.4	13.9 ± 2.5	17.8 ± 5.4	22.5 ± 6.0
Turkish			12.9 ± 8.0	53.1 ± 4.9	12.0 ± 5.2	13.9 ± 5.5
Italian	170	20	38.7 ± 7.7	62.9 ± 5.8	24.7 ± 5.4	68.9 ± 14
Turkish			7.39 ± 1.8	42.3 ± 8.4	10.4 ± 1.9	13.6 ± 5.3
Italian	170	40	17.9 ± 7.3	43.6 ± 10	8.21 ± 1.1	34.5 ± 8.3
Turkish			4.09 ± 4.2	31.3 ± 6.6	6.04 ± 8.5	5.97 ± 8.4

Mean of three replicates ± standard deviation.

2.2. Comparison of Static Hot Air and Infrared Rays Roasting

In a second experimental set, a new roasting technology—infrared rays roasting—was compared to static hot air roasting. An intermediate temperature (140 °C) and two exposure times (20 min and 40 min) were used for both methods.

FAMEs analysis showed a similar composition for both Italian (Table 2) and Turkish (Table 3) hazelnuts, and a total of fourteen fatty acids (FAs) were identified. The main FAs were oleic acid, linoleic acid, palmitic acid, and stearic acid. Monounsaturated fatty acids were always higher than polyunsaturated fatty acids and saturated fatty acids. FAs distribution was preserved after all treatments for both types of hazelnuts, both after static hot air and infrared rays treatment. FAs composition is important for nutritional quality and health benefits offered by MUFAs and PUFAs, flavor, kernel texture, and quality [18]. Some minor changes occurred in the FAs profile after both types of roasting. In particular, the content of SFAs remained low, and the content of UFAs that promote health benefits was always high.

The oleic to linoleic acid (O/L) ratio was considered an important criterion to evaluate the kernel quality, and a greater value indicates better oxidative stability [24]. In TGT hazelnuts, the O/L ratio was reduced after all treatments, but the ratio was higher after infrared rays than after static hot air. The degree of unsaturation expressed as iodine value (IV) was higher after static hot air treatments at both exposure times (88.43 and 88.99 instead of 88.34 of raw hazelnuts). After infrared rays treatments at 140 °C for 20 min, a slight IV reduction occurred (87.97 instead of 88.34 of raw hazelnuts).

Table 2. Fatty acid composition (%), of Tonga Gentile Trilobata (TGT) hazelnuts raw and roasted with static hot air and infrared rays.

Fatty Acid	Raw	140 °C 20 min	140 °C 40 min	Significance
Static Hot Air Treatment				
Myristic (C14:0)	0.02 ± 0.00	0.02 ± 0.00	0.02 ± 0.00	ns
Palmitic (C16:0)	5.79 ± 0.00 [b]	5.92 ± 0.01 [a]	5.74 ± 0.01 [a]	***
Palmitoleic (C16:1)	0.24 ± 0.00 [b]	0.24 ± 0.00 [c]	0.24 ± 0.00 [a]	***
Margaric (C17:0)	0.04 ± 0.00	0.04 ± 0.00	0.04 ± 0.00	ns
Heptadecenoic (C17:1)	0.07 ± 0.00 [a]	0.07 ± 0.00 [b]	0.07 ± 0.00 [b]	***
Stearic (C18:0)	2.32 ± 0.00 [c]	2.36 ± 0.00 [a]	2.27 ± 0.00 [b]	***
Elaidic (C18:1 ω9t)	0.02 ± 0.00	0.02 ± 0.00	0.02 ± 0.00	ns
Oleic (C18:1 ω9c)	84.86 ± 0.00 [c]	84.44 ± 0.02 [b]	84.35 ± 0.03 [a]	***
Linoleic (C18:2 ω6c)	6.27 ± 0.00 [a]	6.53 ± 0.01 [b]	6.88 ± 0.02 [c]	***
Arachidic (C20:0)	0.11 ± 0.00	0.11 ± 0.00	0.11 ± 0.00	ns
Eicosenoic (C20:1)	0.12 ± 0.00 [a]	0.12 ± 0.00 [b]	0.12 ± 0.00 [c]	***
α-Linolenic (C18:3 ω3)	0.08 ± 0.00 [b]	0.08 ± 0.00 [a]	0.08 ± 0.00 [a]	*
Docosanoic (C22:0)	0.02 ± 0.00 [ab]	0.02 ± 0.00 [b]	0.02 ± 0.00 [a]	*
Arachidonic (C20:4 ω6)	0.03 ± 0.00 [a]	0.03 ± 0.00 [a]	0.03 ± 0.00 [b]	*
Σ SFA	8.31 ± 0.00 [c]	8.48 ± 0.01 [a]	8.20 ± 0.01 [b]	***
Σ MUFA	85.31 ± 0.00 [c]	84.89 ± 0.02 [b]	84.80 ± 0.03 [a]	***
Σ PUFA	6.38 ± 0.00 [a]	6.64 ± 0.01 [b]	6.99 ± 0.02 [c]	***
UFA/SFA	11.03 ± 0.00 [a]	10.80 ± 0.01 [c]	11.19 ± 0.02 [b]	***
Oleic/Linoleic (O/L)	13.54 ± 0.01 [c]	12.93 ± 0.02 [b]	12.26 ± 0.03 [a]	***
Iodine value (IV)	88.34 ± 0.00 [a]	88.43 ± 0.01 [b]	88.99 ± 0.01 [c]	***
Infrared Rays Treatment				
Myristic (C14:0)	0.02 ± 0.00	0.02 ± 0.00	0.02 ± 0.00	ns
Palmitic (C16:0)	5.79 ± 0.00 [b]	6.16 ± 0.02 [b]	5.94 ± 0.01 [a]	*
Palmitoleic (C16:1)	0.24 ± 0.00 [b]	0.27 ± 0.00 [a]	0.25 ± 0.00 [b]	***
Margaric (C17:0)	0.04 ± 0.00	0.04 ± 0.00	0.04 ± 0.00	ns
Heptadecenoic (C17:1)	0.07 ± 0.00	0.07 ± 0.00	0.07 ± 0.00	ns
Stearic (C18:0)	2.32 ± 0.00 [b]	2.44 ± 0.00 [a]	2.34 ± 0.00 [b]	*
Elaidic (C18:1 ω9t)	0.02 ± 0.00	0.02 ± 0.00	0.02 ± 0.00	ns
Oleic (C18:1 ω9c)	84.86 ± 0.00 [a]	84.28 ± 0.03 [b]	84.44 ± 0.02 [a]	***
Linoleic (C18:2 ω6c)	6.27 ± 0.00 [b]	6.32 ± 0.01 [a]	6.50 ± 0.01 [b]	***
Arachidic (C20:0)	0.11 ± 0.00	0.12 ± 0.00	0.12 ± 0.00	ns
Eicosenoic (C20:1)	0.12 ± 0.00	0.12 ± 0.00	0.12 ± 0.00	ns
α-Linolenic (C18:3 ω3)	0.08 ± 0.00 [b]	0.08 ± 0.00 [a]	0.08 ± 0.00 [b]	***
Docosanoic (C22:0)	0.02 ± 0.00	0.02 ± 0.00	0.02 ± 0.00	ns
Arachidonic (C20:4 ω6)	0.03 ± 0.00 [b]	0.03 ± 0.00 [a]	0.03 ± 0.00 [b]	***
Σ SFA	8.31 ± 0.00 [b]	8.81 ± 0.02 [a]	8.48 ± 0.01 [b]	*
Σ MUFA	85.31 ± 0.00 [a]	84.76 ± 0.03 [b]	84.91 ± 0.01 [a]	***
Σ PUFA	6.38 ± 0.00 [b]	6.44 ± 0.01 [a]	6.62 ± 0.01 [b]	***
UFA/SFA	11.03 ± 0.00 [a]	10.36 ± 0.02 [b]	10.80 ± 0.01 [a]	*
Oleic/Linoleic (O/L)	13.54 ± 0.01 [a]	13.34 ± 0.03 [b]	12.99 ± 0.02 [a]	***
Iodine value (IV)	88.34 ± 0.00	87.97 ± 0.00 [a]	88.40 ± 0.00 [b]	*

Values are expressed as mean ± standard deviation (n = 3). Means followed by different letters were significantly different at $p < 0.05$. Where letters in columns were not reported, no statistical differences were observed. Significance: * $p < 0.05$; *** $p < 0.001$; ns = not significant. SFA: saturated fatty acid; MUFA: monounsaturated fatty acid; PUFA: polyunsaturated fatty acid.

Turkish hazelnuts (Table 3) showed a similar behavior. The O/L ratio was reduced after all treatments. At 140 °C for 20 min, the O/L ratio was higher after infrared rays than after static hot air (13.34 instead of 12.93). At 140 °C for 40 min, IV was high after static hot air and lower after infrared rays (88.87 and 87.91, respectively, instead of 88.83 of raw hazelnuts).

These results are in agreement with Amaral et al. [25] and Belviso et al. [26], confirming that the lower roasting temperature did not strongly affect the lipid profile and thus the nutritional quality of hazelnuts.

Table 3. Fatty acid composition (%) of Turkish hazelnut raw and roasted with static hot air and infrared rays.

Fatty Acid	Raw	140 °C 20 min	140 °C 40 min	Significance
		Static Hot Air Treatment		
Myristic (C14:0)	0.03 ± 0.00 [b]	0.03 ± 0.00 [a]	0.03 ± 0.00 [b]	***
Palmitic (C16:0)	5.57 ± 0.00 [a]	5.58 ± 0.00 [a]	5.76 ± 0.01 [b]	***
Palmitoleic (C16:1)	0.17 ± 0.00 [a]	0.17 ± 0.00 [b]	0.18 ± 0.00 [c]	***
Margaric (C17:0)	0.04 ± 0.00 [a]	0.04 ± 0.00 [a]	0.04 ± 0.00 [b]	***
Heptadecenoic (C17:1)	0.07 ± 0.00 [a]	0.07 ± 0.00 [b]	0.07 ± 0.00 [a]	***
Stearic (C18:0)	2.21 ± 0.00 [c]	2.04 ± 0.00 [a]	2.17 ± 0.00 [b]	***
Elaidic (C18:1 ω9t)	0.02 ± 0.00	0.02 ± 0.00	0.02 ± 0.00	ns
Oleic (C18:1 ω9c)	84.87 ± 0.01 [b]	85.44 ± 0.01 [c]	84.55 ± 0.01 [a]	***
Linoleic (C18:2 ω6c)	6.64 ± 0.01 [b]	6.25 ± 0.00 [a]	6.81 ± 0.01 [c]	***
Arachidic (C20:0)	0.12 ± 0.00 [b]	0.11 ± 0.00 [a]	0.12 ± 0.00 [b]	***
Eicosenoic (C20:1)	0.15 ± 0.01	0.14 ± 0.00	0.13 ± 0.00	ns
α-Linolenic (C18:3 ω3)	0.06 ± 0.00 [a]	0.06 ± 0.00 [b]	0.07 ± 0.00 [c]	***
Docosanoic (C22:0)	0.02 ± 0.00	0.02 ± 0.00	0.02 ± 0.00	ns
Arachidonic (C20:4 ω6)	0.03 ± 0.00	0.03 ± 0.00	0.03 ± 0.00	ns
Σ SFA	8.00 ± 0.00 [b]	7.83 ± 0.01 [a]	8.15 ± 0.00 [c]	***
Σ MUFA	85.28 ± 0.01 [b]	85.83 ± 0.01 [c]	84.95 ± 0.01 [a]	***
Σ PUFA	6.73 ± 0.01 [b]	6.34 ± 0.00 [a]	6.91 ± 0.01 [c]	***
UFA/SFA	11.51 ± 0.00 [b]	11.78 ± 0.01 [c]	11.27 ± 0.00 [a]	***
Oleic/Linoleic (O/L)	12.78 ± 0.02 [b]	13.68 ± 0.01 [c]	12.42 ± 0.01 [a]	***
Iodine value (IV)	88.83 ± 0.03 [a]	88.64 ± 0.00 [c]	88.87 ± 0.00 [b]	***
		Infrared Rays Treatment		
Myristic (C14:0)	0.03 ± 0.00 [b]	0.03 ± 0.00 [a]	0.03 ± 0.00 [b]	***
Palmitic (C16:0)	5.57 ± 0.00 [c]	5.63 ± 0.01 [b]	5.36 ± 0.04 [a]	***
Palmitoleic (C16:1)	0.17 ± 0.00 [b]	0.17 ± 0.00 [b]	0.16 ± 0.00 [a]	***
Margaric (C17:0)	0.04 ± 0.00 [b]	0.04 ± 0.00 [a]	0.04 ± 0.00 [a]	***
Heptadecenoic (C17:1)	0.07 ± 0.00 [a]	0.07 ± 0.00 [b]	0.07 ± 0.00 [a]	**
Stearic (C18:0)	2.21 ± 0.00 [c]	2.08 ± 0.00 [a]	2.19 ± 0.02 [b]	***
Elaidic (C18:1 ω9t)	0.02 ± 0.00 [a]	0.02 ± 0.00 [ab]	0.03 ± 0.00 [b]	**
Oleic (C18:1 ω9c)	84.87 ± 0.01	85.08 ± 0.02	84.96 ± 0.58	ns
Linoleic (C18:2 ω6c)	6.64 ± 0.01 [b]	6.52 ± 0.00 [c]	6.10 ± 0.04 [a]	***
Arachidic (C20:0)	0.12 ± 0.00 [b]	0.11 ± 0.00 [a]	0.12 ± 0.00 [c]	***
Eicosenoic (C20:1)	0.15 ± 0.01	0.14 ± 0.00	0.14 ± 0.00	ns
α-Linolenic (C18:3 ω3)	0.06 ± 0.00 [a]	0.06 ± 0.00 [c]	0.06 ± 0.00 [b]	***
Docosanoic (C22:0)	0.02 ± 0.00 [b]	0.02 ± 0.00 [a]	0.02 ± 0.00 [b]	***
Arachidonic (C20:4 ω6)	0.03 ± 0.00	0.03 ± 0.00	0.03 ± 0.00	ns
Σ SFA	8.00 ± 0.00 [b]	7.90 ± 0.01 [a]	7.76 ± 0.06 [a]	***
Σ MUFA	85.28 ± 0.01	85.48 ± 0.02	85.35 ± 0.59	ns
Σ PUFA	6.73 ± 0.01 [b]	6.61 ± 0.00 [c]	6.19 ± 0.05 [a]	***
UFA/SFA	11.51 ± 0.00 [a]	11.65 ± 0.02 [b]	11.79 ± 0.01 [a]	***
Oleic/Linoleic (O/L)	12.78 ± 0.02 [b]	13.05 ± 0.01 [a]	13.92 ± 0.01 [c]	***
Iodine value (IV)	88.83 ± 0.03 [b]	88.81 ± 0.01 [c]	87.91 ± 0.61 [a]	***

Values are expressed as mean ± standard deviation (*n* = 3). Means followed by different letters were significantly different at *p* < 0.05. Where letters in columns were not reported, no statistical differences were observed. Significance: ** *p* < 0.01; *** *p* < 0.001; ns = not significant.

During hot air roasting, the moisture was initially removed from the surface, inducing a water diffusion from the interior to the dried surface. Hot air roasting is a convective heat transfer process, and the temperature rises as a function of heat transfer. As the nut temperature approaches 120 °C, the rate of temperature increase slows and the moisture evaporation accelerates [27]. In the present study, static hot air treatment generally induced a higher detoxification with increasing exposure time.

In particular, for Turkish hazelnuts, a higher decontamination of AFB$_1$ and AFG$_1$ was registered after 40 min of treatment (Figure 1). Italian hazelnuts showed a higher concentration range, but AFB$_1$ was always more decontaminated than the other AFs.

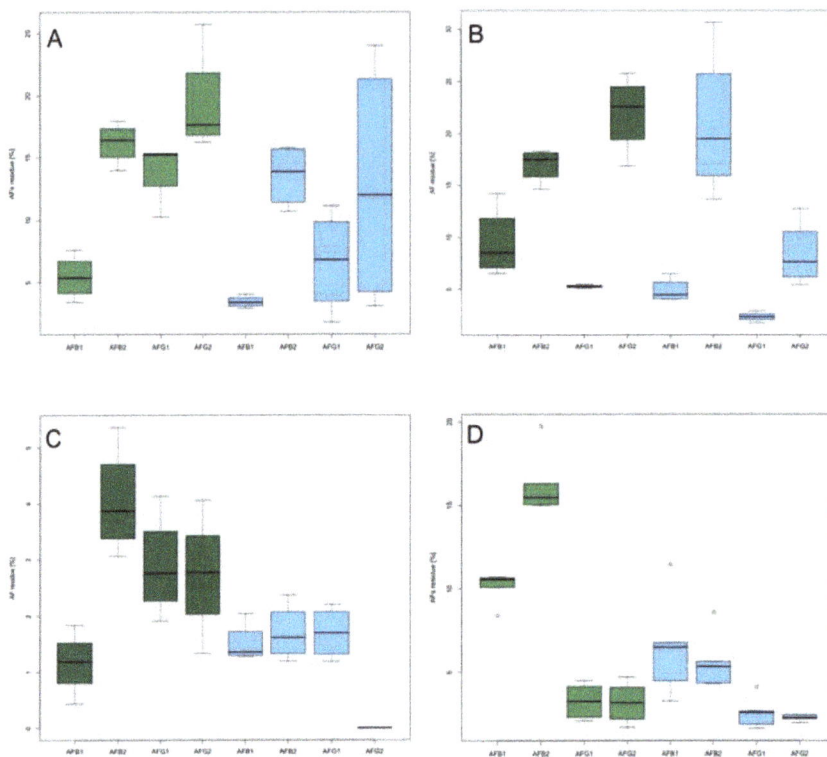

Figure 1. Residual AFB$_1$, AFB$_2$, AFG$_1$, and AFG$_2$ after treatments with static hot air roaster compared with infrared rays roaster. Different colors represent different treatments: green, 140 °C for 20 min; blue, 140 °C for 40 min. The treatments were performed on (**A**) Italian TGT hazelnuts roasted with static hot air roaster; (**B**) Turkish hazelnuts roasted with static hot air roaster; (**C**) Italian TGT hazelnuts roasted with infrared rays roaster; (**D**) Turkish hazelnuts roasted with infrared rays roaster. Boxes represent the interquartile range (IQR) between the first and third quartiles, and the line inside represents the median (second quartile). Whiskers denote the lowest and the highest values within 1.56 IQR from the first and third quartiles, respectively. Circles represent outliers beyond the whiskers.

With infrared rays, heat is transferred by radiation, and the wavelength determines the temperature. During this type of heating, surface irregularities have a small effect on heating transfer [28]. After infrared rays treatments, residual aflatoxins were lower than 5% and 15%, respectively, on Italian and Turkish hazelnuts. Turkish hazelnuts showed a high decontamination of AFG$_1$ for both exposure times and roasting methods, with the lowest value at 40 min treatment. For both methods, a higher reduction of AFB$_1$ and AFG$_1$ compared to AFB$_2$ and AFG$_2$ was confirmed. Infrared rays roasting showed better AFs reduction compared to static hot air roasting, due to higher heat transfer efficiency.

2.3. Analysis of Perisperm

A third set of experiments was realized to understand if the AFs were removed with the perisperm or if they were degraded after exposure to high temperature (Figure 2). The two roasting methods were used only on Italian hazelnuts, and two temperatures (120 °C and 140 °C) were used for 20 and 40 min. After treatments, perisperm was detached from hazelnut and collected separately.

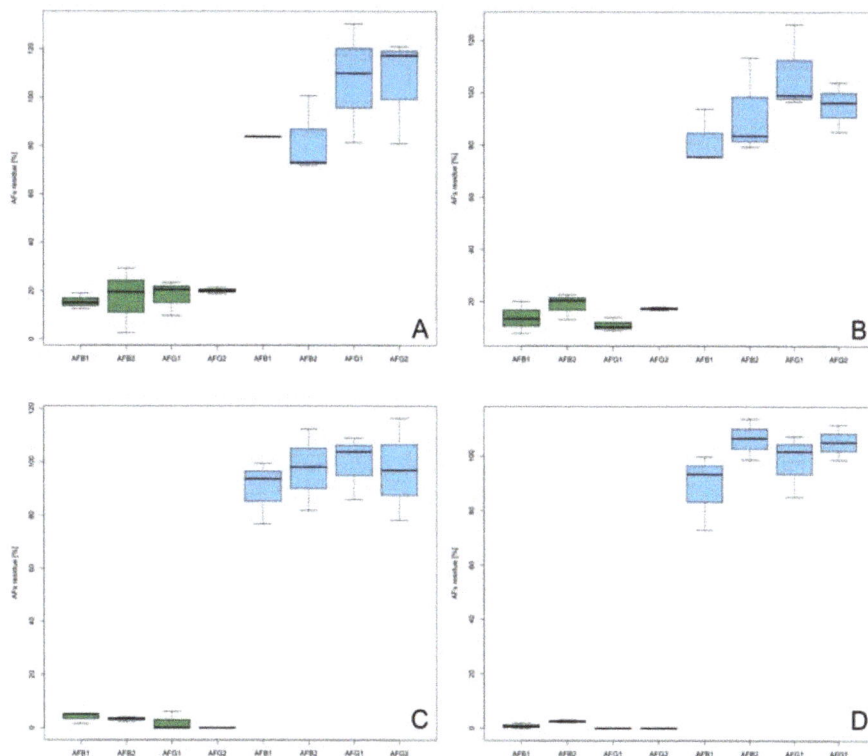

Figure 2. Residual AFB$_1$, AFB$_2$, AFG$_1$, and AFG$_2$ on hazelnuts (green) and perisperm (blue). (**A**) 120 °C for 20 min with static hot air roaster; (**B**) 120 °C for 40 min with static hot air roaster; (**C**) 140 °C for 20 min with infrared rays roaster; (**D**) 140 °C for 40 min with infrared rays roaster. All treatments were applied on TGT hazelnuts. Boxes and whiskers meaning is described in Figure 1.

Infrared rays roasting confirmed a lower level for the four AFs (<10%) similar at 20 and 40 min of exposure. Static hot air treatment also showed a similar trend at both exposure times, but the final residue was higher. The perisperm analysis revealed a high concentration of aflatoxins: 72%–126% for the static hot air method, and 73%–114% for infrared rays roasting.

3. Conclusions

Both treatments were useful for the detoxification of AFs in hazelnuts without affecting the nutritional properties. After roasting, minor changes occurred in FAs, but the FAs profile was preserved, and unsaturated fatty acids were maintained at high levels. Hot static air roasting induced a detoxification in both types of hazelnuts, with some differences due to the temperature. AFB$_1$ and AFG$_1$ were always lower than AFB$_2$ and AFG$_2$ after roasting. When both heating methods were tested under the same conditions, infrared rays roasting induced a higher decontamination. Both thermic treatments were unable to degrade AFs which were accumulated in the perisperm. In conclusion, infrared rays roasting could be considered a promising method for aflatoxin detoxification during hazelnut processing.

4. Materials and Methods

4.1. Chemicals

LC-MS grade methanol, acetonitrile, and acetic acid used as mobile phases and as extraction solvents were purchased from Sigma-Aldrich (St. Louis, MO, USA). NaCl, KCl, Na_2HPO_4, and KH_2PO_4 used to prepare phosphate-buffered saline (PBS) solution were purchased from Merck (Darmstadt, Germany) and dissolved in ultrapure water (Maina, Turin, Italy). Pure standards of AFB_1 (purity $\geq$ 98%), AFB_2 (purity $\geq$ 98%), AFG_1 (purity $\geq$ 98%), and AFG_2 (purity $\geq$ 98%) were purchased from Sigma-Aldrich. AflaClean select immunoaffinity columns (IAC) were obtained from LCTech (Dorfen, Germany). A working solution of AFs at the concentration of 1 mg/mL was prepared in methanol and diluted in order to contaminate hazelnuts and to prepare calibration curves.

4.2. Samples and Contamination Procedure

Hazelnuts used in this work came from Italy and Turkey. The Italian cultivar, "Tonga Gentile Trilobata" (TGT) was harvested in Cortemilia, northern Italy, and provided by La Gentile s.r.l. Turkish hazelnuts were a blend of three major cultivars—"Tombul", "Palaz", and "Kalinkara"—from the Ordu region, northern Turkey. All the hazelnuts used were free from aflatoxins. All samples were divided in two parts—one for detoxification experiments and the other for quality analyses; all the hazelnuts were stored in plastic bags, in the dark, at low relative humidity and a temperature of 4 °C before the experiments.

Dehulled raw hazelnuts were artificially contaminated by spraying at the final concentration of 100 ng/g for each of the four aflatoxins in the first set of experiments. In the second set of experiments, the nuts were contaminated with 8 ng/g of AFB_1 and 2 ng/g of the other three AFs (14 ng/g for total aflatoxins).

4.3. Roasters and Other Equipment

Hazelnuts were roasted at the Brovind–GBV s.r.l. company (Cortemilia, CN, Italy) using different equipment. Known quantities of contaminated hazelnuts were roasted in two ovens at specific temperatures and for selected times.

Hot air static roaster model SD-80. The SD-80 static hot air roaster is a laboratory batch roaster, which does not include product mixing and stirring, and hazelnuts are placed on a perforated grill. Roasting is performed by forced circulation of air heated through electric resistance. Working parameters are set up from the master control panel.

Infrared rays roaster model RI/800. The RI/800 roasting oven is a discontinuous laboratory batch machine equipped with an infrared rays heating system. Through a conical hopper, the raw material is introduced into the roasting chamber. Hazelnuts are heated through a battery of IR lamps, and gentle stirring and mixing is supplied by a vibrating tray, which guarantees the homogeneity of the process. After roasting, the product is expelled through a gate into the cooling unit. Working parameters (temperature, time, and vibration) are set from the master control panel.

Cooler/peeler model PR/1. The PR/1 cooler/peeler is made of a circular tank. Hazelnuts are gently stirred by mixing blades, which guarantee a uniform cooling, and nuts are peeled by friction. Cooling is guaranteed by an aspirating collector and an electro-fan with flux regulator.

Pellicle extractor model CSACK/1. The CSACK/1 pellicle aspirator is used to extract powders and fragments of perisperm with a centrifugal electro-fan and store them in a bag.

4.4. Treatments

In accordance with the commercial hazelnut treatments, in the first experimental set, two temperatures were chosen for the experiments with the static hot air roaster (120 °C and 170 °C), and two exposure times (20 and 40 min) were applied for each temperature.

In the second experimental set, a new roasting technology—infrared rays roasting—was compared to static hot air roasting. One temperature (140 °C) and two exposure times (20 min and 40 min) were used for both methods. First and second experiments were performed on Italian and Turkish hazelnuts.

A third trial was developed with both roasting methods only on Italian hazelnuts, at the same temperatures (120 °C and 140 °C) and exposure times (20 and 40 min) already tested. After treatments, the hazelnut perisperm was collected separately for the analysis of aflatoxin contents.

4.5. Fatty Acid Composition

Fatty acid methyl esters (FAMEs) were determined by gas chromatography according to the method described by Ficarra et al. [29] with slight modification.

Briefly, 50 mg of oil was mixed thoroughly with 1 mL of hexane and 300 μL of 2 M KOH in methanol (*w*/*v*) in a dark tube. The extract was then transferred into a dark glass vial and immediately analyzed by GC. Profiling of the FAMEs was determined using a GC-2010 Shimadzu gas chromatograph (Shimadzu, Kyoto, Japan) equipped with a flame ionization detector, split-splitless injector, an AOC-20i autosampler, and a SP-2560 capillary column (100 m × 0.25 mm id × 0.20 μm, Supelco, Milan, Italy). The following temperature program was used: the initial oven temperature was 165 °C, increasing to 200 °C at 3 °C/min, and then the temperature was held at 200 °C for 45 min. The injector temperature and the detector were 250 °C. Each fatty acid methyl ester was identified and quantified by comparing retention times with Supelco 37 components FAME mix 10 mg/mL. The fatty acid concentration was expressed in relative percentage of each fatty acid, calculated by internal normalization of the chromatographic peak areas. The obtained fatty acid composition was used to calculate the sum of the saturated ($\sum$ SFA), monounsaturated ($\sum$ MUFA), and polyunsaturated ($\sum$ PUFA) fatty acids, as well as the ratio ($\sum$ MUFA + $\sum$ PUFA)/($\sum$ SFA). Moreover, to evaluate the oxidative stability, the ratio of oleic to linoleic (O/L) and the iodine value (IV) were determined. The IV was determined according to the percentages of fatty acids using the following formula: (palmitoleic acid × 1.901) + (oleic acid × 0.899) + (linoleic acid × 1.814) + (linolenic acid × 2.737). Significant differences ($p < 0.05$) among the means were determined using a one-way analysis of variance (ANOVA) with Duncan's test at a fixed level of a = 0.05. All calculations were performed with the STATISTICA software for Windows (Release 7.0; StatSoft Inc., Tulsa, OK, USA).

4.6. Extraction, Clean-Up, and LC-MS/MS Conditions

The AFs were extracted and analyzed using the validated method described by Prelle et al. [30]. The analyses were performed on twenty-five grams of ground hazelnuts. For perisperm analysis, 2 g hazelnut skins, obtained after roasting, were extracted for 2 h on a shaker at 165 rpm with 10 mL of mixture solution (methanol:water; 80:20). Samples were centrifuged at 4000 rpm and subsequently filtered by using a Whatman CA 0.45 μm syringe filter (Maldstone, UK). Filtrate was diluted 1:4 in phosphate buffer solution (PBS) for clean-up using IAC. All samples were analyzed in triplicate.

Analyses were performed using a Varian Model 212-LC micro pumps (Palo Alto, CA, USA) with a Varian autosampler Model 410 Prostar coupled with a Varian 310-MS triple quadrupole mass spectrometer with an electrospray ion source operating in positive ionization mode. Chromatographic separation was performed with a Zorbax Eclipse Plus C18 (100 mm × 4.6 mm, 3.8 μm particle size, Agilent, Santa Clara, CA, USA) column using H_2O and CH_3OH as eluents, both acidified with 0.1% CH_3COOH at the flow rate of 0.2 mL/min. Monitoring reaction mode (MRM) transitions used for quantification were: 313 > 285 (CE 14 V) for AFB_1, 315 > 287 (CE 18 V) for AFB_2, 329 > 243 (CE 18 V) for AFG_1, 331 > 245 (CE 24 V) for AFG_2. AFs quantification was performed using external calibration based on serial dilution of a multi-analyte stock solution.

Supplementary Materials: The following are available online at www.mdpi.com/2072-6651/9/2/72/s1, Table S1: Fatty acid distribution (%) of TGT hazelnuts before and after treatment with static hot air at two temperatures (120 °C and 170 °C) at the two different exposure times (20 and 40 min), Table S2: Fatty acid distribution (%) of

Turkish hazelnuts before and after treatment with static hot air at two temperatures (120 °C and 170 °C) at the two different exposure times (20 and 40 min).

Acknowledgments: Work carried out in the framework of the project "ITACA—Technological Innovation, Automation and new Analytical Controls to improve the quality and safety of the Piedmontese food products" funded by the Piedmont Region (Italy) and "Food Digital Monitoring" (Framework Program MIUR – Piedmont Regione, Action 3 "Smart Factory", Call "Smart Factory Technological Platform". The Authors gratefully acknowledge Ambra Prelle for her support in the chemical analyses.

Author Contributions: D.S. and G.Z. conceived and designed the experiments; I.S. and B.D.B. performed the experiments; I.S., G.Z. and B.D.B. analyzed the data; M.L.G. and G.Z. contributed reagents/materials/analysis tools; I.S., M.L.G. and D.S. wrote the paper.

Conflicts of Interest: The authors declare no conflict of interest. The founding sponsors had no role in the design of the study; in the collection, analyses, or interpretation of data; in the writing of the manuscript, and in the decision to publish the results.

References

1. Kurtzman, C.P.; Horn, B.W.; Hesseltine, C.W. *Aspergillus nomius*, a new aflatoxin-producing species related to *Aspergillus flavus* and *Aspergillus tamarii*. *Antonie van Leeuwenhoek* **1987**, *53*, 147–158. [CrossRef] [PubMed]
2. Prelle, A.; Spadaro, D.; Garibaldi, A.; Gullino, M.L. Co-occurrence of aflatoxins and ochratoxin A in spices commercialized in Italy. *Food Control* **2014**, *39*, 192–197. [CrossRef]
3. Di Stefano, V.; Avellone, G.; Pitonzo, R.; Capocchiano, V.G.; Mazza, A.; Cicero, N.; Dugo, G. Natural co-occurrence of ochratoxin A, ochratoxin B and aflatoxins in Sicilian red wines. *Food Addit. Contam. Part A* **2015**, *32*, 1343–1351. [CrossRef] [PubMed]
4. Siciliano, I.; Spadaro, D.; Prelle, A.; Vallauri, D.; Cavallero, M.C.; Garibaldi, A.; Gullino, M.L. Use of cold atmospheric plasma to detoxify hazelnuts from aflatoxins. *Toxins* **2016**, *8*, 125. [CrossRef] [PubMed]
5. Wogan, G.N. Chemical nature and biological effects of the aflatoxins. *Bacteriol. Rev.* **1966**, *30*, 460–470. [PubMed]
6. Massey, T.E.; Stewart, R.K.; Daniels, J.M.; Liu, L. Biochemical and molecular aspects of mammalian susceptibility to aflatoxin B_1 carcinogenicity. *Proc. Soc. Biol. Med.* **1995**, *208*, 213–227. [CrossRef]
7. IARC Monograph. Available online: http://monographs.iarc.fr/ENG/Monographs/vol82/ (accessed on 20 February 2017).
8. Codex Alimentarius. Codex General Standard for Contaminants and Toxins in Food and Feed. 1995. Available online: http://www.fao.org/fileadmin/user_upload/agns/pdf/CXS_193e.pdf (accessed on 8 July 2016).
9. Abdulkadar, A.H.W.; Al-Ali, A.; Al-Jedah, J. Aflatoxin contamination in edible nuts imported in Qatar. *Food Control* **2000**, *11*, 157–160. [CrossRef]
10. Thuvander, A.; Moller, T.; Barbieri, H.E.; Jansson, A.; Salomonsson, A.C.; Olsen, M. Dietary intake of some important mycotoxins by the Swedish population. *Food Addit. Contam.* **2001**, *18*, 696–706. [CrossRef] [PubMed]
11. Chun, H.S.; Kim, H.J.; Ok, H.E.; Hwang, J.B.; Chung, D.H. Determination of aflatoxin levels in nuts and their products consumed in South Korea. *Food Chem.* **2007**, *102*, 385–391. [CrossRef]
12. Kabak, B.; Dobson, A.D.W.; Var, I. Strategies to prevent mycotoxin contamination of food and animal feed: A review. *Crit. Rev. Food Sci. Nutr.* **2006**, *46*, 593–619. [CrossRef] [PubMed]
13. Chen, R.; Ma, F.; Li, P.; Zhang, W.; Ding, X.; Zhang, Q.; Li, M.; Wang, Y.R.; Xu, B.C. Effect of ozone on aflatoxins detoxification and nutritional quality of peanuts. *Food Chem.* **2014**, *146*, 284–288. [CrossRef] [PubMed]
14. Di Stefano, V.; Pitonzo, R.; Avellone, G. Effect of gamma irradiation on aflatoxins and ochratoxin A reduction in almond samples. *J. Food Res.* **2014**, *3*, 113–118.
15. Mao, J.; He, B.; Zhang, L.; Li, P.; Zhang, Q.; Ding, X.; Zhang, W. A structure identification and toxicity assessment of the degradation products of aflatoxin B_1 in peanut oil under UV irradiation. *Toxins* **2016**, *8*, 332. [CrossRef] [PubMed]
16. Özdemir, M.; Açkurt, F.; Yildiz, M.; Biringen, G.; Gürcan, T.; Löker, M. Effect of roasting on some nutrients of hazelnuts (*Corylus avellana* L.). *Food Chem.* **2001**, *73*, 185–190. [CrossRef]

17. FAOSTAT. Statistic Division. Available online: http://www.fao.org/faostat/en/#data (accessed on 16 January 2017).
18. Alasalvar, C.; Pelvan, E.; Topal, B. Effects of roasting on oil and fatty acid composition of Turkish hazelnut varieties (*Corylus avellana* L.). *Int. J. Food Sci. Nutr.* **2010**, *61*, 630–642. [CrossRef] [PubMed]
19. Kirbaşlar, F.G.; Erkmen, G. Investigation of the effect of roasting temperature on the nutritive value of hazelnuts. *Plant Foods Hum. Nutr.* **2003**, *58*, 1–10. [CrossRef]
20. Rustom, I.Y.S. Aflatoxin in food and feed: Occurrence, legislation and inactivation by physical mathods. *Food Chem.* **1997**, *96*, 57–67. [CrossRef]
21. Pluyer, H.R.; Ahmed, E.M.; Wei, C.I. Destruction of aflatoxin on peanuts by oven- and microwave-roasting. *J. Food Protect.* **1987**, *50*, 504–508. [CrossRef]
22. Yazdanpanah, H.; Mohammadi, T.; Abouhossain, G.; Cheraghali, A.M. Effect of roasting on degradation of aflatoxins in contaminated pistachio nuts. *Food Chem. Toxicol.* **2005**, *43*, 1135–1139. [CrossRef] [PubMed]
23. Soliman, K.M. Incidence, level, and behavior of aflatoxins during coffee bean roasting and decaffenation. *J. Agric. Food Chem.* **2002**, *50*, 5567–5573. [CrossRef]
24. Vujević, P.; Petrović, M.; Vahčić, N.; Milinović, B.; Čmelik, Z. Lipids and minerals of the most represented hazelnut varieties cultivated in Croatia. *Ital. J. Food Sci.* **2014**, *26*, 25–29.
25. Amaral, J.S.; Casal, S.; Seabra, R.M.; Oliveira, B.P.P. Effects of roasting on hazelnut lipids. *J. Agric. Food Chem.* **2006**, *54*, 1315–1321. [CrossRef] [PubMed]
26. Belviso, S.; Dal Bello, B.; Giacosa, S.; Bertolino, M.; Ghirardello, D.; Giordano, M.; Rolle, L.; Gerbi, V.; Zeppa, G. Chemical, mechanical and sensory monitoring of hot air- and infraredroasted hazelnuts (*Corylus avellana* L.) during nine months of storage. *Food Chem.* **2017**, *217*, 398–408. [CrossRef] [PubMed]
27. Ciarmiello, L.F.; Piccirillo, P.; Gerardi, C.; Piro, F.; De Luca, A.; D'Imperio, F.; Rosito, V.; Poltronieri, P.; Santino, A. Microwave irradiation for dry-roasting of hazelnuts and evaluation of microwave treatment on hazelnuts peeling and fatty acid oxidation. *J. Food Res.* **2013**, *2*, 22–35. [CrossRef]
28. Rastogi, N.K. Recent trends and developments in infrared heating in food processing. *Crit. Rev. Food Sci. Nutr.* **2012**, *52*, 737–760. [CrossRef] [PubMed]
29. Ficarra, A.; Lo Fiego, D.P.; Minelli, G.; Antonelli, A. Ultra fast analysis of subcutaneous pork fat. *Food Chem.* **2010**, *121*, 809–814. [CrossRef]
30. Prelle, A.; Spadaro, D.; Garibaldi, A.; Gullino, M.L. Aflatoxin monitoring in Italian hazelnut products by LC-MS. *Food Addit. Contam. Part B* **2012**, *5*, 279–285. [CrossRef] [PubMed]

Article

Plasma-Based Degradation of Mycotoxins Produced by *Fusarium*, *Aspergillus* and *Alternaria* Species

Lars ten Bosch [1,*], Katharina Pfohl [2,*], Georg Avramidis [1,*], Stephan Wieneke [1],
Wolfgang Viöl [1,3] and Petr Karlovsky [2]

[1] University of Applied Sciences and Arts, Faculty N, Von-Ossietzky-Strasse 99/100, 37085 Göttingen,
    Germany; stephan.wieneke@hawk-hhg.de (S.W.); wolfgang.vioel@hawk-hhg.de (W.V.)
[2] Molecular Phytopathology and Mycotoxin Research, Georg-August-University Göttingen,
    Grisebachstrasse 6, 37077 Göttingen, Germany; pkarlov@gwdg.de
[3] Fraunhofer IST Application Centre, Von-Ossietzky-Strasse 100, 37085 Göttingen, Germany
* Correspondence: lars.bosch@hawk-hhg.de (L.t.B.); kpfohl@gwdg.de (K.P.);
    georg.avramidis@hawk-hhg.de (G.A.); Tel.: +49-551-3705-361 (L.t.B.); +49-551-39-13230 (K.P.);
    +49-551-3705-293 (G.A.)

Academic Editor: Ting Zhou
Received: 13 January 2017; Accepted: 7 March 2017; Published: 10 March 2017

Abstract: The efficacy of cold atmospheric pressure plasma (CAPP) with ambient air as working gas for the degradation of selected mycotoxins was studied. Deoxynivalenol, zearalenone, enniatins, fumonisin B1, and T2 toxin produced by *Fusarium* spp., sterigmatocystin produced by *Aspergillus* spp. and AAL toxin produced by *Alternaria alternata* were used. The kinetics of the decay of mycotoxins exposed to plasma discharge was monitored. All pure mycotoxins exposed to CAPP were degraded almost completely within 60 s. Degradation rates varied with mycotoxin structure: fumonisin B1 and structurally related AAL toxin were degraded most rapidly while sterigmatocystin exhibited the highest resistance to degradation. As compared to pure compounds, the degradation rates of mycotoxins embedded in extracts of fungal cultures on rice were reduced to a varying extent. Our results show that CAPP efficiently degrades pure mycotoxins, the degradation rates vary with mycotoxin structure, and the presence of matrix slows down yet does not prevent the degradation. CAPP appears promising for the decontamination of food commodities with mycotoxins confined to or enriched on surfaces such as cereal grains.

Keywords: DBD; atmospheric pressure; low temperature plasma; mycotoxins; degradation

---

1. Introduction

Phytopathogenic fungi infect crops in the field (pre-harvest spoilage) while spoilage fungi colonize harvested commodities during storage (post-harvest spoilage). Besides the reduction of yield and quality, infection with fungal pathogens often leads to contamination with mycotoxins [1,2]. These toxic fungal metabolites have the potential to harm the health of consumers and livestock. Reduction of mycotoxin content in food and feedstuff is therefore an important goal of food and feed safety improvement.

Prevention of fungal contamination is the primary means of agricultural and food industry when it comes to compliance with maximum limits for mycotoxin content. In current production systems, however, even the best agricultural and manufacturing practices cannot fully prevent mycotoxin contamination. Degradation of toxic metabolites may be used to decontaminate food and feed products. Since most mycotoxins exhibit a high chemical stability, development of decontamination methods compatible with food quality standards is a challenging task. Over the last decades chemical, biological and physical strategies for the degradation of mycotoxins and the effect of food processing technologies on mycotoxin content were investigated extensively [3–5]. Among physical treatments mainly heating, irradiation and washing were studied. Mineral and organic mycotoxin binders are established since decades in animal production. More recently chemical and biological decontamination methods were studied. Among chemical methods, successful application of acids, bases, oxidizing agents, chlorinating agents, formaldehyde and ammoniation was described, especially for the decontamination of aflatoxin- and ochratoxin A-contaminated feeds. Although biological and enzymatic strategies have been developed since 1960's [4] physical techniques still offer the most efficient removal of mycotoxins from food and feed [5].

Technical plasma is a novel physical method with a great potential as a post-harvest treatment method for mycotoxin mitigation. Plasma has been successfully used for sterilization and in plasma medicine [6–11]. Recent application of cold atmospheric pressure plasma (CAPP) in breaking seed dormancy and destruction of plant pathogens showed that the technology is suitable for sensitive biological materials [12–15]. Plasma of different types were used in studies of the inhibition of mycotoxin production and mycotoxin degradation. Ouf et al. [16] demonstrated inhibition of the synthesis of fumonisin B2 and ochratoxin A by *A. niger* after treatment with an atmospheric pressure argon plasma jet. Park et al. [17] successfully degraded aflatoxin B1 (AFB1), deoxynivalenol (DON) and nivalenol (NIV) within 5 s using a microwave-induced argon jet at atmospheric pressure.

Physical and chemical treatment of plant products bear a risk of reducing nutritional value and negatively affecting the palatability and sensory quality of the product. Long treatment duration, required by some methods for satisfactory decontamination of large quantities of goods, may increase the risk of these side effects. The effect of physical and chemical decontamination on nutritional value and quality of food commodities has rarely been systematically investigated. Because the energy of free electrons and excited ions and molecular species in CAPP exceeds the dissociation energy of a C-C bond, organic molecules in the discharge are subjected to unspecific degradation. This feature of CAPP was used to degrade chemically stable pollutants in gaseous phase (e.g., [18]). Low penetration depth of CAPP protects nutrients in bulk material from degradation, limiting degradation to a thin surface layer [19]. Kříž et al. [20,21] showed that the content of proteins and fibers as well as residual dry matter, nitrogen-free extract, fat and ash in intact barley grains were not significantly affected by treatment with CAPP while the content of selected mycotoxins was reduced by 20%–70%. We hypothesize that the confinement of mycotoxin contamination to the surface of grains accounted for the selectivity of the degradation in this study.

In the presented study, a dielectric barrier discharge (DBD) operated with ambient air at atmospheric pressure was utilized for the degradation of mycotoxins. Preliminary experiments using a similar setup and comparable dissipated discharge power applied on pea seeds (*Pisum sativum*) revealed no decline of seed germination rate, thus indicating negligible thermal effect due to plasma treatment [22]. The aim of this work was to test the effectiveness of cold atmospheric pressure air plasma based on a dielectric barrier discharge for the degradation of selected mycotoxins important in food and feed safety.

2. Results

2.1. Treatment of Cover Glasses

Mycotoxin solutions were applied on untreated and 5 s pre-treated cover glasses in order to find out whether the activated surface of plasma-pretreated cover glasses (Section 4.3) might chemically affect mycotoxins. Mycotoxin content was determined as described in Section 4.4. As shown in Figure 1, pre-treatment of cover glasses with air plasma did not significantly affect mycotoxins on glass surface.

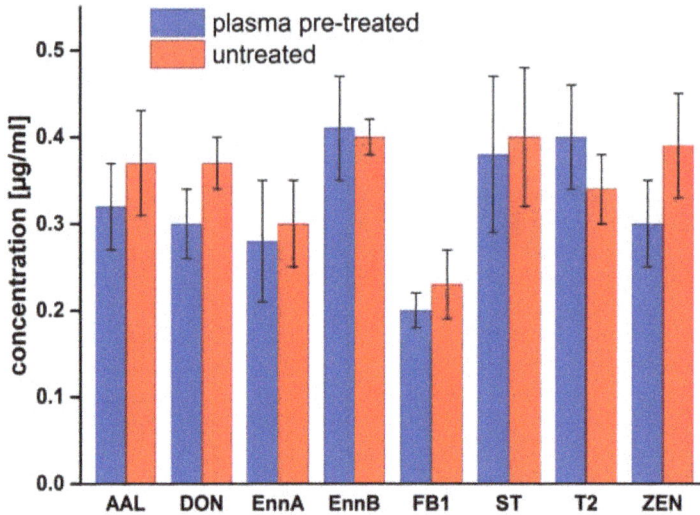

Figure 1. Effect of cover glass pretreatment on mycotoxins. Round cover-glasses were pre-treated with air-plasma for 5 s or not treated (controls). The further sample preparation was executed as described in Section 4.3. Significance of differences between treatments and control was tested by *t*-test at *p* = 0.05 with correction for multiple testing, according to Bonferroni. No significant difference was found.

2.2. Treatment of Pure Mycotoxins and Fungal Extracts

AAL toxin, FB1, DON, ZEN, EnnA, EnnB, T2-toxin and ST as pure compounds covering the surface of cover glasses were subjected to air-plasma for 5 s, 10 s, 20 s, 30 s, and 60 s. HPLC-MS/MS analysis of residues on the glass surface revealed that the plasma treatment led to time-dependent degradation of all mycotoxins (Figure 2a,b). Searching for degradation products by HPLC-MS in a full-scan mode failed to detect any distinct MS signal.

The degradation kinetics seemingly followed an exponential decay. Since distinct differences among the toxins were apparent in their degradation rates, the measured data sets were fitted by an exponential function (Figure 2a,b).

$$C(t) = y_0 + C_0 \cdot e^{-t/\tau} \qquad (1)$$

$C(t)$ is the concentration at time t, C_0 is the initial concentration, y_0 represents the threshold and τ represents the half-life. Half-life values calculated for all toxins by fitting Equation (1) to the data sets (Figure 2) are shown in Table 1.

Table 1. Half-life at $\approx$4 W/cm^2, molecular mass and chemical formula of the mycotoxins.

Mycotoxin	Half-Life τ [s]	Molecular Mass [Da]
Sterigmatocystin	5.0 ± 0.4	324.3
Enniatin A	4.5 ± 0.5	681.9
Zearalenone	4.2 ± 0.5	318.4
Deoxynivalenol	4.0 ± 0.7	296.3
T2-toxin	3.6 ± 0.1	466.5
Enniatin B	3.1 ± 0.2	639.8
AAL-toxin	1.9 ± 0.4	521.6
Fumonisin B1	1.9 ± 0.3	721.8

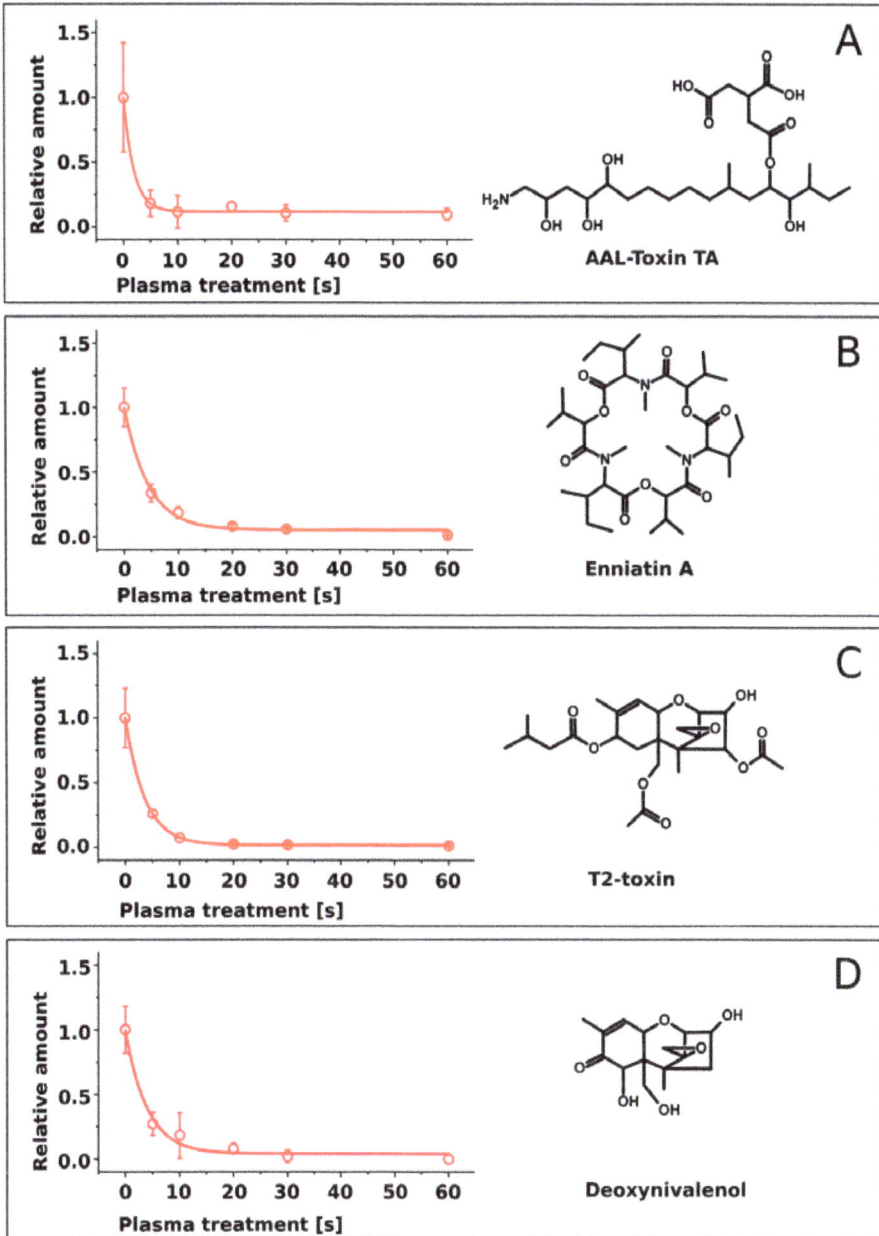

(a)

Figure 2. *Cont.*

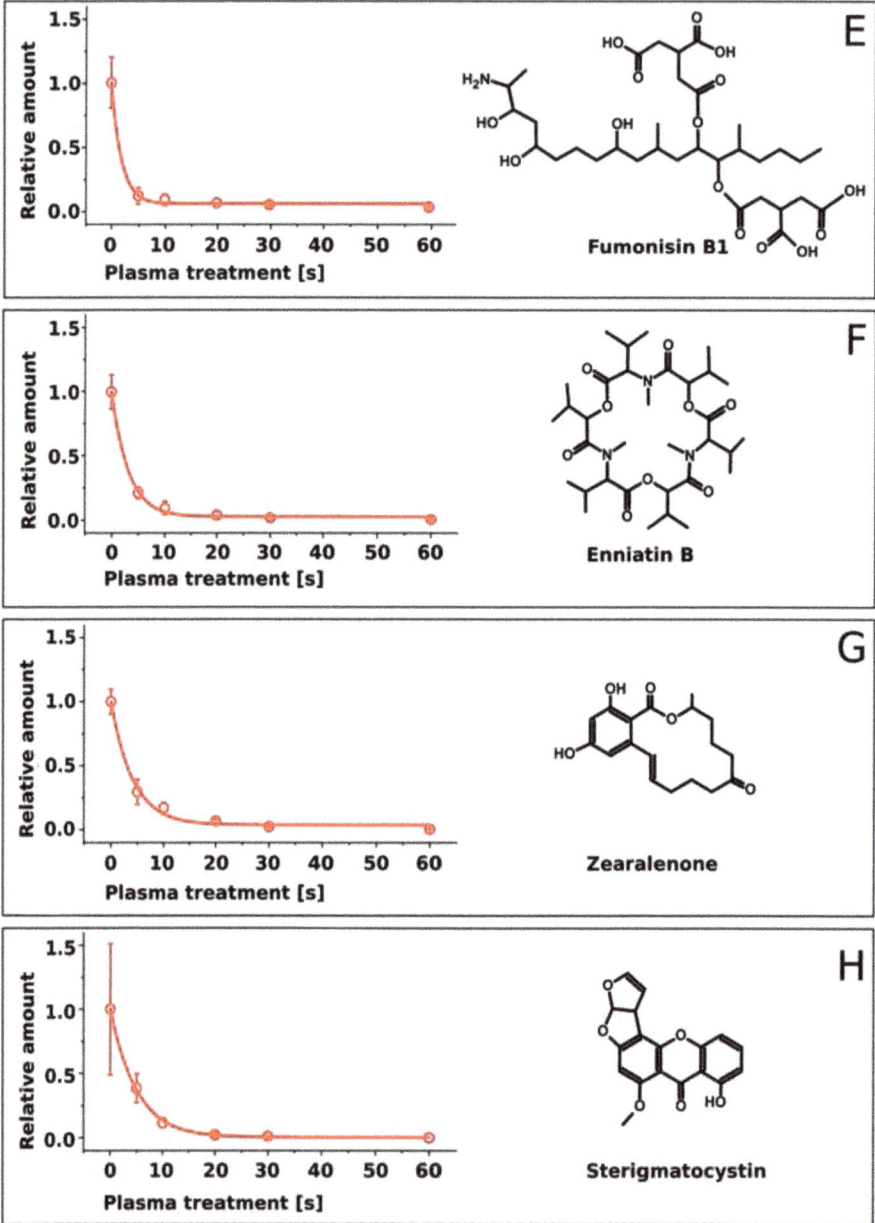

(b)

Figure 2. (a). Time-dependent decay of four pure mycotoxins exposed to air plasma ($N = 5$). (A) AAl-Toxin TA, (B) Enniatin A, (C) T2-toxin, (D) Deoxynivalenol; **(b)**. Time-dependent decay of four pure mycotoxins exposed to air plasma ($N = 5$). (E) Fumonisin B1, (F) Enniatin B, (G) Zearalenone, (H) Sterigmatocystin.

Four mycotoxins (FB1, EnnB, ST and ZEN) were selected to investigate the effect of matrix on mycotoxin degradation by plasma. Extracts of rice cultures of fungal strains producing these mycotoxins (*Fusarium verticillioides*, *Fusarium avenaceum*, *Aspergillus nidulans* and *Fusarium graminearum*), containing approx. 100 µg/mL of each toxin, were exposed to air plasma under the same conditions as pure compounds. The degradation rates were reduced as compared to pure compounds for all four mycotoxins (Figure 3). Particularly strong reduction of degradation rates was observed for FB1 and EnnB; nearly half of these toxins remained intact at the end of the treatment. Moreover, the course of the degradation of FB1 and EnnB was linear, contrasting to exponential decay of pure compounds. Degradation rates of ZEN and ST were reduced to a lesser extent and the progress of degradation of these toxins in matrix followed exponential decay similarly as the degradation of pure compounds (Figure 3).

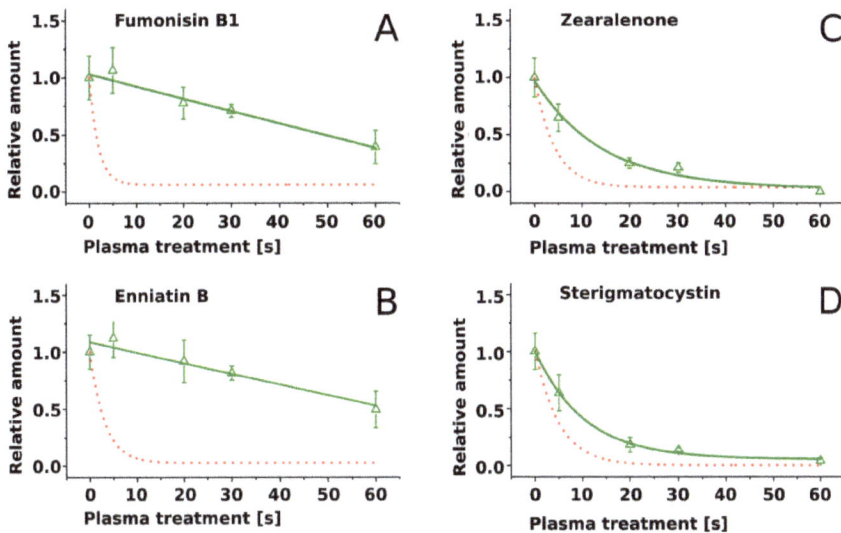

Figure 3. Time-dependent decay of mycotoxins embedded in rice extract (green line) exposed to air plasma (*N* = 5). (**A**) Fumonisin B1; (**B**) Enniatin B; (**C**) Zearalenone; (**D**) Sterigmatocystin. The red dotted line displays the decay slopes of the respective pure mycotoxin standards as shown in Figure 2b.

3. Discussion

All investigated mycotoxins, when present as pure compounds, showed a distinct decay within a few seconds of plasma treatment duration and were reduced by approx. 2 log-ranges within 30 s as shown in Figure 2a,b. It is assumed that the energy dissipated in the discharge gap induces a combination of different degrading mechanisms acting on the toxins such as chemical reactions with reactive species generated in the plasma volume such as O, O_3, OH, and NO_x [23,24] and/or decomposition after collision with electrons and ions [25,26] leading to cleavage of molecular bonds. Further reactions with plasma species can result in fragmentation and generation of volatile compounds. Decomposition of organic compounds into volatile products such as CO, CO_2, and H_2O during exposition of various polymer materials to oxygen-containing plasmas is a known phenomenon that was reported by several authors (e.g., [27–29]).

No stable residues of toxin degradation could be detected with HPLC-MS. Rapid degradation of toxin fragments into volatile products can be expected in analogy with the results of the study of Doraj and Kushner [27] in which degradation of polypropylene was elucidated. We assume that mycotoxin fragments were rapidly converted into volatile compounds which were immediately removed by the gas stream in the discharge gap. Therefore, future work will be dedicated to mass spectrometric investigation of the plasma effluent.

Our results (Figure 2a,b) are in accordance with the degradation of aflatoxin B1, DON, and nivalenol by Park et al. [17] who used a microwave argon plasma jet as well as by Ouf et al. [16] who demonstrated the reduction of fumonisin B2 and ochratoxin A by an atmospheric pressure argon kHz-operated jet. It should be kept in mind that sample temperature measured after the treatment by different approaches differed significantly: 35 °C were estimated after 9 min of treatment by Ouf et al. [16] and 105 °C after 5 s of treatment by Park et al. [17,30]. Elevated temperatures may cause thermal degradation. Since gas temperature and substrate temperature in this study did not exceed 60 °C, no thermal degradation of mycotoxin is expected.

Decay curves of individual toxins showed distinctly different degradation kinetics (Figure 2a,b). These differences were quantified by exponential fitting (Table 1). FB1 and AAL-toxin were degraded at the highest rate ($\tau = 1.9 \pm 0.3$) and ST at the lowest rate ($\tau = 5.0 \pm 0.4$).

The decay rates did not correlate with molecular mass. For example, EnnA with a molecular mass of 681.9 Da showed a similar decay rate as ST with a mass of 324.3 Da. The degradation rate might however be affected by the chemical structure. FB1 and structurally related AAL-toxin with long aliphatic chains were degraded rapidly, while ST with a compact structure of condensed aromatic rings had the highest half life. Most other mycotoxins with intermediate decay rates possessed mixed structures of condensed rings and aliphatic chains.

A relationship between the degradation rate of chemical compounds and their molecular structure was described by Gröning et al. [31] investigating plasma-based decay of polymers. These researchers suggested a buffering effect of aromatic structures on the degradation by low-pressure air-plasma. Klarhöfer et al. [32] reported higher resistance of lignin (which contains aromatic structures) exposed to an air DBD compared to cellulose and ascribed this observations to a similar mechanism. Aromatic structures occurring in mycotoxins might therefore slow down plasma-induced degradation. Future investigations will focus on degradation pathways, especially considering wide-spread toxins lacking aromatic rings. It is desirable to verify the postulated structural effects on degradation and identify reactive species in the plasma accountable for the degradation. Degradation of chemical constituents of solid materials by cold plasma is confined to thin surface layers [19]. Mycotoxins produced by Fusarium species in small-grain cereals are often enriched in the outer layers of grains [33,34]; this circumstance may facilitate selective degradation of mycotoxins in grains by cold plasma with a small loss of nutrients.

Mycotoxins imbedded in extracts of fungal cultures were degraded with lower rates than pure compounds. Presumably, components of the extracts scavenged reactive molecular species in the plasma, shielding mycotoxins from degradation. Activated components of the extracts are expected to rapidly react with other compounds including mycotoxins. Because the presence of matrix significantly reduced the degradation rates of mycotoxins, secondary chemical reactions apparently did not compensate for the loss of reactive molecular species by scavenging. We hypothesize that the effect of matrix on the degradation of other mycotoxins by plasma will be similar. In spite of the protective effects of culture extracts, significant decay of mycotoxins in culture matrix occurred, suggesting that plasma-based methods are promising for mycotoxin degradation in thin surface layers even in the presence of complex matrices.

4. Experimental Setup and Materials

4.1. Mycotoxin Standards

Pure mycotoxin standards of AAL toxin (TA1 + TA2), enniatin A (Enn A), enniatin B (Enn B), fumonisin B1 (FB1), sterigmatocystin (ST), deoxynivalenol, T2-toxin and zearalenone (ZEN) were purchased from Sigma Aldrich (Munich, Germany). Stock solutions were prepared in LC-MS grade methanol (Th. Geyer GmbH, Renningen, Germany).

4.2. Fungal Strains, Rice Cultures and Mycotoxin-Containing Fungal Extracts

Fusarium verticillioides VP2 [35] was kindly provided by Francesca Cardinale (University of Turin, Turin, Italy). *F. avenaceum* DSM 21724 was purchased from Deutsche Sammlung von Mikroorganismen und Zellkulturen (DSMZ, Braunschweig, Germany). *F. graminearum* IFA 66 was kindly provided by Marc Lemmens (Institute of Biotechnology in Plant Production, Tulln, Austria) via Thomas Miedaner (University of Hohenheim, Stuttgart, Germany). *F. graminearum* Fg71 was kindly provided by Thomas Miedaner. *Aspergillus nidulans* RDIT2.3 was kindly provided by Marko Rohlfs (University of Göttingen, Goettingen, Germany).

Fungal spores of *Fusarium* strains were produced according to Bai [36] with modification (Becker et al. [37]). The spores were suspended in sterile tap water and stored at −60 °C. Rice cultures were prepared as described (Nutz et al. [38]). All mycotoxins except fumonisins were extracted from 5 g rice cultures with 40 mL acetonitrile overnight with shaking, while fumonisins were extracted with methanol. An aliquot of 1 mL was dried, redissolved in 1 mL of methanol/water (1:1) and the concentration of mycotoxins was determined by HPLC-MS (see below). Remaining supernatants were dried in vacuum and the residues were dissolved and adjusted to a concentration of 100 µg/mL of ST (*A. nidulans* RDIT2.3), FB1 (*Fusarium verticillioides* VP2), Enn B (*F. avenaceum* DSM 21724) and ZEN (*F. graminearum* Fg71 and IFA 66).

4.3. Sample Preparation

Round cover-glasses (thickness 100 µm, diameter 16 mm) were placed onto microscopy slides and pre-treated with air-plasma for 5 s to facilitate an even spread of the toxin solution on the surface by increased surface tension (see e.g., Gerhard et al. [39]). Subsequently, 1.5 µL mycotoxin solution (100 µg/mL in methanol) were centrally applied onto the cover glass; the solution spread over the entire glass area spontaneously. After solvent evaporation at room temperature, the specimen underwent a plasma treatment at constant input power and varying treatment durations (0 s, 5 s, 10 s, 20 s, 30 s and 60 s). Then, the cover glasses were immersed in 0.5 mL methanol for at least 24 h to dissolve residual mycotoxins for analysis. 5 replicates per mycotoxin and treatment were used.

4.4. Mycotoxin Analyses

High performance liquid chromatography was performed as described (Ratzinger et al. [40]) using C18 column (Kinetex, 50.0 mm × 2.1 mm, particle size 2.6 µm; Phenomenex, Aschaffenburg, Germany); for the quantification of enniatins, 10 µM sodium acetate were added to the mobile phase. The analytes were ionized by electrospray and analyzed either by tandem mass spectrometry for quantification of mycotoxins or in a full scan mode while searching for degradation products using an ion trap detector 500 MS (Varian, Darmstadt, Germany). For the identification of mycotoxins retention time, m/z of molecular ions and fragmentation spectra were used. DON and ZEN were detected in a negative mode while all other mycotoxins were analyzed in positive ionization mode. The m/z values for molecular ions and mass transitions used were 522 > 328 for AAL-toxin, 722 > 686 for FB1, 662 > 549 for Enn B, 704 > 387 for Enn A, 325 > 310 for ST, 489 > 387 for T2-toxin, 355 > 295 for DON and 317 > 175 and 317 > 273 for ZEN. Quantification was carried out based on a linear calibration curve constructed with pure external standards. The estimated limits of quantification were 3 ng/mL for FB1, 5 ng/mL for Enn A/B, ST, AAL-toxin and T2-toxin, 10 ng/mL for DON 1 ng/mL and for ZEN. For all mycotoxins analyzed in positive ionization mode full-scan search for degradation products was carried out as described by Ratzinger et al. [40] followed by pairwise comparison of signal intensities after peak alignment and normalization [41].

4.5. Plasma Device

For plasma treatment, samples were positioned on a glass-insulated (float glass; thickness 4 mm) ground electrode (aluminum) so that the discharge gap between the samples and the upper electrode was 2 mm (Figure 4a). The upper electrode (145 × 85 × 28 mm, filling: bronze powder, dielectric: Al$_2$O$_3$, thickness of dielectric 3 mm) was connected to an alternating high-voltage (≈19 kV peak) pulse generator producing bipolar high-voltage pulses of a duration of approximately 1.9 µs and a repetition frequency of 17 kHz.

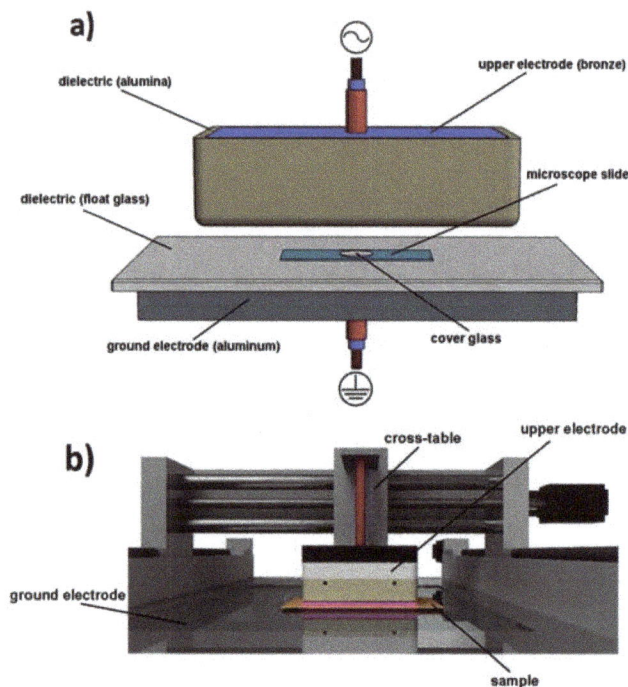

Figure 4. Principle of the electrode configuration (**a**) and scheme of experimental setup (**b**).

Figure 4b depicts the plasma-setup and Table 2 the operational parameters. The upper high voltage electrode is attached to a cross-table and can be moved along 3 axes. One transition over the sample surface results in 5 s of plasma exposition at the chosen movement speed. The treatment of the samples was carried out by several consecutive sample transits by the upper electrode. During treatment compressed air flow was applied in order to cool the system and to homogenize the discharge. The gas temperature, estimation via optical emission spectroscopy (high resolution spectra applying an Echelle spectrograph, see e.g., [42,43]), and substrate temperatures, estimation via thermographic analysis (Fluke TiS Thermal Imager, Fluke Corporation, Everett, WA, USA), did not exceed 330 K or 60 °C after 60 s of treatment.

Table 2. Input parameters for experimental setup.

Input Parameter	Value
power density	≈4 W/cm^2
discharge gap	2 mm
air flow	130 sl/min
appl. voltage	≈38 kV (p-p)
waveform	pulsed sine
gas temperature	$T_{rot} \approx 330$ K

4.6. Measurement of Injected Power

The total energy converted in the gas discharge is an important parameter to characterize dielectric barrier discharges and was determined following the established cyclogram/Lissajous method [44]. The voltage/charge Lissajous figure (Q-U-plot) of the transferred charge and the applied voltage was used to calculate the energy dissipated into the system. The electrical parameters were measured using a high voltage probe (3 pF, 100 MΩ, Tektronix P6015A, Tektronix Inc., Beaverton, OR, USA) and a parallel circuit consisting of a capacitor (200 nF) and a resistor (1 kΩ). This parallel circuit was used for the charge measurement using a Yokokawa DL1740EL Dual 500 MHz oscilloscope (Yokogawa Electric Corp., Musashino, Tokyo, Japan) for digitalization. The voltage applied to the upper electrode was measured directly, whereas the transferred charge was determined by the measurement of the voltage drop on the capacitance (2 $\times$ WIMA FKP1, 100 nF). The calculated power injected to the system was approx. 500 W for all sample treatments. Considering the electrode geometry, a power density of $\approx 4/cm^2$ was calculated as depicted in Table 2.

Acknowledgments: This work was funded by *Niedersächsisches Vorab: Volkswagen Stiftung (ZN2779)*. The authors would like to thank all colleagues involved.

Author Contributions: Georg Avramidis, Petr Karlovsky, Katharina Pfohl and Lars ten Bosch conceived and designed the experiments; Georg Avramidis and Lars ten Bosch performed the experiments; Georg Avramidis, Petr Karlovsky, Katharina Pfohl and Lars ten Bosch analyzed the data; Stephan Wieneke, Wolfgang Viöl and Petr Karlovsky contributed reagents/materials/analysis tools; Georg Avramidis, Petr Karlovsky, Katharina Pfohl and Lars ten Bosch wrote the paper.

Conflicts of Interest: The authors declare no conflict of interest.

References

1. Pestka, J.J. Deoxynivalenol: Mechanisms of action, human exposure, and toxicological relevance. *Arch. Toxicol.* **2010**, *84*, 663–679. [CrossRef] [PubMed]
2. Placinta, C.M.; D'Mello, J.P.F.; Macdonald, A.M.C. A review of worldwide contamination of cereal grains and animal feed with Fusarium mycotoxins. *Anim. Feed Sci. Technol.* **1999**, *78*, 21–37. [CrossRef]
3. He, J.; Zhou, T.; Young, J.C.; Boland, G.J.; Scott, P.M. Chemical and biological transformations for detoxification of trichothecene mycotoxins in human and animal food chains: A review. *Trends Food Sci. Technol.* **2010**, *21*, 67–76. [CrossRef]
4. Karlovsky, P. Biological detoxification of fungal toxins and its use in plant breeding, feed and food production. *Nat. Toxins* **1999**, *7*, 1–23. [CrossRef]
5. Karlovsky, P.; Suman, M.; Berthiller, F.; de Meester, J.; Eisenbrand, G.; Perrin, I.; Oswald, I.P.; Speijers, G.; Chiodini, A.; Recker, T.; et al. Impact of food processing and detoxification treatments on mycotoxin contamination. *Mycotoxin Res.* **2016**, *32*, 179–205. [CrossRef] [PubMed]
6. Helmke, A.; Hoffmeister, D.; Berge, F.; Emmert, S.; Laspe, P.; Mertens, N.; Vioel, W.; Weltmann, K.-D. Physical and microbiological characterisation of *Staphylococcus epidermidis* inactivation by dielectric barrier discharge plasma. *Plasma Process. Polym.* **2011**, *8*, 278–286. [CrossRef]
7. Ehlbeck, J.; Schnabel, U.; Polak, M.; Winter, J.; von Woedtke, T.; Brandenburg, R.; von dem Hagen, T.; Weltmann, K.-D. Low temperature atmospheric pressure plasma sources for microbial decontamination. *J. Phys. D Appl. Phys.* **2011**, *44*. [CrossRef]
8. Fridman, G.; Friedman, G.; Gutsol, A.; Shekhter, A.B.; Vasilets, V.N.; Fridman, A. Applied Plasma Medicine. *Plasma Process. Polym.* **2008**, *5*, 503–533. [CrossRef]
9. Kong, M.G.; Kroesen, G.; Morfill, G.; Nosenko, T.; Shimizu, T.; van Dijk, J.; Zimmermann, J.L. Plasma medicine: An introductory review. *New J. Phys.* **2009**, *11*. [CrossRef]
10. Park, G.Y.; Park, S.J.; Choi, M.Y.; Koo, I.G.; Byun, J.H.; Hong, J.W.; Sim, J.Y.; Collins, G.J.; Lee, J.K. Atmospheric-pressure plasma sources for biomedical applications. *Plasma Sour. Sci. Technol.* **2012**, *21*, 43001. [CrossRef]
11. Halfmann, H.; Bibinov, N.; Wunderlich, J.; Awakowicz, P. A double inductively coupled plasma for sterilization of medical devices. *J. Phys. D Appl. Phys.* **2007**, *40*, 4145–4154. [CrossRef]

12. Bormashenko, E.; Grynyov, R.; Bormashenko, Y.; Drori, E. Cold radiofrequency plasma treatment modifies wettability and germination speed of plant seeds. *Sci. Rep.* **2012**, *2*. [CrossRef] [PubMed]

13. Filatova, I.; Azharonok, V.; Kadyrov, M.; Belyavsky, V.; Gvodzdov, A.; Shik, A.; Antonuk, A. The effect of plasma treatment of seeds of some grain and legumes on their sowing quality and productivity. *Romanian J. Phys.* **2011**, *56*, 139–143.

14. Avramidis, G.; Stüwe, B.; Richard, W.; Bellmann, M.; Stephan, W.; von Tiedemann, A.; Viöl, W. Fungicidal effects of an atmospheric pressure gas discharge and degradation mechanisms. *Surf. Coat. Technol.* **2010**, *205*, 405–408. [CrossRef]

15. Zahoranová, A.; Henselová, M.; Hudecová, D.; Kaliňáková, B.; Kováčik, D.; Medvecká, V.; Černák, M. Effect of cold atmospheric pressure plasma on the wheat seedlings vigor and on the inactivation of microorganisms on the seeds surface. *Plasma Chem. Plasma Process.* **2015**, *36*, 397–414. [CrossRef]

16. Ouf, S.A.; Basher, A.H.; Mohamed, A.-A.H. Inhibitory effect of double atmospheric pressure argon cold plasma on spores and mycotoxin production of *Aspergillus niger* contaminating date palm fruits. *J. Sci. Food Agric.* **2015**. [CrossRef] [PubMed]

17. Park, B.J.; Takatori, K.; Sugita-Konishi, Y.; Kim, I.-H.; Lee, M.-H.; Han, D.-W.; Chung, K.-H.; Hyun, S.O.; Park, J.-C. Degradation of mycotoxins using microwave-induced argon plasma at atmospheric pressure. *Surf. Coat. Technol.* **2007**, *201*, 5733–5737. [CrossRef]

18. Lee, H.M.; Chang, M.B. Abatement of gas-phase p-xylene via dielectric barrier discharges. *Plasma Chem. Plasma Process.* **2003**, *23*, 541–558. [CrossRef]

19. Poll, H.U.; Schladitz, U.; Schreiter, S. Penetration of plasma effects into textile structures. *Surf. Coat. Technol.* **2001**. [CrossRef]

20. Kříž, P.; Olšan, P.; Havelka, Z.; Horáková, M.; Bartoš, P.; Vazdová, P.; Syamkrishna, B.; Špatenka, P. Seed Treatment and Water Purification by the Synergical Effect of Gliding Arc Plasma and Photocatalytic Film. *IEEE* **2014**. [CrossRef]

21. Kříž, P.; Petr, B.; Zbynek, H.; Jaromir, K.; Pavel, O.; Petr, S.; Miroslav, D. Influence of plasma treatment in open air on mycotoxin content and grain nutriments. *Plasma Med.* **2015**, *5*, 145–158. [CrossRef]

22. Bellmann, M.; Avramidis, G.; Wascher, R.; Viöl, W. Accelerated Germination and Altered Surface Characteristics of Pisum Sativum Seeds after Plasma Treatment at Atmospheric Pressure. In Proceedings of the Conference Plasma Surface Engineering, Garmisch-Partenkirchen, Germany, 15 September 2012.

23. Hopfe, V.; Sheel, D.W. Atmospheric-Pressure PECVD Coating and Plasma Chemical Etching for Continuous Processing. *IEEE Trans. Plasma Sci.* **2007**, *35*, 204–214. [CrossRef]

24. Eliasson, B.; Kogelschatz, U. Nonequilibrium volume plasma chemical processing. *IEEE Trans. Plasma Sci.* **1991**, *19*, 1063–1077. [CrossRef]

25. Efremov, A.M.; Kim, D.-P.; Kim, C.-I. Simple Model for Ion-Assisted Etching Using Coupled Plasma: Effect of Gas Mixing Ratio. *IEEE Trans. Plasma Sci.* **2004**, *32*, 1344–1351. [CrossRef]

26. Coburn, J.W.; Winters, H.F. Ion- and electron-assisted gas-surface chemistry—An important effect in plasma etching. *J. Appl. Phys.* **1979**, *50*, 3189–3196. [CrossRef]

27. Dorai, R.; Kushner, M.J. A model for plasma modification of polypropylene using atmospheric pressure discharges. *J. Phys. D Appl. Phys.* **2003**, *36*, 666–685. [CrossRef]

28. Kuvaldina, E.V.; Shikova, T.G.; Smirnov, S.A.; Rybkin, V.V. Surface oxidation and degradation of polyethylene in a mixed argon-oxygen plasma. *High Energy Chem.* **2007**, *41*, 284–287. [CrossRef]

29. Jeong, J.Y.; Babayan, S.E.; Schütze, A.; Tu, V.J.; Park, J.; Henins, I.; Selwyn, G.S.; Hicks, R.F. Etching polyimide with a nonequilibrium atmospheric-pressure plasma jet. *J. Vac. Sci. Technol. A* **1999**, *17*, 2581–2585. [CrossRef]

30. Park, B.J.; Lee, D.H.; Park, J.-C.; Lee, I.-S.; Lee, K.-Y.; Hyun, S.O.; Chun, M.-S.; Chung, K.-H. Sterilization using a microwave-induced argon plasma system at atmospheric pressure. *Phys. Plasmas* **2003**, *10*, 4539–4544. [CrossRef]

31. Gröning, P.; Collaud, M.; Dietler, G.; Schlapbach, L. Plasma modification of polymethylmethacrylate and polyethyleneterephthalate surfaces. *J. Appl. Phys.* **1994**, *76*, 887–892. [CrossRef]

32. Klarhöfer, L.; Viöl, W.; Maus-Friedrichs, W. Electron spectroscopy on plasma treated lignin and cellulose. *Holzforschung* **2010**, *64*, 313–316. [CrossRef]

33. Khatibi, P.A.; Berger, G.; Wilson, J.; Brooks, W.S.; McMaster, N.; Griffey, C.A.; Hicks, K.B.; Nghiem, N.P.; Schmale, D.G. A comparison of two milling strategies to reduce the mycotoxin deoxynivalenol in barley. *J. Agric. Food Chem.* **2014**, *62*, 4204–4213. [CrossRef] [PubMed]

34. Pinotti, L.; Ottoboni, M.; Giromini, C.; Dell'Orto, V.; Cheli, F. Mycotoxin contamination in the EU feed supply chain: A focus on cereal byproducts. *Toxins* **2016**, *8*. [CrossRef] [PubMed]

35. Visentin, I.; Montis, V.; Doll, K.; Alabouvette, C.; Tamietti, G.; Karlovsky, P.; Cardinale, F. Transcription of genes in the biosynthetic pathway for fumonisin mycotoxins is epigenetically and differentially regulated in the fungal maize pathogen *Fusarium verticillioides*. *Eukaryot. Cell* **2012**, *11*, 252–259. [CrossRef] [PubMed]

36. Bai, G.H. Variation in *Fusarium graminearum* and cultivar resistance to wheat scab. *Plant Dis.* **1996**, *80*, 975. [CrossRef]

37. Becker, E.-M.; Herrfurth, C.; Irmisch, S.; Kollner, T.G.; Feussner, I.; Karlovsky, P.; Splivallo, R. Infection of corn ears by *Fusarium* spp. induces the emission of volatile sesquiterpenes. *J. Agric. Food Chem.* **2014**, *62*, 5226–5236. [CrossRef] [PubMed]

38. Nutz, S.; Döll, K.; Karlovsky, P. Determination of the LOQ in real-time PCR by receiver operating characteristic curve analysis: Application to qPCR assays for *Fusarium verticillioides* and *F. proliferatum*. *Anal. Bioanal. Chem.* **2011**, *401*, 717–726. [CrossRef] [PubMed]

39. Gerhard, C.; Roux, S.; Brückner, S.; Wieneke, S.; Viöl, W. Low-temperature atmospheric pressure argon plasma treatment and hybrid laser-plasma ablation of barite crown and heavy flint glass. *Appl. Opt.* **2012**, *51*. [CrossRef] [PubMed]

40. Ratzinger, A.; Riediger, N.; von Tiedemann, A.; Karlovsky, P. Salicylic acid and salicylic acid glucoside in xylem sap of *Brassica napus* infected with *Verticillium longisporum*. *J. Plant Res.* **2009**, *122*, 571–579. [CrossRef] [PubMed]

41. Laurentin, H.; Ratzinger, A.; Karlovsky, P. Relationship between metabolic and genomic diversity in sesame (*Sesamum indicum* L.). *BMC Genom.* **2008**, *9*. [CrossRef] [PubMed]

42. Nassar, H.; Pellerin, S.; Musiol, K.; Martinie, O.; Pellerin, N.; Cormier, J.-M. N2+/N2 ratio and temperature measurements based on the first negative N2+ and second positive N2 overlapped molecular emission spectra. *J. Phys. D Appl. Phys.* **2004**, *37*, 1904–1916. [CrossRef]

43. Paris, P.; Aints, M.; Valk, F.; Plank, T.; Haljaste, A.; Kozlov, K.V.; Wagner, H.-E. Intensity ratio of spectral bands of nitrogen as a measure of electric field strength in plasmas. *J. Phys. D Appl. Phys.* **2005**, *38*, 3894–3899. [CrossRef]

44. Manley, T.C. The Electric Characteristics of the Ozonator Discharge. *Trans. Electrochem. Soc.* **1943**, *84*, 83–96. [CrossRef]

Article

A Structure Identification and Toxicity Assessment of the Degradation Products of Aflatoxin B$_1$ in Peanut Oil under UV Irradiation

Jin Mao [1,2,3,†], Bing He [1,2,†], Liangxiao Zhang [1,2,4], Peiwu Li [1,2,3,4,*], Qi Zhang [1,2,3,4,*], Xiaoxia Ding [1,3,4] and Wen Zhang [1,2,4]

[1]   Oil Crops Research Institute, Chinese Academy of Agricultural Sciences, Wuhan 430062, China;
      maojin106@whu.edu.cn (J.M.); xinxianghebing@163.com (B.H.); liangxiao_zhang@126.com (L.Z.);
      dingdin2355@sina.com (X.D.); zhangwen@oilcrops.cn (W.Z.)
[2]   Laboratory of Quality & Safety Risk Assessment for Oilseed Products (Wuhan), Ministry of Agriculture,
      Wuhan 430062, China
[3]   Key Laboratory of Detection for Mycotoxins, Ministry of Agriculture, Wuhan 430062, China
[4]   Quality Inspection & Test Center for Oilseed Products, Ministry of Agriculture, Wuhan 430062, China
*   Correspondence: peiwuli@oilcrops.cn (P.L.); zhangqi521x@126.com (Q.Z.);
     Tel.: +86-27-8681-2943 (P.L.); +86-27-8671-1839 (Q.Z.)
†   These authors contributed equally to this work.

Academic Editor: Ting Zhou
Received: 20 September 2016; Accepted: 9 November 2016; Published: 12 November 2016

Abstract: Aflatoxins, a group of extremely hazardous compounds because of their genotoxicity and carcinogenicity to human and animals, are commonly found in many tropical and subtropical regions. Ultraviolet (UV) irradiation is proven to be an effective method to reduce or detoxify aflatoxins. However, the degradation products of aflatoxins under UV irradiation and their safety or toxicity have not been clear in practical production such as edible oil industry. In this study, the degradation products of aflatoxin B$_1$ (AFB$_1$) in peanut oil were analyzed by Ultra Performance Liquid Chromatograph-Thermo Quadrupole Exactive Focus mass spectrometry/mass spectrometry (UPLC-TQEF-MS/MS). The high-resolution mass spectra reflected that two main products were formed after the modification of a double bond in the terminal furan ring and the fracture of the lactone ring, while the small molecules especially nitrogen-containing compound may have participated in the photochemical reaction. According to the above results, the possible photodegradation pathway of AFB$_1$ in peanut oil is proposed. Moreover, the human embryo hepatocytes viability assay indicated that the cell toxicity of degradation products after UV irradiation was much lower than that of AFB$_1$, which could be attributed to the breakage of toxicological sites. These findings can provide new information for metabolic pathways and the hazard assessment of AFB$_1$ using UV detoxification.

Keywords: aflatoxin B$_1$; photodegradation product; TQEF-MS/MS; cell viability; furan rings

1. Introduction

Aflatoxins are the secondary metabolites of *Aspergillus flavus* and *Aspergillus parasiticus*, which are found worldwide in air and soil to infest living and dead plants [1,2]. Aflatoxin B$_1$ (AFB$_1$) is the most potent mycotoxin associated with hepatocellular carcinoma, immune-dysfunction, and protein deficiency syndromes, which is classified as a Group I carcinogen by the International Agency for Research in Cancer [3]. It is well known that the double bond in the terminal furan ring was a key active site for its toxic and carcinogenic activities. A chemical molecular structure diagram of AFB$_1$ is shown in Figure 1. Aflatoxins frequently occur in tropical and subtropical regions, and many agricultural commodities and main crops, such as peanut and corn, are susceptible to such contamination in the period of growth and storage. While the contaminated peanuts or corns are used as raw material without appropriate processing measures in the food industry, they will become a serious threat or hazard to human health. The most rational strategy to avoiding hazards associated with aflatoxins is the prevention of aflatoxin contamination by the toxigenic fungi. However, it is difficult to prevent or control the contamination of peanuts by the toxigenic fungi due to that the hazard of fungi is naturally occurring and is not always possible under certain agronomic storage practices [4].

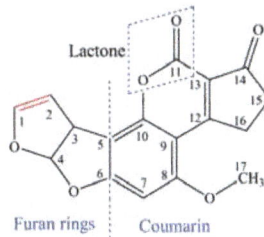

Figure 1. Chemical molecular structure of aflatoxin B_1 (AFB$_1$).

Many kinds of chemical, physical, and biological approaches for the detoxification of aflatoxins have been reported in previous studies [5–8]. Chemical detoxification is an effective measure using strong alkalis or oxidants such as ammoniation, ozone to break the structure of aflatoxins. However, the quality and nutritional compositions in food may be affected by strong alkalis or oxidants during these chemical measures. Alberts et al. reported that a remarkable reduction in AFB$_1$ was observed after 2 h in the presence of *R. erythropolis* extracellular extractions, and only 33.2% residue AFB$_1$ was detected after 72 h by degrading enzymes from extracellular extractions [7]. The biological methods showed the high efficiency and selectivity, but these methods may be difficult to reutilize on a large scale. Ultraviolet (UV) irradiation as a non-thermal technology is widely applied in the food industry for disinfection, which is also considered to be practical and cost-effective method to reduce aflatoxins for its photosensitive properties [8]. As an effective physical method, many studies have been done to investigate the efficiency of UV irradiation, the degradation product of aflatoxins, the safety of degradation products and quality of foods after being irradiated [9–14]. It was found that aflatoxins could be efficiently degraded by UV irradiation, and the degradation efficiency varied with the differences of irradiation conditions [9,10]. For the studies in photodegradation products of aflatoxins, ultra-performance liquid chromatography-quadrupole time-of-flight mass spectrometry (UPLC-Q-TOF MS) was used to identify the photodegradation products. Different degradation products were identified on the basis of low mass error and high matching property in aqueous or acetonitrile solution, and the different AFB$_1$ degradation pathways were proposed [12,14]. Moreover, in other study, it was found that three products of AFB$_1$ formed using UV irradiation for 120 min in the presence of methylene blue [15]. These conclusions indicated that depending upon reaction conditions such as medium or solvent, the degradation products of AFB$_1$ was different under UV irradiation.

In addition, most studies were carried out on the Ames test and cytotoxicity of HepG2 cells of AFB$_1$ products after irradiation [10,13,16,17]. The toxicities of the photodegradation products of AFB$_1$ in water and peanut oil on HepG2 cells were investigated, and it was found that the cytotoxicity of products of AFB$_1$ in water decreased 40%, while that of products of AFB$_1$ in peanut oil reduced about 100% [13]. Moreover, similar results were obtained after being irradiated of AFB$_1$ in peanut oil using a photodegradation reactor in a previous study [16]. This might be because toxicological sites were destroyed by UV irradiation.

From the above studies, it can be determined that the studies on degradation products of AFB$_1$ are mostly in acetonitrile or aqueous media, and the pathway of AFB$_1$ in different solutions are entirely different. However, the identification of degradation products and their toxicity have been poorly investigated in practical production. For instance, the product of AFB_1 in peanut oil under UV irradiation has not yet been reported, which may be due to the complicated compositions in peanut oil. Therefore, the aim of this article is to examine the products of AFB_1 in peanut oil under UV irradiation and the safety or toxicity of degradation products after UV irradiation, and to provide clues to the study of the degradation mechanism of AFB_1 in peanut oil and the assessment of safety issues of the UV method applied in aflatoxins detoxification. In this study, the photodegradation efficiency of AFB_1 in peanut oil were investigated; after optimizing the extract conditions, the photodegradation products were analyzed by Thermo Quadrupole Exactive Focus spectrometry-mass spectrometry/mass spectrometry (TQEF-MS/MS). On the basis of low mass error and high matching property from data of MS/MS, the feasible pathway of AFB_1 in peanut oil was deduced. Moreover, the in vitro toxicity of AFB_1 and its degradation products towards human embryo hepatocytes (L-02 cell) were investigated.

2. Results and Discussions

2.1. Effect of the AFB_1 Initial Concentration on Degradation Performance in Peanut Oil

The effect of the AFB_1 initial concentration on degradation performance in peanut oil was investigated in this study. The result from Figure 2 confirmed that the AFB_1 can be degraded under 365 nm UV irradiation, and there were no obvious changes found in the blank experiment without UV irradiation. The initial concentrations of AFB_1 were 48, 68, 88, 108, and 128 ppb, and it could be seen that the initial concentration of 128 ppb showed only a slightly improved degradation efficiency compared with the others before 20 min, and the value of C_t/C_0 then remained around 4% after 20 min. It was revealed that the effect of the initial concentration of AFB_1 in our selected range on degradation activity was unremarkable, which is in agreement with the result of the previous report [12]. Moreover, after 30 min of irradiation, it was found that there was only about 4% AFB_1 residual in each sample. As we know, many countries have regulated the limitation standard of aflatoxins to protect consumers' health and prevent food safety issues. The Food and Drug Administration (FDA), the World Health Organization (WHO), and the Codex Alimentarius Commission (CAC) have regulated a maximum contamination level for total content of aflatoxin B1, B2, G1, and G2 (15 µg/kg) in food. Moreover, according to national standard (GB 2461-2011) in China, the limitation standard of AFB_1 in peanut oil is 20 µg/kg (ppb). It can be concluded that the contents of AFB_1 in peanut oil with different initial concentrations after 30 min UV treatment in this study are within an acceptable range according to the above limitation standards. However, the European Commission has regulated a maximum contamination level for AFB_1 (2 µg/kg) and total aflatoxin (4 µg/kg) in argo-products according to (EU) No 165/2010, 2010. For this strict restriction in EU, the optimization of the conditions of UV irradiation treatment, such as enhancing the intensity of UV within a defined and accepted range [12,14], may be an effective strategy to reduce and control the content of AFB_1 below the limitation standard of EU.

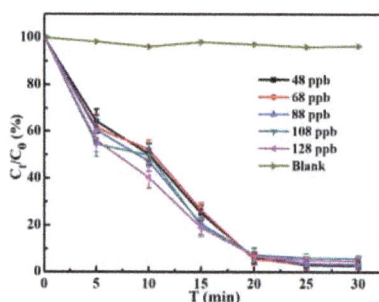

Figure 2. The effect of the AFB_1 initial concentration on degradation performance in peanut oil under UV irradiation from 0 min to 30 min.

2.2. Identification of Photodegradation Products of AFB$_1$ in Peanut Oil

It is well known that the degradation products of AFB$_1$ under UV irradiation in various media are different. Three products, $C_{17}H_{14}O_7$, $C_{16}H_{14}O_6$, and $C_{16}H_{12}O_7$, were formed in aqueous medium and identified by UPLC-Q-TOF-MS [12], while $C_{17}H_{15}O_6$, $C_{16}H_{13}O_5$, and $C_{14}H_{11}O_6$ were obtained in acetonitrile medium by the same identification method [14]. Moreover, in other previous articles, it was reported that the photodegradation products in peanut oil extracted by acetonitrile cannot be detected in UV-treated extracts [9]. In our study, the extraction solvent type and volume were optimized. Various organic reagents were used to extract the photodegradation products of AFB$_1$, and it was found that the products extracted by methyl alcohol-water (10:90) solution could be detected. To study the photodegradation products of AFB$_1$ and identify their structures, the samples ((a) the sample containing AFB$_1$ without UV irradiation; (b) the sample without AFB$_1$ after UV irradiation; and (c) the sample containing AFB$_1$ after UV irradiation) were analyzed by UPLC-QEF-MS/MS. It can be observed distinctly from Figure 3 that AFB$_1$ was degraded within 30 min, leading to the formation of the two new compounds denoted 'P$_1$' and 'P$_2$', which were not detected in the blank experiment a and b. Moreover, these two compounds were considered as the irradiated products of AFB$_1$ in peanut oil. P$_1$, P$_2$, and AFB$_1$ had a retention time of 8.42, 11.34, and 15.68 min, respectively, showing an order of polar character P$_1$ > P$_2$ > AFB$_1$.

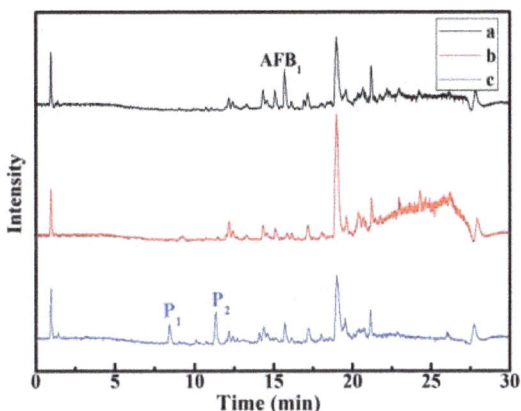

Figure 3. The UPLC chromatogram of peanut oil samples before and after irradiation. ((a) the sample containing AFB$_1$ without UV irradiation, (b) the sample without AFB$_1$ after UV irradiation, and (c) the sample containing AFB$_1$ after UV irradiation).

In a separate QEF MS mode, the ions of P$_1$, P$_2$, and AFB$_1$ were used as precursor compounds, and the ion-filtering function of the quadrupole permitted only ions of $m/z = 340, 227$, and 313 to pass to the collision cell, where 35 eV was applied for 340, 227, and 313. Compared with the standard substance of AFB$_1$, $m/z = 313$ was in agreement with the molecular mass of the parent compound (AFB$_1$). In order to present the structure of these two products, P$_1$ and P$_2$, the MS-MS spectra of degradation products m/z = 340 and m/z = 227 were also recorded. The secondary mass spectrum and fragmentation information of these two precursor compounds were derived, and they are shown in Figure 4. The structural formulas of these photodegradation products were demonstrated on account of the molecular formula calculated from accurate mass analysis. The MS-MS fragmentation formation was supplied by the QEF mode, and illuminated by the tool mass fragment incorporated in the Mass Frontier 7.0 software to verify how much the results matched the corresponding fragment information (Table 1 contains detailed calculation data). From Table 1, it can be seen that the experimental mass of these fragments matched the theoretical mass of these fragments. Moreover, a maximum mass error of 10 ppm for the range of masses was discussed in this study. It can be seen clearly that the mass errors of these fragments of photodegradation products were all below 7 ppm, which indicated at least 93% confidence in the accuracy of the suggested results. From Figure 4a,b, it can be easily found that $m/z = 340$ and $m/z = 227$ have many of the same fragmentations such as $m/z = 209$, $m/z = 114$, $m/z = 96$, $m/z = 79$, etc. Moreover, through accurate calculation and careful molecular model derivation, we finally arrived at the conclusion that the fragmentation of $m/z = 227$ may be the metabolite of $m/z = 340$ under UV irradiation. More important than the above results, and to confirm the relationship between these two products, the time development of the formation of photodegradation products of AFB$_1$ in peanut oil was also investigated in our study. From Figure 5, it can be seen that P$_1$ and P$_2$ were detected during the procedure of UV irradiation, and the content of AFB$_1$ in peanut oil decreased gradually with time. Moreover, by comparing the ratio of the intensities of two photodegradation products at different time under UV irradiation, it was found that the response and intensity of P$_1$ weakened from 10 min to 30 min, while that of P$_2$ increased over time. This result may be due to the fact that some of the P$_1$ translated to P$_2$ during the UV irradiation in peanut oil. These results provided important information for the photodegradation pathway of AFB$_1$ in peanut oil.

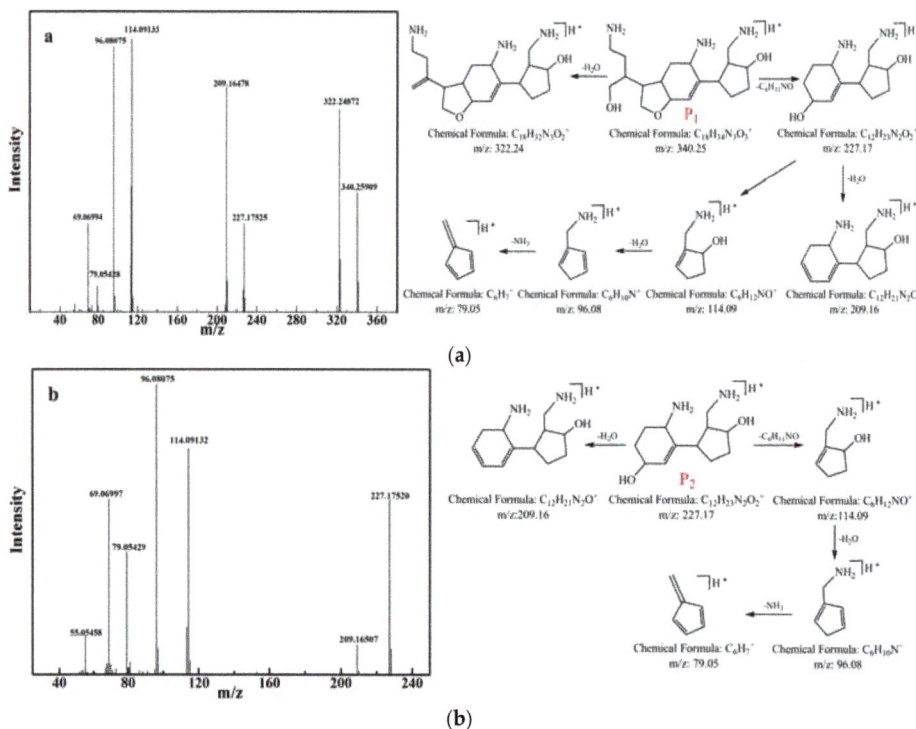

(a)

(b)

Figure 4. The QEF MS/MS spectra and proposed fragmentation of two degradation products under UV irradiation. Proposed fragmentations are shown on the right side of spectra. (a) $m/z = 340$; (b) $m/z = 227$.

Table 1. QEF-MS/MS accurate mass, mass error, and formula of photodegradation product fragments.

Theoretical Mass	Experimental Mass	Fragment Formula	Mass Error (ppm)
79.05478	79.05429	C_6H_7	−6.19823
79.05478	79.05428	C_6H_7	−6.32473
96.08132	96.08075	$C_6H_{10}N$	−5.93247
114.09189	114.09132	$C_6H_{12}NO$	−4.99597
114.09189	114.09133	$C_6H_{12}NO$	−4.90832
209.16539	209.16478	$C_{12}H_{21}N_2O$	−2.91635
209.16539	209.16507	$C_{12}H_{21}N_2O$	−1.52989
227.17595	227.17520	$C_{12}H_{23}N_2O_2$	−3.30141
227.17595	227.17525	$C_{12}H_{23}N_2O_2$	−3.08131
340.26002	340.25909	$C_{18}H_{34}N_3O_3$	−2.73320

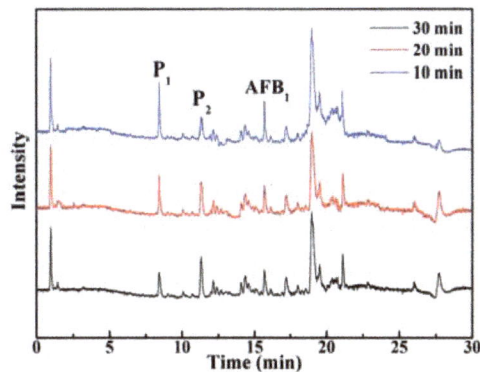

Figure 5. Time development of the formation of photodegradation products of AFB_1 in peanut oil.

2.3. Proposed Pathway of AFB_1 under UV Irradiation in Peanut Oil

From the above results and discussion, the possible photodegradation pathway of AFB_1 in peanut oil was proposed, which was based on the identified chemical structure of the photodegradation products and shown in Figure 6. Because there are kinds of compounds such as proteins, amino acids, polyphenols, and other small molecule compounds in peanut oil [18,19], the photodegradation pathways of AFB_1 in peanut oil might be complicated and accompanied by some complex chemical reactions. As shown in Figure 6, it can be deduced that AFB_1 might lose the C=O of the lactone ring and become to $C_{16}H_{14}O_4$ firstly, because the lactone ring was the active site of AFB_1 [20]. Then, the additional reaction and substitution reaction of small molecules such as $R-NH_2$ and $-NH_2$, which may due to the fact that the concomitant cracking reactions of the nitrogen-containing compound in the peanut oil under UV irradiation occurred on the unsatisfied chemical double bond on the furan ring and the right side five-membered ring of $C_{16}H_{14}O_4$, while the OH groups had been replaced by NH_2. Moreover, the H addition reaction definitely occurred on C=O [21]. After these complex reactions, the structure of $C_{19}H_{33}N_3O_4$ was formed. Thus, the P_1 ($C_{18}H_{33}N_3O_3$) may be the metabolites of $C_{19}H_{33}N_3O_4$ after dropping the methoxy group (OCH_3) under UV irradiation. Finally, as a result of the cracking of the five-membered ring in the middle of the compound of P_1, the compound of P_2 ($C_{12}H_{22}N_2O_2$, molecular weight: 226) formed. Although the proposed photodegradation pathway of AFB_1 in peanut oil was on the base of photochemical principles and identified structural formulas, the detailed and comprehensive degradation pathway need to be proved further in a future study.

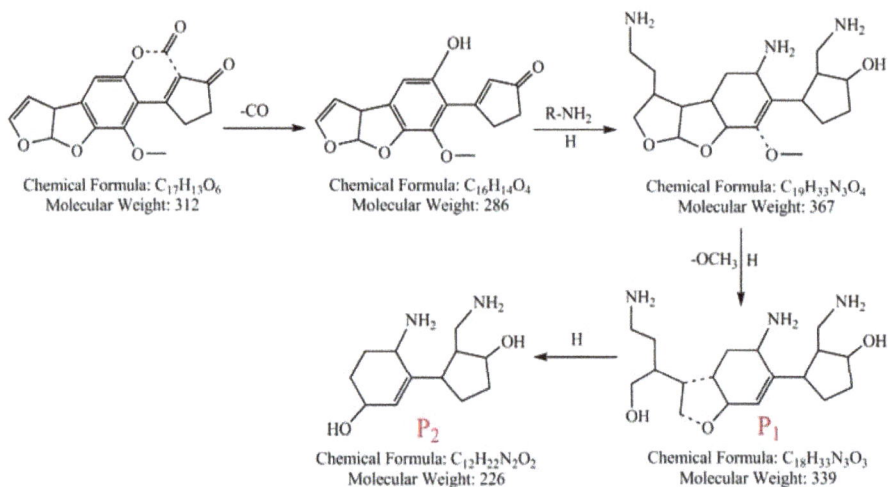

Chemical Formula: $C_{17}H_{13}O_6$
Molecular Weight: 312

Chemical Formula: $C_{16}H_{14}O_4$
Molecular Weight: 286

Chemical Formula: $C_{19}H_{33}N_3O_4$
Molecular Weight: 367

Chemical Formula: $C_{12}H_{22}N_2O_2$
Molecular Weight: 226

Chemical Formula: $C_{18}H_{33}N_3O_3$
Molecular Weight: 339

Figure 6. Possible photodegradation pathway of AFB_1 in peanut oil under 365 nm UV irradiation in this study.

2.4. Cell Viability Assay of AFB_1 and Its Photodegradation Products

In order to investigate the toxicity and hazard of photodegradation of AFB_1 after UV irradiation, the viabilities of human embryo hepatocytes (L-02 liver cells) were assessed using the MTT assay and the CCK-8 assay. As shown in Figure 7a, it can be seen that the consequences of the MTT and CCK-8 assays were similar, and the viability of L-02 liver cells decreased in a linear relation with increasing concentrations of AFB_1 from 48 to 128 ppb after incubation for 24 h. The cell viability treated by the concentration of AFB_1 at 128 ppb decreased more significantly than that of other samples. Compared with a blank control trial, the cell viability of L-02 liver cells was reduced by about 40% after incubation at a concentration of 128 ppb. However, the loss of cell viability of irradiated samples at the same concentration was significantly decreased compared with that of samples without irradiation treatment, which indicated that the toxicity of the degradation products of AFB_1 in peanut oil lowered after UV irradiation. Moreover, it can be seen from Figure 7b that the mortality rates of L-02 liver cells was time-dependent in the presence of AFB_1 while the mortality rates of L-02 liver cells had no obvious changes with time in the presence of the photodegradation products of AFB_1, and the mortality rate of L-02 liver cells in the presence of AFB_1 was up to 70% for 48 h. This investigation further confirmed that the cytotoxicity of AFB_1 was significantly reduced after UV irradiation, which may be due to the changes in the molecular structure of AFB_1. In this present study, the MS/MS date expounded that P_1 ($C_{18}H_{33}N_3O_3$) and P_2 ($C_{12}H_{22}N_2O_2$) were formed after the removal of the double bond in the terminal furan ring and the modification or splitting decomposition of the lactone ring, which may be the main reason for the decrease in toxicity after UV irradiation. Moreover, it is well known that the structure of AFB_1 contained a lactone ring and a furan moiety, and the double bond in the terminal furan ring was the key active site for its toxic and carcinogenic activities [22–24]. This viewpoint is also confirmed by a previous study that the toxicity of radiolytic products was significantly reduced compared with that of AFB_1 because of the addition reaction that occurred on the double bond in the terminal furan ring [25]. Lee et al. found that the lactone ring played a significant role in the fluorescence of AFB_1, and the AFB_1 without the lactone ring became non-florescent with a subsequent decline in toxicity [26]. Therefore, it could be deduced that, due to the reaction on the double bond of the terminal furan ring and the lactone ring for AFB_1 in this study, the toxicity of the two degradation products decreased after UV irradiation treatment compared with that of AFB_1.

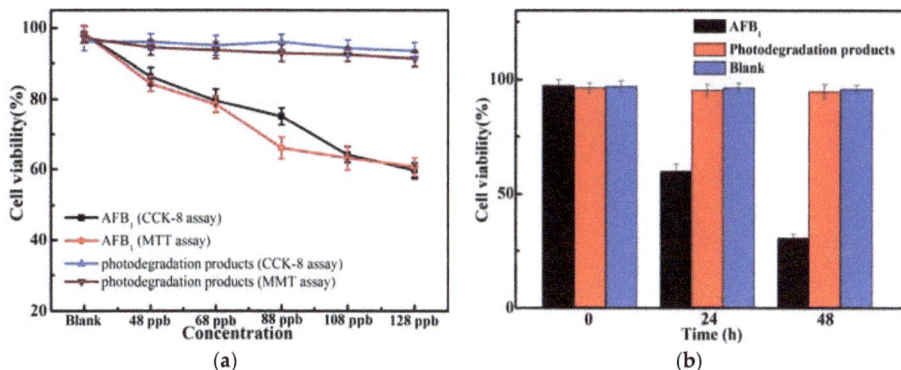

Figure 7. Cell viability assay of AFB$_1$ and its photodegradation products. (**a**) The effect of different concentrations of AFB$_1$ and the same initial concentrations of its photodegradation products in peanut oil on L-02 liver cells viability evaluated by the MTT and CCK-8 assay. (**b**) L-02 liver cells viability exposed to 128 ppb AFB$_1$ and the same concentration photodegradation products for 0, 24, and 48 h evaluated by the CCK-8 assay.

3. Conclusions

To our knowledge, this is the first report to demonstrate the proposed degradation products and pathway of AFB$_1$ in peanuts oil under UV irradiation. The products of AFB$_1$ in peanut oil under UV irradiation were extracted by methyl alcohol and deionized water and analyzed by UPLC-TQEF-MS/MS. It was found that two photodegradation products (P$_1$:C$_{18}$H$_{33}$N$_3$O$_3$ and P$_2$:C$_{12}$H$_{22}$N$_2$O$_2$) were formed. Moreover, reactions of photodegradation mainly occurred on the terminal furan ring and the lactone ring, which were the primary toxicological sites of AFB$_1$. After the resolution of accurate mass spectra and a comparison of UPLC spectra at different times during irradiation, the metabolic pathways of AFB$_1$ in peanut oil were proposed. In addition, the toxicity assessment of degradation products through a human embryo hepatocytes viability assay revealed that the toxicity of degradation products decreased significantly compared with that of AFB$_1$, which reaffirmed that the reactions occurred on the primary toxicological sites of AFB$_1$ under irradiation. The present findings could provide a new idea and means for future studies on metabolic pathways and toxicity assessments of hazardous substances in foodstuffs.

4. Materials and Methods

4.1. Materials and Synthesis

AFB$_1$ (2, 3, 6α, 9α-tetrahydro-4-methoxycyclopenta [c] furo [2, 3:45] furo [2, 3-h] chromene-1, 11-dione; purity >98%) was obtained from Sigma (Sigma-Aldrich, St. Louis, MO, USA). Methyl alcohol, dimethyl sulfoxide (DMSO), and other reagents are analytical reagents and purchased from Sinopharm (Sinopharm Chemical Reagent Co., Ltd., Shanghai, China) Peanut oil was purchased from Wal-Mart supermarket in Wuhan, China.

For high-performance liquid chromatography (HPLC) and TQEF-MS/MS analysis, chromatographic grade acetonitrile and methyl alcohol were purchased from Sigma (St. Louis, MO, USA). The deionized water was obtained from a Milli-Q SP Reagent Water system (Millipore, Bedford, MA, USA).

For cytotoxicity tests, human embryo hepatocytes (L-02 cells) were purchased from Procell (Procell Co. Ltd., Wuhan, China). Culture medium RPMI-1640 and fetal calf serum were purchased from HyClone ((GE-Healthcare, Logan, UT, USA). The 0.05% trypsin solution was from Thermo Fisher (Thermo Fisher Scientific, Waltham, MA, USA).

4.2. Photodegradation Treatment

For UV degradation experiments, 20 g of peanut oil with different initial concentrations of 48 ppb, 68 ppb, 88 ppb, 108 ppb, and 128 ppb AFB_1 were put in glass petri dishes and irradiated under a 100 W ultraviolet lamp (λ = 365 nm, 55–60 mw/cm^2, Changzhou YuYu Electro-Optical Device Co., Ltd., Changzhou, China). The photoreaction was performed in a close reaction box in a temperature-controlled room, and temperature was kept at 26 °C. The distance between the UV lamp and the peanut oil was 40 cm. At certain selected time intervals, irradiated samples were collected and analyzed via HPLC (Agilent 1100, Agilent Technologies, Santa Clara, CA, USA). Each test was repeated five times.

4.2.1. Extraction and Instrumental Analysis for AFB_1 in Peanut Oil

A total of 2.5 g of peanut oil was weighed after UV treatment in a 50 mL centrifuge tube, and 7.5 mL of acetonitrile was then added. The mixture was shaken for 5 min with a platform shaker at room temperature and extracted by sonicator (Nanjing Jiancheng Bioengineering Institute, Nanjing, China) for 10 min. Then, the mixture was centrifuged at 7000 rpm for 15 min, and the precipitate was discarded. A 1.5 mL supernatant with 4 mL of deionized water was collected in a clean test tube. Then, the above mixture was filtered through a 0.22 μm organic ultra-filter membrane (Agilent Technologies, Santa Clara, CA, USA). The filtrate was passed through immunoaffinity columns and washed with 1 mL of methyl alcohol into a glass tube.

HPLC-FLD analysis was performed on an Agilent 1100 system equipped with a fluorescence detector, an auto-injector, and a quaternary solvent-delivery system. The chromatographic column was a 150 mm × 4.6 mm Agilent C_{18} column (Agilent Technologies, Santa Clara, CA, USA) with a 5 μm particle size. The injection volume was 10 μL. The mobile phases were methyl alcohol (A) and aqueous solution (B) in a 45:55 (V_A/V_B) solution with a flow rate of 0.7 mL·min^{-1}. The detection of the excitation and emission wavelengths were at 360 nm and 440 nm, respectively

4.2.2. Test Compounds of Photodegradation Products

The mixtures of photodegradation products in peanut oil was extracted by a methyl alcohol-water solution with a volume ratio of 10:90, and then passed through a 0.22 μm organic ultrafilter membrane (Agilent Technologies, Santa Clara, CA, USA) into a glass tube.

The samples were analyzed and recorded by Ultra Performance Liquid Chromatograph-Thermo Quadrupole Exactive Focus mass spectrometry/mass spectrometry (UPLC-TQEF-MS/MS, Thermo Fisher Scientific, Waltham, MA, USA). UPLC was performed on a Thermo system with a quaternary solvent-delivery system and an autosampler. The injection volume was 10 μL. The chromatographic column was presented on a 100 mm × 2.1 mm Thermo C_{18} column (Thermo Fisher Scientific, Waltham, MA, USA) with a particle size of 5 μm. The mobile phase was a gradient prepared from acetonitrile (A) and a 0.2% formic acid aqueous solution (B). The elution began with 5% A for 0.1 min, and the proportion of A was increased linearly to 100% at 36 min, and then brought back to 5% A at 40 min and kept for 5 min for a total of 45 min. The flow rate of the mobile phase was kept at 200 μL·min^{-1}.

Mass spectrometry was performed on a Thermo Quadrupole Exactive Focus (QEF) system (Thermo Fisher Scientific, Waltham, MA, USA). The products were analyzed under positive-ion (PI) mode. The optimized conditions were Sheath gas 35 L·min^{-1}, Aux gas (Ar) 5 L·min^{-1}, and capillary potentials at 3000 kV. The QEF instrument was operated in full scan mode, and data were collected between m/z = 50 and 500, with a scan accumulation time of 0.2 s. The MS/MS experiments were performed using a collision energy of 35 eV, which was optimized for each product.

4.3. Cell Viability Assay

4.3.1. Cell Culture

L-02 liver cells were serially cultivated in the RPMI-1640 complete medium (15% v/v fetal calf serum, 100 U/mL penicillin, 100 U/mL streptomycin) at 37 °C, with 4% CO_2 in air until an 80% degree of fusion was reached, and then incubated with DMSO, in which the final concentration of DMSO was less than 0.1%.

4.3.2. Cell Viability Assay

A total of 1×10^3 cells were seeded in a 96-well plate, allowing the attachment of cells to the substrate for 6–8 h, and the cells were then divided into different groups (six wells per group). For the sample of photodegradation products, the methyl alcohol-water solution with a volume ratio of 10:90 was used to extract the products of AFB_1 in peanut oil after UV irradiation, and the products were then evaporated under a stream of nitrogen and then collected in DMSO for cell viability assay. Cells were exposed to different concentrations of AFB_1, photodegradation products of AFB_1 and DMSO were used as a control, and cultivation continued for 24 h and 48 h. Cell viability was determined with a MTT assay and a CCK-8 assay. In brief, for the MTT method, 20 μL of 5 mg·mL^{-1} MTT in PBS buffer solution was added to each well, and the plate was incubated at 37 °C for 4 h. After incubation, the medium was then discarded, and the formed crystals were dissolved in 150 μL of DMSO. The absorbance of each well was then measured at 490 nm using a microplate reader (SpectraMax M2e, Molecular Devices, Sunnyvale, CA, USA), and the percentage viability was calculated. Moreover, for the CCK-8 assay, 10 μL of the CCK-8 solution was added to each well, and the plate was incubated at 37 °C for 4 h. After incubation, the absorbance of each well was measured at 450 nm using a microplate reader (SpectraMax M2e, Molecular Devices, Sunnyvale, CA, USA), and the percentage viability was calculated.

Acknowledgments: This work was supported by the Natural Science Foundation of China (31401601), the National Key Project for Agro-product Quality & Safety Risk Assessment, PRC (GJFP2015007), and the Special Fund for Grain-scientific Research in the Public Interest (201513006-02).

Author Contributions: P.L. and Q.Z. conceived and designed the experiments; B.H. and J.M. performed the experiments; X.D., W.Z., and L.Z analyzed the data; J.M. contributed reagents/materials/analysis tools; J.M. and B.H. wrote the paper.

Conflicts of Interest: The authors declare no conflict of interest.

References

1. Wee, J.; Day, D.M.; Linz, J.E. Effects of zinc chelators on aflatoxin production in *Aspergillus parasiticus*. *Toxins* **2016**, *8*, 171. [CrossRef] [PubMed]
2. Torres, A.M.; Barros, G.G.; Palacios, S.A.; Chulze, S.N.; Battilani, P. Review on pre- and post-harvest management of peanuts to minimize aflatoxin contamination. *Food Res. Int.* **2014**, *62*, 11–19. [CrossRef]
3. Monograph on the evaluation of carcinogenic risks to humans. *IARC, International Agency for Research on Cancer (IARC)*; World Health Organization: Lyon, France, 2002; Volume 171, p. 82.
4. Rustom, I.Y.S. Aflatoxin in food and feed: Occurrence, legislation and inactivation by physical methods. *Food Chem.* **1997**, *59*, 57–67. [CrossRef]
5. Kang, F.X.; Ge, Y.Y.; Hu, X.J.; Goikavi, C.; Waigi, M.G.; Gao, Y.Z.; Ling, W.T. Understanding the sorption mechanisms of aflatoxin B_1 to kaolinite, illite, and smectite clays via a comparative computational study. *J. Hazard Mater.* **2016**. [CrossRef] [PubMed]
6. Chen, R.; Ma, F.; Li, P.W.; Zhang, W.; Ding, X.X.; Zhang, Q.; Li, M.; Wang, Y.R.; Xu, B.C. Effect of ozone on aflatoxins detoxification and nutritional quality of peanuts. *Food Chem.* **2014**, *146*, 284–288. [CrossRef] [PubMed]
7. Alberts, J.F.; Engelbrecht, Y.; Steyn, P.S.; Holzapfel, W.H.; van Zyl, W.H. Biological degradation of aflatoxin B_1 by *Rhodococcus erythropolis* culture. *Int. J. Food Microbiol.* **2006**, *109*, 121–126. [CrossRef] [PubMed]

8.	Samarajeewa, U.; Sen, A.C.; Couen, M.D.; Wei, C.I. Detoxification of aflatoxins in foods and feeds by physical and chemical methods. *J. Food Prot.* **1990**, *53*, 489–501.

9.	Liu, R.J.; Jin, Q.Z.; Huang, J.H.; Liu, Y.F.; Wang, X.G.; Mao, W.Y.; Wang, S.S. Photodegradation of aflatoxin B_1 in peanut oil. *Eur. Food Res. Technol.* **2011**, *232*, 843–849. [CrossRef]

10.	Tripathi, S.; Mishra, H.N. Enzymatic coupled with UV degradation aflatoxin B_1 in red chili powder. *J. Food Qual.* **2010**, *33*, 186–203. [CrossRef]

11.	Wang, B.; Mahoney, N.E.; Pan, Z.L.; Khir, R.; Wu, B.G.; Ma, H.L.; Zhao, L.M. Effectiveness of pulsed light treatment for degradation and detoxification of aflatoxin B_1 and B_2 in rough rice and rice bran. *Food Control* **2016**, *59*, 461–467. [CrossRef]

12.	Liu, R.J.; Jin, Q.Z.; Tao, G.J.; Shan, L.; Huang, J.H.; Liu, Y.F.; Wang, X.G.; Mao, W.Y.; Wang, S.S. Photodegradation kinetics and byproducts identification of the aflatoxin B_1 in aqueous medium by ultra-performance liquid chromatography-quadrupole time-of-flight mass spectrometry. *J. Mass. Spectrom.* **2010**, *45*, 553–559. [CrossRef] [PubMed]

13.	Liu, R.J.; Jin, Q.Z.; Tao, G.J.; Huang, J.H.; Liu, Y.F.; Wang, X.G.; Zhou, X.L.; Mao, W.Y.; Wang, S.S. In vitro toxicity of aflatoxin B_1 and its photodegradation products in HepG2 cells. *J. Appl. Toxicol.* **2012**, *32*, 276–281. [CrossRef] [PubMed]

14.	Liu, R.J.; Jin, Q.Z.; Tao, G.J.; Shan, L.; Liu, Y.F.; Wang, X. LC-MS and UPLC-Quadrupole time-of-flight MS for identification of photodegradation products of aflatoxin B_1. *Chromatographia* **2007**, *71*, 107–112. [CrossRef]

15.	Gawade, S.P. Photodynamic studies on aflatoxin B_1 using UV radiation in the presence of methylene blue. *Indian J. Pharm. Educ. Res.* **2010**, *44*, 142–147.

16.	Diao, E.J.; Shen, X.Z.; Zhang, Z.; Ji, N.; Ma, W.W.; Dong, H.Z. Safety evaluation of aflatoxin B_1 in peanut oil after ultraviolet irradiation detoxification in a photodegradation reactor. *Int. J. Food Sci. Tech.* **2015**, *50*, 41–47. [CrossRef]

17.	Netto-Ferreira, J.C.; Heyne, B.; Scaiano, J.C. Photophysics and photochemistry of aflatoxins B_1 and B_2. *Photochem. Photobiol. Sci.* **2011**, *10*, 1701–1708. [CrossRef] [PubMed]

18.	Gosetti, F.; Bolfi, B.; Manfredi, M.; Calabrese, G.; Marengo, E. Determination of eight polyphenols and pantothenic acid in extra-virgin olive oil samples by a simple, fast, high-throughput and sensitive ultra high performance liquid chromatography with tandem mass spectrometry method. *J. Sep. Sci.* **2015**, *38*, 3130–3136. [CrossRef] [PubMed]

19.	Shi, R.; Guo, Y.; Vriesekoop, F.; Yuan, Q.P.; Zhao, S.H.; Liang, H. Improving oxidative stability of peanut oil under microwave treatment and deep fat frying by stearic acid-surfacant-tea polyphenols complex. *Eur. J. Lipid Sci. Technol.* **2015**, *117*, 1008–1015. [CrossRef]

20.	Iram, W.; Anjum, T.; Iqbal, M.; Ghaffar, A.; Abbas, M. Mass spectrometric identification and toxicity assessment of degraded products of aflatoxin B_1 and B_2 by *Corymbia citriodora* aqueous extracts. *Sci. Rep.* **2015**, *5*. [CrossRef] [PubMed]

21.	Suban, P.; Chen, Z. *Elements of Photochemistry*; People's Education Press: Beijing, China, 1982; p. 36.

22.	Wogan, G.N.; Edwards, G.S.; Newberne, P.M. Structure-activity relationships in toxicity and carcinogenicity of aflatoxins and analogs. *Cancer Res.* **1971**, *31*, 1936–1942. [PubMed]

23.	Verma, R.J. Aflatoxin cause DNA damage. *Int. J. Hum. Genet.* **2004**, *4*, 231–236.

24.	Mishra, H.N.; Das, C. A review on biological control and metabolism of aflatoxin. *Crit. Rev. Food Sci.* **2003**, *43*, 245–264. [CrossRef] [PubMed]

25.	Wang, F.; Xie, F.; Xue, X.F.; Wang, Z.D.; Fan, B.; Ha, Y.M. Structure elucidation and toxicity analyses of the radiolytic products of aflatoxin B_1 in methyl alcohol-water solution. *J. Hazard. Mater.* **2011**, *192*, 1192–1202. [CrossRef] [PubMed]

26.	Lee, L.S.J.; Dunn, J.; Delucca, A.J.; Ciegler, A. Role of lactone ring of aflatoxin B_1 in toxicity and mutagenicity. *Experientia* **1981**, *37*, 16–17. [CrossRef] [PubMed]

toxins

Review

Mycotoxin Decontamination of Food: Cold Atmospheric Pressure Plasma versus "Classic" Decontamination

Nataša Hojnik [1,2,*], Uroš Cvelbar [1,2], Gabrijela Tavčar-Kalcher [3], James L. Walsh [4] and Igor Križaj [5,*]

1   Jožef Stefan Institute, Department of Surface Engineering and Optoelectronics, Jamova cesta 39, SI-1000 Ljubljana, Slovenia; uros.cvelbar@ijs.si
2   Jožef Stefan International Postgraduate School, Jamova cesta 39, SI-1000 Ljubljana, Slovenia
3   University of Ljubljana, Veterinary Faculty, Institute of Food Safety, Feed and Environment, Gerbičeva 60, SI-1000 Ljubljana, Slovenia; gabrijela.tavcar-kalcher@vf.uni-lj.si
4   University of Liverpool, Department of Electrical, Engineering and Electronics, Brownlow Hill, Liverpool L69 3GJ, UK; J.L.Walsh@liverpool.ac.uk
5   Jožef Stefan Institute, Department of Molecular and Biomedical Sciences, Jamova cesta 39, SI-1000 Ljubljana, Slovenia
*   Correspondence: natasa.hojnik@ijs.si (N.H.); igor.krizaj@ijs.si (I.K.)

Academic Editor: Ting Zhou
Received: 15 March 2017; Accepted: 26 April 2017; Published: 28 April 2017

Abstract: Mycotoxins are secondary metabolites produced by several filamentous fungi, which frequently contaminate our food, and can result in human diseases affecting vital systems such as the nervous and immune systems. They can also trigger various forms of cancer. Intensive food production is contributing to incorrect handling, transport and storage of the food, resulting in increased levels of mycotoxin contamination. Mycotoxins are structurally very diverse molecules necessitating versatile food decontamination approaches, which are grouped into physical, chemical and biological techniques. In this review, a new and promising approach involving the use of cold atmospheric pressure plasma is considered, which may overcome multiple weaknesses associated with the classical methods. In addition to its mycotoxin destruction efficiency, cold atmospheric pressure plasma is cost effective, ecologically neutral and has a negligible effect on the quality of food products following treatment in comparison to classical methods.

Keywords: cold atmospheric pressure plasma technology; mycotoxins; physical decontamination; chemical decontamination; biological decontamination

---

1. Introduction

Many species of filamentous fungi have the ability to produce toxic secondary metabolites known as mycotoxins. The term mycotoxin is used only for toxic substances produced by fungi related to food products and animal feed; it does not include toxins produced by mushrooms [1]. Today, about 400 structurally different mycotoxins have been discovered and divided into the following main groups: (i) aflatoxins produced by *Aspergillus* species and ochratoxins produced by both *Aspergillus* and *Penicillium* species; (ii) trichothecenes, zearalenone and fumonisins produced by *Fusarium* species; and (iii) ergot alkaloids, produced by *Claviceps* species, and others [2]. Generally, mycotoxins represent a significant threat to human health as they can be carcinogenic, neurotoxic and toxic to the endocrine or immune system [3]. They can appear in the food chain due to infected crops, which are either consumed directly by humans or used as livestock feed, appearing in meat, milk or eggs. Beside this, they can contaminate food such as cereals, fruits, nuts, spices and other by-products as seen from Table 1 [4].

Table 1. Overview of the main characteristics of the most important mycotoxins.

Type	Representatives	Producing Fungi	Contaminated Foods	Structure Type	Toxicity
Aflatoxins (AF)	AFB$_1$, AFB$_2$, AFG$_1$, AFG$_2$, AFM$_1$	*Aspergillus* spp.: *A. flavus* *A. parasiticus* *A. nomius* *A. bombycis* *A. pseudotamari* *A. ochraceoreus*	Crops, cereals, seeds, nuts, spices	Difuranocoumarins	Carcinogenicity
Ochratoxins (OT)	OTA, OTB, OTC	*Aspergillus* spp. and *Penicillium* spp.: *A. ochraceus* *A. aliaceus* *A. auricomus* *A. carbonarius* *A. glaucus* *A. meleus* *A. niger* *P. nordicum* *P. verrucosum*	Crops, fruits, beer, wine, juices, coffee	Polyketide-derived dihydroisocoumarins bound to L-β-phenylalanin by amid bond	Nephrotoxicity, mutagenicy, carinogenicity
Fumonisins	Series A (FA), B (FB), C (FC) and P (FP) with FB being the most common representatives: FB$_1$, FB$_2$, FB$_3$	*Fusarium* spp.: *F. verticillioides* *F. proliferatum* *F. Napiforme* *F. dlamini* *F. nygamai*	Maize and its products	1, 2, 3-propanetricar-boxylic acid	Cytotoxicity, carcinogenicity
Zearalenone (ZEN)	ZEN, a-zearalenol, b-zearalenol	*Fusarium* spp.: *F. graminearum* *F. culmorum* *F. cerealis* *F. equiseti* *F. verticillioides* *F. incarnatum*	Crops, cereals	6-(10-Hydroxy-6-oxo-trans-1-undecenyl)-β-resorcylic acid lactone	Endocrine disruption
Trichothecenes	Deoxynivenol (DON), nivalenol (NIV), T-2 toxin, HT-2 toxin, diacetoxyscirpenol (DAS)	*Fusarium* spp. *Myrothecium* spp. *Phomopsis* spp. *Stachybotrys* spp. *Trychoderma* spp. *Trichotecium* spp. *Verticimonosporium* spp.	Crops	Tetracyclic-12,13-epoxy trichothenes	Inhibition of eucaryotic DNA, RNA and protein synthesis; nausea, vomiting, diarrhea, weight loss and loss of appetite, skin inflammation, vomiting, liver damage
Ergot alkaloids (EAs)	Ergometrine, ergotamine, ergosine, ergocristine, ergocryptine, ergocornine and the corresponding –inine epimers	*Claviceps* spp.: *C. purpurea*	Grains, grass	Tetracyclic ergolines (tryptophan-derived alkaloids)	Neurotoxicity, endocrine disruption
Other mycotoxins	Fusaproliferin (FUS), enniatins (ENNs), beauvericin (BEA), moniliformin (MON), patulin (PAT)	*Fusarium* spp. *Penicillium* spp. *Aspergillus* spp. *Eupenicillium* spp. *Paecilomyces* spp. *Byssochlamys* spp.	Crops, fruits, vegetables, cereals	Sesterterpene cyclic hexadepsipeptides, 3-hydroxycyclobut-3-ene-1,2-dione, 4-hydroxy-4H-furo[3,2-c]pyran-2(6H)-one	Cytotoxicity, abnormal gluconeiogenesis, genotoxicity and mutagenicity

Today, the trend of mycotoxins food contamination is increasing to alarming values with 25% of cereals worldwide already unsuitable for consumption [5]. Undesirable fungal growth and mycotoxin production is usually a result of incorrect agricultural and harvesting practices as well as the low effectiveness of prevention methods [3]. To reduce the potential danger to human health, many countries worldwide adopted strict legislation to control the mycotoxin presence in food and feed. In European Union, the presence of mycotoxins in food and feed is regulated by Regulation (EC) No 1881/2006, Directive 2002/32/EC, Recommendations 2006/576/EC and

2013/165/EU, and their amendments [6–9]. On top of this, recent studies have revealed a correlation between the increased presence of mycotoxins and global climate change [10]. Parameters including elevated temperatures, moisture levels and plant stress-related response stimulate fungal growth and, consequently, production of mycotoxins [10–13]. Furthermore, climate change plays a significant role in the global economy, where food is transported over long distances from producer to consumer, and may be subject to different local climates, transport and prolonged storage times. All these factors may contribute to increased food contamination [14].

Looking forward, it is expected that by the year 2050 the human population will exceed 9.2 billion. This will place an additional and unprecedented burden on the global food supply chain. The combination of modified climatic conditions and a tendency for consumers to eat healthier and fresher foods makes it imperative that new, sustainable and more effective approaches in agriculture, processing, transportation, and storage methods are developed. New mycotoxin-decontamination technologies will play a role in all stages of the supply chain. Beside this, the novel methods will have to preserve the quality of food products, be environmentally benign and economically suitable [14].

Considering the above-mentioned requirements, cold plasma technology represents a promising non-thermal mycotoxin-decontamination approach. Plasma is generally known as the fourth state of matter; a plasma state is reached by increasing the energy level of a substance from a solid state through the liquid and gaseous states of matter, ending in an ionized state of gas, which has unique physical and chemical properties (Figure 1) [15]. In electrically created plasmas, energy is delivered in the form of an electric field from an electrical power source; seed electrons produced by UV or background radiation are accelerated by the applied electric field leading to the excitation, dissociation or ionization of the background gas. Ionization, caused by the collision of an energetic electron with a neutral atom or molecule, results in the production of further electrons which are also accelerated in electric field. These free and energetic electrons subsequently collide with other surrounding molecules and atoms present in the gas, resulting in an avalanche process. Through the simultaneous generation and interaction among electrons, neutrals, metastables and ions, a vast number of reactions occur, yielding a wide variety of reactive chemical species [16,17]. In complex gas mixtures, such as humid air, a large number of reactive chemical species are created which take part in many hundreds of reactions [18]. In addition, molecules or atoms in an exited state can emit photons with wavelengths in the UVC, UVB and UVA range [19].

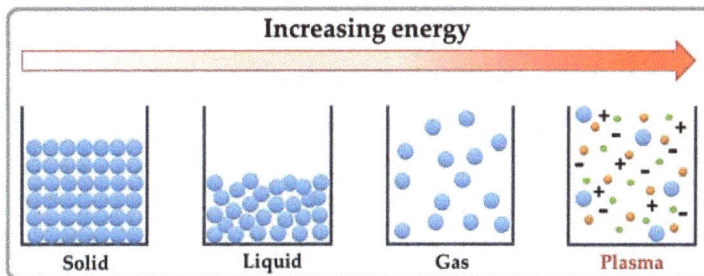

Figure 1. The generation of plasma: by adding energy to material, gas of electrons and ions is eventually produced. This fourth state of matter is referred to as "plasma".

Plasma can be produced under low pressure or even atmospheric pressure conditions. Typically, low-pressure plasma systems require a discharge generator, a gas source and an expensive vacuum system, consisting of pumps and vacuum chamber. Such systems are widely used for applications in material processing. Nevertheless, they are not suitable for materials sensitive to low-pressure conditions including biological material [17]. The use of atmospheric pressure plasma avoids the disadvantages of vacuum systems and enables the treatment of biological materials. Common examples of atmospheric pressure plasma systems include arc, corona and dielectric barrier discharges [20]. The most perspective discharges for the treatment of biological materials are those that are in thermal non-equilibrium, having a gas temperature that is close to room temperature, and are typically referred to as cold plasmas. Recent developments in cold atmospheric pressure plasma (CAP) sources and the ability to tailor discharges to produce highly reactive species in high concentrations, but at temperatures close to room temperature, have paved the way for a wide number of biological applications. Such CAPs are also suitable for use in electronics, surface modifications in polymer and textile industry, synthesis of nanoparticles and degradation of pollutants [21]. The new findings and developments in plasma science through the last decade reveal the great potential of CAP as an innovative technology in the field of biology, moving from the treatment of inanimate materials to living or cellular objects. Such applications include CAP treatments in medicine as well as in agriculture and the food industry. In the area of plasma medicine, research is mostly focused towards CAP use for skin treatment and wound healing, cancer cell and tumour treatment, dental implant sterilization, bone growth and many others [22].

CAP technology is, on the other hand, a newcomer to the field of agriculture and food industry. The main advantage of the CAP treatment of food products refers to its high chemical reactivity, achieved through the reactive species generated, and consequently its ability to deactivate harmful agents such as pathogenic bacteria and toxic pollutants in short processing times and at low temperatures with almost negligible impact on the treated food products. The technology can be applied to different types of food products in both solid and liquid form. In addition, the low energy consumption of such discharges and price-value inputs contribute to CAP being considered as an economically acceptable method [16,21,23,24]. Considering this, CAP technology also has a high potential as a decontamination tool for both mycotoxin-producing fungi and the mycotoxins that they produce. Different set-ups in both low and atmospheric pressure conditions have been used on mycotoxins such as aflatoxin B_1, deoxynivalenol and nivalenol (AFB_1, DON, and NIV, respectively) resulting in high decontamination rates in only a matter of seconds [25–27]. Treatments of the mycotoxin-producing fungi, e.g., one of the main mycotoxin producers *Aspergillus* spp. and *Penicillium* spp., with plasma, demonstrated very promising results as well. Plasma was able to stop or significantly reduce the further growth of fungi on different contaminated food products including corn, bean, cereals, fruits, nuts and many others [28–33]. Perhaps even more importantly, plasma treatments did not significantly influence the organoleptic characteristics of the treated foods or their nutritional properties [27,33–35]. Despite promising results, the use of CAP in the field of mycotoxin decontamination needs further exploration to uncover the CAP-related decontamination chemical processes and to develop optimised plasma systems suitable to meet the requirements of food processing and safety.

The aim of this contribution is to critically review plasma technology as a new food processing approach in the field of mycotoxin decontamination. The main advantages of this method over the classical mycotoxin-decontamination methods are considered and compared.

2. The Background of CAP Decontamination in Agriculture and Food Industry

A key benefit of the use of CAP technology in the field of agriculture and food is its high decontamination efficiency, which can be achieved in short treatment times and in non-thermal conditions. Its applicability has been widely demonstrated by successful decontamination of food with bacterial pathogens (*Escherichia coli*, *Salmonella typhimurium*, *Staphylococcus aureus*, *Listeria monocytogenes*, etc.) and harmful compounds (phenolic compounds, pesticides, azo dyes, etc.) [21,35–43]. The mechanisms of CAP decontamination are attributed primarily to the highly reactive oxygen and nitrogen species (ROS and RNS) created within the plasma as well as UV radiation, which induce highly oxidizing effects [22,24,44–46].

The prevalent primary species in an air plasma include radicals such as OH•, H•, O•, and NO. These radicals can react with each other, and with the ambient/background gas (air), vapour or even liquids, where they create oxygen- and nitrogen-based secondary species such as H_2O_2, NO_x, O_3,

$NO_2{}^-$, $NO_3{}^-$, peroxynitrite, etc. These plasma species can be divided into short- and long-lived species depending on their lifetime [18]. Long-lived species can also exist after the plasma source is removed or turned off resulting in post-discharge reactions, which is named the plasma afterglow [44]. The importance of these species can be observed after primary interaction such as in the case of living cells, where these species first react with the cell plasma membrane, and later can enter the cell and cause damage to intercellular elements, such as organelles and biomolecules such as DNA, RNA and proteins. Similarly, when toxic compounds are exposed to the ROS and RNS produced in the plasma, they are decomposed directly or indirectly through secondary chemical processes with the transformation of toxic substances into less toxic reaction products (Figure 2) [47].

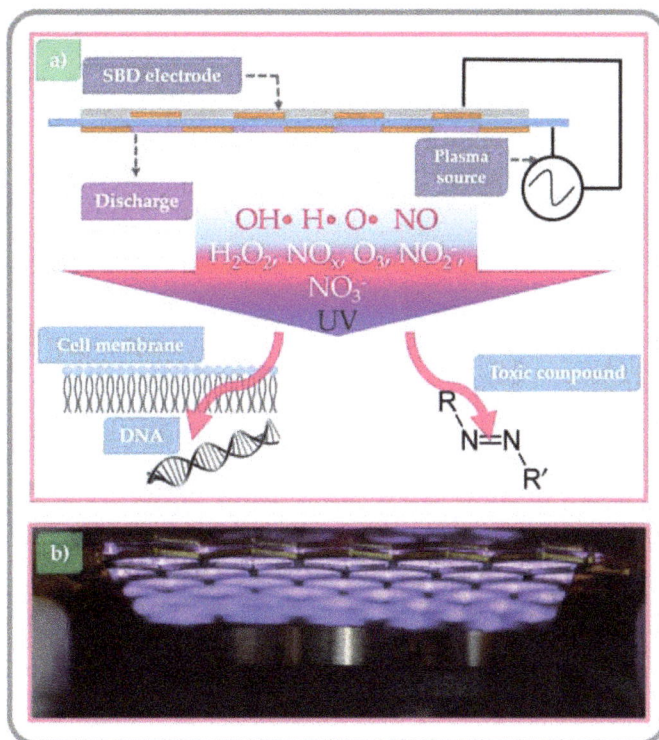

Figure 2. (a) Scheme of an air surface dielectric barrier (SDB) CAP set up; and (b) photo showing the CAP SDB system used in the presented experiments.

Among all plasma species, many studies have highlighted the key role played by atomic Oxygen (O), hydroxyl radical (OH•), ozone (O_3), hydrogen peroxide (H_2O_2) and peroxynitrite in CAP-related decontamination effects, since they all possess a very high oxidative potential. In biological systems such as bacterial and fungal cells, the short-lived O and OH• first react with cell walls and membranes and with all the compounds composing these two structures (lipids, proteins and polysaccharides). The lipids are the most sensitive to oxidation. The mechanism of OH• reaction with lipids refers to its H-abstraction from the unsaturated carbon bonds of the fatty acids, ending in lipid peroxidation [44]. O_3 is also a powerful oxidant; ozonation alone represents one of the most potent sanitizing and detoxifying approaches in the food industry and mycotoxin decontamination. O_3 has high reactivity, penetrability, and spontaneous decomposition into non-toxic oxygen without forming harmful oxygen species. Compared to OH•, O_3 induced reaction kinetics are slower [48]. In addition, the antimicrobial activity of H_2O_2 is well explored. Generally, cytotoxicity caused by H_2O_2 begins with penetration into cells and then transformation to $OH\bullet$ through Fenton's reaction causing intercellular damage [49]. Peroxynitrite has recently been the object of many studies as it has been found to play an important role in oxidative stress and various diseases (neurodegenerative diseases, AIDS, arteriosclerosis, etc.) [50]. It oxidizes biomolecules directly or through H^+- or CO_2-catalysed homolysis. As for direct reactivity, it has affinity on key parts in proteins such as thiols, iron/sulphur centres, and zinc fingers. The lifetime of peroxynitrite is relatively short, nonetheless, it can still cross membranes and reach deep within the cell, which allows it to interact with most of the important biomolecules [51,52]. Regarding the CAP decontamination of toxic compounds, $OH\bullet$ as one of the strongest oxidative species initiates the toxic molecule oxidation, resulting in its degradation. However, other slower reaction pathways such as those caused by O_3 and H_2O_2 are shunted or even bypassed [43].

Plasma species production strongly depends on the CAP system design and its mode of operation. When building a plasma system for food processing, there is a wide range of operating or so-called discharge parameters to choose from, including different gasses (air, O_2, N_2, He, Ar, etc.) and gas flows, discharge types, discharge volumes, electrode setups, etc. The discharge can be generated using high-voltage electrical power sources or intense laser light [46]. In general, these systems can be divided into three groups defined by the position of the treated food product with respect to the point of plasma generation: at some significant distance from the generation point, relatively close to generation point or within the plasma discharge itself. With a change in position of the sample with respect to the plasma, the nature and flux of chemical species varies significantly and result in different surface effects [53]. The first category refers to remote treatment with CAP where the sample is physically separated from the plasma generation point. In this scenario, the plasma generated species are usually transported to the sample by diffusion or by an induced flow. By the time plasma species reach the targeted surface, they are mostly composed of longer-lived plasma species, with a negligible concentration of highly reactive species (Figure 3a) [54]. The second category of system enables a semi-direct treatment with CAP. Then the target is exposed to higher concentrations of short-lived and highly reactive chemical species due to the relatively short distance between plasma generation point and substrate. In this scenario, the flux of UV photons reaching the targeted surface is also relatively high (Figure 3b) [55]. The last category is known as a direct contact system where the sample is placed between the electrodes of the plasma generation system and is consequently bombarded by large fluxes of reactive species and UV light (Figure 3c) [56,57]. While direct treatment should offer the highest possible degradation and decontamination efficacy, its implementation is problematic. The sample forms part of the electrical circuit and its presence can disrupt the discharge leading to the formation of hot spots that can damage the product.

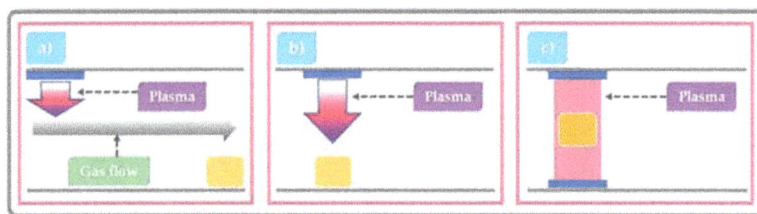

Figure 3. Schematic overview of common CAP systems considered for use in the food industry: (a) remote treatment where the sample is physically separated from the plasma generation point (b) semi-direct exposure, where the sample is placed close to the plasma generating electrodes; and (c) direct-exposure, where the sample is positioned between the plasma generating electrodes.

A wide range of design elements and discharge parameters enable a high degree of flexibility when designing CAP systems for food processing purposes no matter the type, size, and shape of the treated food products [58]. In terms of mycotoxin removal, plasma technology has mostly been used for the treatment of seeds, cereals, crops and fresh products [30,34,58,59]. For example, the use of an atmospheric pressure fluidized bed plasma system with air and nitrogen as a feed gas, which was used for the inactivation of *A. flavus* and *A. parasiticus* contaminated maize resulting in a 5.48 log reduction [28]. Furthermore, the production of fumonisin B_2 and ochratoxin A (FB_2 and OTA, respectively) was inhibited after the exposure of *A. niger* on date palm fruits to an argon CAP source. Oxygen CAP was applied for the treatment of *C. cladosporioides* and *P. citrinum* on the surface of dried filefish fillets reducing the fungi by more than 90% [33]. Moreover, argon and oxygen CAP proved to be efficient against *A. brasiliensis* contaminating pistachios [60]. Siciliano et al. performed CAP treatment with different mixtures of oxygen and nitrogen for decontamination of AFB_1 from dehulled hazelnuts succeeding 70% decontamination rate [27]. One of the most interesting CAP applications is also the in-package treatment of food. In this scenario, strawberries were treated with CAP generated between the electrode gap and inside a sealed package. The background microflora containing fungal species was reduced by 2 log reduction which could significantly prolong the food product expiry date [35].

3. The Comparison of "Classic" Approaches and CAP Technology in the Field of Mycotoxin Decontamination

Actions for preventing the fungal and mycotoxin contamination of feedstuff are performed at critical points before the expected fungal infestation. This may occur at the pre-harvest stage, during the harvest-time or at the post-harvest handling and storage stages [61]. The most effective approach is primary prevention, which should be carried out before the fungal invasion and mycotoxin production occurs. Current approaches include the use of fungicides to inhibit fungal growth, an appropriate scheduling of harvesting, and maintaining the optimum storage conditions after harvest [62,63]. Unfortunately, such techniques are not entirely effective and the efficiency of fungicides varies for different fungal species [64]. For this reason, several recent approaches have focused on the development of fungi-resistant plants [65]. The various physical, chemical and biological methods for the reduction of mycotoxin contamination currently in use or under active investigation for food and feed products are reviewed below, and compared with CAP technology (Figure 4) [61].

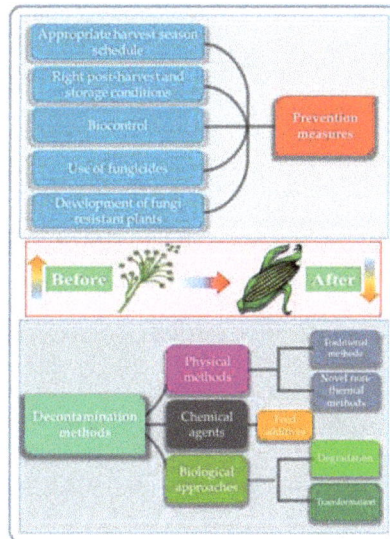

Figure 4. Overview of the currently available mycotoxin prevention and decontamination measures taken before and after fungal and mycotoxin contamination of food.

Methods for the physical decontamination of contaminated food are typically divided into traditional measures and novel non-thermal methods. The first group refers to methods that include sorting, washing, dehulling, density segregation, grain milling and thermal treatment. The principle of their decontamination is mostly based on the removal of the contaminated food parts and consequently the mycotoxins [66]. On the contrary, thermal treatment causes thermal degradation of mycotoxins [67–69]. Most of the traditional methods can reach satisfying decontamination rates for different types of toxins [2,70]. However, the processing time is usually very long, requiring high energy input, and is therefore very expensive. In addition, heat treatment can significantly affect the quality of the treated food products [61]. For these reasons, the food industry is looking to further develop new non-thermal approaches such as UV- and gamma-irradiation, pulsed-light treatment as well as CAP technology. Non-thermal methods typically affect the chemical structure of the mycotoxins leading to their degradation. Their decontamination efficiency depends on the presence of water in the treated food products, the extent of mycotoxin contamination, and the intensity of exposure [66]. For gamma irradiation technology, it has been reported that the decontamination of various mycotoxins was significantly more successful when they were present in solution, reaching up to 90% removal rate. The dosages of gamma irradiation used were from 1 to 20 kGy. Degradation in this case was probably a result of the formation of free radicals which were produced by the radiolysis of water. On the contrary, gamma irradiation decontamination of the mycotoxin-contaminated solids and dry food products in conditions with low moisture values was notably less effective [71]. High doses of gamma irradiation could, however, negatively affect the quality of food products such as grains and seeds, reducing their germination ability for example [72]. UV light irradiation also demonstrated high efficiency in mycotoxin decontamination. Treatments using a wavelength of 365 nm were capable of reducing the content of aflatoxins (AFs) from various types of nuts by more than 90% [73]. The 365 nm light irradiation was capable of removing the AFB_1 in peanut oil almost completely. Moreover, toxicity tests employing human embryo hepatocytes showed a significant reduction of toxicity of the degradation products [74]. When an AFB_1 aqueous solution underwent the UV light irradiation treatment, three major degradation products were observed (Table 2). It was indicated that the UV light irradiation probably interacted with most of the active sites on AFB_1, i.e., C_8-C_9 and O_1-C_{14} bonds. These two bonds have been recognized as being responsible for the AFB_1 toxicity and were transformed to more stable saturated bonds [75]. Interestingly, the structure of AFB_1 degradation products depended significantly on the media. Different degradation products were identified when the treatment was performed in acetonitrile or in peanut oil solution (Table 2) [74,76]. UV light irradiation was also used for patulin (PAT)-contaminated apple juice. Performed at 222 nm, a 90% reduction of mycotoxin content was achieved. However, such treatments lowered the concentration of some other photosensitive substances in the juice, including healthy ascorbic acid [77]. Similar to UV light irradiation, Moreau et al. studied mycotoxin, OTA, ZEN, DON or AFB_1, decontamination with pulsed light. The light flashes used were 300 μs in duration with a broad spectrum of light ranging from 180 to 1100 nm and a light flux 1 J/cm^2. The analysis demonstrated that eight light flashes almost completely removed mycotoxins from the solution. Remaining toxicity was assessed on nematode *Caenhorhabditis elegans*. Degradation products of DON and ZEN were evaluated as not toxic. The mutagenic activity based on an Ames test showed that AFB_1 degradation products were not mutagenic [78]. In a recent study, a pulsed light system of 0.52 J/cm^2/pulse and 360 μs long flashes, with wavelengths ranging from 100 to 1100 nm were used to decontaminate AFB_1 and aflatoxin B_2 (AFB_2) on different rice products. Decontamination efficiency higher than 90% was achieved for AFB_1 after 15 s of treatment in the case of rice bran [79].

An alternate detoxification strategy employs chemical agents, which are able to detoxify mycotoxins when added to a contaminated feedstuff. The effect is achieved by many synthetic and naturally occurring compounds including various organic acids, ammonium hydroxide, calcium hydroxide mono-methylamine, hydrochloric acid, hydrogen peroxide, bisulphite, chlorinating agents, formaldehyde, ammonia, clove oil and many more [63]. Ammoniation is conventionally used for AFs decontamination of feed such as cottonseed and peanut meal. The effectiveness of detoxification with ammonia increases with the quantity of ammonia used, the time of treatment, the temperature and pressure level [80]. Many types of ammoniation are available, with the two most commonly used being high-pressure/high-temperature treatment and atmospheric pressure/moderate temperature treatment. Both methods are able to reduce mycotoxin content up to 90%. For example, the degradation of AFB_1 by ammoniation is accomplished by hydrolysis of the lactone ring, which is followed by decarboxylation to AFD_1 and subsequent loss of cyclopentane ring (Table 2) [81–83]. Ozonation is a rather new way of mycotoxin decontamination in food processing [48]. When AF-contaminated corn flour was exposed to 75 mg/L of ozone for 60 min, the content of AFB_1, aflatoxin G_1 (AFG_1) and AFB_2 decreased from 53.60, 12.08 and 2.42 µg/kg to 11.38, 3.37 and 0.71 µg/kg, respectively [84]. In another study, 89.4% decomposition of AFB_1 was achieved after AFB_1-contaminated peanuts exposure to ozone of 50 mg/L at a flow rate of 5 L/min for 60 min. Following this results, the two most probable AFB_1-degradation pathways were proposed. In the first, the ozone initially reacts with a C_8-C_9 double bond of the furan ring in AFB_1 in electrophilic reaction based on Criegee mechanism, whereas the second degradation pathway starts with oxidation of the AFB_1 benzene methoxy group. Both reaction pathways lead to five final degradation products (Table 2). Generally, the toxicity of most degradation products is reduced under the assumption that a C_8-C_9 double bond represents one of the sites responsible for toxicity of AFB_1 [85]. Wang et al. used ozone to achieve a reduction in toxicity of DON contaminating wheat grains. After 60 min of 100 mg/L ozone treatment, the concentration of DON decreased from 3.89 mg/kg to 0.83 mg/kg, which is under the generally recognized maximum mycotoxin limit in feed [86]. Another method of chemical mycotoxin decontamination is the use of feed additives. These inorganic and organic mycotoxin binders are added to a feedstock when there is an indication of mycotoxin contamination. Typical additives include clays as natural mycotoxin adsorbents made of silicates or aluminosilicates. The level of adsorption of mycotoxins depends on the size and the charge of the mycotoxin with regard to the specific structure of the clay used [61]. The majority of clays are able to bind AFs, but not ZEN, fumonisins and trichothecenes, when added at a concentration of 10 g/kg [87,88]. On the other hand, bentonite was able to adsorb T-2 toxin, but to achieve high binding efficiency much more than 10 g of adsorbent per kg had to be used [89]. Inconveniently, clays are also able to adsorb the micronutrients from feed and disturb the bioavailability of minerals and trace elements. Furthermore, the contamination of clays with dioxins is possible [90]. As the inorganic adsorbents proved to be inefficient removers of the majority of mycotoxins, natural eco-friendly organic binders have been introduced instead, including oath fibres and cell extracts of lactic acid bacteria and *Saccharomyces cerevisiae* [91,92]. The decontamination of food by chemical means may be inexpensive and can achieve good decontamination results; however, most of these methods can present a risk for the environment as well as for human health. A further disadvantage is the long treatment time, which is not good for preservation of high quality foods.

The final category of mycotoxin decontamination measures includes biological methods. These procedures are based on the ability of microorganisms such as bacteria, yeast, moulds, actinomycetes and algae to remove or degrade mycotoxins in food and feed products. A clear advantage of biological decontamination approaches is that no chemicals are involved. The methods are based on biological transformation, enzymatic degradation, or modification of mycotoxins to less toxic substances. Mycotoxins can be thus acetylated, glucosylated, cleaved at their rings, hydrolysed, deaminated or decarboxylated [93]. Microorganisms capable of mycotoxin detoxification include species such as *Bacillus* spp., *Brevibacterium* spp., *Pseudomonas* spp., *Rhodococcus erythropolis*, *Aspergillus* spp., *Rhizopus* spp. and *Trichosporon mycotoxinivorans*. They can efficiently detoxify a wide range of mycotoxins including AFB_1, AFG_1, OTA, ZEN, PAT and DON [94–103]. In addition to reducing bioavailability of mycotoxins, some microorganisms, e.g., probiotic bacteria, are frequently added to feed as an additive with positive effect on the gut flora. The effectiveness of such microorganisms to act anti-mycotoxically largely depends on their ability to remain stable in the gastrointestinal tract. The main representatives of this group of bacteria are lactic acid bacteria [104]. Generally, biological approaches are not expensive and their environmental impact is low. Nevertheless, mycotoxin decontamination processes using microorganisms can be quite time-consuming [93].

In comparison with the methods described previously, CAP mycotoxin decontamination of food overcomes many of the disadvantages and obstacles of physical, chemical and microbial decontamination procedures. As depicted in Table 3, most of the CAP systems used for decontamination of food are environmentally benign, require a low energy input and are economically favourable. Beside this, plasma approaches have proven to have a negligible effect on the quality of many types of treated food. These advantages are based on the reactivity of the plasma species which enable the high decontamination efficiency in a very short time compared to alternative decontamination methods [46,53,105]. To demonstrate the efficiency of the plasma approach, a microwave-induced atmospheric pressure plasma system was used with argon as a carrier gas to treat three different mycotoxins, AFB_1, DON, and NIV dried on glass coverslips. The treatment resulted in the complete decontamination of all three mycotoxins after only 5 s of plasma exposure. Plasma treatment completely eliminated their cytotoxicity as tested on mouse macrophage RAW264.7 cells in vitro [25]. Furthermore, low-temperature radiofrequency plasma was used to degrade AFB_1. After 10 min of treatment, 88.3% AFB_1 was degraded. Analysis of the degradation products indicated that the toxicity should be reduced based on the structure-activity criteria; the degradation pathways indicated the formation of five different decay products (Table 2), where plasma induced the loss of the double bond in the terminal furan ring (C_8-C_9) [26]. AFs were exposed to a dielectric barrier discharge (DBD) plasma system, resulting in the complete destruction of mycotoxins when they were treated alone. Using the same plasma system, a 70% decontamination level was achieved for the treatment of AFB_1 contaminated dehulled hazelnuts [27]. To demonstrate the effectiveness of CAP, our recent experiments consider the use of an air surface barrier discharge (SBD) plasma treatment compared with UV light irradiation or thermal treatment in regards to AFB_1-destruction efficiency. Standard solution of AFB_1 was prepared in the mixture of acetonitrile and deionized water (2:1). One hundred microlitres of AFB_1 standard solution was applied on the glass coverslips and dried for 5 min. Such wet samples were then exposed to CAP, UVC light, or thermal treatment. CAP set-up was similar to the one reported by Ni et al. [40] and was operated at three different discharge powers (P_d): low (10 W), medium (15 W) and high (20 W). Low P_d operated plasma mostly contained ROS whereas RNS were the prevalent species at high P_d conditions. Plasma was observed to achieve more than 80% destruction level after just 15 s of treatment of AFB_1 applied on, regardless of the P_d used. In contrast, no significant transformation of AFB_1 was observed under thermal or UV light treatments, even at the longest exposure times (Figure 5). The ability of plasma to rapidly affect the AFB_1 molecular structure was confirmed by UV-Vis spectrometry. As evident from Figure 6, both major peaks in the UV-Vis spectra of AFB_1 significantly changed after 8 min of exposure of AFB_1 to plasma, independently of the P_d. On the other hand, the AFB_1 UV-Vis spectra remained almost the same following the UV or thermal treatment for the same time period. UV light irradiation treatment is usually efficient in degrading only the mycotoxin molecules, in particular AFs, which are known for their photosensitivity [74–76]. Beside this, UV irradiation represents one of the most commonly used decontamination approaches in food processing [106]. Comparing to CAP, to achieve adequate results, this method requires much longer exposure times (more than 10 min compared to some seconds in the case of CAP). Here, it is worth mentioning that UV requires higher power inputs which further impacts the decontamination efficiency. In addition to the mentioned drawbacks, it has been reported that UV irradiation could even increase the mutagenicity of AFs [107]. The characteristics of mycotoxin plasma treatment can be compared to some extent with ozone treatment, since one of the prevalent plasma-produced long-lived molecular species is ozone [108]. As many other reactive species beside ozone are produced in the plasma, synergistic effects can occur, resulting in the mycotoxin decontamination of food requiring significantly less exposure times than ozone alone [48,84,85]. Despite numerous advantages, CAP technology also has some limitations. One of the major problems is an inability to precisely control the gas phase chemistry when using ambient air, given that it varies with conditions in the surrounding atmosphere (for example increases in humidity). Since CAP contains ROS, it is not suitable for the treatment of high-fat food products. Furthermore, when carried out using very high voltages, additional safety measures are required as well as systems for the destruction and exhaust of potentially harmful long-lived species such as O_3 and NO_2 [46].

Table 2. Degradation products of AFB_1 after treatment with different decontamination methods.

Decontamination Method	Degradation Products	Reference
UV	In aqueous solution:	[75]
	In acetonitrile:	[76]
	In peanut oil:	[74]
Plasma		[26]

Table 2. *Cont.*

Decontamination Method	Degradation Products	Reference
	In acetonitrile:	
Ozone		[85]
Ammoniation		[81]

Table 2. *Cont.*

Decontamination Method	Degradation Products	Reference
P. putida	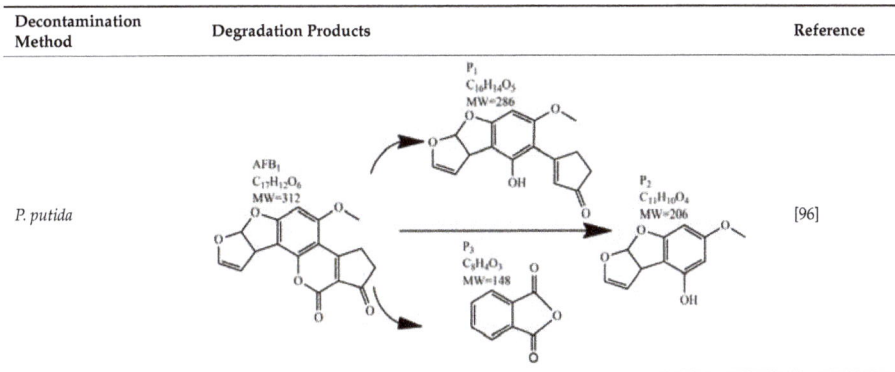	[96]

Table 3. The comparison between mycotoxin decontamination methods.

Decontamination Method	Highest Decontamination Rate Obtained	Food Product	Process Duration	Energy Consumption	Impact on the Food Quality	Reference
Thermal treatment	85–100% (FBs, ZEN, AFs)	Corn	Long	High	Significant	[70,109,110]
Gamma irradiation	90% (mixture)	Grains, seeds	Short	Low	Significant	[71]
UV light irradiation	90% (AFB$_1$, PAT)	Peanut oil; apple juice;	Short	Low	Negligible	[74,77]
Pulsed light technology	90% (AFB$_1$)	Rice products	Short	Low	Negligible	[79]
Ammoniation	90–100% (AFB$_1$)	Rice	Long	High	Significant	[83]
Ozonation	80% (AFs)	Corn flour, peanuts	Long	Low	Negligible	[84,85]
Bacillus spp.	92.5% (OTA)	/	Long	Low	Negligible-significant	[111]
Rhodococcus erythropolis	90% (AFB$_1$)	/	Long	Low	Negligible	[98]
Aspergillus spp.	100% (ZEN)	/	Long	Low	Negligible-significant	[99]
Trichosporon mycotoxinivorans	100% (OTA)	Animal Feed	Long	Low	Negligible	[103]
Lactic acid bacteria	80–100% (FBs)	/	Long	Low	Negligible	[112]
CAP technology	100% (AFs, DON, NIV)	Seeds, crops, cereals	Short	Low	Negligible	[25,27]

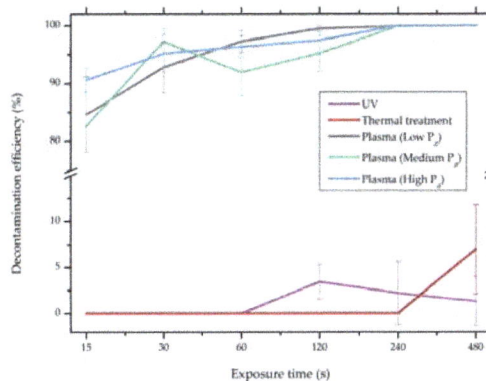

Figure 5. Comparison of decontamination efficiency (%) of aflatoxin B$_1$ (AFB$_1$) between cold atmospheric pressure plasma (CAP) and conventional decontamination approaches, UV light irradiation and thermal treatment; and air surface barrier discharge (SBD) plasma operated with three different discharge powers (P$_d$; low P$_d$, 10 W; med P$_d$, 15 W; and high P$_d$, 20 W). Ambient gas was used as a feed gas.

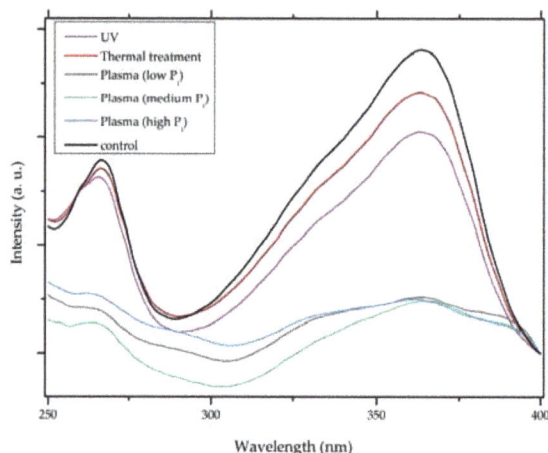

Figure 6. The comparison of aflatoxin B_1 (AFB_1) UV-Vis spectra after 8 min of exposure to heat treatment, UV irradiation and air surface barrier discharge (SBD) plasma operated with three different discharge powers (P_d; low P_d, 10 W; med P_d, 15 W; and high P_d, 20 W). Ambient gas was used as a feed gas.

4. Conclusions

Mycotoxin contaminated food represents a significant and increasing threat to human health and an enormous burden for the global economy. Decontamination methods to tackle this problem are based on physical, chemical and biological principles. In spite of constant improvements, these methods can still suffer from a lack of mycotoxin removal efficiency, they can be environmentally harmful and economically unfavourable. With no doubt, the food industry continuously strives for more effective mycotoxin decontamination approaches.

One of the most promising new procedures to deactivate mycotoxins on food is CAP technology. On the laboratory level, it has been convincingly demonstrated that CAP efficiently kills fungi on the surface of food and destroys the mycotoxins that these organisms secrete. In favour over many of the traditional food decontamination methods, plasma-based decontamination methods are generally lower-cost and ecologically benign. Most importantly, plasma-based mycotoxin decontamination of food has been demonstrated significantly more efficient in both the mycotoxin degradation level and speed of decontamination in comparison to conventional decontamination methods, as presented for the case of one of the most toxic mycotoxins, AFB_1.

Before industrialization of CAP technology can be realised, the molecular mechanisms and kinetics of plasma-based mycotoxin decontamination should be better characterized in order to become standardized. For this reason, additional experimental work is needed to:

-    Draw firm correlations between different plasma operating parameters and the specific reactive chemical species formed.
-    Draw correlations between the composition of the plasma and the structure of the mycotoxin degradation products. As toxicities of the mycotoxin degradation products can be experimentally determined, in this way, the mycotoxin decontamination efficiency would be defined as well.
-    Examine the effects of different plasma treatments on the quality of food products, for example on their nutritional value and organoleptic qualities.
-    Design plasma-forming systems for efficient mycotoxin decontamination of various types and sizes of food products.
-    Test if hybrid plasma-conventional systems for mycotoxin decontamination of food products can be even more effective.

Acknowledgments: Authors acknowledge project funding from NATO grant SPS.984555 and Slovenian Research Agency program P1-0207. JLW acknowledges the support of the UK Engineering and Physical Sciences Research Council (Project EP/N021347/1) and Innovate UK (Project 50769-377232).

Conflicts of Interest: The authors declare no conflict of interest.

References

1. Betina, V. *Mycotoxins: Production, Isolation, Separation and Purification;* Elsevier: Amsterdam, The Netherlands, 1984.
2. Vasanthi, S.; Bhat, R.V. Mycotoxins in foods-occurrence, health & economic significance & food control measures. *Indian J. Med. Res.* **1998**, *108*, 212–224. [PubMed]
3. Marin, S.; Ramos, A.J.; Cano-Sancho, G.; Sanchis, V. Mycotoxins: Occurrence, toxicology, and exposure assessment. *Food Chem. Toxicol.* **2013**, *60*, 218–237. [CrossRef] [PubMed]
4. Sforza, S.; Dall'Asta, C.; Marchelli, R. Recent advances in mycotoxin determination in food and feed by hyphenated chromatographic techniques/mass spectrometry. *Mass Spectrom. Rev.* **2006**, *25*, 54–76. [CrossRef] [PubMed]
5. Bryła, M.; Waśkiewicz, A.; Podolska, G.; Szymczyk, K.; Jędrzejczak, R.; Damaziak, K.; Sułek, A. Occurrence of 26 mycotoxins in the grain of cereals cultivated in poland. *Toxins* **2016**, *8*, 160. [CrossRef] [PubMed]
6. European Commission. Regulation (ec) no 1881/2006 of 19 december 2006 setting maximum levels for certain contaminants in foodstuffs. *Off. J. Eur. Union* **2006**, *364*, 5–24.
7. European Commission. Recommendation 2006/576/ec of 17 august 2006 on the presence of deoxynivalenol, zearalenone, ochratoxin A, T-2 and HT-2 and fumonisins in products intended for animal feeding. *Off. J. Eur. Union* **2006**, *229*, 7–9.
8. European Commission. Recommendation 2013/165/EU of 27 March 2013 on the presence of T-2 and HT-2 toxin in cereals and cereal products. *Off. J. Eur. Union* **2013**, *91*, 12–15.
9. Edgewood Parents and Teachers. Directive 2002/32/EC of 7 May 2002 on undesirable substances in animal feed. *Off. J. Eur. Communities* **2002**, *140*, 10–21.
10. Miraglia, M.; Marvin, H.J.P.; Kleter, G.A.; Battilani, P.; Brera, C.; Coni, E.; Cubadda, F.; Croci, L.; De Santis, B.; Dekkers, S.; et al. Climate change and food safety: An emerging issue with special focus on europe. *Food Chem. Toxicol.* **2009**, *47*, 1009–1021. [CrossRef] [PubMed]
11. Klich, M.A. Environmental and developmental factors influencing aflatoxin production by *Aspergillus flavus* and *Aspergillus parasiticus*. *Mycoscience* **2007**, *48*, 71–80. [CrossRef]
12. Mousa, W.; Ghazali, F.M.; Jinap, S.; Ghazali, H.M.; Radu, S. Modeling growth rate and assessing aflatoxins production by *Aspergillus flavus* as a function of water activity and temperature on polished and brown rice. *J. Food Sci.* **2013**, *78*, M56–M63. [CrossRef] [PubMed]
13. Narasaiah, K.V.; Sashidhar, R.; Subramanyam, C. Biochemical analysis of oxidative stress in the production of aflatoxin and its precursor intermediates. *Mycopathologia* **2006**, *162*, 179–189. [CrossRef] [PubMed]
14. Nellemann, C. *The Environmental Food Crisis: The Environment's Role in Averting Future Food Crises: A Unep Rapid Response Assessment;* UNEP/Earthprint: Nairobi, Kenya, 2009.
15. Fridman, A.; Kennedy, L.A. *Plasma Physics and Engineering;* CRC press: Boca Raton, FL, USA, 2004.
16. Kogelschatz, U. Atmospheric-pressure plasma technology. *Plasma Phys. Control. Fusion* **2004**, *46*, B63. [CrossRef]
17. Lieberman, M.A.; Lichtenberg, A.J. *Principles of Plasma Discharges and Materials Processing;* John Wiley & Sons: Chichester, UK, 2005.
18. Hasan, M.I.; Walsh, J.L. Numerical investigation of the spatiotemporal distribution of chemical species in an atmospheric surface barrier-discharge. *J. Appl. Phys.* **2016**, *119*, 203302. [CrossRef]
19. Locke, B.R.; Lukes, P.; Brisset, J.-L. Elementary chemical and physical phenomena in electrical discharge plasma in gas–liquid environments and in liquids. In *Plasma Chemistry and Catalysis in Gases and Liquids;* Wiley-VCH Verlag GmbH & Co. KGaA: Weinheim, Germany, 2012; pp. 185–241.
20. Schutze, A.; Jeong, J.Y.; Babayan, S.E.; Park, J.; Selwyn, G.S.; Hicks, R.F. The atmospheric-pressure plasma jet: A review and comparison to other plasma sources. *IEEE Trans. Plasma Sci.* **1998**, *26*, 1685–1694. [CrossRef]
21. Misra, N.N. The contribution of non-thermal and advanced oxidation technologies towards dissipation of pesticide residues. *Trends Food Sci. Technol.* **2015**, *45*, 229–244. [CrossRef]
22. Weltmann, K.D.; von Woedtke, T. Plasma medicine-current state of research and medical application. *Plasma Phys. Control. Fusion* **2017**, *59*, 014031. [CrossRef]

23. Laroussi, M.; Kong, M.; Morfill, G. *Plasma medicine: Applications of low-temperature gas plasmas in medicine and biology*; Cambridge University Press: Cambridge, UK, 2012.

24. Scholtz, V.; Pazlarova, J.; Souskova, H.; Khun, J.; Julak, J. Nonthermal plasma—A tool for decontamination and disinfection. *Biotechnol. Adv.* **2015**, *33*, 1108–1119. [CrossRef] [PubMed]

25. Park, B.J.; Takatori, K.; Sugita-Konishi, Y.; Kim, I.H.; Lee, M.H.; Han, D.W.; Chung, K.H.; Hyun, S.O.; Park, J.C. Degradation of mycotoxins using microwave-induced argon plasma at atmospheric pressure. *Surf. Coat. Technol.* **2007**, *201*, 5733–5737. [CrossRef]

26. Wang, S.-Q.; Huang, G.-Q.; Li, Y.-P.; Xiao, J.-X.; Zhang, Y.; Jiang, W.-L. Degradation of aflatoxin B$_1$ by low-temperature radio frequency plasma and degradation product elucidation. *Eur. Food Res. Technol.* **2015**, *241*, 103–113. [CrossRef]

27. Siciliano, I.; Spadaro, D.; Prelle, A.; Vallauri, D.; Cavallero, M.C.; Garibaldi, A.; Gullino, M.L. Use of cold atmospheric plasma to detoxify hazelnuts from aflatoxins. *Toxins* **2016**, *8*, 125. [CrossRef] [PubMed]

28. Dasan, B.G.; Boyaci, I.H.; Mutlu, M. Inactivation of aflatoxigenic fungi (*Aspergillus* spp.) on granular food model, maize, in an atmospheric pressure fluidized bed plasma system. *Food Control* **2016**, *70*, 1–8. [CrossRef]

29. Liang, J.-L.; Zheng, S.-H.; Ye, S.-Y. Inactivation of *Penicillium* aerosols by atmospheric positive corona discharge processing. *J. Aerosol Sci.* **2012**, *54*, 103–112. [CrossRef]

30. Selcuk, M.; Oksuz, L.; Basaran, P. Decontamination of grains and legumes infected with *Aspergillus* spp. and *Penicillum* spp. by cold plasma treatment. *Bioresour. Technol.* **2008**, *99*, 5104–5109. [PubMed]

31. Suhem, K.; Matan, N.; Nisoa, M.; Matan, N. Inhibition of *Aspergillus flavus* on agar media and brown rice cereal bars using cold atmospheric plasma treatment. *Int. J. Food Microbiol.* **2013**, *161*, 107–111. [CrossRef] [PubMed]

32. Ye, S.-Y.; Song, X.-L.; Liang, J.-L.; Zheng, S.-H.; Lin, Y. Disinfection of airborne spores of *Penicillium expansum* in cold storage using continuous direct current corona discharge. *Biosyst. Eng.* **2012**, *113*, 112–119. [CrossRef]

33. Park, S.Y.; Ha, S.-D. Application of cold oxygen plasma for the reduction of *Cladosporium cladosporioides* and *Penicillium citrinum* on the surface of dried filefish (*Stephanolepis cirrhifer*) fillets. *Int. J. Food Sci. Technol.* **2015**, *50*, 966–973. [CrossRef]

34. Basaran, P.; Basaran-Akgul, N.; Oksuz, L. Elimination of *Aspergillus parasiticus* from nut surface with low pressure cold plasma (LPCP) treatment. *Food Microbiol.* **2008**, *25*, 626–632. [CrossRef] [PubMed]

35. Misra, N.N.; Patil, S.; Moiseev, T.; Bourke, P.; Mosnier, J.P.; Keener, K.M.; Cullen, P.J. In-package atmospheric pressure cold plasma treatment of strawberries. *J. Food Eng.* **2014**, *125*, 131–138. [CrossRef]

36. Heo, N.S.; Lee, M.-K.; Kim, G.W.; Lee, S.J.; Park, J.Y.; Park, T.J. Microbial inactivation and pesticide removal by remote exposure of atmospheric air plasma in confined environments. *J. Biosci. Bioeng.* **2014**, *117*, 81–85. [CrossRef] [PubMed]

37. Wang, R.; Nian, W.; Wu, H.; Feng, H.; Zhang, K.; Zhang, J.; Zhu, W.; Becker, K.; Fang, J. Atmospheric-pressure cold plasma treatment of contaminated fresh fruit and vegetable slices: Inactivation and physiochemical properties evaluation. *Eur. Phys. J. D-Atomic Mol. Opt. Plasma Phys.* **2012**, *66*, 1–7. [CrossRef]

38. Edelblute, C.M.; Malik, M.A.; Heller, L.C. Antibacterial efficacy of a novel plasma reactor without an applied gas flow against methicillin resistant *Staphylococcus aureus* on diverse surfaces. *Bioelectrochemistry* **2016**, *112*, 106–111. [CrossRef] [PubMed]

39. Min, S.C.; Roh, S.H.; Niemira, B.A.; Sites, J.E.; Boyd, G.; Lacombe, A. Dielectric barrier discharge atmospheric cold plasma inhibits *Escherichia coli* o157:H7, *Salmonella*, *Listeria monocytogenes*, and tulane virus in romaine lettuce. *Int. J. Food Microbiol.* **2016**, *237*, 114–120. [CrossRef] [PubMed]

40. Ni, Y.; Lynch, M.; Modic, M.; Whalley, R.; Walsh, J. A solar powered handheld plasma source for microbial decontamination applications. *J. Phys. D: Appl. Phys.* **2016**, *49*, 355203. [CrossRef]

41. Doubla, A.; Laminsi, S.; Nzali, S.; Njoyim, E.; Kamsu-Kom, J.; Brisset, J.L. Organic pollutants abatement and biodecontamination of brewery effluents by a non-thermal quenched plasma at atmospheric pressure. *Chemosphere* **2007**, *69*, 332–337. [CrossRef] [PubMed]

42. Li, Z.G.; Hu, Z.; Cao, P.; Zhao, H.J. Decontamination of 2-chloroethyl ethyl sulfide by pulsed corona plasma. *Plasma Sci. Technol.* **2014**, *16*, 1054–1058. [CrossRef]

43. Jiang, B.; Zheng, J.; Qiu, S.; Wu, M.; Zhang, Q.; Yan, Z.; Xue, Q. Review on electrical discharge plasma technology for wastewater remediation. *Chem. Eng. J.* **2014**, *236*, 348–368. [CrossRef]

44. Lukes, P.; Brisset, J.-L.; Locke, B.R. Biological effects of electrical discharge plasma in water and in gas–liquid environments. In *Plasma Chemistry and Catalysis in Gases and Liquids*; Wiley-VCH Verlag GmbH & Co. KGaA: Weinheim, Germany, 2012; pp. 309–352.

45. Machala, Z.; Chládeková, L.; Pelach, M. Plasma agents in bio-decontamination by DC discharges in atmospheric air. *J. Phys. D: Appl. Phys.* **2010**, *43*, 222001. [CrossRef]

46. Niemira, B.A. Cold plasma decontamination of foods. *Annu. Rev. Food Sci. Technol.* **2012**, *3*, 125–142. [CrossRef]

47. Moisan, M.; Barbeau, J.; Crevier, M.-C.; Pelletier, J.; Philip, N.; Saoudi, B. Plasma sterilization. Methods and mechanisms. *Pure Appl. Chem.* **2002**, *74*, 349–358.

48. Karaca, H.; Velioglu, Y.S.; Nas, S. Mycotoxins: Contamination of dried fruits and degradation by ozone. *Toxin Rev.* **2010**, *29*, 51–59. [CrossRef]

49. Laurita, R.; Barbieri, D.; Gherardi, M.; Colombo, V.; Lukes, P. Chemical analysis of reactive species and antimicrobial activity of water treated by nanosecond pulsed DBD air plasma. *Clin. Plasma Med.* **2015**, *3*, 53–61. [CrossRef]

50. Beckman, J.S.; Koppenol, W.H. Nitric oxide, superoxide, and peroxynitrite: The good, the bad, and ugly. *Am. J. Physiol.-Cell Physiol.* **1996**, *271*, C1424–C1437.

51. Naitali, M.; Herry, J.M.; Hnatiuc, E.; Kamgang, G.; Brisset, J.L. Kinetics and bacterial inactivation induced by peroxynitrite in electric discharges in air. *Plasma Chem. Plasma Process.* **2012**, *32*, 675–692. [CrossRef]

52. Naïtali, M.; Kamgang-Youbi, G.; Herry, J.-M.; Bellon-Fontaine, M.-N.; Brisset, J.-L. Combined effects of long-living chemical species during microbial inactivation using atmospheric plasma-treated water. *Appl. Environ. Microbiol.* **2010**, *76*, 7662–7664. [CrossRef] [PubMed]

53. Niemira, B.A.; Gutsol, A. Nonthermal plasma as a novel food processing technology. *Nonthermal Process. Technol. Food* **2011**, 272–288.

54. Chirokov, A.; Gutsol, A.; Fridman, A. Atmospheric pressure plasma of dielectric barrier discharges. *Pure Appl. Chem.* **2005**, *77*, 487–495. [CrossRef]

55. Laroussi, M.; Lu, X. Room-temperature atmospheric pressure plasma plume for biomedical applications. *Appl. Phys. Lett.* **2005**, *87*, 113902. [CrossRef]

56. Smet, C.; Noriega, E.; Rosier, F.; Walsh, J.; Valdramidis, V.; Van Impe, J. Influence of food intrinsic factors on the inactivation efficacy of cold atmospheric plasma: Impact of osmotic stress, suboptimal pH and food structure. *Innovative Food Sci. Emerg. Technol.* **2016**, *38*, 393–406. [CrossRef]

57. Smet, C.; Noriega, E.; Rosier, F.; Walsh, J.; Valdramidis, V.; Van Impe, J. Impact of food model (micro) structure on the microbial inactivation efficacy of cold atmospheric plasma. *Int. J. Food Microbiol.* **2017**, *240*, 47–56. [CrossRef] [PubMed]

58. Ito, M.; Ohta, T.; Hori, M. Plasma agriculture. *J. Korean Phys. Soc.* **2012**, *60*, 937–943. [CrossRef]

59. Zahoranova, A.; Henselova, M.; Hudecova, D.; Kalinakova, B.; Kovacik, D.; Medvecka, V.; Cernak, M. Effect of cold atmospheric pressure plasma on the wheat seedlings vigor and on the inactivation of microorganisms on the seeds surface. *Plasma Chem. Plasma Process.* **2016**, *36*, 397–414. [CrossRef]

60. Pignata, C.; D'Angelo, D.; Basso, D.; Cavallero, M.C.; Beneventi, S.; Tartaro, D.; Meineri, V.; Gilli, G. Low-temperature, low-pressure gas plasma application on *Aspergillus brasiliensis*, *Escherichia coli* and pistachios. *J. Appl. Microbiol.* **2014**, *116*, 1137–1148. [CrossRef] [PubMed]

61. Jouany, J.P. Methods for preventing, decontaminating and minimizing the toxicity of mycotoxins in feeds. *Anim. Feed Sci. Technol.* **2007**, *137*, 342–362. [CrossRef]

62. Abdel-Wahhab, M.; Kholif, A. Mycotoxins in animal feeds and prevention strategies: A review. *Asian J. Anim. Sci.* **2010**, *4*, 113–131. [CrossRef]

63. Beaver, R.W. Decontamination of mycotoxin-containing foods and feedstuffs. *Trends Food Sci. Technol.* **1991**, *2*, 170–173. [CrossRef]

64. Varga, J.; Tóth, B. Novel strategies to control mycotoxins in feeds: A review. *Acta Veterinaria Hungarica* **2005**, *53*, 189–203. [CrossRef] [PubMed]

65. Kabak, B.; Dobson, A.D. Biological strategies to counteract the effects of mycotoxins. *J. Food Prot.* **2009**, *72*, 2006–2016. [CrossRef] [PubMed]

66. Smith, T.; Girish, C. Prevention and control of animal feed contamination by mycotoxins and reduction of their adverse effects in livestock. In *Animal Feed Contamination*; Wooodhead Publishing limited: Cambridge, UK, 2012.

67. Avantaggiato, G.; Havenaar, R.; Visconti, A. Evaluation of the intestinal absorption of deoxynivalenol and nivalenol by an in vitro gastrointestinal model, and the binding efficacy of activated carbon and other adsorbent materials. *Food Chem. Toxicol.* **2004**, *42*, 817–824. [CrossRef] [PubMed]

68. Meister, U.; Springer, M. Mycotoxins in cereals and cereal products—occurrence and changes during processing. *J. Appl. Bot. Food Qual.* **2004**, *78*, 168–173.

69. Park, J.; Scott*, P.; Lau, B.-Y.; Lewis, D. Analysis of heat-processed corn foods for fumonisins and bound fumonisins. *Food Addit. Contam.* **2004**, *21*, 1168–1178. [CrossRef] [PubMed]

70. Scott, P.; Lawrence, G. Stability and problems in recovery of fumonisins added to corn-based foods. *J. AOAC Int.* **1993**, *77*, 541–545.

71. Calado, T.; Venâncio, A.; Abrunhosa, L. Irradiation for mold and mycotoxin control: A review. *Compr. Rev. Food Sci. Food Saf.* **2014**, *13*, 1049–1061. [CrossRef]

72. Kottapalli, B.; Wolf-Hall, C.E.; Schwarz, P.; Schwarz, J.; Gillespie, J. Evaluation of hot water and electron beam irradiation for reducing *Fusarium* infection in malting barley. *J. Food Prot.* **2003**, *66*, 1241–1246. [CrossRef] [PubMed]

73. Jubeen, F.; Bhatti, I.A.; Khan, M.Z.; Zahoor-Ul, H.; Shahid, M. Effect of UVC irradiation on aflatoxins in ground nut (*Arachis hypogea*) and tree nuts (*Juglans regia, Prunus duclus* and *Pistachio vera*). *J. Chem. Soc. Pak.* **2012**, *34*, 1366–1374.

74. Mao, J.; He, B.; Zhang, L.X.; Li, P.W.; Zhang, Q.; Ding, X.X.; Zhang, W. A structure identification and toxicity assessment of the degradation products of aflatoxin B_1 in peanut oil under UV irradiation. *Toxins* **2016**, *8*, 332. [CrossRef] [PubMed]

75. Liu, R.; Jin, Q.; Tao, G.; Shan, L.; Huang, J.; Liu, Y.; Wang, X.; Mao, W.; Wang, S. Photodegradation kinetics and byproducts identification of the aflatoxin B_1 in aqueous medium by ultra-performance liquid chromatography–quadrupole time-of-flight mass spectrometry. *J. Mass Spectrom.* **2010**, *45*, 553–559. [CrossRef] [PubMed]

76. Liu, R.; Jin, Q.; Tao, G.; Shan, L.; Liu, Y.; Wang, X. LC–MS and UPLC–quadrupole time-of-flight MS for identification of photodegradation products of aflatoxin B_1. *Chromatographia* **2012**, *71*, 107–112. [CrossRef]

77. Zhu, Y.; Koutchma, T.; Warriner, K.; Zhou, T. Reduction of patulin in apple juice products by UV light of different wavelengths in the UVC range. *J. Food Prot.* **2014**, *77*, 963–971. [CrossRef] [PubMed]

78. Moreau, M.; Lescure, G.; Agoulon, A.; Svinareff, P.; Orange, N.; Feuilloley, M. Application of the pulsed light technology to mycotoxin degradation and inactivation. *J. Appl. Toxicol.* **2013**, *33*, 357–363. [CrossRef]

79. Wang, B.; Mahoney, N.E.; Pan, Z.L.; Khir, R.; Wu, B.G.; Ma, H.L.; Zhao, L.M. Effectiveness of pulsed light treatment for degradation and detoxification of aflatoxin B_1 and B_2 in rough rice and rice bran. *Food Control* **2016**, *59*, 461–467. [CrossRef]

80. Grenier, B.; Loureiro-Bracarense, A.-P.; Leslie, J.F.; Oswald, I.P. Physical and chemical methods for mycotoxin decontamination in maize. *Mycotoxin Reduct. Grain Chains* **2014**, 116–129.

81. Hoogenboom, L.; Tulliez, J.; Gautier, J.-P.; Coker, R.; Melcion, J.-P.; Nagler, M.; Polman, T.H.; Delort-Laval, J. Absorption, distribution and excretion of aflatoxin-derived ammoniation products in lactating cows. *Food Addit. Contam.* **2001**, *18*, 47–58. [CrossRef] [PubMed]

82. Allameh, A.; Safamehr, A.; Mirhadi, S.A.; Shivazad, M.; Razzaghi-Abyaneh, M.; Afshar-Naderi, A. Evaluation of biochemical and production parameters of broiler chicks fed ammonia treated aflatoxin contaminated maize grains. *Anim. Feed Sci. Technol.* **2005**, *122*, 289–301. [CrossRef]

83. Millán, T.F.; Martinez, Y.A. Efficacy and stability of ammoniation process as aflatoxin B_1 decontamination technology in rice. *Archivos Latinoamer. Nutr.* **2003**, *53*, 287–292.

84. Luo, X.; Wang, R.; Wang, L.; Li, Y.; Wang, Y.; Chen, Z. Detoxification of aflatoxin in corn flour by ozone. *J. Sci. Food Agric.* **2014**, *94*, 2253–2258. [CrossRef] [PubMed]

85. Diao, E.; Hou, H.; Chen, B.; Shan, C.; Dong, H. Ozonolysis efficiency and safety evaluation of aflatoxin B_1 in peanuts. *Food Chem. Toxicol.* **2013**, *55*, 519–525. [CrossRef] [PubMed]

86. Wang, L.; Luo, Y.P.; Luo, X.H.; Wang, R.; Li, Y.F.; Li, Y.N.; Shao, H.L.; Chen, Z.X. Effect of deoxynivalenol detoxification by ozone treatment in wheat grains. *Food Control* **2016**, *66*, 137–144. [CrossRef]

87. Phillips, T.; Clement, B.; Kubena, L.; Harvey, R. Detection and detoxification of aflatoxins: Prevention of aflatoxicosis and aflatoxin residues with hydrated sodium calcium aluminosilicate. *Vet. Hum. Toxicol.* **1989**, *32*, 15–19.

88. Huwig, A.; Freimund, S.; Käppeli, O.; Dutler, H. Mycotoxin detoxication of animal feed by different adsorbents. *Toxicol. Lett.* **2001**, *122*, 179–188. [CrossRef]

89. Carson, M.S.; Smith, T.K. Role of bentonite in prevention of T-2 toxicosis in rats. *J. Anim. Sci.* **1983**, *57*, 1498–1506. [CrossRef] [PubMed]

90. Moshtaghian, J.; Parsons, C.M.; Leeper, R.W.; Harrison, P.C.; Koelkebeck, K.W. Effect of sodium aluminosilicate on phosphorus utilization by chicks and laying hens. *Poult. Sci.* **1991**, *70*, 955–962. [CrossRef]

91. Smith, T.K. Influence of dietary fiber, protein and zeolite on zearalenone toxicosis in rats and swine. *J. Anim. Sci.* **1980**, *50*, 278–285. [CrossRef] [PubMed]

92. Yiannikouris, A.; André, G.; Buléon, A.; Jeminet, G.; Canet, I.; François, J.; Bertin, G.; Jouany, J.-P. Comprehensive conformational study of key interactions involved in zearalenone complexation with β-D-glucans. *Biomacromolecules* **2004**, *5*, 2176–2185. [CrossRef] [PubMed]

93. Hathout, A.S.; Aly, S.E. Biological detoxification of mycotoxins: A review. *Ann. Microbiol.* **2014**, *64*, 905–919. [CrossRef]

94. Schallmey, M.; Singh, A.; Ward, O.P. Developments in the use of *Bacillus* species for industrial production. *Can. J. Microbiol.* **2004**, *50*, 1–17. [CrossRef] [PubMed]

95. Rodriguez, H.; Reveron, I.; Doria, F.; Costantini, A.; De las Rivas, B.; Munoz, R.; Garcia-Moruno, E. Degradation of ochratoxin a by *Brevibacterium* species. *J. Agric. Food Chem.* **2011**, *59*, 10755–10760. [CrossRef] [PubMed]

96. Samuel, M.S.; Sivaramakrishna, A.; Mehta, A. Degradation and detoxification of aflatoxin B_1 by *Pseudomonas putida*. *Int. Biodeterior. Biodegrad.* **2014**, *86*, 202–209. [CrossRef]

97. El-Deeb, B.A. Isolation and characterization of soil bacteria able to degrade zearalenone. *J. Bot.* **2005**, *32*, 3–30.

98. Teniola, O.D.; Addo, P.A.; Brost, I.M.; Färber, P.; Jany, K.-D.; Alberts, J.F.; Van Zyl, W.H.; Steyn, P.S.; Holzapfel, W.H. Degradation of aflatoxin B_1 by cell-free extracts of *Rhodococcus erythropolis* and *Mycobacterium fluoranthenivorans* sp. Nov. Dsm44556 t. *Int. J. Food Microbiol.* **2005**, *105*, 111–117. [CrossRef] [PubMed]

99. Brodehl, A.; Möller, A.; Kunte, H.-J.; Koch, M.; Maul, R. Biotransformation of the mycotoxin zearalenone by fungi of the genera *Rhizopus* and *Aspergillus*. *FEMS Microbiol. Lett.* **2014**, *359*, 124–130. [CrossRef] [PubMed]

100. Kusumaningtyas, E.; Widiastuti, R.; Maryam, R. Reduction of aflatoxin B_1 in chicken feed by using *Saccharomyces cerevisiae*, *Rhizopus oligosporus* and their combination. *Mycopathologia* **2006**, *162*, 307–311. [CrossRef] [PubMed]

101. Chourasia, H.K.; Suman, S.K.; Jha, G.N. Microbial degradation of aflatoxin in maize: A biocontrol approach for management of preharvest aflatoxin contamination. *J. Mycol. Plant Pathol.* **2011**, *41*, 408.

102. Garda-Buffon, J.; Kupski, L.; Badiale-Furlong, E. Deoxynivalenol (DON) degradation and peroxidase enzyme activity in submerged fermentation. *Food Sci. Technol. (Camp.)* **2011**, *31*, 198–203. [CrossRef]

103. Molnar, O.; Schatzmayr, G.; Fuchs, E.; Prillinger, H. *Trichosporon mycotoxinivorans* sp. nov., a new yeast species useful in biological detoxification of various mycotoxins. *Syst. Appl. Microbiol.* **2004**, *27*, 661–671. [CrossRef] [PubMed]

104. Sangsila, A.; Faucet-Marquis, V.; Pfohl-Leszkowicz, A.; Itsaranuwat, P. Detoxification of zearalenone by *Lactobacillus pentosus* strains. *Food Control* **2016**, *62*, 187–192. [CrossRef]

105. Misra, N.N.; Schlüter, O.; Cullen, P.J. Chapter 1 - plasma in food and agriculture. In *Cold Plasma in Food and Agriculture*; Academic Press: San Diego, CA, USA, 2016; pp. 1–16.

106. Fellows, P.J. *Food Processing Technology: Principles and Practice*; Elsevier: Amsterdam, The Netherlands, 2009.

107. Stark, A.-A.; Gal, Y.; Shaulsky, G. Involvement of singlet oxygen in photoactivation of aflatoxins B_1 and B_2 to DNA-binding forms in vitro. *Carcinogenesis* **1990**, *11*, 529–534. [CrossRef] [PubMed]

108. Olszewski, P.; Li, J.; Liu, D.; Walsh, J. Optimizing the electrical excitation of an atmospheric pressure plasma advanced oxidation process. *J. Hazard. Mater.* **2014**, *279*, 60–66. [CrossRef] [PubMed]

109. Castells, M.; Marin, S.; Sanchis, V.; Ramos, A. Fate of mycotoxins in cereals during extrusion cooking: A review. *Food Addit. Contam.* **2005**, *22*, 150–157. [CrossRef] [PubMed]

110. Hale, O.; Wilson, D. Performance of pigs on diets containing heated or unheated corn with or without aflatoxin. *J. Anim. Sci.* **1979**, *48*, 1394–1400. [CrossRef] [PubMed]

111. Petchkongkaew, A.; Taillandier, P.; Gasaluck, P.; Lebrihi, A. Isolation of *Bacillus* spp. from thai fermented soybean (Thua-nao): Screening for aflatoxin B_1 and ochratoxin A detoxification. *J. Appl. Microbiol.* **2008**, *104*, 1495–1502. [CrossRef] [PubMed]

112. Niderkorn, V.; Morgavi, D.P.; Aboab, B.; Lemaire, M.; Boudra, H. Cell wall component and mycotoxin moieties involved in the binding of fumonisin B_1 and B_2 by lactic acid bacteria. *J. Appl. Microbiol.* **2009**, *106*, 977–985. [CrossRef] [PubMed]

toxins

MDPI

Article

Ameliorative Effects of Neutral Electrolyzed Water on Growth Performance, Biochemical Constituents, and Histopathological Changes in Turkey Poults during Aflatoxicosis

Denise Gómez-Espinosa [1,†], Francisco Javier Cervantes-Aguilar [2,†],
Juan Carlos Del Río-García [3,†], Tania Villarreal-Barajas [4], Alma Vázquez-Durán [3] and
Abraham Méndez-Albores [3,*]

[1] National Autonomous University of Mexico-Superior Studies Faculty at Cuautitlan (UNAM-FESC),
    Master in Animal Production and Health Sciences, Cuautitlan Izcalli 54714, Mexico;
    denisegoes1985@gmail.com
[2] National Autonomous University of Mexico-Superior Studies Faculty at Cuautitlan (UNAM-FESC),
    Veterinary Medicine and Animal Husbandry, Department of Biological and Livestock Sciences,
    Cuautitlan Izcalli 54714, Mexico; fjca_03@hotmail.com
[3] National Autonomous University of Mexico-Superior Studies Faculty at Cuautitlan (UNAM-FESC),
    Campus 4, Multidisciplinary Research Unit L14 (Food, Mycotoxins and Mycotoxicosis),
    Cuautitlan Izcalli 54714, Mexico; delriog@unam.mx (J.C.D.R.-G.);
    almavazquez@comunidad.unam.mx (A.V.-D.)
[4] Esteripharma SA de CV, Atlacomulco 50450, Mexico; tvillarreal@esteripharma.com.mx
[*] Correspondence: albores@unam.mx; Tel.: +52-55-5623-1999 (ext. 39434)
[†] These authors contributed equally to this work.

Academic Editor: Ting Zhou
Received: 9 February 2017; Accepted: 11 March 2017; Published: 14 March 2017

Abstract: Different in vitro and in silico approaches from our research group have demonstrated that neutral electrolyzed water (NEW) can be used to detoxify aflatoxins. The objective of this investigation was to evaluate the ability of NEW to detoxify B-aflatoxins (AFB_1 and AFB_2) in contaminated maize and to confirm detoxification in an in vivo experimental model. Batches of aflatoxin-contaminated maize were detoxified with NEW and mixed in commercial feed. A total of 240 6-day-old female large white Nicholas-700 turkey poults were randomly divided into four treatments of six replicates each (10 turkeys per replicate), which were fed ad libitum for two weeks with the following dietary treatments: (1) control feed containing aflatoxin-free maize (CONTROL); (2) feed containing the aflatoxin-contaminated maize (AF); (3) feed containing the aflatoxin-contaminated maize detoxified with NEW (AF + NEW); and (4) control feed containing aflatoxin-free maize treated with NEW (NEW). Compared to the control groups, turkey poults of the AF group significantly reduced body weight gain and increased feed conversion ratio and mortality rate; whereas turkey poults of the AF + NEW group did not present significant differences on productive parameters. In addition, alterations in serum biochemical constituents, enzyme activities, relative organ weight, gross morphological changes and histopathological studies were significantly mitigated by the aflatoxin-detoxification procedure. From these results, it is concluded that the treatment of aflatoxin-contaminated maize with NEW provided reasonable protection against the effects caused by aflatoxins in young turkey poults.

Keywords: maize; aflatoxins; neutral electrolyzed water; detoxification; turkey

1. Introduction

Among the many types of microorganisms that cause food-borne disease outbreaks, toxigenic fungi and their products (mycotoxins) threaten the health and economy of the poultry industry by contaminating feed materials [1–4]. Mycotoxins are widespread food and feed compounds capable of causing innumerable damaging effects to humans and animals; these contaminants are a chemical assemblage of fungal-originated toxins that may be found naturally in grains and cereal products. Aflatoxins are a special group of mycotoxins compelling investigation due to their toxic, mutagenic, teratogenic and carcinogenic potential. Some species of fungi of the *Aspergillus* genus, for instance *Aspergillus flavus* and *Aspergillus parasiticus*, produce aflatoxins. These are secondary metabolites easily found in fungal-contaminated peanuts, maize, pistachios and rice, among others, and include four major structures: AFB_1, AFB_2, AFG_1, and AFG_2 [5–8].

Aflatoxins have been reported to cause severe economic losses in the poultry industry due to their various health effects. Specific responses of aflatoxicosis include reduced growth performance, increased relative pancreas weight, decreased relative liver weight, immunosuppression, hemorrhage, alteration in digestion absorption, nutrient metabolism, serum enzyme activities, biochemical and hematological values and possibly death [9,10]. Specifically, turkeys fed with aflatoxins rapidly incorporate the metabolite in the small intestine, thus mainly damaging the liver. Reduction of protein production, increase in hepatic enzyme activity and coagulopathies indicate fat degeneration and proliferation of biliary ducts [11]. Turkey poults generally exhibit aflatoxicosis as erratic walk, inappetence, reduced activity, anemia, and even death. Therefore, in the poultry industry, production parameters such as body weight gain, feed consumption and mortality rate may be affected by aflatoxin poisoning [12].

Developing strategies to detoxify mycotoxin contamination is crucial for the food and animal feed industries. The best option to prevent mycotoxicosis is to avoid the use of contaminated materials, followed by fungi disinfection [13,14], and finally inactivation of the mycotoxin. A diversified portfolio of strategies to detoxify food and feed has been proposed to reduce the adverse effects of aflatoxin contamination, including physical selection of the contaminated commodity, thermal inactivation, irradiation and supplementation with nano-clay adsorbent, biological degradation by microbial fermentation, and chemical treatment with acids, bases, organic solvents, and gases [1,15]. Many control methods are impractical because they also alter the quality of the food and feed materials; moreover, some of them are expensive and environmental unfriendly. Hence, research into safe and effective alternatives of decontamination has gained much attention recently.

Electrolyzed oxidizing water (EOW) has notable biocidal activity mainly due to three physicochemical properties: pH, oxidation-reduction potential (ORP) and available chlorine concentration (ACC). EOW is produced by electrolysis of pure water added with sodium chloride in an electrolysis apparatus. Water molecules and chloride ions are transformed into chlorine oxidants such as hypochlorous acid (HOCl), hypochlorite ions (ClO$^-$) and chlorine (Cl_2) [16]. During the electrolysis, neutral electrolyzed water (NEW) is generated and it is stable in terms of loss of chlorine oxidants and ORP. Since NEW is non-toxic, non-corrosive and safe due to its physicochemical properties and a reverting capacity to ordinary water when diluted with tap water [17], NEW could be used for the application of safer, healthier and more acceptable methods for aflatoxin detoxification. Mounting evidence suggests that EOW has strong antifungal activity [18,19] and it is also effective in detoxifying AFB_1 in peanuts and maize [20,21]; however, there are some challenges faced in the development of this technology, such as the quick loss of activity in the presence of organic matter. Nevertheless, the application of NEW to detoxify aflatoxin-contaminated maize and the confirmation of the detoxification process in an in vivo experimental model have not yet been reported. Consequently, the purpose of this study was to evaluate the ability of NEW to detoxify B-aflatoxins in contaminated maize and confirm detoxification in turkey poults.

2. Results and Discussion

2.1. Physicochemical Properties of NEW

Regarding the main physicochemical properties of NEW, after electrolysis and up to the end of the experiment, the pH value (pH 7.03 ± 0.02), ORP (862 ± 3.9 mV) and ACC (54 ± 1 mg/L) were completely stable. It has been established that the most important factor in AFB$_1$ transformation is the high level of ACC, which is totally dependent on the electrolysis conditions. Taking into account that NEW contains primarily hypochlorous acid and the hypochlorite ions [16], their concentrations were also determined spectrophotometrically. Ultraviolet spectrum demonstrated two absorption peaks, one at 237 nm and one at 293 nm (Figure 1). Thus, using Lambert-Beer's law equation, NEW contained 43.01 ± 1.7 mg/L and 10.87 ± 0.71 mg/L of HOCl and OCl$^-$, respectively. The Ultra High Range Chlorine photometer uses the N,N-diethyl-p-phenylenediamine ferrous ethylenediammonium sulfate (DPD-FEAS) method, which is a titrimetric procedure for determining free available chlorine; however, care should be taken with this procedure as it cannot differentiate between the HOCl and OCl$^-$ concentrations. Consequently, ultraviolet determination could be considered as a practical, inexpensive and environmentally friendly technique for chlorine measurement that allows separate determinations of the two free chlorine species in NEW without the use of chemicals [22].

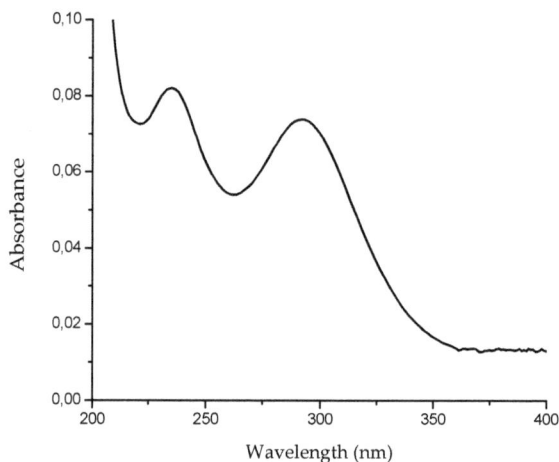

Figure 1. Ultraviolet absorption spectrum of neutral electrolyzed water (NEW) at 25 °C.

2.2. Aflatoxin Analyses and NEW Detoxifying Capacity

The aflatoxin analyses indicated that artificially contaminated feed contained 430 ± 17 ng/g. The Ultra Performance Liquid Chromatography (UPLC) chromatogram confirmed that the toxins were only AFB$_2$ and AFB$_1$. The retention time (Rt) values for these toxins were 1.60 and 2.00 min, respectively. AFB$_1$ was the most abundant toxin, making up 77% ($\approx$331 ng/g) of the total aflatoxin content (Figure 2). In this research, the presence of AFB$_2$ in the contaminated feed was considered insignificant, since this molecule is approximately 200-fold less toxic than AFB$_1$ [23]. Moreover, no T-2/HT-2 toxins or fumonisins (B$_1$, B$_2$ and B$_3$) were detected in the ration. After treatment with NEW, the aflatoxin content in the detoxified ration did not significantly decrease (determined with both methods via fluorescence detection); in consequence, the aflatoxin fluorescence strength of NEW-detoxified feed was almost similar to the aflatoxin-contaminated feed samples, since it has been reported that this particular detoxification procedure does not reduce the aflatoxin content, measured as loss of fluorescence [20]. Recently, our research group demonstrated that –Cl and –OH groups of the HOCl were added to the C$_8$ and C$_9$ atoms in the terminal furan ring of the AFB$_1$ molecule to yield 8-chloro-9-hydroxy-aflatoxin

B_1. Data of the compound indicate an important reduction in the cytotoxic and genotoxic potential, as demonstrated by using different in vitro and in silico assays [20,24]. The most important factor in AFB_1 detoxification is the high concentration of HOCl [24]; in this study, NEW contained 43.01 mg/L; however, the post-reaction concentration was <1 mg/L. This reduction represents about 98% decrease in HOCl concentration. In addition, the pH and ORP values were also significantly reduced, reaching values of 5.99 and 232 mV, respectively.

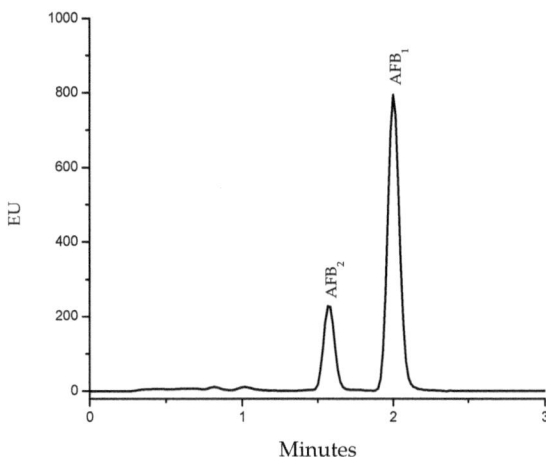

Figure 2. Representative Ultra Performance Liquid Chromatography (UPLC) profile of the aflatoxin-contaminated turkey feed. The retention time values for AFB_2 and AFB_1 were 1.60 and 2.00 min, respectively.

2.3. Poult Performance

The effects of providing the aflatoxin-contaminated and NEW-detoxified feed on poult performance and mortality rate are summarized in Table 1. At the end of week 1 (13 days-old), there were no statistical differences in body weight gain among the four treatments. However, by the end of week 2 (20 days-old), body weight gain was significantly reduced in poults of the AF group when compared to the CONTROL and NEW groups, respectively. The poults receiving the aflatoxin-contaminated diet had a 33.6% reduction in body weight gain as compared to the controls. This notable decrement in body weight gain caused by the aflatoxins was significantly ameliorated by the detoxification procedure of the aflatoxins with NEW. Moreover, cumulative data (6 to 20 days) showed that poults of the AF group had −24% deviation in body weight gain in comparison with the CONTROL and NEW groups, respectively. However, a moderate deviation in body weight gain (−11%) was observed in poults of the AF + NEW group, which means that NEW treatment offers a reasonable protection against the harmful effects in body weight gain caused by the aflatoxins. It is also important to mention that performance was not affected by treatment with NEW alone.

It is well known that growth performance of turkey poults can be modified significantly by the presence of aflatoxins in the diet in a dose-dependent manner [4]. McKenzie et al. [2] reported 23% reduction in body weight gain in turkey poults receiving approximately 560 ng/g AFB_1 in the diet by the end of week 3. Rauber et al. [4] showed growth rate depressions of 16% and 39% in initial-phase turkey poults fed a diet containing 500 ng/g and 1000 ng/g total aflatoxins (AFB_1 = 75.2%, AFB_2 = 1.4%, AFG_1 = 22.6%, AFG_2 = 0.8%), respectively. The current study demonstrated that a dose of 331 ng/g of AFB_1 in the diet significantly reduced the body weight gain in turkey poults from 6 to 20 days of age. This finding is in close agreement with the results found by Hamilton et al. [25] and Giambrone et al. [26], who reported that 250 ng/g and 400 ng/g are the minimum contents of AFB_1

that can significantly affect body weight gain in turkey poults. On the other hand, feed conversion ratio (g feed:g body weight gain) was only affected in poults of the AF group, presenting values up to 2.841 (Table 1). Lala et al. [1] reported a feed conversion rate of 2.930 in starter turkeys fed a diet containing 110 ng/g AFB_1, which is consistent with our findings. In this investigation, the observed mortality during the 2-week period was as follows: 13 poults in the AF group (22%), 2 poults in the group AF + NEW (3%) and 0 mortality in the CONTROL and NEW groups, respectively. It is important to mention that mortality in the AF group occurred within the first days of the second week of the trial and that the second poult in the AF + NEW group died from chronic staphylococcal peritonitis. The high mortality rate registered in poults of the AF group could be attributed to the formation of the intermediate product 2,3-dihydro-AFB_1 in the liver, resulting in the rapid onset of necrosis, breakdown of the immune system and eventually death [3]. Rauber et al. [4] reported mortality rates of 18.7% and 37.5% in turkey poults receiving 200 ng/g or 1000 ng/g total aflatoxins. Data shown by these researchers are consistent with our results. Aflatoxin sensibility varies among species, with turkeys being the most sensitive due to a combination of efficient activation and deficient detoxification of this mycotoxin in the liver, the primary target organ [27]. Turkeys that consume aflatoxins in their diet generally develop inappetence, reduced activity, unsteady gait, recumbence and ultimately death [28]. During the experiment, turkey poults of the AF group showed these markedly clinical signs of aflatoxicosis plus poor fathering, apathy and convulsive crises, signs that were completely alleviated by the detoxification treatment with the use of NEW.

Table 1. Effects of aflatoxin-contaminated and NEW-detoxified feed on body weight gain, feed conversion ratio and mortality rate in turkey poults.

Treatment	Body Weight Gain (g)			Deviation from CONTROL (%)	FCR (g Feed: g Gain)	MR (%)
	6 to 13 days old	13 to 20 days old	6 to 20 days old			
CONTROL	51.3 ± 4.2 [a]	131.9 ± 4.5 [a]	183.2	0	2.021 [a]	0
AF	52.2 ± 1.9 [a]	87.6 ± 5.8 [b]	139.8	−24	2.841 [b]	22
AF + NEW	50.5 ± 3.3 [a]	111.8 ± 7.5 [a]	162.3	−11	2.184 [a]	3
NEW	52.2 ± 3.0 [a]	131.8 ± 11.9 [a]	184.0	0	2.149 [a]	0

Mean of six replicates of ten poults each per treatment (minus mortality) ± standard error. [a,b] Means, within the same column, not sharing a common superscript differ significantly (Dunnett test $p < 0.05$). FCR = feed conversion ratio, MR = mortality rate.

2.4. Biochemical Constituents

Table 2 shows the summarized serum biochemical results. In general, aflatoxins caused a significant decrease in the total protein and albumin concentrations; reductions of 35% and 38% in these parameters were observed in poults of the AF group in comparison to the CONTROL group. McKenzie et al. [2] reported decrements of 59% and 53% in the serum concentrations of total protein and albumin in turkey poults fed a diet containing 560 ng/g AFB_1 during a 3-week period. Quist et al. [12] reported reductions of 19% and 27% in the total protein and albumin concentrations in 4-month old wild turkey poults fed a diet containing 400 ng/g total aflatoxins during 14 days. It has been reported that biochemical changes in the nitrogenous compounds in poults are maximal during the first three weeks after aflatoxin exposure, with hypoalbuminemia and the resultant hypoproteinemia being the most sensitive indicators of aflatoxin intoxication [29]. These effects are compatible with the compromised hepatic function seen in turkey poults of the AF group. Furthermore, an increment of 49% in the creatinine concentration was recorded in poults of the same experimental group (Table 2). Mathuria and Verma [30] reported a dose-dependent increase in creatinine concentration in young inbred Swiss strain male albino mice administered with aflatoxins for a 45-day period. Authors reported that low doses of aflatoxins (750 µg/kg body weight) caused a 60% increment in the creatinine concentration as compared to the vehicle control. The heightened appearance of creatinine in the serum of poults in the AF group indicates the increased transformation of phosphocreatinine to creatinine in the muscle, which might be due to a lesser utilization of phosphocreatinine during muscular contraction. Since the kidney rapidly excretes creatinine, the significant increase in creatinine concentration in the serum could be due to increased release from muscles and/or decreased excretion from the kidney, suggesting that aflatoxins cause adverse changes in both skeletal muscle and kidney at very early stages [30]. Furthermore, there were no significant differences in total bilirubin and uric acid concentrations among all four treatments (Table 2). In general, it can be concluded that the detoxification procedure with the use of NEW ameliorates aflatoxin-induced effects in the serum parameters as compared to the aflatoxin alone treated group.

Table 2. Effects of aflatoxin-contaminated and NEW-detoxified feed on some biochemical constituents in turkey poults at the end of the trial.

Treatment	Total Protein	Albumin	Total Bilirubin	Creatinine	Uric Acid
	(g/dL)			(mg/dL)	
CONTROL	2.42 ± 0.64 [a]	0.69 ± 0.16 [a]	0.28 ± 0.32 [a]	0.35 ± 0.04 [a]	6.98 ± 1.88 [a]
AF	1.57 ± 0.07 [b]	0.43 ± 0.02 [b]	0.22 ± 0.15 [a]	0.52 ± 0.10 [b]	4.09 ± 0.92 [a]
AF + NEW	1.97 ± 0.33 [a]	0.61 ± 0.23 [a]	0.25 ± 0.04 [a]	0.35 ± 0.12 [a]	4.24 ± 0.17 [a]
NEW	2.38 ± 0.51 [a]	0.79 ± 0.09 [a]	0.25 ± 0.19 [a]	0.39 ± 0.08 [a]	6.26 ± 1.70 [a]

Mean of six replicates of three poults each per treatment ($n = 18$) ± standard error. [a,b] Means, within the same column, not sharing a common superscript differ significantly (Dunnett test $p < 0.05$).

2.5. Serum Enzyme Activities

Conspicuous symptoms of aflatoxin toxicity were detected in poults of the AF group by the serum aspartate aminotransferase (AST) activity level, which increased by 1.34-fold in comparison to the CONTROL group (Table 3). On the other hand, there were no significant differences in alanine aminotransferase (ALT), gamma glutamiltranspeptidase (GGT) and alkaline phosphathase (AP) activities among all four treatments, which is in close agreement with data reported by Rauber et al. [4] in turkey poults fed seven different doses of aflatoxins in their diet. It is well known that AST is present in the cytosol, and whenever liver hepatocytes are damaged, the enzyme is released into the blood stream; consequently, a significant increase in its activity indicates damage. The results obtained in the present study showed a significant increase in the AST activity in poults of the AF group; similar results have been reported by different researchers [2,12,31]. However, this effect was completely mitigated by the treatment of the aflatoxin molecules with NEW.

Table 3. Effects of aflatoxin-contaminated and NEW-detoxified feed on serum enzyme activities in turkey poults at the end of the trial.

Treatment	AST	ALT	GGT	AP
	(IU/L)			
CONTROL	478.00 ± 8.49 [a]	49.55 ± 3.66 [a]	7.60 ± 0.75 [a]	2360.91 ± 26.35 [a]
AF	644.32 ± 5.60 [b]	54.82 ± 9.83 [a]	7.60 ± 0.61 [a]	2136.66 ± 59.52 [a]
AF + NEW	369.01 ± 2.50 [a]	67.32 ± 5.55 [a]	6.66 ± 0.78 [a]	2189.73 ± 33.80 [a]
NEW	536.44 ± 7.05 [a]	50.60 ± 7.17 [a]	9.67 ± 0.42 [a]	2481.40 ± 85.24 [a]

Mean of six replicates of three poults each per treatment ($n = 18$) ± standard error. [a,b] Means, within the same column, not sharing a common superscript differ significantly (Dunnett test $p < 0.05$). AST = aspartate aminotransferase, ALT = alanine aminotransferase, GGT = gamma glutamiltranspeptidase, AP = alkaline phosphathase.

2.6. Relative Organ Weight

As shown in Table 4, no significant differences were noted among all treatments in the relative spleen and bursa of Fabricius relative weights. However, when compared to the CONTROL group, the relative weight of the liver decreased significantly (20%), whereas the relative weight of the kidney increased (37%), in poults of the AF group. The reduced relative liver weight and the increments in the relative kidney weight in poults of the AF group are consistent with previous studies [2,4,12].

In general, the aflatoxin detoxification procedure with the use of NEW protected the poults from significant relative organ weight changes.

Table 4. Effects of aflatoxin-contaminated and NEW-detoxified feed on relative organ weight in turkey poults at the end of the trial.

Treatment	Liver	Kidney	Spleen	Bursa of Fabricius
CONTROL	3.18 ± 0.27 [a]	1.02 ± 0.12 [a]	0.09 ± 0.01 [a]	0.18 ± 0.03 [a]
AF	2.55 ± 0.10 [b]	1.40 ± 0.07 [b]	0.08 ± 0.01 [a]	0.14 ± 0.02 [a]
AF + NEW	2.91 ± 0.08 [a]	1.07 ± 0.04 [a]	0.08 ± 0.01 [a]	0.14 ± 0.02 [a]
NEW	3.16 ± 0.04 [a]	0.97 ± 0.05 [a]	0.08 ± 0.01 [a]	0.14 ± 0.01 [a]

Mean of six replicates of three poults each per treatment ($n = 18$) ± standard error. [a,b] Means, within the same column, not sharing a common superscript differ significantly (Dunnett test $p < 0.05$).

2.7. Gross Morphological Changes and Histopathological Studies

No gross morphological alterations were observed in the liver, kidney, spleen or bursa of Fabricius of the poults in the CONTROL, AF + NEW and NEW groups throughout the course of the experiment. However, livers of the AF group were pale, friable and appeared smaller in size compared to those of the CONTROL group. These lesions were also accompanied by hemorrhagic streaks and accentuated lobular patterns. On the other hand, the AF group kidneys showed no gross lesions but were swollen and had light colored areas on the surface. Furthermore, histological studies provided additional evidence of the beneficial effect of NEW treatment on suppressing the toxicity induced by the aflatoxins. In general, aflatoxin exposure caused extensive liver damage in all poults of the AF group; lesions observed were composed of variable areas of hepatic degeneration, massive bile duct epithelial necrosis as well as severe hyperplasia of the epithelium and bile duct proliferation (Figure 3, profile b). In contrast, only a minimal degree of bile duct epithelial necrosis and minimal bile duct proliferation was seen in the livers of AF + NEW group (Figure 3, profile c). Histopathological examination of the bursa of Fabricius indicated lymphoid depletion, abundant apoptotic figures in the germinal centers and hyperplasia of the epithelium—all of them in severe grade—in poults of the AF group (Figure 4, profile b). On the contrary, only a minimal degree of lymphoid depletion and hyperplasia of the epithelium was seen in the bursa of Fabricius of poults of the AF + NEW group (Figure 4, profile c). Lesions on the spleen were more discrete; the changes seen in poults of the AF and AF + NEW groups, but not seen in the CONTROL and NEW groups, were severe and minimal lymphoid depletion, respectively (Figure 5, profiles b,c). Finally, no significant pathological changes on the kidney were found in poults of the four experimental groups (not shown). It has been reported that the principal target organ for aflatoxicosis is the liver; consequently, all microscopic lesions caused by the aflatoxins agree with our previous studies [32]. On the contrary, the severity of lesions in the liver was significantly reduced in the AF + NEW treatment group, suggesting that the detoxification procedure with the use of NEW is effective in reducing the toxicity of aflatoxins in the liver of turkey poults. Aflatoxins are also potent immunosuppressive agents causing immunosuppression through impairment of humoral and cellular immune responses [26]; thus, resistance to infectious diseases is generally dependent on antibody production which is dependent in turn on the bursa of Fabricius. In this research, aflatoxins caused severe depletion of lymphoid cells in the bursa of Fabricius and spleen, indicating that lymphoid organs are very sensitive to aflatoxins. However, similar but less severe changes in the bursa of Fabricius and spleen were seen in turkey poults of the AF + NEW group when compared with the AF group. These findings provide further evidence of the effectiveness of the detoxification treatment in reduction of aflatoxin toxicity.

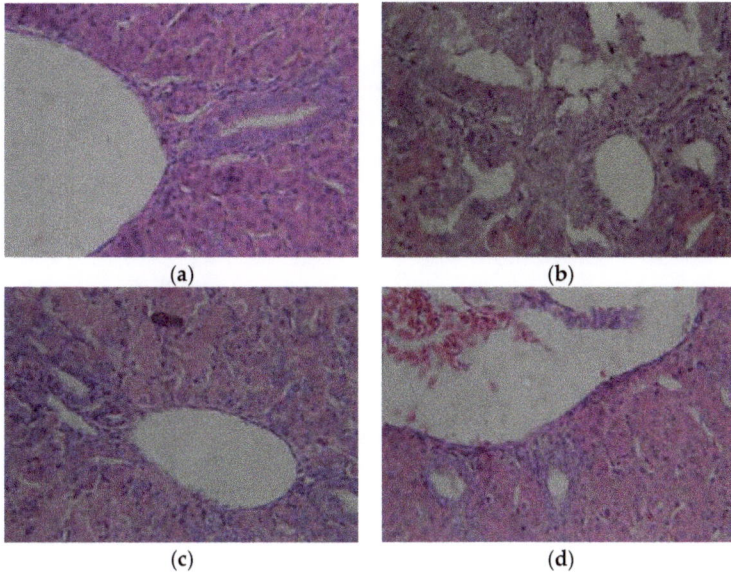

Figure 3. Comparative histological changes in the liver (20×, H & E stain). The normal histological structure is similar in the CONTROL (profile **a**) and NEW (profile **d**) groups. Massive bile duct epithelial necrosis as well as severe hyperplasia of the epithelium and bile duct proliferation is observed in the AF group (profile **b**), in contrast with a minimal degree of bile duct epithelial necrosis and minimal bile duct proliferation seen in the AF + NEW group (profile **c**).

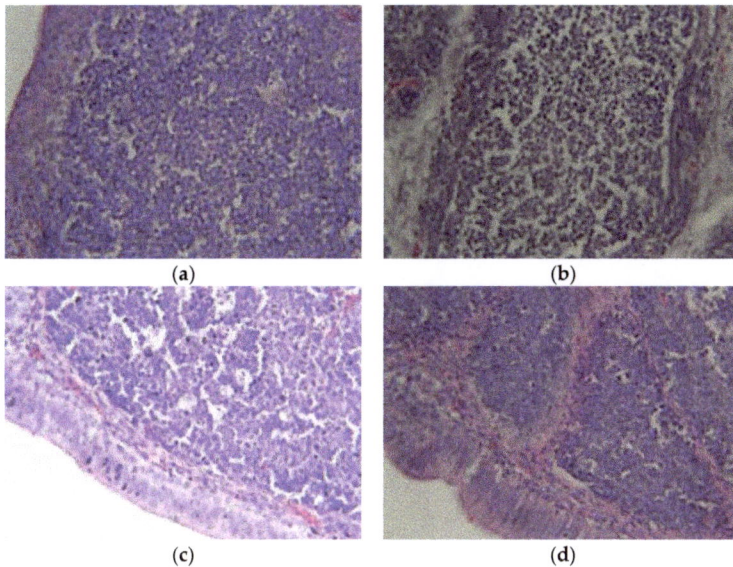

Figure 4. Comparative histological changes in the bursa of Fabricius (20×, H & E stain). The normal structure is similar in the CONTROL (profile **a**) and NEW (profile **d**) groups. Severe lymphoid depletion and hyperplasia of the epithelium is clear in the AF group (profile **b**), in comparison with minimal degree of lymphoid depletion and hyperplasia of the epithelium in the AF + NEW group (profile **c**).

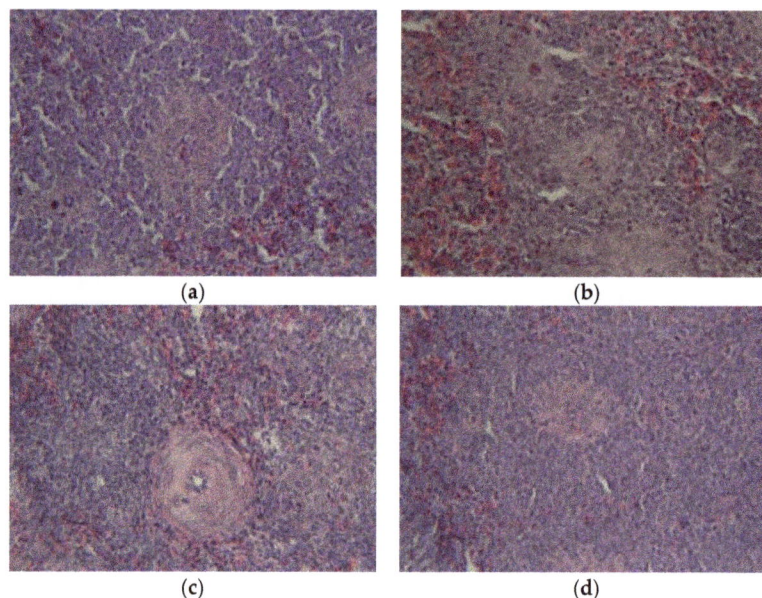

Figure 5. Comparative histological changes in the spleen (20×, H & E stain). The normal histological structure in the CONTROL (profile **a**) is similar to that of group NEW (profile **d**). Severe and minimal lymphoid depletion was observed in the AF and AF + NEW groups (profiles **b**,**c**), respectively.

To the best of our knowledge, the present study is the first one to demonstrate that treatment of aflatoxin-contaminated maize with NEW significantly diminishes AFB_1-toxicity in young turkey poults. This effect can be related to the reaction of hypochlorous acid with the double bond in the terminal furan ring of the AFB_1 molecule to yield 8-chloro-9-hydroxy-aflatoxin B_1. Nevertheless, further studies on the complete growing cycle of other animal species need to be conducted. Finally, to accurately determine the risk of using chlorine compounds in the animal feed industry, the available chlorine concentration used and the reactivities of the chlorine species with feed components such as carbohydrates, lipids, and proteins should be extensively evaluated.

3. Materials and Methods

3.1. Safety Precautions

Procedures used for handling and decontaminating aflatoxin-contaminated materials were adopted from recommendations published by the International Agency for Research on Cancer [33].

3.2. Animal Ethics

This study was conducted according to the Internal Committee for Care and Use of Experimental Animals (CICUAE), approved by the National Autonomous University of Mexico. Ethical approval code: CICUAE-29102009. Date of approval: 29 October 2009. Project identification code: C16_02. Date of approval: 8 April 2016. Ethics committee: Internal Committee for Care and Use of Experimental Animals (CICUAE, from its abbreviation in Spanish).

3.3. Chemicals and Reagents

B-aflatoxins, tween-80, sodium chloride, formaldehyde, xylenes and hematoxylin-eosin staining solution were purchased from Sigma-Aldrich Co. (St. Louis, MO, USA). Acetonitrile (HPLC grade), ethanol, methanol, and methanol (HPLC grade) were obtained from J.T Baker, Mallinckrodt Baker (Ecatepec, Mexico). Paraffin for tissue embedding was procured from Leica Biosystems Richmond Inc. (Richmond, IL, USA). Malt extract-sodium chloride-agar (MSA) medium was obtained from DIBICO SA de CV (Mexico, DF, Mexico). Saline solution was purchased from Becton Dickinson Co. (Mexico, DF, Mexico). All other chemicals used were analytical reagent grade.

3.4. Preparation of Neutral Electrolyzed Water (NEW)

NEW was electrochemically generated using two patented generators from Esteripharma SA de CV (Atlacomulco, Mexico). This method of NEW production has been previously described in detail by Jardon-Xicotencatl et al. [20]. NEW was used immediately after elaboration. Three physicochemical properties were verified: the pH and ORP values were measured with a combination pH/ORP/temperature meter PC-45 (Conductronic, Mexico, DF, Mexico), and the ACC was determined using an Ultra High Range Chlorine photometer model HI96771C (Hanna Instruments, Melrose, MA, USA). The concentrations of HOCl (the main available chlorine form in NEW) and OCl^- were further determined using a diode-array system spectrophotometer Cary-8454 (Agilent Technologies, Santa Clara, CA, USA). Ultraviolet spectra was collected in the range of 200–400 nm at room temperature in 1-cm quartz cell and the absorbance at 237 nm and 293 nm was used to determine the concentration using Lambert-Beer's law equation with 100 1/M·cm and 350 1/M·cm as the molar absorption coefficients for HOCl and OCl^-, respectively [34].

3.5. Maize Grain

White maize grains of the commercial hybrid AS-900 provided by Aspros Comercial SA de CV (Cortazar, Mexico) were used. The aflatoxin content of the grain was below the immunoaffinity column minimum detection limit (0.5 ng/g). Contents of T-2/HT-2 toxins and total fumonisins (FB_1, FB_2 and FB_3) were also determined using monoclonal antibody columns [35]. Moisture content was determined by drying replicate portions of 5 to 10 g each of whole grain at 103 °C for 72 h, with percentages calculated on a wet-weight basis.

3.6. Fungal Isolate

The *Aspergillus flavus* Link strain UNIGRAS-1231 (Culture Collection of the Grain and Seed Research Unit of the National Autonomous University of Mexico) was plated into Petri dishes containing MSA medium (%: malt extract, 2; sodium chloride, 6; and agar, 2) for 7 days at 25 °C. This strain has high ability to produce AFB_1 and AFB_2 [36].

3.7. Aflatoxins Production

Aflatoxins were produced according to the technique proposed by Méndez-Albores et al. [36]. The inoculated maize was incubated during 37 days to obtain an approximate aflatoxin content of 13,500 ng/g. The aflatoxin-contaminated maize was stored at 4 °C prior to treatment.

3.8. Aflatoxin Analyses

3.8.1. Using Immunoaffinity Columns (IACs)

The aflatoxin content was determined according to the 991.31 AOAC method [37] using antibody-based IACs for AFB_1 and AFB_2. Samples (50 g) were extracted by blending with 100 mL methanol-water (80:20, v/v) and 5 g of NaCl using a laboratory blender model 51BL30 (Waring, CT, USA). The mixture was filtered through a Whatman 1 filter paper and a portion of 5 mL was diluted with 20 mL of distilled water. The diluted preparation was filtered through a micro-fiber filter, and 10 mL were passed through the IACs (Afla B, VICAM Science Technology, Watertown, MA, USA). Subsequently, the column was washed twice with 10 mL of distilled water and dried with sterile air flow. The toxins were then eluted with 1 mL of HPLC grade methanol and quantified in a fluorometer

VICAM Series-4EX (VICAM Source Scientific, Irvine, CA, USA) after reaction with 1 mL of 0.002% aqueous bromine. The detection limit for aflatoxins via fluorescence measurement is approximately 0.5 ng/g.

3.8.2. Using Ultra Performance Liquid Chromatography (UPLC)

The aflatoxin identification was carried out according to the technique proposed by Jardon-Xicontencatl et al. [20], using a Waters ACQUITY UPLC H-Class System equipped with a quaternary solvent manager and an ACQUITY UPLC BEH C18 phase reverse column (2.1 × 100 mm, 1.7 μm). Standards, as well as samples collected from the IACs (1 μL), were injected and eluted with a single ternary mixture of 64:18:18 water/methanol/acetonitrile (all HPLC grade) at a flow rate of 400 μL/min. Aflatoxins were fluorometrically detected and identified using a UPLC-optimized fluorescence detector (Waters, Milford, MA, USA). The excitation and emission wavelengths were 365 and 429 nm, respectively. Aflatoxins were identified by their Rt and compared with those of a pure aflatoxin standard solution. The estimated detection limits are 0.58 and 2.01 ng/kg for AFB_2 and AFB_1, respectively.

3.9. Detoxification of the Aflatoxin-Contaminated Maize with NEW and Diet Formulation

Batches of the aflatoxin-contaminated maize (1 kg) were milled (Molinos Pulvex, Mexico, DF, Mexico) using a hammer head and a 0.5 mm mesh screen, and subjected to detoxification treatment using standardized conditions previously outlined by Jardon-Xicotencatl et al. [20]. Subsequently, the aflatoxin-detoxified material was placed into flat stainless steel pans for 2 h in a sterile air flow chamber. The ground maize was mixed in a commercial turkey feed procured from Nutricion Tecnica Animal SA de CV (Queretaro, Mexico) containing 26% protein, 12.64 MJ/kg (metabolizable energy) and met or exceeded levels of other nutrients recommended by the National Research Council [38]. The commercial feed was free of any detectable mycotoxins (aflatoxins, T-2/HT-2 toxins and fumonisins) and contained no antibiotic, nor anticoccidial drugs or even growth promoters. Feed batches (25 kg) were artificially contaminated with B-aflatoxins (430 ng/g) using 27 g of the milled maize per kilogram of feed. In order to assure the proper distribution of the aflatoxins, the contaminated feed was mixed for 20 min in a Ribbon Blender Mixer model MH-7050 (Molinos Pulvex, Mexico, DF, Mexico).

3.10. Birds and Housing

For the experiment, 240 6-day-old female large white Nicholas-700 turkey poults (obtained from a commercial hatchery) were individually weighed at placement and randomly distributed in four pens at the Poultry Research Station of the National Autonomous University of Mexico. Six replicates of ten poults (n = 60 per treatment) were grouped based on the following four dietary treatments: (1) control feed containing aflatoxin-free maize (CONTROL); (2) feed containing the aflatoxin-contaminated maize (AF); (3) feed containing the aflatoxin-contaminated maize detoxified with NEW (AF + NEW); and (4) control feed containing aflatoxin-free maize treated with NEW (NEW). The temperature, lighting and ventilation programs were followed according to standard recommendations of the supplier. Feed and water were provided ad libitum during the whole period of the experiment (2 weeks).

3.11. Collection of Samples and Measurements

Poults were individually weighed (on a weekly basis) and feed consumption for each replicate was measured weekly until the end of the experiment (2 weeks). Body weight gain, feed intake, feed conversion ratio (FCR) and mortality rate (MR) were calculated. Feed intake and FCR were adjusted for mortalities when necessary. At 20 days of age, blood was drawn by cardiac puncture under anesthesia (the turkey was exposed for 1 minute to 40% carbon dioxide, 30% oxygen, and 30% nitrogen) from 18 randomly selected birds from each treatment (3 birds per replicate), and serum was prepared. The following analyses were performed spectrophotometrically using commercially available kits (BioSystems, Barcelona, Spain): total protein (545 nm), albumin (630 nm), total bilirubin

(540 nm), creatinine (500 nm), and uric acid (520 nm). The serum aspartate aminotransferase (AST), alanine aminotransferase (ALT), gamma glutamiltranspeptidase (GGT), and alkaline phosphathase (AP) activities were also determined spectrophotometrically at 340 nm, 340 nm, 410 nm and 405 nm, respectively. The bled poults were then exposed to 80% carbon dioxide, 5% oxygen, and 15% nitrogen for euthanasia [39]. All efforts were made to minimize unnecessary pain and suffering. Liver plus gall bladder, kidney, spleen, and bursa of Fabricius were excised and washed in cold saline and their relative percentages estimated. For histopathological examination, all specimens were fixed in 10% neutral-buffered formalin for 48 h, routinely embedded in paraffin, cut into 4 μm thick sections and processed for hematoxylin and eosin (H & E) staining.

3.12. Statistical Analysis

Data were subjected to analysis of variance (ANOVA) using the General Linear Model (GLM) procedure in Statistical Analysis System software version 8.0 [40], and means were separated by the Dunnett procedure and judged to be significantly different if $p < 0.05$.

Acknowledgments: The authors are grateful to Consejo Nacional de Ciencia y Tecnologia (CONACYT) for the financial support for this research through the Grant PROINNOVA-230213. D.G.-E. also acknowledges CONACYT for the MSc scholarship (615804).

Author Contributions: D.G.-E. and F.J.C.-A. carried out the experiments and analyzed the data; A.V.-D. performed the experimental design and statistical analysis; J.C.D.R.-G. conducted the histopathological examination; T.V.-B. took part in discussion and helped in editing the manuscript; A.M.-A. conceived and designed the experiment and wrote the paper. All authors read and approved the final version of the document.

Conflicts of Interest: The authors declare that there are no conflicts of interest.

References

1. Lala, A.; Oso, A.; Ajao, A.; Idowu, O.; Oni, O. Effect of supplementation with molecular or nano-clay adsorbent on growth performance and haematological indices of starter and grower turkeys fed diets contaminated with varying dosages of aflatoxin B$_1$. *Livest. Sci.* **2015**, *178*, 209–215. [CrossRef]
2. McKenzie, K.; Kubena, L.; Denvir, A.; Rogers, T.; Hitchens, G.; Bailey, R.; Harvey, R.; Buckley, S.; Phillips, T. Aflatoxicosis in turkey poults is prevented by treatment of naturally contaminated corn with ozone generated by electrolysis. *Poult. Sci.* **1998**, *77*, 1094–1102. [CrossRef] [PubMed]
3. Miazzo, R.; Peralta, M.; Magnoli, C.; Salvano, M.; Ferrero, S.; Chiacchiera, S.; Carvalho, E.; Rosa, C.; Dalcero, A. Efficacy of sodium bentonite as a detoxifier of broiler feed contaminated with aflatoxin and fumonisin. *Poult. Sci.* **2005**, *84*, 1–8. [CrossRef] [PubMed]
4. Rauber, R.; Dilkin, P.; Giacomini, L.; de Almeida, C.A.; Mallmann, C. Performance of turkey poults fed different doses of aflatoxins in the diet. *Poult. Sci.* **2007**, *86*, 1620–1624. [CrossRef] [PubMed]
5. Asao, T.; Buchi, G.; Abdel-Kader, M.; Chang, S.; Wick, E.L.; Wogan, G. Aflatoxins B and G. *J. Am. Chem. Soc.* **1963**, *85*, 1706–1707. [CrossRef]
6. Boyacioglu, D.; Gönül, M. Survey of aflatoxin contamination of dried figs grown in Turkey in 1986. *Food Addit. Contam.* **1990**, *7*, 235–237. [CrossRef] [PubMed]
7. Njapau, H.; Muzungaile, E.M.; Changa, R.C. The effect of village processing techniques on the content of aflatoxins in corn and peanuts in Zambia. *J. Sci. Food Agric.* **1998**, *76*, 450–456. [CrossRef]
8. Rustom, I.Y. Aflatoxin in food and feed: Occurrence, legislation and inactivation by physical methods. *Food Chem.* **1997**, *59*, 57–67. [CrossRef]
9. Del Bianchi, M.; Oliveira, C.; Albuquerque, R.; Guerra, J.; Correa, B. Effects of prolonged oral administration of aflatoxin B$_1$ and fumonisin B$_1$ in broiler chickens. *Poult. Sci.* **2005**, *84*, 1835–1840. [CrossRef] [PubMed]
10. Yang, J.; Bai, F.; Zhang, K.; Bai, S.; Peng, X.; Ding, X.; Li, Y.; Zhang, J.; Zhao, L. Effects of feeding corn naturally contaminated with aflatoxin B$_1$ and B$_2$ on hepatic functions of broilers. *Poult. Sci.* **2012**, *91*, 2792–2801. [CrossRef] [PubMed]
11. Fernandez, A.; Verde, M.; Gomez, J.; Gascon, M.; Ramos, J. Changes in the prothrombin time, haematology and serum proteins during experimental aflatoxicosis in hens and broiler chickens. *Res. Vet. Sci.* **1995**, *58*, 119–122. [CrossRef]

12. Quist, C.; Bounous, D.; Kilburn, J.; Nettles, V.; Wyatt, R. The effect of dietary aflatoxin on wild Turkey poults. *J. Wildl. Dis.* **2000**, *36*, 436–444. [CrossRef] [PubMed]

13. Neucere, J.N. Inhibition of *Aspergillus flavus* growth by silk extracts of resistant and susceptible corn. *J. Agric. Food Chem.* **1996**, *44*, 1982–1983. [CrossRef]

14. Zeringue, H.; Brown, R.; Neucere, J.; Cleveland, T. Relationships between C_6–C_{12} alkanal and alkenal volatile contents and resistance of maize genotypes to *Aspergillus flavus* and aflatoxin production. *J. Agric. Food Chem.* **1996**, *44*, 403–407. [CrossRef]

15. Milićević, D.R.; Škrinjar, M.; Baltić, T. Real and perceived risks for mycotoxin contamination in foods and feeds: Challenges for food safety control. *Toxins* **2010**, *2*, 572–592. [CrossRef] [PubMed]

16. Guentzel, J.L.; Lam, K.L.; Callan, M.A.; Emmons, S.A.; Dunham, V.L. Reduction of bacteria on spinach, lettuce, and surfaces in food service areas using neutral electrolyzed oxidizing water. *Food Microbiol.* **2008**, *25*, 36–41. [CrossRef] [PubMed]

17. Xiong, K.; Liu, H.J.; Li, L.T. Product identification and safety evaluation of aflatoxin B_1 decontaminated by electrolyzed oxidizing water. *J. Agric. Food Chem.* **2012**, *60*, 9770–9778. [CrossRef] [PubMed]

18. Audenaert, K.; Monbaliu, S.; Deschuyffeleer, N.; Maene, P.; Vekeman, F.; Haesaert, G.; De Saeger, S.; Eeckhout, M. Neutralized electrolyzed water efficiently reduces *Fusarium* spp. in vitro and on wheat kernels but can trigger deoxynivalenol (DON) biosynthesis. *Food Control* **2012**, *23*, 515–521. [CrossRef]

19. Suzuki, T.; Noro, T.; Kawamura, Y.; Fukunaga, K.; Watanabe, M.; Ohta, M.; Sugiue, H.; Sato, Y.; Kohno, M.; Hotta, K. Decontamination of aflatoxin-forming fungus and elimination of aflatoxin mutagenicity with electrolyzed NaCl anode solution. *J. Agric. Food Chem.* **2002**, *50*, 633–641. [CrossRef] [PubMed]

20. Jardon-Xicotencatl, S.; Díaz-Torres, R.; Marroquín-Cardona, A.; Villarreal-Barajas, T.; Méndez-Albores, A. Detoxification of aflatoxin-contaminated maize by neutral electrolyzed oxidizing water. *Toxins* **2015**, *7*, 4294–4314. [CrossRef] [PubMed]

21. Zhang, Q.; Xiong, K.; Tatsumi, E.; Liu, H.-J. Elimination of aflatoxin B_1 in peanuts by acidic electrolyzed oxidizing water. *Food Control* **2012**, *27*, 16–20. [CrossRef]

22. Len, S.-V.; Hung, Y.-C.; Erickson, M.; Kim, C. Ultraviolet spectrophotometric characterization and bactericidal properties of electrolyzed oxidizing water as influenced by amperage and pH. *J. Food Prot.* **2000**, *63*, 1534–1537. [CrossRef] [PubMed]

23. Méndez-Albores, A.; Arambula-Villa, G.; Loarca-Piña, M.; Castano-Tostado, E.; Moreno-Martínez, E. Safety and efficacy evaluation of aqueous citric acid to degrade B-aflatoxins in maize. *Food Chem. Toxicol.* **2005**, *43*, 233–238. [CrossRef] [PubMed]

24. Escobedo-González, R.; Méndez-Albores, A.; Villarreal-Barajas, T.; Aceves-Hernández, J.M.; Miranda-Ruvalcaba, R.; Nicolás-Vázquez, I. A theoretical study of 8-chloro-9-hydroxy-aflatoxin B_1, the conversion product of aflatoxin B_1 by neutral electrolyzed water. *Toxins* **2016**, *8*. [CrossRef] [PubMed]

25. Hamilton, P.; Tung, H.-T.; Harris, J.; Gainer, J.; Donaldson, W. The effect of dietary fat on aflatoxicosis in turkeys. *Poult. Sci.* **1972**, *51*, 165–170. [CrossRef] [PubMed]

26. Giambrone, J.; Diener, U.; Davis, N.; Panangala, V.; Hoerr, F. Effects of aflatoxin on young turkeys and broiler chickens. *Poult. Sci.* **1985**, *64*, 1678–1684. [CrossRef] [PubMed]

27. Klein, P.J.; Buckner, R.; Kelly, J.; Coulombe, R.A. Biochemical basis for the extreme sensitivity of turkeys to aflatoxin B_1. *Toxicol. Appl. Pharm.* **2000**, *165*, 45–52. [CrossRef] [PubMed]

28. Arafa, A.S.; Bloomer, R.J.; Wilson, H.R.; Simpson, C.F.; Harms, R.H. Susceptibility of various poultry species to dietary aflatoxin. *Brit. Poult. Sci.* **1981**, *22*, 431–436. [CrossRef] [PubMed]

29. Gumbmann, M.; Williams, S.; Booth, A.; Vohra, P.; Ernst, R.; Bethard, M. Aflatoxin susceptibility in various breeds of poultry. *Exp. Biol. Med.* **1970**, *134*, 683–688. [CrossRef]

30. Mathuria, N.; Verma, R.J. Ameliorative effect of curcumin on aflatoxin-induced toxicity in serum of mice. *Acta Pol. Pharm.* **2008**, *65*, 339–343. [PubMed]

31. Kubena, L.; Edrington, T.; Kamps-Holtzapple, C.; Harvey, R.; Elissalde, M.; Rottinghaus, G. Effects of feeding fumonisin B_1 present in Fusarium moniliforme culture material and aflatoxin singly and in combination to Turkey poults. *Poult. Sci.* **1995**, *74*, 1295–1303. [CrossRef] [PubMed]

32. Méndez-Albores, A.; Del Río-García, J.; Moreno-Martinez, E. Decontamination of aflatoxin duckling feed with aqueous citric acid treatment. *Anim. Feed Sci. Technol.* **2007**, *135*, 249–262. [CrossRef]

33. Castegnaro, M.; Hunt, D.; Sansone, E.; Schuller, P.; Siriwardana, M.; Telling, G.; Van Egmond, H.; Walker, E. *Laboratory Decontamination and Destruction of Aflatoxins B1, B2, G1, G2 in Laboratory Wastes*; IARC Scientific Publications; WHO Publications Centre: Lyon, France, 1980.

34. Morris, J.C. The acid ionization constant of HOCl from 5 to 35. *J. Phys. Chem.* **1966**, *70*, 3798–3805. [CrossRef]

35. Mendez-Albores, A.; Cardenas-Rodriguez, D.A.; Vazquez-Duran, A. Efficacy of microwave-heating during alkaline processing of fumonisin-contaminated maize. *Iran. J. Public Health* **2014**, *43*, 147–155. [PubMed]

36. Pérez-Flores, G.; Moreno-Martínez, E.; Méndez-Albores, A. Effect of Microwave Heating during Alkaline-Cooking of Aflatoxin Contaminated Maize. *J. Food Sci.* **2011**, *76*, T48–T52. [CrossRef] [PubMed]

37. Horwitz, W. *Official Methods of Analysis of AOAC International*; AOAC International: Gaithersburg, MD, USA, 2000.

38. National Research Council. *Nutrient Requirements of Poultry*, 9th ed.; National Academy Press: Washington, DC, USA, 1994.

39. Coenen, A.; Smit, A.; Zhonghua, L.; Van Luijtelaar, G. Gas mixtures for anaesthesia and euthanasia in broiler chickens. *World Poult. Sci. J.* **2000**, *56*, 226–234. [CrossRef]

40. SAS/STAT User's Guide. Version 8. Available online: http://www.okstate.edu/sas/v8/saspdf/stat/pdfidx.htm (accessed on 7 January 2017).

Review

Mitigation of Patulin in Fresh and Processed Foods and Beverages

J. David Ioi [1,2], Ting Zhou [1], Rong Tsao [1,*] and Massimo F. Marcone [2]

[1]   Guelph Research and Development Centre, Agriculture and Agri-Food Canada, Guelph, ON N1G5C9, Canada; jioi@uoguelph.ca (J.D.I.); ting.zhou@agr.gc.ca (T.Z.)

[2]   Department of Food Science, University of Guelph, Guelph, Ontario, N1G 2W1, Canada; mmarcone@uoguelph.ca

*   Correspondence: rong.cao@agr.gc.ca; Tel.: +1-226-217-8108

Academic Editor: Gerhard Adam

Received: 1 February 2017; Accepted: 3 May 2017; Published: 11 May 2017

Abstract: Patulin is a mycotoxin of food safety concern. It is produced by numerous species of fungi growing on fruits and vegetables. Exposure to the toxin is connected to issues neurological, immunological, and gastrointestinal in nature. Regulatory agencies worldwide have established maximum allowable levels of 50 µg/kg in foods. Despite regulations, surveys continue to find patulin in commercial food and beverage products, in some cases, to exceed the maximum limits. Patulin content in food can be mitigated throughout the food processing chain. Proper handling, storage, and transportation of food can limit fungal growth and patulin production. Common processing techniques including pasteurisation, filtration, and fermentation all have an effect on patulin content in food but individually are not sufficient safety measures. Novel methods to remove or detoxify patulin have been reviewed. Non-thermal processing techniques such as high hydrostatic pressure, UV radiation, enzymatic degradation, binding to microorganisms, and chemical degradation all have potential but have not been optimised. Until further refinement of these methods, the hurdle approach to processing should be used where food safety is concerned. Future development should focus on determining the nature and safety of chemicals produced from the breakdown of patulin in treatment techniques.

Keywords: patulin; mycotoxin; mitigation; decontamination; food and beverage; processing

1. Introduction

Mycotoxins are toxic secondary metabolites produced by fungi that present a potential hazard regrading food safety. Patulin is a mycotoxin and is known to be produced by more than 60 species of fungi belonging to greater than 30 genera [1,2]. Although typically associated with *Penicillium expansum*, patulin is also known to be produced by other fungi, including *P. claviforme*, *P. urticae*, *P. patulum*, *Aspergillus clavatus*, *A. giganteus*, *Byssoclamys fulva*, *B. nivea*, and *Alternaria alternata* [3]. Chemically speaking, patulin (4-hydroxy-4-H-furo[3,2-c]pyran-2[6H]-one) is an unsaturated heterocyclic lactone with a molecular weight of 154 (Figure 1) [4,5]. Discovered in the 1940s, patulin was initially investigated for its potential as an antibiotic due to its strong activity against gram-positive and gram-negative bacteria such as *Mycobacterium tuberculosis* [6] and more than 75 other different bacterial species [7]. In the 1960s, it was found to be toxic not only to bacterial cells but to animal and plant cells as well and was subsequently reclassified as a mycotoxin [8].

OH

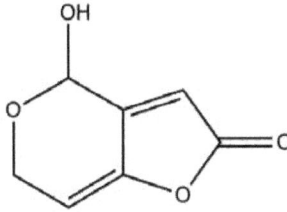

Figure 1. Chemical structure of patulin.

Patulin has a broad spectrum of toxicities that include both acute and chronic effects. Some examples include congestion and edema of blood vessels and tissues [9]. Formation of sarcomas has been observed when large doses of the mycotoxin were injected into animals, causing concerns of carcinogenicity in humans [10]. Other effects seen in animals include teratogenicity, liver, spleen and kidney damage, lung and brain edema, and immune system toxicity [11]. In humans, the primary reported acute symptoms include gastrointestinal issues, nausea, and vomiting, but there is no conclusive evidence as to the nature of the chronic effects [3]. The LD_{50} of patulin ranges from 15 to 25 mg/kg and is dependent on the characteristics of exposure as well as the route of ingestion [5]. Patulin is a highly reactive molecule, able to interact with proteins to form intramolecular and intermolecular crosslinks with specific amino acids, causing it to behave as an enzyme inhibitor [12,13]. It has also been shown that patulin can form intermolecular links with DNA molecules [14]. These properties may explain the reported teratogenic and carcinogenic effects. No reports are available on the possible toxicity of patulin due to inhalation of the toxin in powdered form [15]. Children are more at risk for toxicities from patulin as they often consume more potentially contaminated products. Information taken from a study by the USDA has shown that children have a very high consumption of apple products as compared to adults [16]. During the first year of life, children were found to consume on average 6.4 g/kg body weight/day of apples while adults consumed only 0.4 g/kg body weight/day [16]. This means that it is of particular importance to be cautious of the potential danger that patulin and other mycotoxins present in baby foods.

Due to the potential negative health effects of consuming patulin, regulatory agencies from around the world have instituted limits regarding the maximum amount of patulin that can be in food products. Many organizations such as Health Canada, The United States Food and Drug Administration, and the Codex Committee on Food Additives and Contaminants have all set limits of 50 µg/kg patulin [17–19]. The World Health Organization has suggested a limit of 0.4 µg/kg body weight and the European Union has set a much lower maximum limit of 25 µg/kg for solid products and 10 µg/kg for any food marketed towards infants [20,21].

Despite the presence of these regulations, patulin continues to be found in food products around the world. Table 1 summarizes the patulin contamination that has been quantified in various food products. Patulin is typically associated with apples and apples products; however it has also been found in other fruits such as pears, figs, and tomatoes [22–24]. Scientific surveys have also discovered patulin contamination in vegetables such as bell peppers, grains like wheat, rice and corn, and some cheeses [25–27].

In some cases patulin has been found in commercial food products exceeding regulatory limits. An examination of apple and pear products in Tunisia for patulin contamination found that 50% of samples were contaminated [28]. The level of patulin ranged from 2 µg/kg to 889 µg/kg, with 22% of contaminated samples exceeding the limit for the toxin set by the European Union. Another study based in Turkey found that dried figs contained patulin in levels as high as 151 µg/kg [23].

Other studies have surveyed food products for the presence of spores of patulin producing fungi. The concern is that, should fungi capable of producing mycotoxins survive processing, they may under the right circumstances continue to produce patulin later. Fungi such as *P. expansum*, *B. nivea*, and

A. terreus have been isolated from cereals and a variety of fruits [29]. Strains of patulin producing fungi have been found in other products such as peanuts, pecans, and hazelnuts [30].

The presence of fungi does not necessarily indicate the presence of patulin within a food. Likewise the absence of visible fungal growth is not a guarantee of safety. The mycotoxin has been shown to diffuse away from mold in food products. In apples, patulin has been found 1–2 cm away from the infected flesh [31,32]. Similar diffusion was seen in pears in a later study [33]. In tomatoes, likely due to the lower viscosity of their interiors, patulin can diffuse throughout the entirety of the fruit [31]. The same study also found that patulin can diffuse through wheat based products by as much as 3–4 cm.

Given the potential toxicity of patulin and the continued occurrence of it in commercial food products, there has been research done to look at the effect of different treatments on patulin. This review will examine the methods used to mitigate the threat of patulin in food through the commercial food chain with a focus on techniques devised for the specific function of mycotoxin reduction.

Table 1. Recent surveys of the occurrence of patulin in food commodities.

Food Commodity	Location	Range (µg/kg)	Positive (%)	Reference
Apples	Spain	0	0	[34]
Apples	Portugal	1–70.6	ND	[24]
Apples	United States of America	8.8–417.6	40.8	[35]
Figs	Turkey	39.3–151.6	ND	[23]
Tomatoes	Portugal	21.29	ND	[24]
Tomatoes	Belgium	ND	10.8	[25]
Bell Peppers	Belgium	ND	11.4	[25]
Hazelnuts	Turkey	16.6–92.4	ND	[36]
Cereal Based Foods	Portugal	0–4.5	75	[26]
Apple Juice	Italy	1.6–55.4	47	[37]
Apple Juice	Turkey	7–376	100	[38]
Apple Juice	Brazil	3–7	3	[39]
Apple Juice	Tunisia	2–889	64	[28]
Apple Juice	Portugal	1.2–42	41	[40]
Apple Juice	Belgium	2.5–39	81	[41]
Apple Juice	Spain	0–36.5	45	[42]
Apple Juice	South Korea	9.9–30.9	12.5	[43]
Apple Juice	Spain	2.5–6	7.1	[30]
Apple Juice	South Africa	5–45	24	[44]
Apple Juice	United States of America	8.8–2700.4	22.7	[35]
Apple Puree	Argentina	22–221	50	[22]
Apple Puree	Portugal	1.2–5.7	7	[40]
Apple Puree	Spain	0–50.3	13	[42]
Apple Puree	Italy	1.92	-	[45]
Apple Puree	South Africa	5–20	35	[44]
Apple Products	Argentina	17–39	16	[22]
Apple Products	China	1.2–94.7	12.6	[46]
Pear Products	Argentina	25	17	[22]
Pear Products	Italy	0.79	ND	[45]
Tomato Products	Italy	7.15	ND	[45]
Fruit Jam	Tunisia	2–554	20	[28]
Pear Juice	Tunisia	5–231	47.6	[28]
Orange Juice	South Korea	9.9–30.9	8	[43]
Grape Juice	South Korea	5.2–14.5	17	[43]
Semi-hard cheese	Italy	15–460	28	[27]

ND = No data available.

2. Pre-Processing Control of Patulin

The degree of patulin contamination in a food product can be managed at all levels in the food processing chain. There are common pre-processing steps that are completed regardless of the end product. These can have a significant effect on the patulin content of the finished food. These steps include storage, removal of the fungi from fresh fruit and vegetables, and the application of fungicides.

2.1. Storage

Prevention of fungal growth is the first step in the mitigation of mycotoxins. The conditions that foods face directly after harvest and before processing can have a large effect on the final quality of the produce. *P. expansum* is known to possess psychrotrophic characteristics; it is able to grow and produce patulin under refrigeration temperatures [47]. This fungus is able to produce patulin between 4 and 25 °C for 20 to 90 days [48]. The implication is that refrigerated storage is only suitable for relatively short periods of storage. The Food and Agriculture Organization of the United Nations (FAO) suggests keeping storage to <10 °C or to store fresh produce for <48 h in order to prevent the risk of patulin. Another study determined that the 48 h mark is only important for produce stored at 20 °C or higher (open deck storage); it found that in refrigerated storage the 48 h mark was not a critical time slot [49].

The use of a modified atmosphere is a second control option used for the storage of food products. The application of a high carbon dioxide and/or nitrogen atmosphere with low oxygen content has been shown to be a potential means of controlling mold growth and rot in apples [50,51]. Apples packaged in polyethylene have shown high degrees of inhibition of fungal growth and patulin production, while polypropylene is not an effective inhibiting material [52]. Without the use of a modified atmosphere, polyethylene limited growth by 68% and toxin production by 99.5%. Inhibition was further increased with the use of a modified atmosphere with CO_2 in a dose dependent manner on both patulin production and fungal growth.

2.2. Fungicides

Another means for controlling mold growth and mycotoxins in the field and in storage is the application of fungicides. There are a large number of fungicides that have been shown to possess varying levels of effectiveness. Benzimidazole fungicides used to be a common form of post-harvest treatment for fruit to deter fungal growth but have seen a steep decline in use due to increased fungal resistance [53]. Ripening makes fruits more sensitive to contamination by mold; ripened fruit put into storage had a higher risk of contamination in the absence of fungicide use. In contrast, when treated with fungicide, the more ripened fruit had smaller amounts of fungal contamination than the unripened ones, suggesting that the efficacy of fungicides is dependent on the characteristics of the fruit and the degree of ripening at the onset of storage [54]. Fludioxonil is another conventional fungicide that has been found to be effective in controlling *Penicillium* growth in apples [55]. While the use of conventional fungicides is an effective control measure, increased regulatory and health concerns and fungal resistance to the fungicides have led to research into finding alternative control agents [47]. A naturally occurring volatile compound trans-2-hexenal has been shown to be an effective fumigant in controlling *P. expansum* growth and patulin reduction in apples during storage [56]. A 3% solution of sodium hypochlorite (NaOCl) also effectively inhibited the growth of a number of fungi, including *P. expansum*, *A. alaternata*, and *Fusarium* sp., on apples [57]. Essential oils such as lemon and orange oils were also tested in the same study, and it was found that the production of patulin by *P. expansum* in apples was completely inhibited by a 0.2% solution of lemon oil and by >90% with the use of 0.05% lemon oil and 0.2% orange oil solutions.

2.3. Physical Removal of Fungi and Infected Tissue

The overall quality of food products is highly dependent on the quality of the raw ingredients. Ingredients with a high concentration of mycotoxin contamination will inherently cause a high level of contamination in the finished product. It is a recommended practice to sort out any damaged produce prior to processing and storage [58]. This not only decreases the patulin content in the finished product but also helps to reduce the possibility of cross contamination. Unfortunately the highest quality produce are typically sorted out for sale as fresh produce, whereas it is common to use bruised damaged fruit or even windfalls for the creation of juices or purees [59,60]. While this is purely a choice to lower the cost and make use of lower quality fruits, it can decrease the safety of food products.

There are other steps that can be taken in order to reduce the final patulin content and increase product quality other than sorting out infected fruit. These will typically either take the form of a washing or trimming process.

Washing typically involves either immersion into a tumbling water tank/bath or application of a high pressure water stream [61]. The primary purpose of the washing step is to remove debris, including dirt, plant matter, bugs, and mold/fungi [60]. Patulin is a water soluble molecule, and, by including a washing step, a portion of the patulin content in produce can be solubilized and removed [62]. Studies on the effects of washing treatments on the patulin level in apples found that this was one of the most critical stages of processing and could remove up to 54% of patulin from infected apples [61,63]. Researchers found that use of a high pressure water spray was more effective than a rotary wash tub as the spray would also aid in the physical removal of infected tissue for an improved reduction of both patulin and pathogenic fungi. The efficacy of washing can vary, ranging from 10% to 100% reductions [64].

Research has also been done into examining whether a chemical wash solution would be more effective than water at reducing patulin. Both the United States Food and Drug Administration (USFDA) and the Canadian Food Inspection Agency (CFIA) have recommended the use of a 100–150 ppm chlorine wash solution for processed fruit and vegetable products [65,66]. The use of a chlorine solution is primarily for the purpose of reducing microbial levels and not for the destruction of patulin. The effectiveness of chlorine as an antimicrobial is highly dependent on pH [67]. Research has been conducted to compare the use of a chlorine solution as opposed to pure water to reduce patulin content in apples [64], but, other than a larger impact on mold/fungi counts, no significant difference was found in patulin reduction by chlorine treatment. Chlorine levels need to be closely monitored as chlorine may cause the possible formation of chemical by-products and increased wear and corrosion of the equipment [68]. Other options for treating fruit include chlorine dioxide [68], hydrogen peroxide [69], acetic acid vapour [70], ozonated water [58,71], calcium salts [72], and electrolyzed oxidizing water [73]. A 200 ppm NaOCl solution inhibited the growth of *P. expansum* but was not able to completely prevent the production of patulin [67]. The same study found that the use of an acetic acid solution of 2–5% for >1 min was able to completely inhibit the growth of mold and prevent the production of patulin.

While an effective technique for removing patulin, a washing step alone is not sufficient to guarantee the safety of fresh food products. There is also the potential for the wash water to spread the patulin and cause cross contamination. Moreover, if not properly handled and disposed of, the wash solution containing patulin becomes a potential source of contamination for the entire processing facility. Washing is not a sufficient standalone mitigation technique and should be applied as a carefully monitored critical control point with other patulin reduction methods following it.

Another means of patulin reduction by fungi removal is the trimming of tissue from produce that is infected with fungi or is otherwise damaged. Regarding apples, patulin has been shown to concentrate around the area of infection and thus the removal of just this section of the apple will significantly reduce its patulin content [47]. Trimming damaged tissue was found to be capable of removing more than 93–99% of the total patulin content from apples [74].

Trimming of the fruit is an economically useful means of patulin reduction as, unlike sorting, only small portions of the material is discarded. The danger of this method is that the contaminated material must be handled properly and disposed of following its removal from the apple. This allows for possible cross contamination as well as a potential increased cost to processors. The other limitation is that, as previously discussed, patulin can diffuse from the infected area to other areas of the food products [31]. For products such as tomatoes, trimming will have no significant effect on patulin reduction.

3. Effect of Processing Steps on Patulin

Commercially sold food products have been found to contain patulin, in some cases exceeding the regulatory limit. As has been discussed, pre-processing methods can have a significant impact on the reduction of the patulin content. There is a limit to the efficacy of such techniques, and in instances of high patulin contents these methods are not able to reduce patulin to the regulatory limit. Studies have been conducted to determine the effect of conventional processing on patulin, with the intention of combining these techniques with pre-processing methods in an integrated system to increase safety.

3.1. Clarification/Filtration

Clarification or filtration techniques are commonly used in the manufacture of fruit juices. The purpose of these steps is primarily to remove solid particles such as pectins or proteins from solution [61]. This category of processing techniques has been studied for their potential to also reduce the patulin content from liquid apple products.

A system of depectinisation, clarification (with gelatine and bentonite), and filtration using a rotary vacuum pre-coat filter could reduce patulin content by 39% [61]. The depectinisation step caused negligible change to the patulin content, and the clarification step was found to be responsible for the majority of the reduction in patulin. A different study examining filtration media found that bentonite gave a patulin reduction of 8.5% and diatomaceous earth only 3.4% [75]. This same study also examined the use of centrifugation either alone or combined with other filtration methods. This type of separation works based on the difference in force applied to substances as determined by their individual masses [76]. Centrifugation alone reduced patulin content by 20.5% [75].

The removal of patulin by activated carbon/charcoal as a filtration medium has been extensively studied [77–80]. An early work compared stirring powdered activated carbon in the juice with using an activated carbon filter screen [77] and found that both methods were equally effective in reducing patulin in apple juice by 98.15–100%. It should to be noted that follow up studies did not find such high levels of reduction by activated carbon [78]. Nevertheless, activated carbon was found to be the most effective clarification means to reduce patulin, although it was also found to cause the highest reduction in pH, colour, sugar content, and possibly other nutrients. Different types of activated carbon showed different effects on patulin removal. Steam activated carbon was more effective than chemically activated carbon [44]. The same study also suggested that more activated carbon was required to reduce patulin in juices with higher solids contents.

Enzymatic depectinisation is a common juice/concentrate clarification method that uses pectinase enzymes to break down the pectin that surrounds protein particles in the solution, causing the protein to sediment out [60]. Depectinisation has been studied for its potential to also remove patulin from apple juice with varying results. Enzyme treatment only reduced patulin content by <5% in previous studies [61,75,80]; however, a study using apple juice samples from an industrial processing facility found that the use of enzymatic depectinisation removed 28% of the patulin content [81]. It is theorized that the patulin removal ability of this method is due to patulin binding to the solid particles that are removed from the solution [75]. Given these varying results, enzyme treatment does not appear to be a practical means of controlling patulin contamination.

Microfiltration and ultrafiltration have been used in recent years as an additional means of liquid product clarification. Membrane filtration separates molecules based on the molecular weight of solutes suspended in a solution [82]. Microfiltration typically filters out molecules in the size range of 0.1–2 μm, and ultrafiltration in the range of 0.001–0.1 μm. These types of systems are advantageous over conventional clarifying methods in that they cut down on enzyme consumption and the use of other filtering aids and are readily adapted to a continuous flow process [81]. The two methods are distinguished primarily by difference in membrane pore size; thus the maximum molecular weights they will let through [76]. Both have been studied for their ability to filter out patulin from apple juice [61,78]. Microfiltration could reduce the patulin content by 20.1% at different stages of apple juice processing [81]. Ultrafiltration was less effective in patulin removal from apple juice, with a

3–12% reduction [61,78]. The ability of membrane filtration to remove patulin is likely dependent on its binding to larger molecules that can be removed by the membranes rather than filtering out patulin itself [81]. This could explain the disparity of results, as the individual experimental parameters and particular composition of the apple juices used may have had a greater effect on the results than the difference in filter size.

3.2. Heat Treatment

Heat treatments such as pasteurization and distillation are common preservation methods in food processing [76]. Pasteurization is a mild heat treatment used to extend the shelf life and increase the safety of foods by destroying detrimental microorganisms. Evaporation and distillation are processes that use heat to remove water and/or volatile components from liquid foods.

The use of thermal treatments to reduce the risk of patulin has been questioned due to its resistance to heat [74,83–85]. Pasteurization has historically shown differing and contrasting results with regard to patulin reduction. These are summarized briefly in Table 2. Pasteurization for 20 min at 80 °C of apple juice spiked with patulin was found to cause a reduction of up to 50% [86]; however a later study reported that patulin was stable at 80 °C for 30 min in apple juice, and a more severe treatment of 120 °C for 30 min was required in order to achieve patulin reduction [87]. A high temperature short time (HTST) system was found to cause greater reduction in patulin (18.8% reduction at 90 °C) compared to a batch pasteurization treatment [88]. Re-examination of the two methods found that longer time pasteurization of 20 min at 90°C and 100 °C could lead to 18.81% and 25.99% patulin reduction, respectively [89]. Conflicting results continue to be generated by different investigators when comparing HTST and batch pasteurization [81,89].

Distillation, on the other hand, was found to reduce patulin by 24% in apple juice; however this was likely due to the degradation of patulin into breakdown products by heat as opposed to removal as a volatile phase [87]. This was confirmed in a later study [90].

The ability of heat treatments to remove patulin is highly dependent on the treatment parameters as well as the characteristics of the apple product. It has been indicated that, for a juice that has an initial patulin level equal to or greater than 200 µg/kg, no variation of heat treatment can reduce the patulin to below the regulated level of 50 µg/kg [89]. Other components of the food matrices may also play a role in how patulin is affected. As in filtration, it is possible that the presence of other large molecules such as fibre, protein, or sugars may cause binding to patulin and an increased resistance to destruction [81]. The pH of the product may also have had an effect on patulin reduction. Patulin is stable in acidic conditions; it has shown resistance to heat at a pH range of 3.5–5.5 at 125 °C [85].

Table 2. Patulin degradation during heat treatment of liquid food products.

Processing Temperature (°C)	Processing Time (min)	Initial PAT (µg/kg)	PAT Reduction (%)	Reference
80	20	4	55	[86]
80	30	ND	NS	[87]
90	0.17	96.5	13.4	[80]
90	0.17	20	19	[88]
90	0.5	433	39.6	[81]
90	7	1500	60	[32]
90	10	140	12.1	[91]
90	20	220	18.8	[89]
90	20	1000	NS	[92]
100	20	220	26	[89]

NS = No significant reduction in patulin found; ND = No data available; PAT = patulin.

3.3. Fermentation

Fermentation is a process in which yeast converts sugars into alcohols, gases, and/or acids [76]. Numerous studies have been reported on the contamination of unfermented apple juice by patulin; however few surveys have documented patulin contamination in alcoholic cider [93,94]. Yeast fermentation was found to be able to reduce patulin by up to 90% [95]. Another study used eight strains of yeast and three different types of fermentation processes to assess the effect on patulin in apple juice spiked with 15 mg/L patulin and found that all treatments were able to reduce the patulin content by 99% over a two week fermentation period [96]. Similarly, in an examination of the ability of three different strains of *Saccharomyces cerevisiae* to degrade patulin in apple juice, it was found that all three were effective at reducing the patulin content during fermentation but were ineffective during periods of aerobic growth [94]. This suggests that the reduction of patulin seen is a result of degradation by fermentation as opposed to adsorbing to the yeast cells themselves. This study also examined the breakdown products of patulin by fermentation and whether they were adsorbed by the yeast cells [94]. They found that numerous decomposition products were generated by fermentation and that they remained in the juice after fermentation. Of note, one of these products was E-ascadiol, which is a mycotoxin itself, though considered to be less toxic than patulin [97].

4. Patulin Reduction Techniques

Conventional processing has been shown to have an effect on patulin content in food products [47,61]. The extent to which the patulin content can be reduced by these means is unclear, with results dependent on the parameters and the initial patulin content [22,37]. This suggests that further treatment may be required to reduce the mycotoxin content. Methods of processing specifically for the reduction of patulin have been proposed to replace traditional methods or for inclusion in the production process as part of a hurdle approach. These include biological, chemical, and physical methods to either bind patulin or degrade it.

4.1. Biological Control Agents

Biological control of patulin refers to any method which uses microorganisms to reduce the patulin content in a product or to prevent the production of patulin. These methods differ from fermentation in that they do not necessarily contribute to the characteristics or chemical properties of the food; they are studied for the purpose of patulin reduction. This control method falls under two general categories based on the mode of patulin reduction; detoxification and adsorption. Detoxification refers to processes that chemically modify the mycotoxin to inactivate or reduce toxicity, while adsorption refers to the means by which the mycotoxin is bound and removed from solution. Similar to the fermentative removal of patulin by the yeast *S. cerevisiae*, microorganisms such as lactic acid bacteria (LAB), other yeasts, and fungi have been investigated for their ability to degrade patulin in aqueous solution [98–102]. Microorganisms, including LAB and yeast cells have also been found to be capable of patulin reduction in food products through the mechanism of adsorption onto their cell walls [103–108].

LAB are microorganisms of significance due to their high usage in the food industry as both a means to process food and as additives. They are used as probiotics in maintaining gastrointestinal health and have been shown to offer protection against some toxic compounds [109]. Research has found that significant amounts of patulin were removed by 10 strains of LAB, resulting in a 47 to 80% reduction of patulin [106]. The result also suggests that this reduction of patulin by LAB is highly strain specific. The increased surface area and cell wall volume of the bacteria showed a higher ability to adsorb patulin from aqueous solution [108]. Functional groups including C-O, OH, and NH were found to be involved in adsorbing patulin, suggesting that polysaccharides and proteins that are rich in these groups may play important roles. The effect of cell viability on degree of patulin removal has also been assed [107]. The highest reduction of patulin was achieved by a strain of *Bifidobacterium bifidum* at 51.1% for viable cells and 54.1% for nonviable cells.

Yeast cells or cell wall components have also been studied for their ability to bind patulin. Two types of inactivated yeast powder were examined for their capacity to adsorb the mycotoxin [105]. Patulin was reduced to below 4.6 µg/kg after 36 h. Cell powders of eight or 10 yeast strains were shown to be able to reduce patulin by greater than 50% over a 24 h period [104]. The ability of two *Enterococcus faecium* strains to remove patulin from aqueous solution has also been investigated [103]. The strains were able to remove between 15% and 45% of the patulin over a 48 h period. The viability of the strains did not have a significant effect on the ability to remove patulin. It was also found that patulin forms a stable complex with the bacteria, and the patulin-bacteria complexes could reasonably be maintained through further processing.

Microorganisms have also been studied for their ability to detoxify patulin. Two strains of *Metschnikowia pulcherrima* yeast were able to degrade patulin in a liquid media spiked with 5, 7.5, 10, and 15 µg/mL patulin [101]. It was found that one of the yeast strains was able to reduce patulin levels by 100% within 48 h and the other within 72 h at all concentrations. No patulin was found in the cell walls after degradation, suggesting that the ability of these yeast strains to remove patulin is unrelated to cell wall binding, as was the case in the other form of biological control of patulin. Researchers also noted that the presence of patulin did not influence the concentration of yeast cells during growth, suggesting that the yeast was immune to the toxic effects of the mycotoxin. It has not been determined whether all yeast strains possess patulin resistance. However, using a pre-treatment of low amounts of patulin prior to fermentation, it was found that patulin resistance and degradation abilities could be induced in *Sporobolomyces* sp. cells [102]. A variety of yeast strains have been examined for their ability to degrade patulin. The yeast *P. ohmeri* was able to degrade more than 83% of patulin after two days at 25 °C, and after 15 days patulin was degraded below the detectable limit [99]. *S. cerevisiae* degraded 96% of patulin in apple juice that had an initial concentration of 4.5 µg/mL after 6 d at 25 °C. However only 90% was degraded when the initial patulin content was 7.0 µg/mL, suggesting that the rate of degradation is concentration dependent. A strain of marine yeast, *Kodameae ohmeri*, has been reported to have a high tolerance to patulin and to have the ability to significantly reduce patulin content in apple juice. It has been suggested that the ability of yeast to detoxify patulin is enzymatic in nature [110]. It was found that the yeast *Rhodosporidium paludignum* could significantly reduce the patulin content in apples and pears. A potential hazard concerning the use of *R. paludignum* has also been noted; however the application of this yeast in high concentrations actually increased the patulin concentration in infected fruits [111]. This is possibly caused by a triggering of stress responses of patulin-producing fungi. Another concern with patulin degradation is the potential toxicity of the breakdown products. While not all of the degradation products have been assessed, some have been identified as E-ascladiol, Z-ascladiol, and deoxypatulinic acid [98,100,112]. E-ascladiol and Z-ascladiol have been found to exhibit no signs of toxicity towards human cell lines derived from the intestinal tract, kidney, liver, and immune system [113].

4.2. Chemical Additives

Several methods of chemical degradation of patulin have been proposed. Some of these are novel methods, while others are additives used for other purposes in apple production but that have found subsequent use as patulin reducers. Of the variety of chemicals studied, the most promising include ascorbic acid, ammonia, potassium permanganate, sulfur dioxide, ozone, and some of the B vitamins [95,114–117].

Ascorbic acid and ascorbate (Vitamin C) have been studied to reduce patulin in apple products. Slight reductions of patulin by ascorbic acid have been reported, with one study noting only 5% losses after 3 h and 36% after 44 h [115]. Similarly, the degradation of patulin during storage was observed; the patulin content of apple juice with ascorbic acid added was decreased by 70% but only by 30% in juice without added ascorbic acid after a 34 d period [118].

The degradation of patulin by ammoniation and by oxidation with potassium permanganate in an acidic and a basic environment has also been studied as a control measure [115]. Both treatments were effective and able to reduce patulin by more than 99.9% in a standard aqueous solution. Under acidic conditions, treatment with potassium permanganate was found to produce potentially mutagenic and harmful compounds, limiting its potential for use in foodstuffs.

Patulin has been shown to be unstable in the presence of sulfur containing compounds [119]. For this reason there has been study on the effect of sulfur dioxide to degrade patulin in solution. One study found that a solution with 200 ppm of sulfur dioxide was able to reduce patulin by 12% after 24 h and 90% after two days [95]. A later study found that just 100 ppm sulfur dioxide could reduce patulin content by 50% in 15 min. The differences noted here could be attributed to differences in the composition of the sample solutions, as interfering components may have been present. Patulin is also thought to react with a number of other sulfur compounds to produce less toxic compounds [9,120]. It has been shown that patulin will form adducts with a number of sulfur containing compounds such as cysteine, *N*-acetylcysteine, and glutathione [12]. This is because patulin has a strong affinity for binding covalently to sulfhydryl groups as well as amino, thiol, and NH_2 groups [121]. While still possessing some toxicity, it has been determined that these adducts are 100 times less toxic than patulin itself [122].

Ozone is a strong oxidant, capable of reacting with numerous chemical groups and is thought to be able to detoxify patulin [116,121,123]. Patulin treated with a 10% solution of ozone degraded from 5 ppm to below detectable levels in 15 s [116]. Metal ions in general did not affect patulin degradation by ozone oxidation although iron and manganese both significantly reduced the effect. Ozone alone was able to degrade patulin by up to 98% in 1 min [123]. Ozone treatment is highly effective and it does not have a significant effect on the quality parameters of food products [121].

Among thiamine hydrochloride, pyridoxine hydrochloride, and calcium D-pantothenate, the latter i.e. calcium D-pantothenate was the most effective at reducing patulin, being able to reduce it by up to 94.3% over six months with no significant loss in quality characteristics, compared to 35.8% with no addition of any substance [117]. While effective, this length of time may be impractical for many products and the toxicity of the adducts formed requires further study.

4.3. Physical Treatments

Patulin is known to be resistant to degradation by heat treatment [85]. Furthermore some of the treatments used in apple processing that have been suggested as possibilities for the reduction of patulin are known to have a negative effect on some of the quality characteristics of the food product such as pH, clarity, colour, sugars, and °Brix [78,79,91]. Development in the field of non-thermal food processing techniques has opened up the potential for unconventional processing methods to play a role in apple processing and the reduction of patulin.

4.3.1. Ultraviolet Radiation

Ultraviolet (UV) radiation is an approved non-thermal method for the preservation of fruit juices in both Canada and the United States [124,125]. Typically used for the destruction of microorganisms, it has also been studied as a means of degrading patulin [126,127]. UV radiation with an exposure range of 14.2–99.4 mJ/cm^2 on apple cider was able to cause reductions ranging from 9.4 to 43.4% with higher exposures causing higher patulin reductions [128]. No loss of chemical components or sensory properties was found including pH, °Brix, and total acids. A later study on the effect of UV radiation (253.7 nm) on patulin in apple juice, apple cider, and a model aqueous solution showed that it was highly effective in all but apple cider [127,129]. It was suggested that apple cider might contain components that are interfering with the breakdown reaction. The increased turbidity of apple cider is considered to hamper the action of UV. The use of filtration/clarification processes therefore can increase the effectiveness of this technique on apple cider. Other studies have shown that the effect of UV radiation can be significantly hindered by the presence of large amounts of ascorbic acid [126]. Ascorbic acid is a common additive to apple juice for its anti-browning and antioxidant properties [130]. In order for UV radiation to be an effective treatment, any addition of ascorbic acid would have to occur afterwards. The particular wavelength of UV light used was also shown to affect patulin reduction [131]. It was determined that patulin reduction in apple juice or cider was most effective at 222 nm as opposed to 254 and 282 nm. No significant changes to pH, soluble solids, or the colour of the apple juice were found during this treatment. Despite this apparent lack of change to other components of apple juice, a trained sensory panel found that UV radiation treated juice was significantly different from conventionally treated juice [126]. Further research should examine not only the effect of UV on patulin but also on flavour compounds.

4.3.2. Pulsed Light

The use of pulsed light is another processing technique that has been proposed for the destruction of patulin in food products [132]. Pulsed light is a non-thermal food preservation technique that involves the use of short (1 μs–0.1 s) bursts of broad spectrum light with wavelengths ranging from 200 to 1100 nm [133]. Patulin was degraded by pulsed light in McIlvaine buffer, apple juice, and apple puree [132]. Treatment with pulsed light was able to reduce the patulin content by 85–95% in the buffer, 22% in the juice, and 51% in the apple puree. The effectiveness was not dependent on the initial patulin content.

4.3.3. High Hydrostatic Pressure

High hydrostatic pressure processing (HPP) is a non-thermal food processing method originally designed for the reduction of microorganisms that has also been studied as a means of reducing mycotoxins in foods [134]. HPP is a food preservation treatment that uses high pressure to inactivate microorganisms and proteins [135,136]. HPP treatment has been found to reduce up to 56.24% of patulin in apple juice contaminated with 100 ppb of the mycotoxin, depending on the operating conditions [134]. Pressures ranged from 30 to 500 Mpa, and temperatures ranged from (30–50 °C). No clear trend as to the optimal pressure/temperature combination has been concluded, suggesting that further study is required to refine this technology. A higher pressure at 600 MPa for 300 s was found to reduce patulin in juice by 31% [137]. It has been shown that HPP primarily works on hydrophobic and electrostatic interactions, not the covalent ones found in patulin molecules [138]. The reductions in patulin content have been attributed to the formation of adducts with compounds containing sulphhydryl groups such as glutathione or cysteine [134,139]. These adducts have been shown to be 100 times less toxic than patulin itself [122].

5. Conclusions

Patulin is a mycotoxin of threat to human health. While the precise nature of the long term effects in humans are uncertain, the evidence from animal cases is sufficient to justify concern. A number of regulatory agencies have laid out limitations as to the allowable patulin content in food products. Despite these regulations, patulin maintains a presence in foods produced around the world. Means of removing or detoxifying patulin in food are necessary considerations in the processing chain to increase safety.

A combination of temperature control, the use of modified atmosphere storage, and the application of fungicides can significantly decrease fungal growth and the production of patulin in fresh fruits. However not every storage facility has access to all of these technologies, and there are issues with the over use of chemical fungicides. Improving the quality of the fruit to be processed by means of washing, trimming, and sorting are all very useful means for controlling patulin. These processes may not be economically feasible for all producers as they also significantly increase raw material waste. The waste itself presents a contamination issue and requires special consideration for handling and disposal. Typical pre-processing techniques are not sufficient and further patulin reduction is necessary. Conventional processing methods using heat, fermentation, and/or chemical binders have similarly been shown to have an effect on patulin. However the evidence is inconclusive as to the actual efficacy of these techniques and so cannot be relied upon to guarantee safety. Furthermore, there is concern with the potential toxicity of compounds produced by the degradation of patulin by some of these methods. Most of these products have yet to be identified, and some that have been identified are known to be toxic.

Due to the resistance of patulin to acidic conditions and heat treatment, alternative methods have been proposed for the removal or detoxification of patulin using modified or novel processing techniques. For these methods to be considered for this purpose they must be safe and effective. The FAO has requirements for a decontamination process that state that the method must do the following:

1. Destroy, inactivate, or otherwise remove the mycotoxins.
2. Not leave or create any products that possess toxic/mutagenic/carcinogenic properties.
3. Be practical in so far as it is technologically and economically feasible.
4. Prevent the re-occurrence of mycotoxins by destroying any fungal spores or mycelium.

The development of non-thermal physical processing techniques has led to some promising possibilities for techniques to reduce the patulin content of food products. UV and pulsed light radiation can effectively remove patulin from solution; however they may also destroy beneficial components of the food. HPP can be an effective means of patulin degradation; however the optimal conditions have yet to be determined and may depend in part on the properties of the food product. Regarding the FAO requirements, the primary concern for any of these treatments is that current research is inadequate to fully explain the mechanism of action and the potential toxicity of the breakdown products. These methods would also require highly specialized equipment, which represents a major cost factor and may not be allowed for use in food products.

Chemical degradation of patulin is another potential solution, given that the method is easily accommodated into traditional processing streams. However, the use of chemical agents solely for the purpose of mycotoxin reduction is currently not permitted, though many of the chemicals listed here can also serve other purposes; for example, ascorbic acid (Vitamin C) is commonly used to prevent oxidation or browning [60,140]. Despite this potential, further research is still required to validate some of the inconsistencies found and to determine the optimal processing conditions. The exact nature of the reactions and, more importantly, the reaction products for many of these cases are still unknown, which provides some health and safety concerns.

Biologically based methods can eliminate or reduce the patulin content in food products. Both enzymatic degradation and the adsorptive removal of patulin by microorganisms like LAB or yeast have a significant impact on patulin content. Furthermore, it has been found that the application of biological control methods causes no significant impact to the juice quality characteristics of °Brix, acidity, colour, and clarity [104]. Various microorganisms have the ability to either detoxify or bind to patulin in an aqueous environment, though it is unclear what the optimal processing conditions are. It has been shown that strain, pH, temperature, incubation time, concentration of microorganisms, and patulin level all play an important role, but further study is required [106]. Some breakdown products have been identified, but not all are known [98,100,112]. More work is required to more fully understand the processes involved and to determine how to safely incorporate these methods into food processing.

There are numerous methods that have an effect on fungi and patulin; however there does not appear to be any singular method that can reliably prevent patulin in food products. Instead the most practical course of action appears to be the application of the hurdle approach to food safety. That is to say that multiple safety checks and measures should be implemented throughout the entire growth, harvest, and production cycle. Work on the refinement of these techniques to determine the optimal processing conditions is necessary to increase the practicality of usage. The main concern at this point and the area to which future research focus must be turned is in understanding and controlling the degradation compounds that are produced by a number of these methods.

Acknowledgments: This work is funded by the A-Base Project (# J-000048.001.09) of Agriculture and Agri-Food Canada.

Conflicts of Interest: The authors declare no conflict of interest.

References

1. Barug, D.; Bhatnagar, D.; Van Egmond, H.P.; van der Kamp, J.W.; Van Osenbruggen, W.A.; Visconti, A. *The Mycotoxin Factbook*; Wageningen Academic Publishers: Wageningen, The Netherlands, 2006.
2. Lai, C.; Fuh, Y.; Shih, D. Detection of mycotoxin patulin in apple juice. *J. Food Drug Anal.* **2000**, *2*, 85–96.
3. Drusch, S.; Ragab, W. Mycotoxins in fruits, fruit juices, and dried fruits. *J. Food Prot.* **2003**, *66*, 1514–1527. [CrossRef] [PubMed]
4. Stott, W.; Bullerman, L.B. Patulin: A mycotoxin of potential concern in foods. *J. Food Prot.* **1975**, *38*, 695–705. [CrossRef]
5. Yang, J.; Li, J.; Jiang, Y.; Duan, X.; Qu, H.; Yang, B.; Chen, F.; Sivakumar, D. Natural occurrence, analysis, and prevention of mycotoxins in fruits and their processed products. *Crit. Rev. Food Sci. Nutr.* **2014**, *54*, 64–83. [CrossRef] [PubMed]
6. Paterson, R.R.; Venancio, A.; Lima, N. Solutions to *Penicillium* taxonomy crucial to mycotoxin research and health. *Res. Microbiol.* **2004**, *155*, 507–513. [CrossRef] [PubMed]
7. Ciegler, A.; Detroy, R.; Lillehoj, E. Patulin, penicillic acid, and other carcinogenic lactones. *Microb. Toxins* **1971**, *6*, 409–434.
8. Puel, O.; Galtier, P.; Oswald, I.P. Biosynthesis and toxicological effects of patulin. *Toxins* **2010**, *2*, 613–631. [CrossRef] [PubMed]
9. Ciegler, A.; Vesonder, R.; Jackson, L. Production and biological activity of patulin and citrinin from *Penicillium expansum*. *Appl. Environ. Microbiol.* **1977**, *4*, 1004–1006.
10. Dickens, F.; Jones, H. Carcinogenic activity of a series of reactive lactones and related substances. *Br. J. Cancer* **1961**, *15*, 85–100. [CrossRef] [PubMed]
11. Llewellyn, G.C.; McCay, J.A.; Brown, R.D. Immunological evaluation of the mycotoxin patulin in female B6C3F$_1$ mice. *Food Chem. Toxicol.* **1998**, *36*, 1107–1115. [CrossRef]
12. Fliege, R.; Metzler, M. Electrophilic properties of patulin. Adduct structures and reaction pathways with 4-bromothiophenol and other model nucleophiles. *Chem. Res. Toxicol.* **2000**, *13*, 363–372. [CrossRef] [PubMed]
13. Baert, K.; Devlieghere, F.; Flyps, H.; Oosterlinck, M.; Ahmed, M.M.; Rajkovic, A.; Verlinden, B.; Nicolai, B.; Debevere, J.; De Meulenaer, B. Influence of storage conditions of apples on growth and patulin production by *Penicillium expansum*. *Int. J. Food Microbiol.* **2007**, *119*, 170–181. [CrossRef] [PubMed]
14. Schumacher, D.M.; Müller, C.; Metzler, M.; Lehmann, L. DNA-DNA cross-links contribute to the mutagenic potential of the mycotoxin patulin. *Toxicol. Lett.* **2006**, *166*, 268–275. [CrossRef] [PubMed]
15. Fung, F.; Clark, R.F. Health effects of mycotoxins: A toxicological overview. *J. Toxicol. Clin. Toxicol.* **2004**, *42*, 217–234. [CrossRef] [PubMed]
16. Plunkett, L.; Turnbull, D.; Rodricks, J. Differences between adults and children affecting exposure assessment. In *Similarities and Differences between Children and Adults, Implications for Risk Assessment*; Guzelian, P., Henry, C., Olin, S., Eds.; ILSI Press: Washington, DC, USA, 1992; pp. 79–94.
17. CODEX. *Maximum Level for Patulin in Apple Juice and Apple Juice Ingredients and Other Beverages*; Codex Alimentarius Commission, Ed.; 235; Codex Alimentarius Commission: Rome, Italy, 2003.
18. Food and Drug Administration (FDA). Compliance policy guidance for fda staff. Sec. 510.150 Apple juice, apple juice concentrates, and apple juice products—Adulteration with patulin. In *Compliance Policy Guide*; U.S. Food and Drug Administration: Silver Spring, MD, USA, 2004.
19. Health Canada. Canadian standards for various chemical contaminants in foods. *Food and Drug Regulations*. Health Canada, Ed.; 2014. Available online: http://www.hc-sc.gc.ca/fn-an/securit/chem-chim/contaminants-guidelines-directives-eng.php (accessed on 20 October 2016).
20. World Health Organization. Evaluation of certain food additives and contaminants. *Tech. Rep. Ser.* **1995**, *859*, 36–38.
21. European Commission. Commission regulation (EC) No 1881/2006 of 19 december 2006 setting maximum levels for certain contaminants in foodstuffs. *Off. J. Eur. Union* **2006**, *364*, 5–24.

22. Funes, G.J.; Resnik, S.L. Determination of patulin in solid and semisolid apple and pear products marketed in argentina. *Food Control* **2009**, *20*, 277–280. [CrossRef]

23. Karaca, H.; Nas, S. Aflatoxins, patulin and ergosterol contents of dried figs in turkey. *Food Addit. Contam.* **2006**, *23*, 502–508. [CrossRef] [PubMed]

24. Cunha, S.C.; Faria, M.A.; Pereira, V.L.; Oliveira, T.M.; Lima, A.C.; Pinto, E. Patulin assessment and fungi identification in organic and conventional fruits and derived products. *Food Control* **2014**, *44*, 185–190. [CrossRef]

25. Van de Perre, E.; Jacxsens, L.; Van Der Hauwaert, W.; Haesaert, I.; De Meulenaer, B. Screening for the presence of patulin in molded fresh produce and evaluation of its stability in the production of tomato products. *J. Agric. Food Chem.* **2014**, *62*, 304–309. [CrossRef] [PubMed]

26. Assuncao, R.; Martins, C.; Dupont, D.; Alvito, P. Patulin and ochratoxin a co-occurrence and their bioaccessibility in processed cereal-based foods: A contribution for portuguese children risk assessment. *Food Chem. Toxicol.* **2016**, *96*, 205–214. [CrossRef] [PubMed]

27. Pattono, D.; Grosso, A.; Stocco, P.P.; Pazzi, M.; Zeppa, G. Survey of the presence of patulin and ochratoxin a in traditional semi-hard cheeses. *Food Control* **2013**, *33*, 54–57. [CrossRef]

28. Zouaoui, N.; Sbaii, N.; Bacha, H.; Abid-Essefi, S. Occurrence of patulin in various fruit juice marketed in tunisia. *Food Control* **2015**, *51*, 356–360. [CrossRef]

29. Deshpande, S.S. *Handbook of Food Toxicology*; Marcel Dekker Inc.: New York, NY, USA, 2002; p. 920.

30. Marín, S.; Mateo, E.M.; Sanchis, V.; Valle-Algarra, F.M.; Ramos, A.J.; Jiménez, M. Patulin contamination in fruit derivatives, including baby food, from the spanish market. *Food Chem.* **2011**, *124*, 563–568. [CrossRef]

31. Rychlik, M.; Schieberle, P. Model studies on the diffusion behavior of the mycotoxin patulin in apples, tomatoes, and wheat bread. *Eur. Food Res. Technol.* **2001**, *212*, 274–278. [CrossRef]

32. Taniwaki, M.; Hoenderboom, C.; De Almeida Vitali, A.; Firoa, M. Migration of patulin in apples. *J. Food Prot.* **1992**, *55*, 902–904. [CrossRef]

33. Laidou, I.A.; Thanassoulopoulos, C.C.; Liakopoulou-Kyriakides, M. Diffusion of patulin in the flesh of pears inoculated with four post-harvest pathogens. *J. Phytopathol.* **2001**, *149*, 457–461. [CrossRef]

34. Marin, S.; Morales, H.; Hasan, H.A.; Ramos, A.J.; Sanchis, V. Patulin distribution in Fuji and golden apples contaminated with *Penicillium expansum*. *Food Addit. Contam.* **2006**, *23*, 1316–1322. [CrossRef] [PubMed]

35. Harris, K.; Bobe, G.; Bourquin, L. Patulin surveillance in apple cider and juice marketed in michigan. *J. Food Prot.* **2009**, *72*, 1255–1261. [CrossRef] [PubMed]

36. Ekinci, R.; Otag, M.; Kadakal, C. Patulin & ergosterol: New quality parameters together with aflatoxins in hazelnuts. *Food Chem.* **2014**, *150*, 17–21. [PubMed]

37. Spadaro, D.; Ciavorella, A.; Frati, S.; Garibaldi, A.; Gullino, M.L. Incidence and level of patulin contamination in pure and mixed apple juices marketed in italy. *Food Control* **2007**, *18*, 1098–1102. [CrossRef]

38. Gokmen, V.; Acar, J. Incidence of patulin in apple juice concentrates produced in turkey. *J. Chromatogr. A* **1998**, *815*, 99–102. [CrossRef]

39. Iha, M.H.; Sabino, M. Incidence of patulin in brazilian apple-based drinks. *Food Control* **2008**, *19*, 417–422. [CrossRef]

40. Barreira, M.J.; Alvito, P.C.; Almeida, C.M.M. Occurrence of patulin in apple-based-foods in portugal. *Food Chem.* **2010**, *121*, 653–658. [CrossRef]

41. Tangni, E.K.; Theys, R.; Mignolet, E.; Maudoux, M.; Michelet, J.Y.; Larondelle, Y. Patulin in domestic and imported apple-based drinks in belgium: Occurrence and exposure assessment. *Food Addit. Contam.* **2003**, *20*, 482–489. [CrossRef] [PubMed]

42. Pique, E.; Vargas-Murga, L.; Gomez-Catalan, J.; Lapuente, J.; Llobet, J.M. Occurrence of patulin in organic and conventional apple-based food marketed in catalonia and exposure assessment. *Food Chem. Toxicol.* **2013**, *60*, 199–204. [CrossRef] [PubMed]

43. Cho, M.S.; Kim, K.; Seo, E.; Kassim, N.; Mtenga, A.B.; Shim, W.-B.; Lee, S.-H.; Chung, D.-H. Occurrence of patulin in various fruit juices from South Korea: An exposure assessment. *Food Sci. Biotechnol.* **2010**, *19*, 1–5. [CrossRef]

44. Leggott, N.; Shephard, G. Patulin in south african commercial apple products. *Food Control* **2001**, *12*, 73–76. [CrossRef]

45. Sarubbi, F.; Formisano, G.; Auriemma, G.; Arrichiello, A.; Palomba, R. Patulin in homogenized fruit's and tomato products. *Food Control* **2016**, *59*, 420–423. [CrossRef]

46. Yuan, Y.; Zhuang, H.; Zhang, T.; Liu, J. Patulin content in apple products marketed in northeast China. *Food Control* **2010**, *21*, 1488–1491. [CrossRef]
47. De Souza Sant'Ana, A.; Rosenthal, A.; de Massaguer, P.R. The fate of patulin in apple juice processing: A review. *Food Res. Int.* **2008**, *41*, 441–453. [CrossRef]
48. Taniwaki, M.; Bleinroth, E.; De Martin, Z. Bolores produtores de patulina em macã e suco industrializado. *Colet. Inst. Tecnol. Alimentos* **1989**, *19*, 42–49.
49. Morales, H.; Marin, S.; Centelles, X.; Ramos, A.J.; Sanchis, V. Cold and ambient deck storage prior to processing as a critical control point for patulin accumulation. *Int. J. Food Microbiol.* **2007**, *116*, 260–265. [CrossRef] [PubMed]
50. Johnsonn, D.; Stow, J.; Dover, C. Prospect for the control of fungal rotting in cox's orange pippin apples by low oxygen and low ethylene storage. *Acta Hortic.* **1993**, *343*, 334–336. [CrossRef]
51. Paster, N.; Huppert, D.; Barkai-Golan, R. Production of patulin by different strains of *Penicillium expansum* in pear and apple cultivars stored at different temperatures and modified atmospheres. *Food Addit. Contam.* **1995**, *155*, 507–513.
52. Moodley, R.S.; Govinden, R.; Odhav, B. The effect of modified atmospheres and packaging on patulin production in apples. *J. Food Prot.* **2002**, *65*, 867–871. [CrossRef] [PubMed]
53. Rosenberger, D. Control of *Penicillium Expansum* During Apple Harvest Storage. In Proceedings of the Patulin Technical Symposium, Kissimmee, FL, USA, 18–19 February 2003.
54. Morales, H.; Marin, S.; Rovira, A.; Ramos, A.J.; Sanchis, V. Patulin accumulation in apples by *Penicillium expansum* during postharvest stages. *Lett. Appl. Microbiol.* **2007**, *44*, 30–35. [CrossRef] [PubMed]
55. Errampalli, D. Effect of fludioxonil on germination and growth of *Penicillium expansum* and decay in apple cvs. Empire and gala. *Crop Prot.* **2004**, *23*, 811–817. [CrossRef]
56. Neri, F.; Mari, M.; Menniti, A.M.; Brigati, S.; Bertolini, P. Control of *Penicillium expansum* in pears and apples by trans-2-hexenal vapours. *Postharvest Biol. Technol.* **2006**, *41*, 101–108. [CrossRef]
57. Hasan, H.A. Patulin and aflatoxin in brown rot lesion of apple fruits and their regulation. *World J. Microbiol. Biotechnol.* **2000**, *16*, 607–612. [CrossRef]
58. CODEX. *Code of Practice for the Prevention and Reduction of Patulin Contamination in Apple Juice and Apple Juice Ingredients in other Beverages*; Codex Alimentarius Commission, Ed.; CA/RCP-2003; Food and Agriculture Organization: Rome, Italy, 2003.
59. Moake, M.M.; Padilla-Zakour, O.I.; Worobo, R.W. Comprehensive review of patulin control methods in foods. *Compr. Rev. Food Sci. Food Saf.* **2005**, *1*, 8–21. [CrossRef]
60. Root, W.H.; Barrett, D.M. Apples and apple processing. In *Processing Fruits*; Barrett, D.M., Somogyi, L., Ramaswamy, H., Eds.; CRC Press: Danvers, MA, USA, 2005; pp. 455–479.
61. Acar, J.; Gokmen, V.; Taydas, E.E. The effects of processing technology on the patulin content of juice during commercial apple juice concentrate production. *Z. Lebensm. Unters. Forsch. A* **1998**, *207*, 328–331. [CrossRef]
62. Cole, R.; Jarvis, B.B.; Schweikert, M.A. *Handbook of Secondary Fungal Metabolites*; Academic Press: San Diego, CA, USA, 2003.
63. Sydenham, E.; Vismer, H.; Marasas, W.; Brown, N.; Schlechter, M.; van der Westhuizen, L.; Rheeder, J. Reduction of patulin in apple juice samples—Influence of initial processing. *Food Control* **1995**, *6*, 195–200. [CrossRef]
64. Jackson, L.S.; Beacham-Bowden, T.; Keller, S.E.; Adhikari, C.; Taylor, K.T.; Chirtel, S.J.; Merker, R.I. Apple quality, storage, and washing treatments affect patulin levels in apple cider. *J. Food Prot.* **2003**, *66*, 618–624. [CrossRef] [PubMed]
65. Food and Drug Administration (FDA). *Guide to Minimize Microbial Food Safety Hazards for Fresh Fruits and Vegetables*; Food Safety Initiative Staff, HFS-32, U.S. Food and Drug Administration, Ed.; Center for Food Safety and Applied Nutrition: Washington, DC, USA, 1998.
66. Canadian Food Inspection Agency (CFIA). *Code of Practice for Minimally Processed Ready-To-Eat Vegetables*; Canadian Food Inspection Agency, Ed.; Canadian Food Inspection Agency: Ottawa, ON, Canada, 2009.
67. Chen, L.; Ingham, H.; Ingham, S.C. Survival of *Penicillium expansum* and patulin production on stored apples after wash treatments. *J. Food Sci.* **2004**, *69*, 669–675. [CrossRef]
68. Roberts, R.G.; Reymond, S.T. Chlorine dioxide for reduction of postharvest pathogen inoculum during handling of tree fruits. *Appl. Environ. Microbiol.* **1994**, *60*, 2864–2868. [PubMed]

69. Baldry, M. The bactericidal, fungicidal and sporicidal properties of hydrogen peroxide and peracetic acid. *J. Appl. Bacteriol.* **1983**, *54*, 417–423. [CrossRef] [PubMed]

70. Sholberg, P.; Haag, P.; Hocking, R.; Bedford, K. The use of vinegar vapor to reduce postharvest decay of harvested fruit. *Hortic. Sci.* **2000**, *35*, 898–903.

71. Spotts, R.; Cervantes, L. Effect of ozonated water on postharvest pathogens of pear in laboratory and packinghouse tests. *Plant Dis.* **1992**, *76*, 256–259. [CrossRef]

72. Conway, W.; Lanisiewicz, W.; Klein, I.; Sams, C. Strategy for combining heat treatment, calcium infiltration, and biological control to reduce postharvest decay of "gala" apples. *Hortic. Sci.* **1999**, *34*, 700–704.

73. Okull, D.; LaBorde, L. Activity of electrolyzed oxidizing water against *Penicillium expansum* in suspension and on wounded apples. *J. Food Sci.* **2004**, *69*, 23–27. [CrossRef]

74. Lovett, J.; Thompson, R.; Boutin, B. Patulin production in apples stored in a controlled atmosphere. *J. Assoc. Off. Anal. Chem.* **1975**, *58*, 912–914. [PubMed]

75. Bisseur, J.; Permaul, K.; Odhav, B. Reduction of patulin during apple juice clarification. *J. Food Prot.* **2001**, *64*, 1216–1219. [CrossRef]

76. Fellows, P.J. *Food Processing Technol*, 3rd ed.; Woodland Publishing: Boca Raton, FL, USA, 2009.

77. Sands, D.C.; McIntyre, J.L.; Walton, G.S. Use of activated charcoal for the removal of patulin from cider. *Appl. Environ. Microbiol.* **1976**, *32*, 388–391. [PubMed]

78. Gökmen, V.; Artık, N.; Acar, J.; Kahraman, N.; Poyrazoğlu, E. Effects of various clarification treatments on patulin, phenolic compound and organic acid compositions of apple juice. *Eur. Food Res. Technol.* **2001**, *213*, 194–199. [CrossRef]

79. Kadakal, C.; Nas, S. Effect of activated charcoal on patulin, fumaric acid, and some other properties of apple juice. *Nahr. Food* **2002**, *46*, 31–33. [CrossRef]

80. Kadakal, C.; Sebahattin, N.; Poyrazoğlu, E.S. Effect of commercial processing stages of apple juice on patulin, fumaric acid and hydroxymethylfurfural (HMF) levels. *J. Food Qual.* **2002**, *25*, 359–368. [CrossRef]

81. Welke, J.E.; Hoeltz, M.; Dottori, H.A.; Noll, I.B. Effect of processing stages of apple juice concentrate on patulin levels. *Food Control* **2009**, *20*, 48–52. [CrossRef]

82. Fukumoto, L.; Delaquis, P.; Girard, B. Microfiltration and ultrafiltration ceramic membranes for apple juice clarification. *J. Food Sci.* **1998**, *63*, 845–866. [CrossRef]

83. Wiesner, B. Bactericidal effects of *Aspergillus clavatus*. *Nature* **1942**, *149*, 356–357. [CrossRef]

84. Heatley, N.; Philpot, F. The routine examination for antibiotic produced by moulds. *J. Gen. Microbiol.* **1947**, *1*, 232–237. [CrossRef] [PubMed]

85. Lovett, J.; Peeler, J. Effect of ph on the thermal destruction kinetics of patulin in aqueous solution. *J. Food Sci.* **1973**, *38*, 1094–1095. [CrossRef]

86. Scott, P.; Somers, E. Stability of patulin and penicillic acid in fruit juices and flour. *J. Agric. Food Chem.* **1968**, *16*, 483–485. [CrossRef]

87. Kubacki, S. The analysis and occurrence of patulin in apple juice. In Proceedings of the 6th International IUPAC Symposium on Mycotoxins Phycotoxins, Pretoria, South Africa, 22–25 July 1986; pp. 293–304.

88. Wheeler, J.L.; Harrison, M.A.; Koehler, P.E. Presence and stability of patulin in pasteurized apple cider. *J. Food Sci.* **1987**, *52*, 479–780. [CrossRef]

89. Kadakal, C.; Nas, S. Effect of heat treatment and evaporation on patulin and some other properties of apple juice. *J. Sci. Food Agric.* **2003**, *83*, 987–990. [CrossRef]

90. Kryger, R.A. Volatility of patulin in apple juice. *J. Agric. Food Chem.* **2001**, *49*, 4141–4143. [CrossRef] [PubMed]

91. Janotová, L.; Čížková, H.; Pivoňka, J.; Voldřich, M. Effect of processing of apple puree on patulin content. *Food Control* **2011**, *22*, 977–981. [CrossRef]

92. Woller, R.; Majerus, P. Patulin in obsterzeugnissen-egenschaften, bildung und vorkommen. *Flussiges Obst* **1982**, *49*, 564–570.

93. Harwig, J.; Chen, Y.; Kennedy, P.; Scott, P. Occurrence of patulin and patulin producing strains of *Penicillium expansum* in natural rots of apples in canada. *J. Can. Inst. Food Sci. Technol.* **1973**, *6*, 22–25. [CrossRef]

94. Moss, M.O.; Long, M.T. Fate of patulin in the presence of the yeast *Saccharomyces cerevisiae*. *Food Addit. Contam.* **2002**, *19*, 387–399. [CrossRef] [PubMed]

95. Burroughs, L. Stability of patulin to sulfur dioxide and to yeast fermentation. *J. Assoc. Off. Anal. Chem.* **1977**, *60*, 100–103. [PubMed]

96. Stinson, E.E.; Osman, S.F.; Huhtanen, C.N.; Bills, D.D. Disappearance of patulin during alcoholic fermentation of apple juice. *Appl. Environ. Microbiol.* **1978**, *36*, 620–622. [PubMed]

97. Suzuki, T.; Takeda, M.; Tanabe, H. A new mycotoxin produced by *Aspergillus clavatus*. *Chem. Pharm. Bull.* **1971**, *19*, 1786–1788. [CrossRef] [PubMed]

98. Ricelli, A.; Baruzzi, F.; Solfrizzo, M.; Morea, M.; Fanizzi, F.P. Biotransformation of patulin by *Gluconobacter oxydans*. *Appl. Environ. Microbiol.* **2007**, *73*, 785–792. [CrossRef] [PubMed]

99. Coelho, A.; Celli, M.; Sataque Ono, E.; Hoffmann, F.; Pagnocca, F.; Garcia, S.; Sabino, M.; Harada, K.; Wosiacki, G.; Hirooka, E. Patulin biodegradation using *Pichia ohmeri* and *Saccharomyces cerevisiae*. *World Mycotoxin J.* **2008**, *1*, 325–331. [CrossRef]

100. Fuchs, S.; Sontag, G.; Stidl, R.; Ehrlich, V.; Kundi, M.; Knasmuller, S. Detoxification of patulin and ochratoxin a, two abundant mycotoxins, by lactic acid bacteria. *Food Chem. Toxicol.* **2008**, *46*, 1398–1407. [CrossRef] [PubMed]

101. Reddy, K.R.; Spadaro, D.; Gullino, M.L.; Garibaldi, A. Potential of two *metschnikowia pulcherrima* (yeast) strains for in vitro biodegradation of patulin. *J. Food Prot.* **2011**, *74*, 154–156. [CrossRef] [PubMed]

102. Ianiri, G.; Pinedo, C.; Fratianni, A.; Panfili, G.; Castoria, R. Patulin degradation by the biocontrol yeast *sporobolomyces* sp. Is an inducible process. *Toxins* **2017**, *9*, 61. [CrossRef] [PubMed]

103. Topcu, A.; Bulat, T.; Wishah, R.; Boyaci, I.H. Detoxification of aflatoxin B1 and patulin by *enterococcus faecium* strains. *Int. J. Food Microbiol.* **2010**, *139*, 202–205. [CrossRef] [PubMed]

104. Yue, T.; Dong, Q.; Guo, C.; Worobo, R.W. Reducing patulin contamination in apple juice by using inactive yeast. *J. Food Prot.* **2011**, *74*, 149–153. [CrossRef] [PubMed]

105. Guo, C.; Yue, T.; Hatab, S.; Yuan, Y. Ability of inactivated yeast powder to adsorb patulin from apple juice. *J. Food Prot.* **2012**, *75*, 585–590. [CrossRef] [PubMed]

106. Hatab, S.; Yue, T.; Mohamad, O. Reduction of patulin in aqueous solution by lactic acid bacteria. *J. Food Sci.* **2012**, *77*, M238–M241. [CrossRef] [PubMed]

107. Hatab, S.; Yue, T.; Mohamad, O. Removal of patulin from apple juice using inactivated lactic acid bacteria. *J. Appl. Microbiol.* **2012**, *112*, 892–899. [CrossRef] [PubMed]

108. Wang, L.; Yue, T.; Yuan, Y.; Wang, Z.; Ye, M.; Cai, R. A new insight into the adsorption mechanism of patulin by the heat-inactive lactic acid bacteria cells. *Food Control* **2015**, *50*, 104–110. [CrossRef]

109. Knasmüller, S.; Steinkellner, H.; Hirschl, A.M.; Rabot, S.; Nobis, E.C.; Kassie, F. Impact of bacteria in dairy products and of the intestinal microflora on the genotoxic and carcinogenic effects of heterocyclic aromatic amines. *Mutat. Res.* **2001**, *480*, 129–138. [CrossRef]

110. Zhu, R.; Yu, T.; Guo, S.; Hu, H.; Zheng, X.; Karlovsky, P. Effect of the yeast *Rhodosporidium paludigenum* on postharvest decay and patulin accumulation in apples and pears. *J. Food Prot.* **2015**, *78*, 157–163. [CrossRef] [PubMed]

111. Zhu, R.; Feussner, K.; Wu, T.; Yan, F.; Karlovsky, P.; Zheng, X. Detoxification of mycotoxin patulin by the yeast *Rhodosporidium paludigenum*. *Food Chem.* **2015**, *179*, 1–5. [CrossRef] [PubMed]

112. Castoria, R.; Mannina, L.; Duran-Patron, R.; Maffei, F.; Sobolev, A.P.; De Felice, D.V.; Pinedo-Rivilla, C.; Ritieni, A.; Ferracane, R.; Wright, S.A. Conversion of the mycotoxin patulin to the less toxic desoxypatulinic acid by the biocontrol yeast *Rhodosporidium kratochvilovae* strain LS11. *J. Agric. Food Chem.* **2011**, *59*, 11571–11578. [CrossRef] [PubMed]

113. Tannous, J.; Snini, S.P.; El Khoury, R.; Canlet, C.; Pinton, P.; Lippi, Y.; Alassane-Kpembi, I.; Gauthier, T.; El Khoury, A.; Atoui, A.; et al. Patulin transformation products and last intermediates in its biosynthetic pathway, E- and Z-ascladiol, are not toxic to human cells. *Arch. Toxicol.* **2016**. [CrossRef] [PubMed]

114. Brackett, R.E.; Marth, E.H. Ascorbic acid and ascorbate cause disappearance of patulin from buffer solutions and apple juice. *J. Food Prot.* **1979**, *42*, 864–866. [CrossRef]

115. Fremy, J.M.; Castegnaro, M.J.; Gleizes, E.; De Meo, M.; Laget, M. Procedures for destruction of patulin in laboratory wastes. *Food Addit. Contam.* **1995**, *12*, 331–336. [CrossRef] [PubMed]

116. McKenzie, K.S.; Sarr, A.B.; Mayura, K.; Bailey, R.H.; Miller, D.R.; Rogers, T.D.; Norred, W.P.; Voss, K.A.; Plattner, R.D.; Kubena, L.F.; et al. Oxidative degradation and detoxification of mycotoxins using a novel source of ozone. *Food Chem. Toxicol.* **1997**, *35*, 807–820. [CrossRef]

117. Yazici, S.; Velioglu, Y.S. Effect of thiamine hydrochloride, pyridoxine hydrochloride and calcium-d-pantothenate on the patulin content of apple juice concentrate. *Nahrung/Food* **2002**, *46*, 256–257. [CrossRef]

118. Drusch, S.; Kopka, S.; Kaeding, J. Stability of patulin in a juice-like aqueous model system in the presence of ascorbic acid. *Food Chem.* **2007**, *100*, 192–197. [CrossRef]

119. Pohland, A.; Allen, R. Stability studies with patulin. *J. AOAC* **1970**, *53*, 688–691.
120. Cavallito, C.; Bailey, J. Preliminary note on the inactivation of antibiotics. *Science* **1944**, *100*, 390. [CrossRef] [PubMed]
121. Wu, T.S.; Liao, Y.C.; Yu, F.Y.; Chang, C.H.; Liu, B.H. Mechanism of patulin-induced apoptosis in human leukemia cells (HL-60). *Toxicol. Lett.* **2008**, *183*, 105–111. [CrossRef] [PubMed]
122. Lindroth, S.; von Wright, A. Detoxification of patulin by adduct formation with cysteine. *J. Environ. Pathol. Toxicol. Oncol.* **1990**, *10*, 254–259. [PubMed]
123. Karaca, H.; Sedat Velioglu, Y. Effects of some metals and chelating agents on patulin degradation by ozone. *Ozone Sci. Eng.* **2009**, *31*, 224–231. [CrossRef]
124. Food and Drug Administration (FDA). *Irradiation in the Production, Processing and Handling of Food*; U.S. Food and Drug Administration, Ed.; Code of Federal Regulations: Washington, DC, USA, 2000; pp. 71056–71058.
125. Health Canada. Ultraviolet Light Treatment of Apple Juice/Cider Using the Cidersure 3500. Available online: http://www.hc-sc.gc.ca/fn-an/gmf-agm/appro/dec85_rev_nl3-eng.php (accessed on 7 May 2017).
126. Assatarakul, K.; Churey, J.J.; Manns, D.C.; Worobo, R.W. Patulin reduction in apple juice from concentrate by UV radiation and comparison of kinetic degradation models between apple juice and apple cider. *J. Food Prot.* **2012**, *75*, 717–724. [CrossRef] [PubMed]
127. Tikekar, R.V.; Anantheswaran, R.C.; LaBorde, L.F. Patulin degradation in a model apple juice system and in apple juice during ultraviolet processing. *J. Food Process. Preserv.* **2014**, *38*, 924–934. [CrossRef]
128. Dong, Q.; Manns, D.C.; Feng, G.; Yue, T.; Churey, J.J.; Worobo, R.W. Reduction of patulin in apple cider by UV radiation. *J. Food Prot.* **2010**, *1*, 69–74. [CrossRef]
129. Zhu, Y.; Koutchma, T.; Warriner, K.; Shao, S.; Zhou, T. Kinetics of patulin degradation in model solution, apple cider and apple juice by ultraviolet radiation. *Food Sci. Technol. Int.* **2012**, *19*, 291–303. [CrossRef] [PubMed]
130. Sapers, G.M.; Hicks, K.B.; Philips, J.G.; Garzarella, L.; Pondish, D.L.; Matulaitis, R.M.; McCormack, T.J.; Sondey, S.M.; Seib, P.A.; El-Atawy, Y.S. Control of enzymatic browning in apple with ascorbic acid derivatives, polyphenol oxidase inhibitors, and complexing agents. *J. Food Sci.* **1989**, *54*, 997–1002. [CrossRef]
131. Zhu, Y.; Koutchma, T.; Warriner, K.; Zhou, T. Reduction of patulin in apple juice products by UV light of different wavelengths in the UVC range. *J. Food Prot.* **2014**, *77*, 963–971. [CrossRef] [PubMed]
132. Funes, G.J.; Gómez, P.L.; Resnik, S.L.; Alzamora, S.M. Application of pulsed light to patulin reduction in mcilvaine buffer and apple products. *Food Control* **2013**, *30*, 405–410. [CrossRef]
133. Gomez-Lopez, V.; Ragaert, P.; Debevere, J.; Devlieghere, V. Pulsed light for food decontamination: A review. *Trends Food Sci. Technol.* **2007**, *18*, 464–473. [CrossRef]
134. Avsaroglu, M.D.; Bozoglu, F.; Alpas, H.; Largeteau, A.; Demazeau, G. Use of pulsed-high hydrostatic pressure treatment to decrease patulin in apple juice. *High Press. Res.* **2015**, *35*, 214–222. [CrossRef]
135. San Martin, M.F.; Barbosa-Canovas, G.V.; Swanson, B.G. Food processing by high hydrostatic pressure. *Crit. Rev. Food Sci. Nutr.* **2002**, *42*, 627–645. [CrossRef] [PubMed]
136. Rendueles, E.; Omer, M.K.; Alvseike, O.; Alonso-Calleja, C.; Capita, R.; Prieto, M. Microbiological food safety assessment of high hydrostatic pressure processing: A review. *LWT Food Sci. Technol.* **2011**, *44*, 1251–1260. [CrossRef]
137. Hao, H.; Zhou, T.; Koutchma, T.; Wu, F.; Warriner, K. High hydrostatic pressure assisted degradation of patulin in fruit and vegetable juice blends. *Food Control* **2016**, *62*, 237–242. [CrossRef]
138. Patterson, M.F. Microbiology of pressure-treated foods. *J. Appl. Microbiol.* **2005**, *98*, 1400–1409. [CrossRef] [PubMed]
139. Schebb, N.H.; Faber, H.; Maul, R.; Heus, F.; Kool, J.; Irth, H.; Karst, U. Analysis of glutathione adducts of patulin by means of liquid chromatography (HPLC) with biochemical detection (BCD) and electrospray ionization tandem mass spectrometry (ESI-MS/MS). *Anal. Bioanal. Chem.* **2009**, *394*, 1361–1373. [CrossRef] [PubMed]
140. Park, D.L.; Troxell, T.C. U.S. Perspective on mycotoxin regulatory issues. *Adv. Exp. Med. Biol.* **2002**, *504*, 277–285. [PubMed]

![toxins logo] *toxins*

MDPI

Article

Beyond Ribosomal Binding: The Increased Polarity and Aberrant Molecular Interactions of 3-*epi*-deoxynivalenol

Yousef I. Hassan, Honghui Zhu, Yan Zhu and Ting Zhou *

Guelph Research and Development Centre, Agriculture and Agri-Food Canada, Guelph, ON N1G5C9, Canada; yousef.hassan@agr.gc.ca (Y.I.H.); honghui.zhu@agr.gc.ca (H.Z.); yan.zhu@agr.gc.ca (Y.Z.)
* Correspondence: ting.zhou@agr.gc.ca; Tel.: +1-226-217-8084; Fax: +1-519-829-2600

Academic Editor: Deepak Bhatnagar
Received: 3 August 2016; Accepted: 29 August 2016; Published: 8 September 2016

Abstract: Deoxynivalenol (DON) is a secondary fungal metabolite and contaminant mycotoxin that is widely detected in wheat and corn products cultivated around the world. Bio-remediation methods have been extensively studied in the past two decades and promising ways to reduce DON-associated toxicities have been reported. Bacterial epimerization of DON at the C3 carbon was recently reported to induce a significant loss in the bio-toxicity of the resulting stereoisomer (3-*epi*-DON) in comparison to the parental compound, DON. In an earlier study, we confirmed the diminished bio-potency of 3-*epi*-DON using different mammalian cell lines and mouse models and mechanistically attributed it to the reduced binding of 3-*epi*-DON within the ribosomal peptidyl transferase center (PTC). In the current study and by inspecting the chromatographic behavior of 3-*epi*-DON and its molecular interactions with a well-characterized enzyme, *Fusarium graminearum* Tri101 acetyltransferase, we provide the evidence that the C3 carbon epimerization of DON influences its molecular interactions beyond the abrogated PTC binding.

Keywords: deoxynivalenol; epimer; polarity; Tri101; molecular; interactions

---

1. Introduction

Deoxynivalenol (DON) is a secondary fungal metabolite and contaminant mycotoxin that is wildly detected in wheat and corn products cultivated around the world. It is estimated that more than four billion people are exposed to high levels of this metabolite especially in developing countries [1]. In mono-gastric animals, DON presence within feed is associated with adverse effects ranging from low animal-productivity, feed-refusal, to decreased weight-gain [2–4]. Similarly, DON causes both chronic and acute symptoms in humans with changes observed at the molecular and cellular levels [5,6]. Intestinal inflammations, increased susceptibility to infections, negative effects on the active transport of many nutrients [2,7], and brain homeostasis [8] are among the reported outcomes of DON exposure. Finally, DON is considered a major factor in the pathogenicity of Fusarium species against plants [9].

DON initial toxicity emerges from its ability to bind eukaryotic ribosomes hence inhibits protein biosynthesis [2,5,10]. DON binds to the 60S subunit and inside the A-site of the peptidyl transferase center (PTC) of the wildtype ribosome forming three hydrogen bonds: (a) The first is between the hydrogen of the uracil U2873 sugar and the oxygen of the epoxy group on C12 in DON; (b) The second is between the hydrogen of the guanine G2403 and the oxygen of the C15 group CH_2OH in DON; (c) Finally, the oxygen of uracil U2869 interacts with hydrogen of the C3 group in DON [11,12]. The attenuation or abrogation of such interactions as in the case of ribosomal protein L3 (RPL3) mutants lead to a semi-dominant resistance toward this toxin in plants and yeast [13–15].

The chemical structure of DON influences both its potency and the associated sensitivity-levels endured in different animal species [2]. Changes to DON structure (such as de-epoxidation, acetylation, and hydroxylation) can substantially influence its toxicity [2,16]. For example, the C15 carbon acetylation is report to increase the toxicological potency of DON in Caco-2 cells while modifications that take place at the C3 carbon are reported to decrease its toxicity by a number of orders of magnitude [2].

Many microorganisms isolated in the past two decades including bacteria, filamentary fungi, and yeasts were shown to reduce DON toxicity [17–19] either by physically binding DON (hence reducing its absorption) or through introducing enzymatic/chemical modifications at specific side groups [20–22]. Recently, a soil bacterium identified as *Devosia mutans* Strain 17-2-E-8, was reported to epimerize the -OH group at the C3 carbon within DON leading to a significant abrogation of DON cellular toxicity [23,24]. The diminished toxicity of this epimer, 3-*epi*-DON, was confirmed in different cell lines including Caco-2 and 3T3 fibroblasts in addition to B6C3F1 mouse model [25]. Other teams have reported similar results for 3-*epi*-DON [26,27].

In spite of the confirmed reduction of toxicity for 3-*epi*-DON using different in vivo models, a clear mechanistic explanation for the drastic reduction in potency connected with just one (-OH) group rotation is lacking. A recent study compared the toxicological profiles of DON, 3-*epi*-DON, and deepoxy-deoxynivalenol (DOM-1) and confirmed the toxicity reduction for both 3-*epi*-DON and DOM-1 attributing the reduction of 3-*epi*-DON cellular toxicity to the reduced hydrogen bonding of 3-*epi*-DON within the ribosomal peptidyl transferase center [12].

In this study, we first compare the chromatographic behavior of DON and 3-*epi*-DON and later explore the ability of both compounds to interact with a recombinant *Fusarium graminearum* Tri101 acetyltransferase [28] by tracking acetyl-groups transfer as an indicative marker of such interactions. While the reduced toxicity of 3-*epi*-DON can be plausibly attributed to the attenuated binding of 3-*epi*-DON to ribosomal peptidyl transferase centers, the obtained data show that the increased polarity of 3-*epi*-DON in addition to its aberrant molecular interactions are indeed factors that can contribute to the overall toxicity reduction possibly in a synergistic fashion.

2. Results

2.1. Fusarium graminearum Tri101 Acetyltransferase Expression and Purification

Expression levels of His-tagged FgTri101 in *E. coli* whole cell lysates are shown in Figure 1a. A clear band of the induced enzyme with the expected molecular weight (52 kDa) was present in the third lane after 0.5 mM IPTG addition overnight.

It was critical to start the induction within the OD_{600} = 0.6–0.8 range as higher values led to protein degradation and much lower yields. The induction took place also at 16 °C overnight even though such low temperatures were not always necessary and the induction could possibly proceed at 37 °C. The outcome of His-FgTri101 purification is shown in Figure 1b. The second elution led to the least contaminated yet highest enzyme yield. Final protein preparations were subdivided in aliquots (10–20 μL) and stored at −20 °C until usage.

2.2. Chromatographic Elution Patterns Suggested the Increased Polarity of 3-epi-DON

The epimerization of DON at the C3 group rendered 3-*epi*-DON to a higher polarity spectrum compared to the parental compound, DON. Under the same isocratic separation conditions (buffer composition, temperature, reverse phase column), DON was originally eluting at 13.2 min. (Figure 2A) however after the bacterial epimerization, 3-*epi*-DON shifted to the 8.9 min. (Figure 2B). The noted change in elution time, despite the similarity in structures/molecular weights/ionization status, was an indicative of changes of the polarity index associated with electron rearrangements. Figure 2C evidently presents DON and 3-*epi*-DON separation patterns for *Devosia mutans* 17-2-E-8 LB broth sample under the same analytical conditions (isocratic) and using a reverse-phase C12 Jupiter 4μ Proteo 90A HPLC column (commonly used for the purification of peptide mixtures) which reflected an altered interactions of 3-*epi*-DON with the column matrix. While chromatographic separations are not conventionally used to draw conclusions about polarity changes of mycotoxins, the data presented here collectively supported the notions that the observed changes in elution times are due to the increased polarity of 3-*epi*-DON. Our experimental approach was similar to the one that was suggested earlier by Namjesnik-Dejanovic and Cabaniss [29].

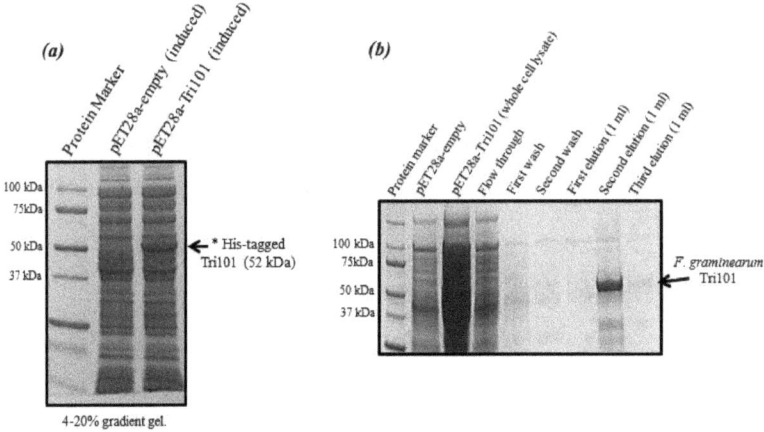

Figure 1. The induction and purification of His-FgTri101 expressed in *E. coli* BL21(DE3) cells. (**a**) His-FgTri101 from *Fusarium graminearum* was induced and overexpressed in BL21(DE3) cells; (**b**) The enzyme was purified with Clontech's His-TALON gravity columns by overnight incubation at 4 °C. All washes and subsequent elution steps were conducted at room temperature.

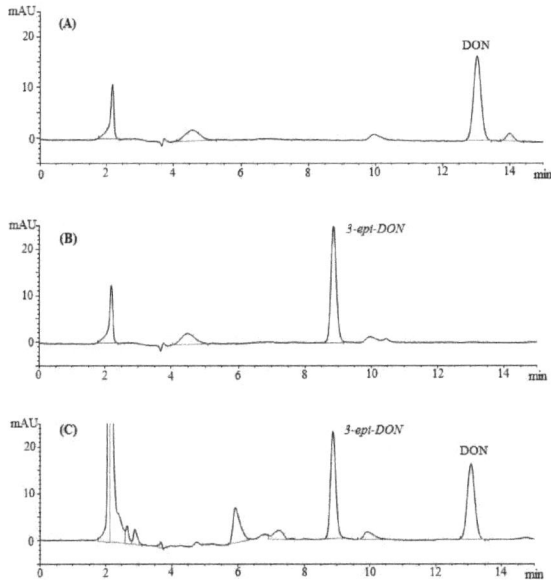

Figure 2. Chromatographic separation of DON and 3-*epi*-DON. Both DON and 3-*epi*-DON were injected on Jupiter 4μ Proteo 90A HPLC column. The results indicated that DON (less-polar) eluted in a longer time frame (**A**) in comparison to 3-*epi*-DON (**B**) under the same isocratic separation conditions. Section (**C**) represents the injection of both compounds at the same run.

2.3. Fusarium graminearum Tri101 Carried DON Acetylation Both in Vitro and in Vivo but Not 3-epi-DON

As mentioned below in the Materials and Methods section, FgTri101 was used to track the molecular interactions of DON and 3-*epi*-DON as it provides means of measuring such interactions indirectly and quantitatively. Both in vivo and in vitro models were used to minimize any aberrant interpretations resulting from DON and 3-*epi*-DON abilities to adsorb to bacterial cell walls/proteins hence reduce their availability within cellular matrixes/reaction mixes.

In short, FgTri101 was able to interact with DON both in vivo and in vitro as indicated by the C3 carbon acetylation. The in vivo interactions between DON and *E. coli* BL21(DE3) cells expressing His-FgTri101 led to a significant decrease of DON recovery (Figure 3a) in test tubes in comparison to the tubes that harbored only the LB broth supplemented with 50 µg/mL DON (control) or the empty pET-28a(+) vector (negative control) (Figure 3a).

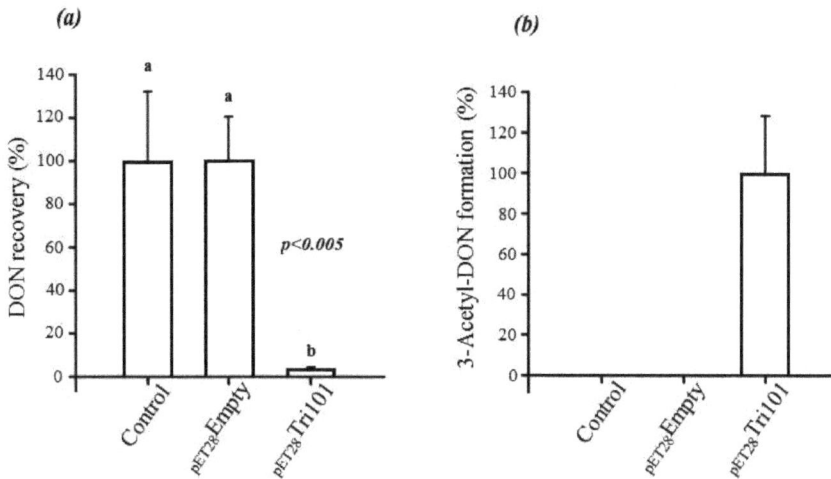

Figure 3. Changes in DON concentrations in *E. coli* BL21(DE3) cells overexpressing *Fusarium graminearum* Tri101. (**a**) A significant decrease in DON recovery was noted for the construct that encoded for the Tri101 enzyme in comparison to the control or the empty vector. The noted decrease in DON was confirmed to be due to the acetylation of DON by His-FgTri101 (**b**). The above tendency was observed in two separate experiments, each with three replications, and the different letters signify significant differences between the means according to Fisher's Least Significant Difference (LSD) test ($p < 0.005$).

Furthermore, FgTri101 presence with DON led to the accumulation of the 3ADON metabolite as shown in Figure 3b. No 3ADON was detected in tubes harboring LB broth (with 50 µg/mL DON) nor the empty pET-28a(+) vector (negative control) alone indicating that the accumulation was the result of specific DON/FgTri101 interactions (acetylation).

On the other hand, incubation of 3-*epi*-DON with either *E. coli* BL21(DE3) cells expressing His-FgTri101 or BL21(DE3) cells with empty pET-28a(+) vector did not lead to any changes in 3-*epi*-DON recovery levels after methanol-extractions (Figure 4a) or to the detection of 3ADON presence in these reactions (data not shown) as a result of 3-*epi*-DON and FgTri101 interactions.

In order to assure that *E. coli* BL21(DE3) cells incubated with both DON and 3-*epi*-DON were all expressing FgTri101, a SDS-PAGE separation of total cell-lysates was conducted and the results are shown in Figure 4b. The three triplicates of DON/BL21(DE3)-FgTri101 cells expressed His-tagged FgTri101 in an equal fashion to 3-*epi*-DON/BL21(DE3)-FgTri101 cells offering the chance for both metabolites to interact with FgTri101 equally and forming 3ADON (if possible).

(a)   (b)

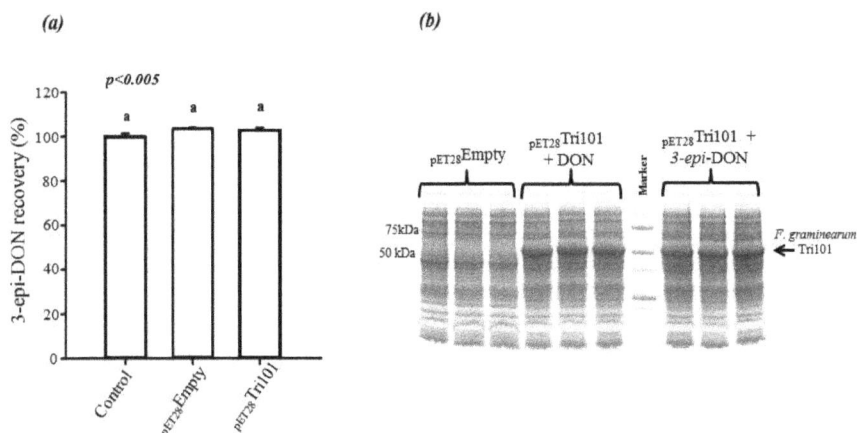

Figure 4. Changes in 3-*epi*-DON concentrations in *E. coli* BL21(DE3) cells overexpressing *Fusarium graminearum* Tri101. (**a**) No changes in 3-*epi*-DON recoveries were noticed for the construct that encoded the Tri101 enzyme in comparison to the control or empty vector. The above tendency was observed in two separate experiments, each with three replications, and the different letters signify significant differences between the means according to Fisher's Least Significant Difference (LSD) test ($p < 0.005$); (**b**) To confirm that the above cells where all over-expressing His-FgTri101, bacterial pellets were collected, lysed, and ran on SDS-PAGE gels (4%–20%) before Coomassie Blue staining. The overexpression levels of His-FgTri101 were all evident in both DON and 3-*epi*-DON treatments.

In an identical fashion, the in vitro reactions assembled as stated later within the Materials and Methods section yielded parallel observations (Table 1). No 3ADON was detected using LC-MS/MS analytical method when 3-*epi*-DON was incubated with purified his-tagged FgTri101 (tube #6) nor when either DON or 3-*epi*-DON were incubated with FgTri101 in the absence of acetyl-CoA (tubes #2 and #5). The 3ADON metabolite was only detectable in the presence of DON, FgTri101, and acetyl-CoA (tube #3).

Table 1. The in vitro testing of FgTri101 ability to interact with DON and 3-*epi*-DON, respectively. Reaction mixes were assembled in triplicates as mentioned under the Materials and Methods section. Methanol-extracted aliquots were analyzed for 3ADON using the LC-MS/MS approach. Tri101 supported the acetylation of DON but not 3-*epi*-DON.

Component/Tube	(1)	(2)	(3)	(4)	(5)	(6)
DON	+	+	+	−	−	−
3-*epi*-DON	−	−	−	+	+	+
recombinant His-FgTri101	−	+	+	−	+	+
acetyl-CoA	−	−	+	−	−	+
3ADON formation/detection	No	No	Yes	No	No	No

2.4. 3-epi-DON Cross-Reactivity with DON-Monoclonal Antibodies Suggested Similar Structural Configurations/Side-Group Rearrangements of Both Compounds

A commercial monoclonal antibody that was generated against DON and is widely used to specifically capture this toxin from complex food matrixes/biological samples was tested for its ability to interact with 3-*epi*-DON. The assumption was that, if both compounds were captured and retained equally on the DONtest immunoaffinity column, then it is plausible to hypothesize that the two metabolites display similar epitopes and share close structural configurations needed for such monoclonal antibody recognition/interactions. In other words, despite the structural stress induced by the C3 carbon epimerization of DON (and the correlated electrons rearrangements to stabilize the resulting isoform), the two chemicals can be assumed to endure a similar epitope/3D chemical configuration in case they were both recognized/bound by the same monoclonal antibodies.

Satisfyingly enough, the DONtest immunoaffinity column showed the ability to bind 3-*epi*-DON in a comparable fashion to DON. When the methanol eluted compounds were analyzed using HPLC methods described below, both metabolites were detectable as indicated in Figure 5.

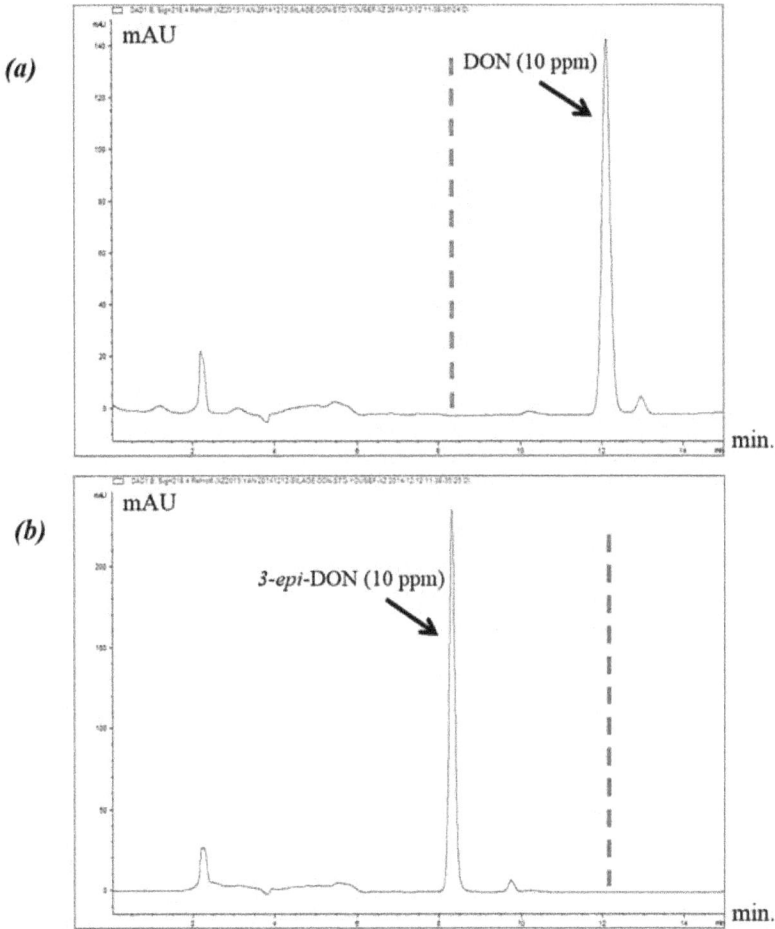

Figure 5. Both DON and 3-*epi*-DON interacted with the same conjugated commercial DON-monoclonal antibody in a similar fashion. The chromatographic separation of DON and 3-*epi*-DON after the affinity column purification step (DONtest) and injection on Jupiter 4μ Proteo 90A HPLC column is shown. The results indicated that DON (**a**) was captured in a similar fashion to 3-*epi*-DON (**b**).

The obtained results suggested that most of the 3D structural features of DON were still indeed conserved after carbon C3 epimerization. Such conserved features were responsible for the observed affinity between the tested DON-monoclonal antibody and 3-*epi*-DON in a similar fashion to the ones noticed earlier between DON with that exact monoclonal antibody.

2.5. Molecular Visualization of FgTri101/DON Interactions Highlighted the Role of Some Non-Polar Amino Acids

As mentioned in the Materials and Methods section, the FgTri101 enzyme crystal structure was solved by Garvey et al. [28] showing the key amino acids that occupy the active pocket of the enzyme. Among these amino acids are: Leu16, TRP380, SER378, SER379, and the catalytic His156 (Figure 6).

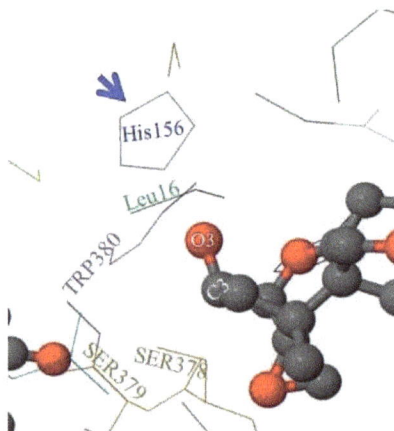

Figure 6. The molecular visualization of DON within *F. graminearum* Tri101 binding pocket. The C3 carbon (and attached -OH group) in DON (less-polar) is facing a non-polar amino acid (Leu16). In the case of 3-*epi*-DON (higher-polarity), the same group will be facing a matrix of non-polar (Leu16) and neutral (Trp380) amino acid residues.

When DON occupies the enzyme's active pocket (and taking the steric hindrance in consideration), the C3-O3 group extends to establish the interactions with the nearby Leu16 and His156 and shows higher affinity toward the non-polar Leu16 residue (Figure 6). In contrast, when 3-*epi*-DON occupied the same space: (a) first, it elicited a reduce affinity toward Leu16 (an extension to the previously observed chromatographic behavior on the reverse phase column) and (b) second, the C3-O3 bond was slightly leaning away in the opposite direction. Altogether, this leads to decreased dwelling/interaction times between FgTri101 and 3-*epi*-DON in comparisons with the original interactions noticed for FgTri101 and DON. Such reduced interactions between 3-*epi*-DON and other molecular targets (including ribosomal binding sites) were reported to be the force that dictates 3-*epi*-DON reduced toxicity [12].

Furthermore, the molecular modeling and visualization of 3-*epi*-DON and FgTri101 interactions indicated while this metabolite do indeed fit within the Tri101 active site without any major conformational perturbations, the 3-OH group in 3-*epi*-DON is sufficiently disoriented to prevent the acetylation reaction from proceeding.

3. Discussion

Food and feed research is driven nowadays by a strong support from current consumers to adapt green biotechnological solutions [21,30]. The field is increasingly exploring new avenues within the bioremediation arena to mitigate the growing problems of mycotoxin contamination. DON presence in the food/feed chain is still considered as a challenging task to address as no effective biological means of control are optimized yet. Most recently, a number of research teams have reported the use of bacterial isolates to bio-transform DON to other metabolites with a significant reduction in the associated toxicity [2,18,31–34]. Among the reported bacterial metabolites within this regard were DOM-1, 16-hydroxy-deoxynivalenol, and 3-*epi*-DON. A clear mechanistic understanding of the phenomena behind the diminished toxicity of such metabolites is a must in order to avoid any unpredicted scenarios such as the case of previously reported masked mycotoxins [35].

To fully comprehend the reduction of DON toxicity upon C3 carbon epimerization [24,26,33], we tested the ability of 3-*epi*-DON to adapted/interact with the active pocket of a recombinant Tri101 enzyme cloned from *F. graminearum*. The Tri101 enzyme which represents a large group of acetyltransferases that naturally interact with DON, was expressed and purified in an exogenous host, namely *E. coli* BL21(DE3). The ability of the recombinant FgTri101 enzyme to transfer acetyl groups from acetyl-CoA to DON was used in this study as an indicative of such interactions while the absence of acetylated by-products, judged by LC-MS/MS analysis, hinted toward aberrant enzyme-substrate interactions. Other parameters including substrate polarity and 3D structural rearrangements were investigated too.

The obtained observations here collectively supported the highlighted role of C3 carbon in establishing the molecular interactions of DON with different cellular targets. This carbon plays a pivotal role in DON toxicity which emerges mainly from DON's ability to bind to ribosomal RNA hence inhibiting protein bio-synthesis and suppressing host-defense mechanisms. A decent amount of scientific literature was dedicated for understanding the multifaceted relations between DON interactions with cellular targets including ribosomes, and overall bio-toxicity [2,9]. The conserved role of C3 carbon (modification and orientation) in DON bio-potency is not surprising. In fact, some of the mitigation approaches that were suggested earlier to reduce DON efficacy as plant immuno-inhibitor where to enzymatically modify this group through acetylation and/or glycosylation [27,36,37].

By far the most prominent interaction between DON and the PTC is mediated by the 3-OH group, which coordinates a structural Mg-Ion. Modeling 3-*epi*-DON/PTC shows that the epimerized 3-OH group is not in place to form reasonable interactions, and the strongest interaction (with the Mg-Ion) is broken. While the epoxide group is still in close contact with U2873 of the rRNA but due to the isomeric changes, 3-*epi*-DON cannot establish any bonds with U2869 [12].

In Tri101, 3-*epi*-DON could be modelled quite convincingly into the DON binding site, and there were no obvious clashes. The epimerized 3-OH group was out of reach for acetylation. This in addition to the increased polarity clearly shows why Tri101 cannot work with 3-*epi*-DON as a substrate.

Furthermore, the obtained data indicated that 3-*epi*-DON was still capable of interacting with a commercial DON-specific monoclonal antibody rolling out the assumption that a mere steric hindrance/global 3D structural rearrangements are the base for the observed reduced interactivity (and toxicity) associated with C3 epimerization. Collectively, the provided data suggested that the reduced toxicity of 3-*epi*-DON is due to a combination of reduced interactions and increased polarity index which both affect the observed interactions of 3-*epi*-DON at the molecular level as noted also by Pierron et al. [12].

The reported differences in polarity can be advantageous from the analytical point of view as they allow to validly distinguish between the two stereoisomers based on their separation patterns. The interactions of DON and 3-*epi*-DON with reverse phase HPLC columns packed with non-polar stationary phases clearly can be utilized to analyze both metabolites under isocratic conditions. The retention times of less-polar eluents (DON in this case) are substantially longer than the more-polar molecules (such as 3-*epi*-DON).

Since chemical modifications do in some cases affect the ability of certain chemicals of crossing cellular membranes as in the case of 15-acetyldeoxynivalenol (15ADON) which shows a higher permeability in human Caco-2 intestinal cells in comparison to DON or 3ADON [38] and in order to minimize the chances of aberrant interpretations, we used both in vitro and in vivo approaches to address the molecular intractability questions with Tri101. Both models yielded identical outcomes.

In conclusion, the presented work here provides the base for a better understanding of the correlations between DON, 3-*epi*-DON, role of C3 carbon epimerization, and host-toxicity at the molecular level. Further studies that securitize the effect of C3 carbon epimerization on the transport/absorbability of 3-*epi*-DON in polarized eukaryotic cells (such as Caco-2) can also provide great insights about the kinetics of this microbial metabolite in animal/human gastrointestinal tracts.

4. Materials and Methods

4.1. Bacterial Strains, Plasmids, and Chemicals

LB broth (Cat. #244620) was obtained from BD (Franklin Lakes, NJ, USA). Kanamycin (Cat. #60615) and acetyl coenzyme A sodium salt (Cat. #A2056) were ordered from Sigma-Aldrich (Oakville, ON, Canada). Deoxynivalenol (Cat. #51481-10-8) was obtained from TripleBond (Guelph, ON, Canada) and stored at −20 °C freezer until usage. 3-*epi*-DON was prepared and purified as described earlier [39]. BL21(DE3) competent cells (Cat. #69450-3) and pET-28a(+) vector (Cat. #69864-3) were both purchased from Novagen-EMD Millipore (Etobicoke, ON, Canada).

4.2. Recombinant F. graminearum Tri101 Acetyltransferase Cloning, Expression, and Purification

Fusarium graminearum Tri101 acetyltransferase (FgTri101) (GenBank # AB000874, 1356 bp) was selected to investigate DON and 3-*epi*-DON interactions for multiple reasons: (a) first, substrate-enzyme interactions can be measured indirectly through the determination of the acetylated-C3-DON in proportion to the un-modified original toxin; (b) second, the enzyme crystal structure was solved earlier by Garvey et al. [28] and clear insights about DON-binding site and the catalytic mechanism are already established; (c) third, the reported FgTri101 enzyme shows structural plasticity, particularly related to substrate-binding within the catalytic pocket. This plasticity is manifested by the ability of T-2 toxin to bind to the same site where DON was shown to bind [28]. Based on the bulkiness of T-2 structure (Figure 7a) (type A trichothecene) in comparison to DON or 3-*epi*-DON (Figure 7b,c), it can be logically assumed with confidence that steric hindrance that results from DON to 3-*epi*-DON epimerization should not be the sole factor that dictates 3-*epi*-DON interactions with FgTri101.

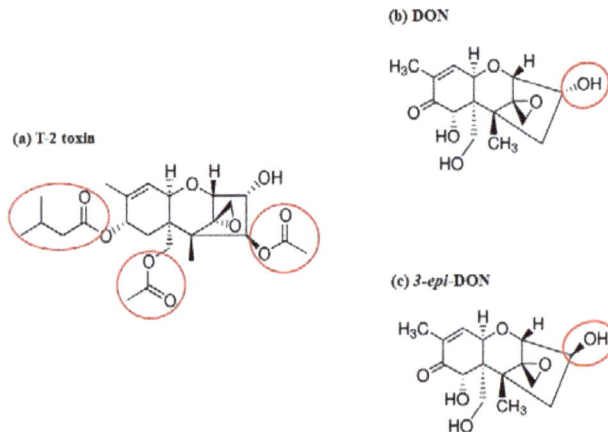

Figure 7. T-2 (**a**) and Deoxynivalenol (DON) (**b**) toxins were both reported to complex with FgTri101 despite the bulkiness of T-2 toxin in comparison to DON [28]. Due to the noted plasticity of FgTri101 active pocket, it is logical to assume that 3-*epi*-DON (**c**) can also bind within the enzyme's catalytic cavity without any major structural restrain.

FgTri101 was cloned in pET-28a(+) vector after linearizing the vector with *Nde*I and *Bam*HI restriction digests. gBlock fragments encoding FgTri101 sequence (optimized for *E. coli* expression) were ordered from IDT-DNA (Integrated DNA Technologies, Inc.; Coralville, IA, USA) and In-Fusion cloning reactions (In-Fusion HD Cloning Plus CE; Cat. #638916; TaKaRa-Clontech; Mountain View, CA, USA) were used to assemble the fragments according to the supplier's protocol. His-tag was fused to the enzyme's N-terminus. The resulting pET-28a(+)-His-FgTri101 kanamycin resistant vector was transformed into TOP10 competent cells (Life Technologies Inc., Burlington, ON, Canada) and selected on *Kan*+ plates before sending for sequencing at the University of Guelph Laboratory Services.

The sequence-verified plasmid, pET-28a(+)-His-FgTri101, was transformed into *E. coli* BL21(DE3) competent cells which were grown at 37 °C for 5 h (OD_{600}= 0.6–0.8) then transferred to 300 mL LB broth supplemented with kanamycin before inducing overnight with 0.5 mM IPTG (as final concentration). The induced culture was kept at 16 °C and 150 rpm until harvesting by spinning at 4500 rpm in swing-bucket type centrifuge. Collected pellets were subjected to a gentle protein extraction protocol using Qproteome bacterial protein preparation kit (Cat. #37900; Qiagen, Toronto, ON, Canada) according to the supplier's recommendations. The His-tagged enzyme was purified by binding overnight to HisTALON Gravity columns (Cat. #ST0165; TaKaRa-Clontech), washing, and elution in 1 mL buffer increments. The purified enzyme was sub-aliquoted and stored at −20 °C until usage.

4.3. The in Vivo Acetyl-Transfer Assays

In the in vivo assays, *E. coli* BL21(DE3) cells either harboring empty pET-28a(+) or pET-28a(+)-His-FgTri101 vectors were incubated with either DON or 3-*epi*-DON separately. The resulting 3-acetyl-DON (3ADON) formation (if any) and reductions of DON/3-*epi*-DON concentrations were tracked using LC-MS/MS approach as described later. In short, tubes (in triplicates) containing LB broth supplemented with 50 ppm DON or 50 ppm 3-*epi*-DON were inoculated with 50 µL of overnight cultures of *E. coli* BL21(DE3) cells either harboring empty pET-28a(+) or pET-28a(+)-His-FgTri101. The cultures were kept growing on an orbital shaker (150 rpm) at 37 °C for 5 h (culture OD_{600} = 0.6–0.8) then were induced using IPTG (0.5 mM as final concentration) at 16 °C for another 16–20 h before spinning at 10,000 rpm and collecting the depleted broth and bacterial pellets for later analysis. Bacterial pellets were subject to SDS-PAGE analysis and Coomassie Brilliant Blue staining to confirm protein induction/expression while the depleted culture broth was extracted with 95% HPLC-grade methanol, filter-sterilized, and used as described later for DON, 3-*epi*-DON, and 3ADON analysis.

4.4. The in Vitro Acetyltransferase Enzymatic Assay

The purified His-FgTri101 enzyme prepared as described above was used as a model cellular target to investigate DON and 3-*epi*-DON in vitro interactions. In essence, working reaction mixes were prepared as suggested by Garvey et al. [28] through combining acetyl-CoA (0.62 mM as final concentration), 100 µg/mL DON or 3-*epi*-DON, 3.25 µg of His-FgTri101 enzyme in 50 mM Tris-HCl buffer (pH = 8). Mixes (*n* = 5) were kept at 28 °C overnight and reactions were terminated by adding absolute methanol in an equal volume. DON, 3-*epi*-DON, and 3ADON concentrations were measured using LC-MS/MS as described below. Both DON and 3-*epi*-DON recovery percentages were calculated in addition to 3ADON accumulation (if any).

4.5. Using Immunoaffinity Binding to Delineate DON and 3-epi-DON Structural Overlaps

Monoclonal antibodies can characteristically recognize a single, specific epitope that is displayed on a protein or a chemical. Such highly specific antibodies can be used generally speaking to track structural changes [40]. In the following approach we made use of this concept to tentatively test the ability of 3-*epi*-DON to bind to the same commercial monoclonal antibody generated against DON in order to gain insights about any structural changes associated with the epimerization of C3 carbon. An immunoaffinity column, DONtest HPLC obtained from VICAM (Cat. #G1005; MA, USA), packed with a DON-specific proprietary monoclonal antibody was tested for the ability to bind 3-*epi*-DON. In short, 1 mL of DON standard (1 µg/mL), 3-*epi*-DON (1 µg/mL), or combinations were loaded onto the pre-set columns and pressure was applied to achieve 1–2 drops/second flow rates. Columns were then washed with 5 mL of pure water and eluted with 1 mL of HPLC-grade methanol according to the supplier's recommendations. The eluted DON and 3-*epi*-DON were dried either under nitrogen-streams or by using a SpeedVac concentrator (SPD2010, Thermo Scientific, Waltham, MA, USA) to be re-suspended later in 0.5 mL of 10% acetonitrile in water. Finally, samples were passed through 0.45 µm filters (Whatman, Florham Park, NJ, USA; Cat. #6765-1304) and analyzed using HPLC as described later.

4.6. HPLC Analysis

An Agilent HPLC system (1200 Series; Palo Alto, CA, USA) equipped with a quaternary pump, an inline degasser, an auto-sampler, and a diode array detector (DAD) was used in this study. A C12 reverse phase Phenomenex Jupiter 4μ Proteo 90A column (250 × 4.6 mm, Cat. #00G-4396-E0) coupled with a C18 guard column (Torrance, CA, USA) was used for the reported separation. A mobile phase consisting of two solvents (A: acetonitrile and B: H_2O) mixed in 1:9 ratio by volume was used for elution. Isocratic conditions with flow rates controlled at 1 mL/min. were utilized with an injection volumes set to 50 μL. The total running time was 15 minutes while the detection wavelength was set at 218 nm.

4.7. LC-MS/MS Determination of DON and 3-epi-DON

Since mass spectrometry (MS) detectors have much better sensitivity levels, it was the preferred method to track DON, 3-*epi*-DON, and their acetylated products (if any). Liquid chromatography-tandem mass spectrometry (LC-MS/MS) was used to simultaneously determine DON and 3-*epi*-DON in methanol-extracted samples similar to what was described earlier by Young et al. [19]. In short, the system was composed of a Shimadzu HPLC system (Shimadzu, Kyoto, Japan) equipped with an UV/VIS photodiode array detector connected to a triple quadrupole IONICS 3Q Molecular Analyzer (IONICS, Bolton, ON, Canada). An Agilent (Agilent Technologies, Santa Clara, CA, USA) ZORBAX SB-C18 column (2.1 × 100 mm, 3.5 μm) was used for separation. Two mobile phases, solvent A (99.9% H_2O + 0.1% formic acid) and solvent B (99.9% MeOH + 0.1% formic acid) were used. The chromatographic elution conditions were as the following: 0−1 min, isocratic 10% B; 1–10 min, gradient 10% to 80% B; 10–12 min, isocratic 80% B; 12–13 min gradient 80% to 10% B; 13–18 min, isocratic 10% B. The column temperature was set to 25 °C, the flow rate was controlled at 0.4 mL/min, and the injection volume was 10 μL, respectively. The ESI positive mode was used for data collection. Before carrying out the actual analyses, the system was optimized using standards at very low concentrations. Quantification was accomplished at multiple reaction monitoring (MRM) mode by monitoring the transition pairs of m/z 296.52 (molecular ion)/ 249.13 (fragment ion) for DON and 3-*epi*-DON and 338.83/231.13 for 3ADON. The fragment ion of each compound was selected based on the ion which has the highest sensitivity among the fragment ions of that compound. The dwelling time for MRM data collection was 200 ms. Peak areas obtained at the MRM mode were used to establish the calibration curves through least-squares regressions with R^2 values ≥ 0.99.

4.8. Statistical Analysis and Molecular Visualization

All statistical analyses were conducted using SigmaPlot (version 12.5, Systat Software, Inc.; San Jose, CA, USA). One-way analysis of variance (ANOVA) was applied to the reported results followed by Fisher's Least Significant Difference (LSD) test. Crystal structures of *F. graminearum* TRI101 complexed with Coenzyme A, Deoxynivalenol (#3B2S) and T-2 mycotoxin (#2RKV) were obtained from RCSB-Protein Data Bank. Both PyMOL (The PyMOL Molecular Graphics System, Version 1.7.4; Schrödinger LLC; Cambridge, MA, USA) and JSmol (https://sourceforge.net/projects/jsmol/) were used to visualize and manipulate the 3D structure of the reported complexes.

Acknowledgments: Authors would like to thank Agriculture and Agri-Food Canada (AAFC) for the financial support of this research work and Xiu-Zhen Li (AAFC) and Marcus Hartmann (Department of Protein Evolution, Max Planck Institute for Developmental Biology, Germany) for their valuable technical advice and in-depth scientific discussions.

Author Contributions: Design of the work: Y.I.H. and T.Z. Conducting experiments: Y.I.H., H.Z., and Y.Z. Interpretation of data: Y.I.H., H.Z., Y.Z., and T.Z. Drafting the work: Y.I.H. and T.Z. Final approval: Y.I.H., H.Z., Y.Z., and T.Z.

Conflicts of Interest: The authors declare no conflict of interest.

References

1.	Sobrova, P.; Adam, V.; Vasatkova, A.; Beklova, M.; Zeman, L.; Kizek, R. Deoxynivalenol and its toxicity. *Interdisciplin. Toxicol.* **2010**, *3*, 94–99. [CrossRef] [PubMed]

2. Hassan, Y.I.; Watts, C.; Li, X.Z.; Zhou, T. A novel peptide-binding motifs inference approach to understand deoxynivalenol molecular toxicity. *Toxins* **2015**, *7*, 1989–2005. [CrossRef] [PubMed]

3. Ghareeb, K.; Awad, W.A.; Bohm, J.; Zebeli, Q. Impacts of the feed contaminant deoxynivalenol on the intestine of monogastric animals: Poultry and swine. *J. Appl. Toxicol.* **2015**, *35*, 327–337. [CrossRef] [PubMed]

4. Pinton, P.; Oswald, I.P. Effect of deoxynivalenol and other Type B trichothecenes on the intestine: A review. *Toxins* **2014**, *6*, 1615–1643. [CrossRef] [PubMed]

5. Pestka, J.J.; Zhou, H.R.; Moon, Y.; Chung, Y.J. Cellular and molecular mechanisms for immune modulation by deoxynivalenol and other trichothecenes: Unraveling a paradox. *Toxicol. Lett.* **2004**, *153*, 61–73. [CrossRef] [PubMed]

6. Wang, Z.; Wu, Q.; Kuca, K.; Dohnal, V.; Tian, Z. Deoxynivalenol: Signaling pathways and human exposure risk assessment—an update. *Arch. Toxicol.* **2014**, *88*, 1915–1928. [CrossRef] [PubMed]

7. Maresca, M. From the gut to the brain: Journey and pathophysiological effects of the food-associated trichothecene mycotoxin deoxynivalenol. *Toxins* **2013**, *5*, 784–820. [CrossRef] [PubMed]

8. Razafimanjato, H.; Benzaria, A.; Taieb, N.; Guo, X.J.; Vidal, N.; Di Scala, C.; Varini, K.; Maresca, M. The ribotoxin deoxynivalenol affects the viability and functions of glial cells. *Glia* **2011**, *59*, 1672–1683. [CrossRef] [PubMed]

9. Audenaert, K.; Vanheule, A.; Hofte, M.; Haesaert, G. Deoxynivalenol: A major player in the multifaceted response of fusarium to its environment. *Toxins* **2014**, *6*, 1–19. [CrossRef] [PubMed]

10. Pan, X.; Whitten, D.A.; Wilkerson, C.G.; Pestka, J.J. Dynamic changes in ribosome-associated proteome and phosphoproteome during deoxynivalenol-induced translation inhibition and ribotoxic stress. *Toxicol. Sci.* **2014**, *138*, 217–233. [CrossRef] [PubMed]

11. Garreau de Loubresse, N.; Prokhorova, I.; Holtkamp, W.; Rodnina, M.V.; Yusupova, G.; Yusupov, M. Structural basis for the inhibition of the eukaryotic ribosome. *Nature* **2014**, *513*, 517–522. [CrossRef] [PubMed]

12. Pierron, A.; Mimoun, S.; Murate, L.S.; Loiseau, N.; Lippi, Y.; Bracarense, A.F.; Schatzmayr, G.; He, J.W.; Zhou, T.; Moll, W.D.; et al. Microbial biotransformation of DON: Molecular basis for reduced toxicity. *Sci. Rep.* **2016**, *6*, 29105. [CrossRef] [PubMed]

13. Afshar, A.S.; Mousavi, A.; Majd, A.; Renu; Adam, G. Double mutation in tomato ribosomal protein l3 cdna confers tolerance to deoxynivalenol (DON) in transgenic tobacco. *Pak. J. Biol. Sci.* **2007**, *10*, 2327–2333. [CrossRef] [PubMed]

14. Adam, G.; Mitterbauer, R.; Raditschnig, A.; Poppenberger, B.; Karl, T.; Goritschnig, S.; Weindorfer, H.; Glossl, J. Molecular mechanisms of deoxynivalenol resistance in the yeastsaccharomyces cerevisiae. *Mycotoxin Res.* **2001**, *17* (Suppl. 1), 19–23. [CrossRef] [PubMed]

15. Mitterbauer, R.; Poppenberger, B.; Raditschnig, A.; Lucyshyn, D.; Lemmens, M.; Glossl, J.; Adam, G. Toxin-dependent utilization of engineered ribosomal protein L3 limits trichothecene resistance in transgenic plants. *Plant Biotechnol. J.* **2004**, *2*, 329–340. [CrossRef] [PubMed]

16. Danicke, S.; Hegewald, A.K.; Kahlert, S.; Kluess, J.; Rothkotter, H.J.; Breves, G.; Doll, S. Studies on the toxicity of deoxynivalenol (DON), sodium metabisulfite, don-sulfonate (DONS) and de-epoxy-don for porcine peripheral blood mononuclear cells and the intestinal porcine epithelial cell lines IPEC-1 and IPEC-J2, and on effects of DON and DONS on piglets. *Food Chem. Toxicol.* **2010**, *48*, 2154–2162. [PubMed]

17. Hassan, Y.I.; Bullerman, L.B. Cell-surface binding of deoxynivalenol to *Lactobacillus paracasei* subsp. *tolerans* isolated from sourdough starter culture. *J. Microbiol. Biotechnol. Food Sci.* **2013**, *2*, 2323–2325.

18. Islam, R.; Zhou, T.; Young, J.C.; Goodwin, P.H.; Pauls, K.P. Aerobic and anaerobic de-epoxydation of mycotoxin deoxynivalenol by bacteria originating from agricultural soil. *World J. Microbiol. Biotechnol.* **2012**, *28*, 7–13. [CrossRef] [PubMed]

19. Young, J.C.; Zhou, T.; Yu, H.; Zhu, H.; Gong, J. Degradation of trichothecene mycotoxins by chicken intestinal microbes. *Food Chem. Toxicol.* **2007**, *45*, 136–143. [CrossRef] [PubMed]

20. Volkl, A.; Vogler, B.; Schollenberger, M.; Karlovsky, P. Microbial detoxification of mycotoxin deoxynivalenol. *J. Basic Microbiol.* **2004**, *44*, 147–156. [CrossRef] [PubMed]

21. Zhu, Y.; Hassan, Y.I.; Watts, C.; Zhou, T. Innovative technologies for the mitigation of mycotoxins in animal feed and ingredients—A review of recent patents. *Anim. Feed Sci. Technol.* **2016**, *216*, 19–29.

22. Vanhoutte, I.; Audenaert, K.; de Gelder, L. Biodegradation of mycotoxins: Tales from known and unexplored worlds. *Front. Microbiol.* **2016**, *7*, 561. [CrossRef] [PubMed]

23. Hassan, Y.I.; Lepp, D.; He, J.; Zhou, T. Draft genome sequences of devosia sp. Strain 17-2-E-8 and devosia riboflavina strain IFO13584. *Genome Announc.* **2014**, *2*, e00994-14. [CrossRef] [PubMed]

24. He, J.W.; Hassan, Y.I.; Perilla, N.; Li, X.Z.; Boland, G.J.; Zhou, T. Bacterial epimerization as a route for deoxynivalenol detoxification: The influence of growth and environmental conditions. *Front. Microbiol.* **2016**, *7*, 572. [CrossRef] [PubMed]

25. He, J.W.; Bondy, G.S.; Zhou, T.; Caldwell, D.; Boland, G.J.; Scott, P.M. Toxicology of 3-*epi*-deoxynivalenol, a deoxynivalenol-transformation product by devosia mutans 17-2-e-8. *Food Chem. Toxicol.* **2015**, *84*, 250–259. [CrossRef] [PubMed]

26. Ikunaga, Y.; Sato, I.; Grond, S.; Numaziri, N.; Yoshida, S.; Yamaya, H.; Hiradate, S.; Hasegawa, M.; Toshima, H.; Koitabashi, M.; et al. Nocardioides sp. Strain wsn05-2, isolated from a wheat field, degrades deoxynivalenol, producing the novel intermediate 3-*epi*-deoxynivalenol. *Appl. Microbiol. Biotechnol.* **2011**, *89*, 419–427. [CrossRef] [PubMed]

27. Karlovsky, P. Biological detoxification of the mycotoxin deoxynivalenol and its use in genetically engineered crops and feed additives. *Appl. Microbiol. Biotechnol.* **2011**, *91*, 491–504. [CrossRef] [PubMed]

28. Garvey, G.S.; McCormick, S.P.; Rayment, I. Structural and functional characterization of the TRI101 trichothecene 3-*O*-acetyltransferase from *Fusarium sporotrichioides* and *Fusarium graminearum*: Kinetic insights to combating fusarium head blight. *J. Biol. Chem.* **2008**, *283*, 1660–1669. [CrossRef] [PubMed]

29. Namjesnik-Dejanovic, K.; Cabaniss, S.E. Reverse-phase hplc method for measuring polarity distributions of natural organic matter. *Environ. Sci. Technol.* **2004**, *38*, 1108–1114. [CrossRef] [PubMed]

30. He, J.; Zhou, T. Patented techniques for detoxification of mycotoxins in feeds and food matrices. *Recent Pat. Food Nutr. Agric.* **2010**, *2*, 96–104. [CrossRef]

31. Fuchs, E.; Binder, E.M.; Heidler, D.; Krska, R. Structural characterization of metabolites after the microbial degradation of type A trichothecenes by the bacterial strain BBSH 797. *Food Addit. Contam.* **2002**, *19*, 379–386. [CrossRef] [PubMed]

32. Ito, M.; Sato, I.; Ishizaka, M.; Yoshida, S.; Koitabashi, M.; Yoshida, S.; Tsushima, S. Bacterial cytochrome P450 system catabolizing the *Fusarium* toxin deoxynivalenol. *Appl. Environ. Microbiol.* **2013**, *79*, 1619–1628. [CrossRef] [PubMed]

33. Sato, I.; Ito, M.; Ishizaka, M.; Ikunaga, Y.; Sato, Y.; Yoshida, S.; Koitabashi, M.; Tsushima, S. Thirteen novel deoxynivalenol-degrading bacteria are classified within two genera with distinct degradation mechanisms. *FEMS Microbiol. Lett.* **2012**, *327*, 110–117. [CrossRef] [PubMed]

34. Zhao, L.; Li, X.; Ji, C.; Rong, X.; Liu, S.; Zhang, J.; Ma, Q. Protective effect of devosia sp. Ansb714 on growth performance, serum chemistry, immunity function and residues in kidneys of mice exposed to deoxynivalenol. *Food Chem. Toxicol.* **2016**, *92*, 143–149. [CrossRef] [PubMed]

35. Stoev, S.D. Foodborne mycotoxicoses, risk assessment and underestimated hazard of masked mycotoxins and joint mycotoxin effects or interaction. *Environ. Toxicol. Pharmacol.* **2015**, *39*, 794–809. [CrossRef] [PubMed]

36. Ohsato, S.; Ochiai-Fukuda, T.; Nishiuchi, T.; Takahashi-Ando, N.; Koizumi, S.; Hamamoto, H.; Kudo, T.; Yamaguchi, I.; Kimura, M. Transgenic rice plants expressing trichothecene 3-*O*-acetyltransferase show resistance to the *Fusarium* phytotoxin deoxynivalenol. *Plant Cell Rep.* **2007**, *26*, 531–538. [CrossRef] [PubMed]

37. Poppenberger, B.; Berthiller, F.; Lucyshyn, D.; Sieberer, T.; Schuhmacher, R.; Krska, R.; Kuchler, K.; Glossl, J.; Luschnig, C.; Adam, G. Detoxification of the fusarium mycotoxin deoxynivalenol by a udp-glucosyltransferase from *Arabidopsis thaliana*. *J. Biol. Chem.* **2003**, *278*, 47905–47914. [CrossRef] [PubMed]

38. Kadota, T.; Furusawa, H.; Hirano, S.; Tajima, O.; Kamata, Y.; Sugita-Konishi, Y. Comparative study of deoxynivalenol, 3-acetyldeoxynivalenol, and 15-acetyldeoxynivalenol on intestinal transport and 1L-8 secretion in the human cell line Caco-2. *Toxicol. In Vitro* **2013**, *27*, 1888–1895. [CrossRef] [PubMed]

39. He, J.W.; Yang, R.; Zhou, T.; Boland, G.J.; Scott, P.M.; Bondy, G.S. An epimer of deoxynivalenol: Purification and structure identification of 3-*epi*-deoxynivalenol. *Food Addit. Contam.* **2015**, *32*, 1523–1530. [CrossRef] [PubMed]

40. Iram, S.H.; Cole, S.P. Differential functional rescue of Lys[513] and Lys[516] processing mutants of mrp1 (ABCC1) by chemical chaperones reveals different domain-domain interactions of the transporter. *Biochim. Biophys. Acta* **2014**, *1838*, 756–765. [CrossRef] [PubMed]

![toxins logo] *toxins*   MDPI

Article

Identification of the Anti-Aflatoxinogenic Activity of *Micromeria graeca* and Elucidation of Its Molecular Mechanism in *Aspergillus flavus*

Rhoda El Khoury [1,2], Isaura Caceres [1], Olivier Puel [1], Sylviane Bailly [1], Ali Atoui [3], Isabelle P. Oswald [1], André El Khoury [2] and Jean-Denis Bailly [1,*]

[1] Toxalim, Université de Toulouse, INRA, ENVT, INP Purpan, UPS, Toulouse F-31027, France; rhodakhoury@gmail.com (R.E.K.); isauracaceres@hotmail.com (I.C.); olivier.puel@inra.fr (O.P.); s.bailly@envt.fr (S.B.); isabelle.oswald@inra.fr (I.P.O.)

[2] Laboratoire de Mycologie et Sécurité des Aliments (LMSA), Département des sciences de la vie et de la terres - Biochimie, Faculté des Sciences, Université Saint-Joseph, P.O. Box 17-5208, Mar Mikhael Beirut 1104 2020 Lebanon; andre.khoury@usj.edu.lb

[3] Laboratory of Microbiology, Department of Natural Sciences and Earth, Faculty of Sciences I, Lebanese University, Hadath Campus, P.O. Box 5, Beirut, Lebanon; a.atoui@cnrs.edu.lb

*   Correspondence: jd.bailly@envt.fr; Tel.: +33-5-6119-3229

Academic Editor: Ting Zhou
Received: 18 January 2017; Accepted: 24 February 2017; Published: 1 March 2017

Abstract: Of all the food-contaminating mycotoxins, aflatoxins, and most notably aflatoxin B_1 (AFB$_1$), are found to be the most toxic and economically costly. Green farming is striving to replace fungicides and develop natural preventive strategies to minimize crop contamination by these toxic fungal metabolites. In this study, we demonstrated that an aqueous extract of the medicinal plant *Micromeria graeca*—known as hyssop—completely inhibits aflatoxin production by *Aspergillus flavus* without reducing fungal growth. The molecular inhibitory mechanism was explored by analyzing the expression of 61 genes, including 27 aflatoxin biosynthesis cluster genes and 34 secondary metabolism regulatory genes. This analysis revealed a three-fold down-regulation of *aflR* and *aflS* encoding the two internal cluster co-activators, resulting in a drastic repression of all aflatoxin biosynthesis genes. Hyssop also targeted fifteen regulatory genes, including *veA* and *mtfA*, two major global-regulating transcription factors. The effect of this extract is also linked to a transcriptomic variation of several genes required for the response to oxidative stress such as *msnA*, *srrA*, *catA*, *cat2*, *sod1*, *mnsod*, and *stuA*. In conclusion, hyssop inhibits AFB$_1$ synthesis at the transcriptomic level. This aqueous extract is a promising natural-based solution to control AFB$_1$ contamination.

Keywords: Aflatoxin B_1; *Aspergillus flavus*; hyssop; inhibition; oxidative stress

---

1. Introduction

Aspergillus flavus, a saprophytic fungus that develops on many crops including maize, oilseed, dried fruit, and spices [1], is the main producer of aflatoxin B_1 (AFB$_1$), the most potent naturally occurring carcinogen. AFB$_1$ is associated with several pathologies mainly targeting the liver [2]. This mycotoxin also has major economic impacts as it contributes to considerable amounts of crop and livestock losses occur [3], endangering food and feed security. Globalization of food trade and global climate changes have further exacerbated the situation [4,5].

Many methodologies have been developed to limit AFB$_1$ contamination in crops. First, the implementation of good agricultural practices is undoubtedly a key factor to reduce undesirable growth of fungi. Fungal growth and mycotoxin production closely depend on temperature and humidity [6] and since these meteorological parameters are impossible to control, contamination cannot be completely avoided. The massive use of fungicides in crops over the last decades led to an accumulation of toxic chemical residues in food products but also in water and soil and resulted in the development of resistant pathogen populations [7].

Recent studies have pointed out the use of physio-chemical approaches to counteract aflatoxin contamination [8,9] while others were based on the detoxification properties or the protective physiological effect of bacterial metabolites or natural extracts used as feed supplements [10–13]. Although proven to be efficient, these approaches remain strictly restricted to animal feed. Nonetheless, such alternative strategies could be used at a much earlier phase, postharvest, to prevent aflatoxin production in crops. As an example, several bio-control approaches were developed relying on the use of microorganisms, such as lactic acid bacteria strains [14] or atoxinogenic fungi [15,16]. These strains displayed the ability to inhibit aflatoxin production or fungal growth to a certain extent [17,18]. Prevention strategies could also rely on the use of natural substances like plant extracts or essential oils. As plants grow, they produce many metabolites that serve as a defense against a number of environmental stresses. Therefore, plant extracts have long been studied as protective bioactive agents and it was demonstrated that some of them have antifungal or anti-toxinogenic properties [19,20].

Micromeria graeca, commonly known as hyssop, is an herbaceous plant belonging to the *Lamiaceae* family. It is a widespread species in the Mediterranean basin and is frequently used for medicinal purposes and as a condiment [21]. The composition of essential oil, the toxicity, and the antimicrobial effect of different species of hyssop were previously described [22–27], but there is little data on the composition and toxicity of the aqueous extract of hyssop [21,28], and none on its antifungal or anti-mycotoxinogenic effect. The purpose of this study was to test the aqueous extract of *M. graeca* for its ability to prevent aflatoxin's biosynthesis. We observed that it inhibits AFB_1's production by *A. flavus* strains without interfering with fungal growth. The molecular mechanism of action of this extract involved down-regulation of all the AFB_1 biosynthetic cluster as well as modulation in the expression of 15 secondary metabolite regulating genes.

2. Results

2.1. Effect of Aqueous Extract of Hyssop on the Production of AFB_1 and the Development of A. flavus

When *A. flavus* strain NRRL 62477 was grown in a hyssop-supplemented medium, a dose-dependent decrease in AFB_1 production was observed. The downward trend started at the lowest concentration (0.0195 mg/mL) and was statistically significant (28.5%, *p*-value 0.0032) starting at 0.078 mg/mL. Inhibition reached 99.2% at 10 mg/mL and AFB_1 was no longer detectable at 15 mg/mL (Figure 1). Further experiments were then conducted supplementing the culture medium with 10 mg/mL of hyssop extract, this being the lowest to present a quasi-total inhibition of AFB_1 synthesis.

At 10 mg/mL, aflatoxin inhibition by hyssop was accompanied by a mild increase of the colony diameter (4.4 ± 0.03 vs. 4.25 ± 0.03 cm for treated and control respectively, *p*-value 0.0213). However, no significant increase in the total spore count or in the spore density was observed following the addition of hyssop. Besides, no further change was observed in mycelium weights or for the germination delay in the presence of hyssop in the medium (Table 1).

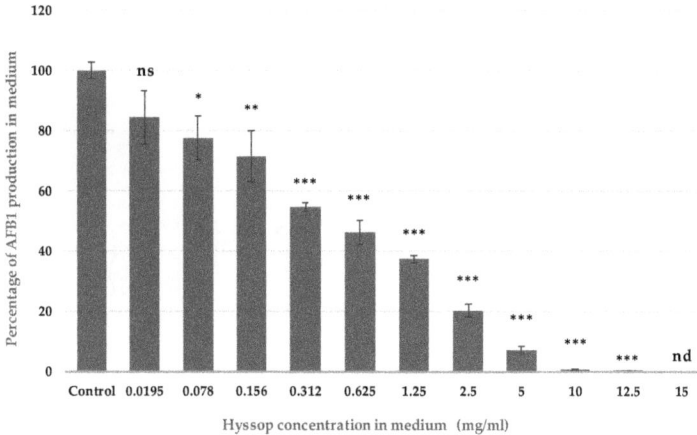

Figure 1. Aflatoxin B$_1$ (AFB$_1$) production as a function of hyssop concentration. Malt extract agar (MEA) medium was supplemented with increasing concentrations of hyssop extract ranging from 0.0195 to 15 mg/mL and cultivated at 27 °C, in the dark, for 8 days. AFB$_1$ concentrations were quantified through HPLC/FLD. Results are expressed as mean % ± SEM ($n = 3$). ns = no significant changes; nd = not detectable; * p-value < 0.05; ** p-value < 0.01; *** p-value < 0.001.

Table 1. The effect of the addition of 10 mg/mL hyssop to the culture medium on the development of *A. flavus* (i) colony diameter was measured in length and width; (ii) weight was measured after a 48 h-drying at 60 °C; (iii) germinating conidia were counted by observation under stereo-microscope after 16 h incubation at 27 °C; (iv) total spore count is estimated following a complete wash of conidia and a Malassez-cell count of proper dilutions and (v) spore density was calculated based on the total spore count related to the colony surface. Results are expressed as mean ± SEM ($n = 3$).

	Observed Parameters	MEA	MEA + Hyssop 10 mg/mL
Growth	Colony diameter (cm)	4.25 ± 0.03	4.4 ± 0.03
	Mycelium dry weight (g)	0.16 ± 0.03	0.15 ± 0.02
	Germinating conidia after 16 h (%)	96.5 ± 8.5%	101.5 ± 4%
Sporulation	Total spore count	$8.1 \times 10^8 \pm 4.5 \times 10^7$	$1.1 \times 10^9 \pm 9.9 \times 10^7$
	Spore density (conidia/cm^2)	$5.7 \times 10^7 \pm 2.6 \times 10^6$	$7 \times 10^7 \pm 5.6 \times 10^6$

Following addition of hyssop in medium, *A. flavus* colonies presented numerous macro and microscopic modifications. The major noticeable macroscopic morphological change was the development of an abundant aerial mycelium covering the entire surface of the colony. This latter also displayed fasciculation on the edge. The presence of these numerous floccose tufts also increased the depth of the colony (Figure 2).

Under microscope, classic *A. flavus* structures were present in the basal mycelium of hyssop-treated cultures: long, coarse, un-branched conidiophores and radiate biseriate conidial heads. However, in the aerial mycelium, conidiophores, vesicles, and conidia presented an atypical morphology and organization: (1) an increased number of short conidiophores bearing small columnar heads in relation with the abundant aerial mycelium; (2) phialides developing anarchically on hyphae and on conidiophores in the absence of a vesicle; (3) and the presence of conidiophores with two, and less frequently three, sporulated vesicles (Figure 3).

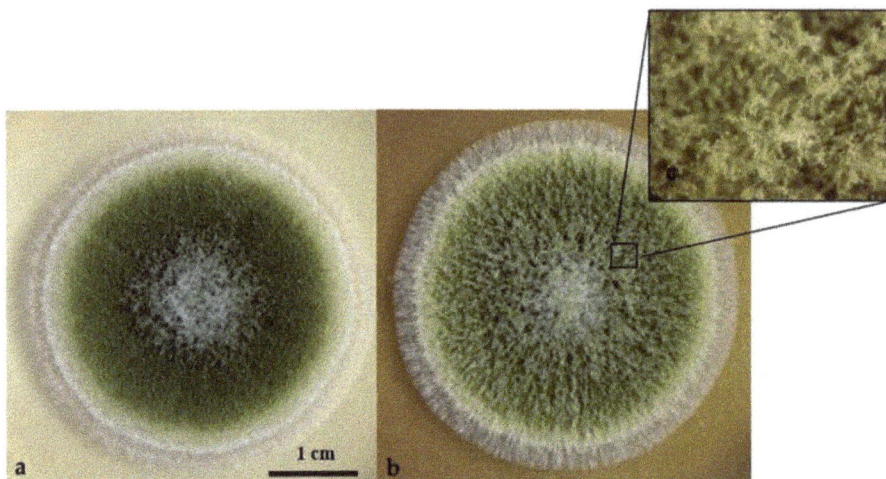

Figure 2. Phenotype of *A. flavus* strain NRRL62477 after four days of culture at 27 °C in MEA medium or MEA medium supplemented with 10 mg/mL of hyssop aqueous solution. (**a**) Control culture grown on a regular MEA medium; (**b**) MEA medium was supplemented with 10 mg/mL of aqueous solution of hyssop; (**c**) Magnification of the aerial mycelium covering the hyssop treated culture.

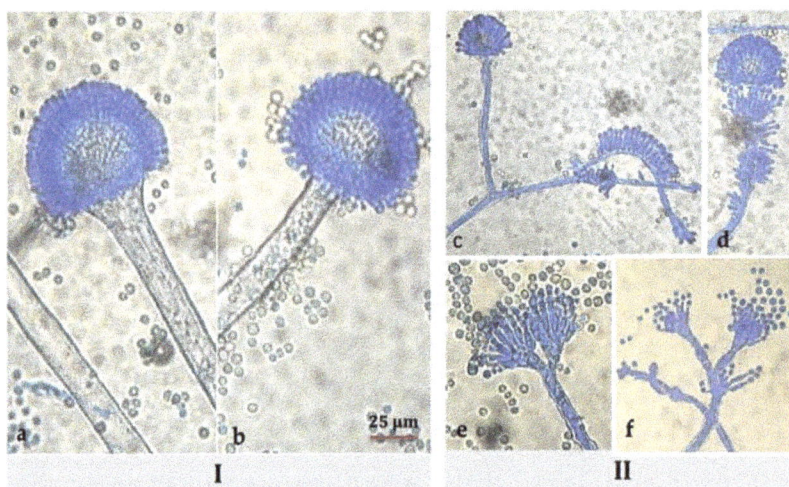

Figure 3. Microscopic views (x400) of *A. flavus* NRRL 62477 conidiophores in the (I) basal mycelium on (**a**) MEA medium and (**b**) MEA supplemented with 10 mg/mL of hyssop extract, and the (II) aerial mycelium showing the development of anarchic philalides when strain was grown on a hyssop-treated MEA medium; (**c**) and (**d**) development of anarchic philalides; (**e**) conidiophore bearing two vesicles and (**f**) presence of short conidiophores with columnar heads.

At the dose of 10 mg/mL, we observed a 77.7% and 70.8% inhibition of the production of AFB_1 in two other *A. flavus* strains (E28 and E71respectively), without alteration of fungal growth. Similar morphological changes were also observed for these two strains (data not shown).

2.2. Aqueous Extract of Hyssop Down-Regulated the Expression of AFB₁ Cluster Genes

The biosynthesis of AFB_1 is the result of a well-described cascade of more than 20 enzymatic reactions. This cascade is governed by 27 clustered genes encoding the corresponding enzymes as well as two internal regulators, AflR encoding a Gal4 zinc finger transcription factor and its co-activator AflS. The expression of all of these genes was analyzed in order to determine whether the inhibition of AFB_1 synthesis occurred at a transcriptomic level. Inhibition of AFB_1 production in hyssop-supplemented media was accompanied by a decrease in the expression of both of *aflR* and *aflS* genes by 3.2 and 2.8 times respectively (p-value < 0.0001). Apart from *aflT* (encoding a MFS-family transporter), which is not regulated by the AflR/AflS complex [29] and was down regulated only by 2.3 times (p-value < 0.0001), the expression of the remaining cluster genes was severely repressed (Figure 4).

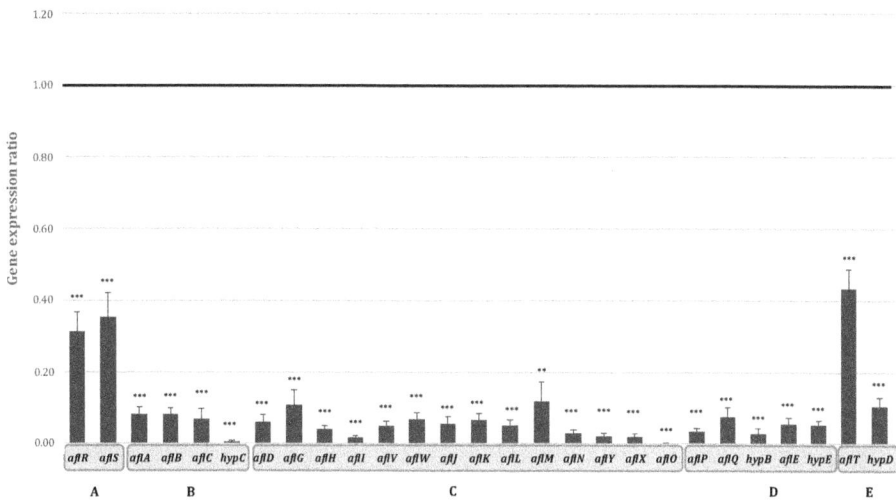

Figure 4. Expression of genes belonging to AFB_1 cluster genes in the presence of 10 mg/mL of hyssop aqueous extract; (**A**) internal cluster regulators (**B**) genes involved in the earlier steps of AFB_1 enzymatic cascade leading to the formation of norsolorinic acid (**C**) genes involved in the middle steps of the AFB_1 enzymatic cascade converting norsolorinic acid into sterigmatocystin (**D**) genes involved in the final steps of the cascade leading to AFB_1 synthesis (**E**) genes with uncharacterized functions. Black line represents the expression level of control. ** p-value < 0.01; *** p-value < 0.001.

Genes undergoing the most drastic inhibitions were *hypC*, *aflI*, and *aflO*, encoding enzymes respectively intervening at the beginning, middle, and end of the biosynthetic pathway and with corresponding fold-changes of 167.2, 60.7, and 468.8 with p-values < 0.001. For the genes encoding enzymes involved in the first steps of the cascade leading to the polyketide structure, *aflA*, *aflB*, and *aflC*, expression was decreased by 12.2, 12.3, and 14.7 respectively. The least impacted genes were *aflM*, *aflG*, and *hypD* with expression levels decreased by 8.4, 9.3, and 9.4 times respectively. For the remaining AFB_1 cluster genes, the same downward trend was observed with expression levels decreased by 14 to 50 folds (Supplementary Material Table S1). This repression of the entire aflatoxin gene cluster could then be directly linked to the inhibition of the production of this mycotoxin.

2.3. Transcriptomic Effect of Hyssop Extract on the Expression of Genes Coding for Regulators of Secondary Metabolites

The expression of AFB$_1$ cluster genes is subjected to the control exerted by regulatory factors encoded by genes outside of the cluster. In order to investigate a possible genetic relationship between those regulators and the transcriptional inhibition of AFB$_1$ cluster genes after the addition of hyssop extract, we conducted a study on the regulatory network affecting secondary metabolism that includes 34 genes involved in several fungal functional pathways. Among these, a total of 15 genes involved in diverse cellular mechanisms were modulated by adding hyssop to the culture medium (Figure 5). The modulated genes could be grouped in five categories based on their cellular functions in other ascomycetes [30]. The first one includes global regulating factors such as *veA*, *mtfA*, *nsdC* that were affected with expression levels respectively increased by 3.8, 1.9, and 1.5 folds (*p*-values < 0.0001, 0.0001 and 0.0122). Secondly, genes encoding enzymes involved in cellular protection from oxidative stress such as superoxide dismutases (*sod1* and *mnsod*) and catalases (*catA* and *cat2*) had their expression decreased by 1.6, 2, 2.2, and 3 folds respectively (*p*-values 0.013, 0.0007, 0.004, and < 0.0001). Other genes intervening in the oxidative stress response and encoding transcription factors, notably *msnA* and *srrA*, had their expression levels increased by 3.2 and 1.4 times with *p*-values of 0.0126 and 0.0017, respectively. The third category involves *gprK* and *gprH*, encoding G-protein receptors involved in relaying external signals with *gprK* expression increased by 2 folds (*p*-value < 0.0001) and *gprH* decreased by 2.1 folds (*p*-value 0.0006). Moreover, *ppoC* encoding a fatty acid dioxygenase involved in oxylipin production also presented an expression decreasing by 1.5 times (*p*-value 0.003). The conidial developmental factor *stuA* whose expression increased by 1.8 times (*p*-value 0.0012) constitutes the fourth category. Finally, two genes encoding environmentally influenced transcription factors whose expression is respectively modulated by nitrogen and medium pH, *areA* and *pacC* were also triggered by the addition of hyssop in the medium and their expression levels were respectively increased by 1.7 and 1.6 folds with *p*-values of 0.0215 and < 0.0001.

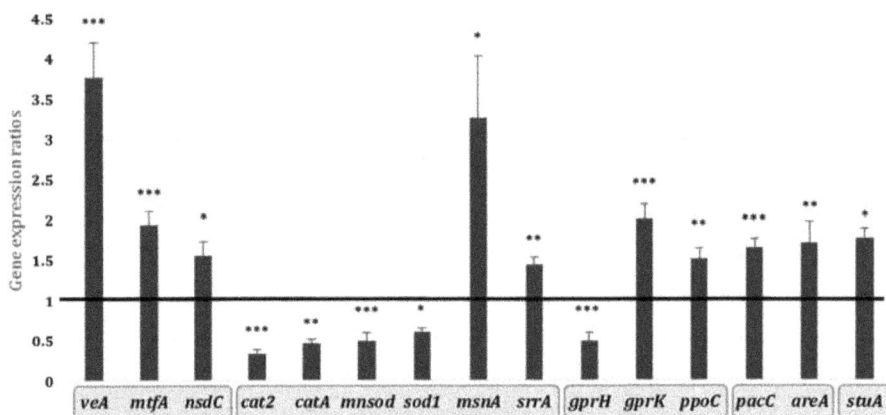

Figure 5. Schematic representation of gene expression ratios of the different regulatory genes affected upon 10 mg/mL hyssop supplementation of MEA media. Genes are grouped into the five categories described above. The black line represents the expression level of genes in control cultures. * *p*-value < 0.05; ** *p*-value < 0.01; *** *p*-value < 0.001.

3. Discussion

3.1. Hyssop Leads to an Inhibition of AFB₁ Synthesis in A. flavus by a Transcriptomic Regulation of AFB₁ Cluster Genes

3.1. Hyssop Leads to an Inhibition of AFB_1 Synthesis in A. flavus by a Transcriptomic Regulation of AFB_1 Cluster Genes

The inhibition of AFB_1 synthesis was dose-dependent (Figure 1) and the same impact was observed using several strains of *A. flavus*. Another recent study by Omidpanah et al. (2015) [31] pointed out that some aqueous extracts including thyme and mint had fungicidal effect on *A. flavus* at concentrations of 0.2 and 0.8 mg/mL respectively, yet without determining aflatoxin inhibition at any of the concentration range used. At a comparable concentration of 0.625 mg/mL, hyssop's extract was able to inhibit the production of aflatoxin by 52% without restraining the growth of *A. flavus*. A slight trend for an increase in total spore count and density was observed upon addition of hyssop, but these results remain statistically non-significant and do not affect the anti-aflatoxinogenic property of the extract.

AFB_1 biosynthesis in aflatoxinogenic fungi is the result of a coordinated cascade of enzymatic reactions. The enzymes catalyzing these reactions are encoded by 27 genes and grouped into a cluster located in the telomeric region of the 3rd chromosome of aflatoxinogenic species [32]. The function of almost all of these genes was elucidated and described in previous works [33–35]. We demonstrated that AFB_1 inhibition in *A. flavus* by *M. graeca* extract is consistent with a drastic repression of all AFB_1 cluster genes following a significant decrease in the expression levels of the two internal cluster regulators *aflR* and *aflS* (previously annotated as *aflJ*) (Figure 4). Furthermore, the down-regulation of these two latter genes in AF-repressive conditions has been previously described as associated with the repression of the entire AF cluster genes [36]. It could also be noted that the inhibition level of the cluster genes does not depend on their chronological intervention in the enzymatic cascade leading to AFB_1 synthesis, contrary to what has been described for other anti-aflatoxinogenic agents such as eugenol [30].

3.2. The Implication of VeA and MtfA: Two Leading Transcriptional Regulators

VeA is a global regulating transcription factor involved in primary and secondary metabolism [37] and recruiting other factors such as LaeA and VelB to form the trimeric velvet complex. The activity of this complex affects fungal development, conidiation, and secondary metabolism [38]. In hyssop-treated cultures, transcripts of *laeA*, *velB*, and *vosA*, the latter being an interacting partner of *velB* [39], were not affected (Table S1). This result further highlights the independent role of VeA in multiple other cellular mechanisms [40]. The presence of VeA is necessary for the expression of secondary metabolite genes; however, it can also act as a repressor of some of these genes and thus inhibit the production of the concerned metabolite. VeA is essential for the transcription of AF cluster genes, including the transcription factor *aflR* and others (*aflD*, *aflM*, and *aflP*) regulating production of aflatoxin. Deletion of the *veA* gene led to the repression of AFB_1 cluster genes in *A. flavus* [41]. However, according to our current study and to another recent one [30], a repression of all AFB_1 cluster genes can also coincide with a *veA*-over-expression profile (Figure 5). Such is the case of penicillin, produced by *A. nidulans* where an OE:*veA* led to the repression of *acvA*, the penicillin biosynthesis gene and subsequent inhibition of penicillin production [42].

VeA can also interact with another conserved global transcription factor, MtfA. The latter has a major role in regulating the development and the secondary metabolism in filamentous fungi [43] and it is linked to AFB_1's biosynthesis as well as to *aflR*'s expression. An over-expression of *mtfA* in *A. flavus* has drastically inhibited the production of AFB_1 following a down-regulation of *aflR*, whereas the effect of an *mtfA* deletion was less important [44]. The concurring over-expression of *mtfA* and *veA* in our conditions (Figure 5), which further highlights the interaction between these two factors, could then be responsible for down-regulating AF-cluster genes and inhibiting the production of aflatoxin.

3.3. Implication of Other Regulatory Factors in AF Inhibition by M. graeca—Hyssop Extract

Stressful environmental conditions may lead fungi to establish several defense lines to limit cellular damage. For example, when subjected to oxidative stress, fungi might respond by producing secondary metabolites such as aflatoxins. The oxidative stress generated by enhanced lipid peroxidation and the free radical generation was proven to be a prerequisite for aflatoxin production in *A. parasiticus* [45]. Moreover, the anti-aflatoxinogenic effect of antioxidant agents, such as eugenol, seems to be linked to an alleviation of oxidative stress and lipid peroxidation as well as a modulation of the expression of a number of genes involved in the oxidative stress response [30,46].

It has been demonstrated that VeA contributes to a positive transcriptomic modulation of stress-tolerance genes such as *msnA* and *srrA* under induced oxidative stress conditions [39]. The expression of these two transcription factors is then highly dependent on that of *veA*. Therefore, their over-expression in a hyssop-treated medium might be the outcome of an over-expressed *veA* (Figure 5). The developmental factor StuA has also been associated to stress-response in fungi, yet there isn't a clear view on its contribution [47]. However, its dependence on *msnA* was shown since its expression levels were modulated in *A. flavus* as well as in *A. parasiticus msnA*-deleted strains [48]. MsnA is also known for regulating the expression of the catalases (CAT)- and superoxide dismutases (SOD)-encoding genes [48]. Those antioxidant enzymes along with aflatoxin formation are suggested as part of the fungus defense mechanism against reactive oxygen species (ROS) damages [49]. When the medium was supplemented with *M. graeca* extract, *A. flavus* responded by decreasing the expression of SOD- and CAT-encoding genes such as *sod1*, *mnsod*, *catA* and *cat2* as levels of *msnA* increased thus resulting in an AF-biosynthesis repression, possibly related to an alleviation of environmental oxidative stress.

Two other regulatory factors that were not linked to oxidative stress were also modulated by hyssop's addition to medium. The first regulator is PacC, which is a pH-dependent factor whose over-expression in aflatoxin-repressive conditions is to be investigated. In fact, PacC is usually activated under alkaline conditions [50] whereas hyssop addition does not change the pH of the medium (data not shown).

The second regulatory factor is NsdC, known to be a developmental regulator whose alteration causes several morphological aberrances such as shorter-stipe conidiophores presenting abnormal conidial-head formations. Similar to VeA, NsdC modulation could participate to the morphological modifications observed in hyssop-treated cultures (Figure 3). It has also been linked to aflatoxin cluster-gene expression [51] as well as being a global regulator of secondary metabolism [52]. More data is yet to be collected on the individual and possibly collaborative roles of both of these factors in the regulation of secondary metabolism.

3.4. Morphological Modifications of Conidiophores and Vesicles of A. flavus in Hyssop-Supplemented Media

Besides its role in secondary metabolite regulation, VeA is also a developmental factor in regulating morphogenesis as alterations in its expression levels can result in morphological abnormalities. For example, a reduction in fungal aerial hyphae was noted in both *A. flavus* and *Fusarium graminaerum veA* deleted strains [41,53]. Therefore, the modulation of *veA* expression could contribute to morphological abnormalities observed upon hyssop exposure (Figure 3).

Furthermore, a single previous study has described the modification of the aerial hyphae in an AF-inhibiting profile in *A. parasiticus* in the presence of n-decyl aldehyde, a corn-derived volatile compound [54]. However, another study conducted on *A. flavus* mutant strains described the appearance of morphological abnormalities, notably on phialide formation, associated to a cessation of AFB$_1$ production [51].

Alterations in morphology, such as the development of aerial hyphae, were also associated to an imbalance in the G-protein signal transduction pathway [55,56]. This pathway is governed by the binding of signaling molecule to G-protein coupled receptors (GPCR) such as those encoded by *gprK* and *gprH*, both affected by the addition of hyssop to the media (Figure 5), and tuned by regulators of the G-protein signaling cascade (RGS), whose roles and implication in AFB_1 synthesis are being investigated in *A. flavus* [57]. G-protein signaling pathway is also linked to oxylipins that are hormonal-like signaling molecules [58] produced by fatty-acid-oxygenases such as PpoC. Moreover, oxylipins regulation has also been described as VeA-dependent [37]. However, fungal signal perception and transduction pathways are a very complex loop due to the diversity of signals that might initiate them and most importantly to the numerous acting factors involved downstream. Since *M. graeca* extract is a complex extract containing many signal-provoking agents such as polysaccharides, amino acids, minerals, phenolic compounds and many others, it remains possible that morphological modifications have no direct link to AFB_1's inhibition.

4. Conclusions

This study demonstrates the efficiency of *M. graeca* aqueous extract in limiting AFB_1 contamination without altering fungal growth. Such an effect could ensure food safety without affecting biodiversity. Indeed, *A. flavus* is a very competitive crop-contaminating agent; therefore, the use of fungistatic agents could favor the emergence of other, possibly uncontrollable microorganisms. According to our results, inhibition by hyssop extract occurs at a transcriptomic level as expression ratios of all of aflatoxin cluster-genes were severely decreased. Nonetheless, hyssop extract triggered a response in several fungal cellular mechanisms including cellular signaling, global transcription factors, and conidial development, as well as factors acting in the oxidative stress response. Massive transcriptomic analyses, such as RNAseq or Microarray assays, would be a good complement to this study since they could provide a broader vision of cellular functions affected by the addition of hyssop to the medium. Nonetheless, being as complex as it is, this extract may shelter several bioactive compounds [59] contributing in a complementary way to its anti-aflatoxinogenic activity. To ascertain more accurately the inhibitory mechanism of action, the content of this extract needs to be deciphered in order to determine and purify its active molecules as well as the inhibition extent of each of the isolated compounds.

5. Materials and Methods

5.1. Solvents and Standards

All solvents were HPLC grade and purchased from ThermoScientific Fisher (Villebon-Sur-Yvette, France). Lyophilized aflatoxin B_1 standard was purchased from Sigma Aldrich (St. Louis, MO, USA). Stock solutions of each of the standards were prepared in methanol and stored at 4 °C in the dark. Calibration curves were prepared beforehand by diluting stock solutions with mobile phase used for HPLC analysis.

5.2. Preparation of the Aqueous Solution of Hyssop

Dried hyssop (*M. graeca*) was commercially purchased from Tyr, Lebanon. Hyssop species was kindly confirmed by Prof. Marc Beyrouthy (Department of Agricultural Sciences, USEK—Lebanon). Leaves were ground with an electrical grinder and ten grams of ground hyssop were added to 80 mL of bi-distilled water and placed on a horizontal shaking table at 220 rpm for 24 h. Extracts were then filtered through cotton gauze before being centrifuged for 10 min at 3500 rpm. Filtrates were centrifuged once again, at 4700 rpm for 30 min and autoclaved at 121 °C for 15 min. Final sterile extracts were stored at +4 °C until their use.

5.3. Fungal Strains and Growth Conditions

A referenced toxinogenic *Aspergillus flavus* strain NRRL 62477 isolated from paprika samples harvested from a Moroccan market [59] was used to evaluate aflatoxin inhibition by aqueous solution of hyssop as well as the molecular mechanism of inhibition. Further analysis of total aflatoxin inhibition by the aqueous hyssop solution was conducted on two other *A. flavus* strains (E28 and E71) that were previously isolated from white pepper and paprika samples from Morocco [60]. Strains were cultivated on a malt extract agar (MEA) medium (30 g malt extract and 15 g agar-agar per liter) (Biokar Diagnostics, Allone, France), supplemented at 2% v/v with the autoclaved aqueous hyssop solution, whereas 2% v/v water-supplemented media were used for control cultures. The media for RNA isolation and dry weight measurement were layered with 8.5 cm diameter cellophane disks (Hutchinson, Chalette-sur-Loing, France) before inoculation in order to allow separation of mycelium from the culture medium. Spore suspensions were prepared in Tween 80 (0.05% in water) from a one-week-old MEA culture. Spores were counted on a Malassez cell and 10^3 spores were inoculated in the center of the medium. Cultures destined for RNA isolation were incubated in 6 replicates per condition for 4 days. For AFB$_1$ quantification assays, cultures were incubated for 8 days and were in triplicates. All cultures were incubated at 27 °C. The media pH was measured before and after inoculation and after incubation using a H199161 food pH-meter (Hanna Instruments, Tanneries, France).

5.4. Examination of Cultural Parameters

5.4.1. Effect on Growth

The final growth mean was estimated by the measurement of culture diameters in length and width at day 4.

5.4.2. Mycelium Dry Weight

Following a four-day incubation period, cellophane disks were peeled off and placed in new petri dishes that were incubated for 48 h at 60 °C. Dried mycelium films were allowed to cool in a desiccator before being weighed on an analytical balance. Final weight was calculated by subtracting the mean weight of four desiccated control cellophane disks.

5.4.3. Total Spore Quantification

Colonies were cut out of MEA media, 1 mm beyond the mycelium border, placed in a stomacher bag with 50 mL of Tween 0.05% and spores were gently manually scraped off of culture without tearing the media. The bag was then placed in a stomacher for 90 s. The supernatant was filtered through cotton gauze that was then rinsed with 3×20 mL Tween 0.05%. Spore solutions were homogenized by thorough vortex and subsequent dilutions were prepared in Tween 0.05% for counting on a Malassez cell in order to determine the total spore count (SC). Spore density (SD) was calculated as $SD = SC/(\pi r^2)$, r = average colony radius.

5.4.4. Delay to Germination

Two hundred spores were inoculated in the center of the media and germinating spores were counted after a 16-h incubation period at 27 °C by stereo-microscopic examination.

5.4.5. Fungal Morphological Features

Macroscopic (e.g., color of conidial areas, thallus margin and texture, aspect of conidial heads and colony reverse) and microscopic (e.g., conidiophore, shape of vesicles, number of sterigmata, shape of conidia and ornamentation) characters were observed under stereomicroscope SZX9—X12-120 (Olympus, Rungis, France) and optical microscope CX41—X400 and X1000 (Olympus, Rungis, France) respectively.

5.5. RNA Isolation and Reverse Transcription

Cellophane disks along with the four-day mycelium were peeled off from the medium, finely grinded with liquid nitrogen and a maximum of 100 mg were used for total RNA purification through a RNeasy Plus Minikit (Qiagen, Hilden, Germany), which includes an on-column genomic

DNA clean-up, following the manufacturer's instructions. RNA integrity and purity were checked with agarose gel electrophoresis and a NanoDrop ND1000 (Labtech, Palaiseau, France) that also determined its concentration. First-strand cDNA synthesis reaction was primed using RevertAid Reverse Transcriptase (MBI Fermentas, UK), RNase Inhibitor (Applied Biosystems, Warrington, UK) and an anchored oligo(dT) Bys 3′ Primer (5′-GCTGTCAACGATACGCTAACGTAACGGCATGAC AGTGTTTTTTTTTTTTTTTTTT-3′). An RT minus sample, where no reverse transcriptase reaction takes place, and a sterile water sample were added as negative controls in order to verify the absence of undesirable genomic DNA contamination and primer complementation, respectively.

5.6. Real-Time PCR Expression Profile Analysis of Genes Regulating AFB₁ Biosynthesis in A. flavus

The genome of *A. flavus* strain NRRL 3357 (GenBank accession number EQ963478) served as a matrix for all of the primer used in this study. Gene selection and the corresponding primer pair sequences were adapted from a previous work [29] and primer sequences of the *stuA* gene (AFLA_046990) were added in this study (stuA_F: GATAAACGGAACCAAACTGCTCAA; stuA_R: CACGCTCAAATGGGATCCAA). In total, the expression of 61 genes was simultaneously analyzed, 27 of which corresponded to the AFB_1 cluster and 34 to regulatory factors. Regulatory factors were grouped into 6 categories: (1) environmental transcription factors (Area, CreA, MeaB and PacC); (2) oxidative stress response factors (AtfA, AP-1, CatA, Cat2, MnSOD, MsnA, and SrrA); (3) the velvet complex (LaeA, VeA, VelB and VosA); (4) factors belonging to different cellular signaling families such as the Ras-family (RasA), the G-protein signaling family (FadA and FlbA) as well as the G-protein receptors (GprK, GprA, GprH, GprP, and GprG) and oxylipin enzymes (PpoA, PpoB, and PpoC); (5) regulators of development and conidiation (FluG, BrlA, StuA, and AbaA) and (6) global secondary-metabolism-regulating transcription factors (NsdC, MtfA, and Fcr3). Primer pairs design was based on the coding sequence of the corresponding genes, with at least one of the primers extending on an exon/exon junction in order to avoid undesirable genomic DNA amplification. Primer-dimers or self-complementarities were evaluated using the PrimerExpress 2.0 software (Applied Biosystems, Courtaboeuf, France). All primers were synthesized by Sigma Aldrich (Saint-Quentin Fallavier, France). Following RNA extractions and reverse transcriptase reactions, real-time PCR assays were performed on 15 ng cDNA in a 5 µL reaction volume per well, using Power SYBR® Green PCR Master Mix (Applied Biosystems, Warrington, UK) as a fluorescent dye for cDNA quantification. Master mixes and diluted cDNA samples were prepared separately on 96-well Framestar Jupe plates (Dominique Dutscher, Issy-les-Moulineaux, France) and mixed in 384-well plates by an Agilent Bravo Automated Liquid Handling Platform (Agilent Technologies, Santa Clara, CA, USA). All real time amplification reactions were carried out on a ViiA7 Real-Time PCR System (Applied Biosystems, Warrington, UK), as described by Tannous et al. [61].

5.7. Aflatoxin Extraction and HPLC Quantification

Media of four- and eight-day cultures were entirely retrieved and their AFB_1 content determined after extraction with 25 and 40 mL chloroform respectively. Extracts were held for 2 h on a horizontal shaking table at 200 rpm and were then filtered through a Whatman 1PS phase separator filter (GE Healthcare Life Sciences, Vélizy-Villacoublay, France, 150 mm diameter). Filtrates were evaporated to dryness and dissolved in 1 mL of a water-acetonitrile-methanol mixture (65:17.5:17.5; $v/v/v$). Extracts were filtered using 0.45 µm porosity disks (Thermo Scientific Fisher, Villebon-Sur-Yvette, France) before analysis. HPLC analysis was performed using a Dionex Ultimate 3000 UHPLC (Thermo Scientific, France) using a 125 × 2 mm, 5 µm, 100 Å, Luna® C18(2) LC column (Phenomenex, Torrance, CA, USA). Aflatoxins were separated using the program described by Fu, Huang, & Min, 2008, with minor modifications [62]. A mixture of water (acidified with 0.2% acetic acid)-acetonitrile (79:21, v/v) is eluent A and methanol is eluent B. Separation program consists of a 30 min A:B (82.5:17.5) isocratic flow at 0.2 mL/min. Aflatoxins were detected using a fluorescent detector at wavelengths of 365/430 nm (excitation/emission). UV Spectra were confirmed by an additional diode array detector (DAD) coupled to the apparatus. Sample concentrations were calculated based on a standard calibration curve.

5.8. Statistics

All experiments were performed in triplicate. For gene expression assays, six biological replicates were used each time for each gene. Data presented for the dose-dependent effect of hyssop extract were analyzed using a One-way ANOVA followed by a post hoc Dunnett's test ($\alpha = 0.05$) and conducted with XLSTAT (Addinsoft, Paris). Gene expression and growth and sporulation data were analyzed using Student's t-test in Excel statistics. Differences were considered statistically significant when *p*-value was lower than 0.05.

Supplementary Materials: The following are available online at www.mdpi.com/2072-6651/9/3/87/s1, Table S1: Gene expression ratios of AFB_1 cluster genes upon hyssop addition. Ratios are obtained in comparison to control values.

Acknowledgments: The authors would like to thank Marc Beyrouthy (Department of Agricultural Sciences, USEK—Lebanon) for identifying the species of the hyssop plant used in this study and Françoise Michaud (National Veterinary School of Toulouse) and Amy Smith for English editing. This work was funded by the projects Aflafree (ANR-11-ALID-0003), Aflared (joint project, 001-2012 STDF-AIRD), and ToxinFree (Campus France PHC cèdre 32763). The authors would also like to thank the Agence Nationale de la Recherche (ANR) and the Research Council of Saint-Joseph University (Lebanon), the two organizations funding the doctoral studies of Rhoda El Khoury. Doctoral studies of Isaura Caceres were funded by the Consejo Nacional de Ciencia y Tecnología (CONACYT) México. The funders had no role in study design, data collection and analysis, decision to publish, or preparation of the manuscript.

Author Contributions: Isabelle P. Oswald, Olivier Puel, Jean-Denis Bailly, Ali Atoui, and André El Khoury conceived, supervised, and designed the experiments. Rhoda El Khoury and Isaura Caceres performed the experiments, analyzed the data, and contributed to experiment design. Sylviane Bailly performed morphological analysis. Rhoda El Khoury and Jean-Denis Bailly wrote the paper.

Conflicts of Interest: The authors declare no conflict of interest. The funding sponsors had no role in the design of the study; in the collection, analyses, or interpretation of data; in the writing of the manuscript; or in the decision to publish the results.

References

1. Klich, M.A. *Aspergillus flavus*: The major producer of aflatoxin. *Mol. Plant Pathol.* **2007**, *8*, 713–722. [CrossRef] [PubMed]
2. Cano, P.; Puel, O.; Oswald, I.P. Mycotoxins: Fungal Secondary Metabolites with Toxic Properties. In *Fungi Applications and Management Strategies*; Deshmukh, S.K., Misra, J.K., Tewari, J.P., Papp, T., Eds.; CRC Press: Boca Raton, FL, USA, 2016; pp. 318–371.
3. Bryden, W.L. Mycotoxin contamination of the feed supply chain: Implications for animal productivity and feed security. *Anim. Feed Sci. Technol.* **2012**, *173*, 134–158. [CrossRef]
4. Wu, F.; Guclu, H. Aflatoxin regulations in a network of global maize trade. *PLoS ONE* **2012**, *7*, e45151. [CrossRef] [PubMed]
5. Medina, A.; Rodriguez, A.; Magan, N. Effect of climate change on *Aspergillus flavus* and aflatoxin B_1 production. *Front. Microbiol.* **2014**, *5*, 348. [CrossRef] [PubMed]
6. Magan, N.; Aldred, D. Post-harvest control strategies: Minimizing mycotoxins in the food chain. *Int. J. Food Microbiol.* **2007**, *119*, 131–139. [CrossRef] [PubMed]
7. Da Cruz Cabral, L.; Fernández Pinto, V.; Patriarca, A. Application of plant derived compounds to control fungal spoilage and mycotoxin production in foods. *Int. J. Food Microbiol.* **2013**, *166*, 1–14. [CrossRef] [PubMed]
8. Mao, J.; He, B.; Zhang, L.; Li, P.; Zhang, Q.; Ding, X.; Zhang, W. A Structure identification and toxicity assessment of the degradation products of aflatoxin B_1 in peanut oil under UV irradiation. *Toxins (Basel)* **2016**, *8*, 332. [CrossRef] [PubMed]
9. Diao, E.; Li, X.; Zhang, Z.; Ma, W.; Ji, N.; Dong, H. Ultraviolet irradiation detoxification of aflatoxins. *Trends Food Sci. Technol.* **2015**, *42*, 64–69. [CrossRef]

10. Zhang, L.; Ma, Q.; Ma, S.; Zhang, J.; Jia, R.; Ji, C.; Zhao, L. Ameliorating effects of *Bacillus subtilis* ANSB060 on growth performance, antioxidant functions, and aflatoxin residues in ducks fed diets contaminated with aflatoxins. *Toxins (Basel)* **2016**, *9*, 1. [CrossRef] [PubMed]

11. Zhang, N.-Y.; Qi, M.; Zhao, L.; Zhu, M.-K.; Guo, J.; Liu, J.; Gu, C.-Q.; Rajput, S.; Krumm, C.; Qi, D.-S.; et al. Curcumin prevents aflatoxin B$_1$ hepatoxicity by inhibition of cytochrome P450 isozymes in chick liver. *Toxins (Basel)* **2016**, *8*, 327. [CrossRef] [PubMed]

12. Xu, L.; Eisa Ahmed, M.; Sangare, L.; Zhao, Y.; Selvaraj, J.; Xing, F.; Wang, Y.; Yang, H.; Liu, Y. Novel aflatoxin-degrading enzyme from *Bacillus shackletonii* L7. *Toxins (Basel)* **2017**, *9*, 36. [CrossRef] [PubMed]

13. Dellafiora, L.; Galaverna, G.; Reverberi, M.; Dall'Asta, C. Degradation of aflatoxins by means of laccases from *Trametes versicolor*: An in silico insight. *Toxins (Basel)* **2017**, *9*, 17. [CrossRef] [PubMed]

14. Saladino, F.; Luz, C.; Manyes, L.; Fernández-Franzón, M.; Meca, G. In vitro antifungal activity of lactic acid bacteria against mycotoxigenic fungi and their application in loaf bread shelf life improvement. *Food Control* **2016**, *67*, 273–277. [CrossRef]

15. Mauro, A.; Battilani, P.; Cotty, P.J. Atoxigenic *Aspergillus flavus* endemic to Italy for biocontrol of aflatoxins in maize. *Biocontrol* **2015**, *60*, 125–134. [CrossRef]

16. Odhiambo, B.O.; Murage, H.; Wagara, I.N. Screening for Atoxigenic *Aspergillus* Species and Evaluating their Inhibitory Potential against Growth and Sporulation of Aflatoxigenic *Aspergillus* Species. *Egert. J. Sci. Technol.* **2014**, *14*, 61–80.

17. Dalié, D.K.D.; Deschamps, A.M.; Richard-Forget, F. Lactic acid bacteria—Potential for control of mould growth and mycotoxins: A review. *Food Control* **2010**, *21*, 370–380. [CrossRef]

18. Jane, C.; Kiprop, E.K.; Mwamburi, L.A. Biocontrol of Aflatoxins in Corn using Atoxigenic *Aspergillus flavus*: Review. *Int. J. Sci. Res.* **2014**, *3*, 1954–1958.

19. Kohiyama, C.Y.; Yamamoto Ribeiro, M.M.; Mossini, S.A.G.; Bando, E.; Bomfim, N.D.S.; Nerilo, S.B.; Rocha, G.H.O.; Grespan, R.; Mikcha, J.M.G.; Machinski, M. Antifungal properties and inhibitory effects upon aflatoxin production of *Thymus vulgaris* L. by *Aspergillus flavus* Link. *Food Chem.* **2015**, *173*, 1006–1010. [CrossRef] [PubMed]

20. Chulze, S.N. Strategies to reduce mycotoxin levels in maize during storage: A review. *Food Addit. Contam. Part A Chem. Anal. Control. Expo. Risk Assess.* **2010**, *27*, 651–657. [CrossRef] [PubMed]

21. Abu-Gharbieh, E.; Shehab, N.G.; Khan, S.A. Anti-inflammatory and gastroprotective activities of the aqueous extract of *Micromeria fruticosa* (L.) *Druce ssp Serpyllifolia* in mice. *Pak. J. Pharm. Sci.* **2013**, *26*, 799–803. [PubMed]

22. Džamić, A.M.; Soković, M.D.; Novaković, M.; Jadranin, M.; Ristić, M.S.; Tešević, V.; Marin, P.D. Composition, antifungal and antioxidant properties of *Hyssopus officinalis* L. subsp. *pilifer* (Pant.) Murb. essential oil and deodorized extracts. *Ind. Crops Prod.* **2013**, *51*, 401–407. [CrossRef]

23. Michalczyk, M.; Macura, R.; Tesarowicz, I.; Banaś, J. Effect of adding essential oils of coriander (*Coriandrum sativum* L.) and hyssop (*Hyssopus officinalis* L.) on the shelf life of ground beef. *Meat Sci.* **2012**, *90*, 842–850. [CrossRef] [PubMed]

24. Soylu, E.M.; Kurt, S.; Soylu, S. In vitro and in vivo antifungal activities of the essential oils of various plants against tomato grey mould disease agent *Botrytis cinerea*. *Int. J. Food Microbiol.* **2010**, *143*, 183–189. [CrossRef] [PubMed]

25. Tisserand, R. Essential oil safety II. metabolism, neurotoxicity, reproductive toxicity. *Int. J. Aromather.* **1996**, *7*, 26–29. [CrossRef]

26. Alwan, S.; El Omari, K.; Soufi, H.; Zreika, S.; Sukarieh, I.; Chihib, N.-E.; Jama, C.; Hamze, M. Evaluation of the antibacterial activity of *Micromeria barbata* in Lebanon. *J. Essent. Oil Bear. Plants* **2016**, *19*, 321–327. [CrossRef]

27. Formisano, C.; Oliviero, F.; Rigano, D.; Saab, A.M.; Senatore, F. Chemical composition of essential oils and in vitro antioxidant properties of extracts and essential oils of *Calamintha origanifolia* and *Micromeria myrtifolia*, two *Lamiaceae* from the Lebanon flora. *Ind. Crops Prod.* **2014**, *62*, 405–411. [CrossRef]

28. Skotti, E.; Anastasaki, E.; Kanellou, G.; Polissiou, M.; Tarantilis, P.A. Total phenolic content, antioxidant activity and toxicity of aqueous extracts from selected Greek medicinal and aromatic plants. *Ind. Crops Prod.* **2014**, *53*, 46–54. [CrossRef]

29. Chang, P.-K.; Yu, J.; Yu, J.-H. *aflT*, a MFS transporter-encoding gene located in the aflatoxin gene cluster, does not have a significant role in aflatoxin secretion. *Fungal Genet. Biol.* **2004**, *41*, 911–920. [CrossRef] [PubMed]

30. Caceres, I.; El Khoury, R.; Medina, Á.; Lippi, Y.; Naylies, C.; Atoui, A.; El Khoury, A.; Oswald, I.P.; Bailly, J.-D.; Puel, O. Deciphering the anti-aflatoxinogenic properties of eugenol using a large-scale q-PCR approach. *Toxins (Basel)* **2016**, *8*, 123. [CrossRef] [PubMed]

31. Omidpanah, S.; Sadeghi, H.; Mohamadian, M.; Manayi, A. Evaluation of antifungal activity of aqueous extracts of some medicinal plants against *Aspergillus flavus*, pistachio aflatoxin producing fungus in vitro. *Drug Dev. Ther.* **2015**, *6*, 66–69.

32. Georgianna, D.R.; Payne, G.A. Genetic regulation of aflatoxin biosynthesis: From gene to genome. *Fungal Genet. Biol.* **2009**, *46*, 113–125. [CrossRef] [PubMed]

33. Yu, J.; Chang, P.-K.; Ehrlich, K.C.; Bhatnagar, D.; Cleveland, T.E.; Payne, G.A.; Linz, J.E.; Woloshuk, C.P.; Bennett, J.W. Clustered pathway genes in aflatoxin biosynthesis. *Appl. Environ. Microbiol.* **2004**, *70*, 1253–1262. [CrossRef] [PubMed]

34. Ehrlich, K.C. Predicted roles of the uncharacterized clustered genes in aflatoxin biosynthesis. *Toxins (Basel)* **2009**, *1*, 37–58. [CrossRef] [PubMed]

35. Ehrlich, K.C.; Yu, J.; Cotty, P.J. Aflatoxin biosynthesis gene clusters and flanking regions. *J. Appl. Microbiol.* **2005**, *99*, 518–527. [CrossRef] [PubMed]

36. Holmes, R.A.; Boston, R.S.; Payne, G.A. Diverse inhibitors of aflatoxin biosynthesis. *Appl. Microbiol. Biotechnol.* **2008**, *78*, 559–572. [CrossRef] [PubMed]

37. Calvo, A.M. The VeA regulatory system and its role in morphological and chemical development in fungi. *Fungal Genet. Biol.* **2008**, *45*, 1053–1061. [CrossRef] [PubMed]

38. Bayram, O.; Braus, G.H. Coordination of secondary metabolism and development in fungi: The velvet family of regulatory proteins. *FEMS Microbiol. Rev.* **2012**, *36*, 1–24. [CrossRef] [PubMed]

39. Bayram, O.; Krappmann, S.; Ni, M.; Bok, J.W.; Helmstaedt, K.; Valerius, O.; Braus-Stromeyer, S.; Kwon, N.-J.; Keller, N.P.; Yu, J.-H.; et al. VelB/VeA/LaeA complex coordinates light signal with fungal development and secondary metabolism. *Science* **2008**, *320*, 1504–1506. [CrossRef] [PubMed]

40. Baidya, S.; Duran, R.M.; Lohmar, J.M.; Harris-Coward, P.Y.; Cary, J.W.; Hong, S.-Y.; Roze, L.V.; Linz, J.E.; Calvo, A.M. VeA is associated with the response to oxidative stress in the aflatoxin producer *Aspergillus flavus*. *Eukaryot. Cell* **2014**, *13*, 1095–1103. [CrossRef] [PubMed]

41. Duran, R.M.; Cary, J.W.; Calvo, A.M. Production of cyclopiazonic acid, aflatrem, and aflatoxin by *Aspergillus flavus* is regulated by *veA*, a gene necessary for sclerotial formation. *Appl. Microbiol. Biotechnol.* **2007**, *73*, 1158–1168. [CrossRef] [PubMed]

42. Sprote, P.; Brakhage, A.A. The light-dependent regulator *velvet A* of *Aspergillus nidulans* acts as a repressor of the penicillin biosynthesis. *Arch. Microbiol* **2007**, *188*, 69–79. [CrossRef] [PubMed]

43. Ramamoorthy, V.; Dhingra, S.; Kincaid, A.; Shantappa, S.; Feng, X.; Calvo, A.M. The putative C2H2 transcription factor MtfA is a novel regulator of secondary metabolism and morphogenesis in *Aspergillus nidulans*. *PLoS ONE* **2013**, *8*, e74122. [CrossRef] [PubMed]

44. Zhuang, Z.; Lohmar, J.M.; Satterlee, T.; Cary, J.W.; Calvo, A.M. The master transcription factor *mtfA* governs aflatoxin production, morphological development and pathogenicity in the fungus *Aspergillus flavus*. *Toxins (Basel)* **2016**, *8*, 29. [CrossRef] [PubMed]

45. Jayashree, T.; Subramanyam, C. Oxidative stress as a prerequisite for aflatoxin production by *Aspergillus parasiticus*. *Free Radic. Biol. Med.* **2000**, *29*, 981–985. [CrossRef]

46. Jayashree, T.; Subramanyam, C. Antiaflatoxigenic activity of eugenol is due to inhibition of lipid peroxidation. *Lett. Appl. Microbiol.* **1999**, *28*, 179–183. [CrossRef] [PubMed]

47. Hong, S.-Y.; Roze, L.V.; Linz, J.E. Oxidative stress-related transcription factors in the regulation of secondary metabolism. *Toxins (Basel)* **2013**, *5*, 683–702. [CrossRef] [PubMed]

48. Chang, P.-K.P.; Scharfenstein, L.L.L.; Luo, M.; Mahoney, N.; Molyneux, R.J.; Yu, J.; Brown, R.L.; Campbell, B.C. Loss of *msnA*, a putative stress regulatory gene, in *Aspergillus parasiticus* and *Aspergillus flavus* increased production of conidia, aflatoxins and kojic acid. *Toxins (Basel)* **2011**, *3*, 82–104. [CrossRef]

49. Hong, S.-Y.; Roze, L.V.; Wee, J.; Linz, J.E. Evidence that a transcription factor regulatory network coordinates oxidative stress response and secondary metabolism in aspergilli. *Microbiologyopen* **2013**, *2*, 144–160. [CrossRef] [PubMed]

50. Ke, R.; Haynes, K.; Stark, J. Modelling the activation of alkaline pH response transcription factor PacC in *Aspergillus nidulans*: Involvement of a negative feedback loop. *J. Theor. Biol.* **2013**, *326*, 11–20. [CrossRef] [PubMed]

51. Cary, J.W.; Harris-Coward, P.Y.; Ehrlich, K.C.; Mack, B.M.; Kale, S.P.; Larey, C.; Calvo, A.M. NsdC and NsdD affect *Aspergillus flavus* morphogenesis and aflatoxin production. *Eukaryot. Cell* **2012**, *11*, 1104–1111. [CrossRef] [PubMed]

52. Gilbert, M.K.; Mack, B.M.; Wei, Q.; Bland, J.M.; Bhatnagar, D.; Cary, J.W. RNA sequencing of an *nsdC* mutant reveals global regulation of secondary metabolic gene clusters in *Aspergillus flavus*. *Microbiol. Res.* **2015**, *182*, 150–161. [CrossRef] [PubMed]

53. Jiang, J.; Liu, X.; Yin, Y.; Ma, Z. Involvement of a velvet protein FgVeA in the regulation of asexual development, lipid and secondary metabolisms and virulence in *Fusarium graminearum*. *PLoS ONE* **2011**, *6*, e28291. [CrossRef] [PubMed]

54. Wright, M.S.; Greene-Mcdowelle, D.M.; Zeringue, H.J.; Bhatnagar, D.; Cleveland, T.E. Effects of volatile aldehydes from *Aspergillus*-resistant varieties of corn on *Aspergillus parasiticus* growth and aflatoxin biosynthesis. *Toxicon* **2000**, *38*, 1215–1223. [CrossRef]

55. Yang, Q.; Borkovich, K.A. Mutational activation of a Gα(i) causes uncontrolled proliferation of aerial hyphae and increased sensitivity to heat and oxidative stress in *Neurospora crassa*. *Genetics* **1999**, *151*, 107–117. [PubMed]

56. Han, K.H.; Seo, J.A.; Yu, J.H. Regulators of G-protein signalling in *Aspergillus nidulans*: RgsA downregulates stress response and stimulates asexual sporulation through attenuation of GanB (Gα) signalling. *Mol. Microbiol.* **2004**, *53*, 529–540. [CrossRef] [PubMed]

57. Affeldt, K.; Carrig, J.; Amare, M.G.; Keller, N. Global survey of canonical *Aspergillus flavus* G protein-coupled receptors. *MBio* **2014**, *5*, 1501–1514. [CrossRef] [PubMed]

58. Brodhagen, M.; Keller, N.P. Signalling pathways connecting mycotoxin production and sporulation. *Mol. Plant Pathol.* **2006**, *7*, 285–301. [CrossRef] [PubMed]

59. Atoui, A.K.; Mansouri, A.; Boskou, G.; Kefalas, P. Tea and herbal infusions: Their antioxidant activity and phenolic profile. *Food Chem.* **2005**, *89*, 27–36. [CrossRef]

60. EL Mahgubi, A.; Bailly, S.; Tadrist, S.; Querin, A.; Ouadia, A.; Oswald, I.P.; Bailly, J.-D. Distribution and toxigenicity of *Aspergillus* section *Flavi* in spices marketed in Morocco. *Food Control* **2013**, *32*, 143–148. [CrossRef]

61. Tannous, J.; El Khoury, R.; Snini, S.P.; Lippi, Y.; El Khoury, A.; Atoui, A.; Lteif, R.; Oswald, I.P.; Puel, O. Sequencing, physical organization and kinetic expression of the patulin biosynthetic gene cluster from *Penicillium expansum*. *Int. J. Food Microbiol.* **2014**, *189*, 51–60. [CrossRef] [PubMed]

62. Fu, Z.; Huang, X.; Min, S. Rapid determination of aflatoxins in corn and peanuts. *J. Chromatogr. A* **2008**, *1209*, 271–274. [CrossRef] [PubMed]

Article

Protective Effects of Sporoderm-Broken Spores of *Ganderma lucidum* on Growth Performance, Antioxidant Capacity and Immune Function of Broiler Chickens Exposed to Low Level of Aflatoxin B_1

Tao Liu [1], Qiugang Ma [1,*], Lihong Zhao [1], Ru Jia [2], Jianyun Zhang [1], Cheng Ji [1] and Xinyue Wang [1]

[1] State Key Laboratory of Animal Nutrition, College of Animal Science and Technology, China Agricultural University, Beijing 100193, China; liuyitao2004@126.com (T.L.); lihongzhao100@126.com (L.Z.); jyzhang@cau.edu.cn (J.Z.); jicheng@cau.edu.cn (C.J.); xinyuewang2016@126.com (X.W.)

[2] College of Life Science, Shanxi University, Taiyuan 030006, Shanxi, China; jiaru.bjdk071@163.com

* Correspondence: maqiugang@cau.edu.cn; Tel.: +86-10-62731998; Fax: +86-10-62732774

Academic Editor: Ting Zhou

Received: 16 August 2016; Accepted: 21 September 2016; Published: 24 September 2016

Abstract: This study was conducted to investigate the toxic effects of aflatoxin B_1 (AFB_1) and evaluate the effects of sporoderm-broken spores of *Ganoderma lucidum* (SSGL) in relieving aflatoxicosis in broilers. A total of 300 one-day-old male Arbor Acre broiler chickens were randomly divided into four dietary treatments; the treatment diets were: Control (a basal diet containing normal peanut meal); AFB_1 (the basal diet containing AFB_1-contaminated peanut meal); SSGL (basal diet with 200 mg/kg of SSGL); AFB_1+SSGL (supplementation of 200 mg/kg of SSGL in AFB_1 diet). The contents of AFB_1 in AFB_1 and AFB_1+SSGL diets were 25.0 μg/kg in the starter period and 22.5 μg/kg in the finisher period. The results showed that diet contaminated with a low level of AFB_1 significantly decreased ($p < 0.05$) the average daily feed intake and average daily gain during the entire experiment and reduced ($p < 0.05$) serum contents of total protein IgA and IgG. Furthermore, a dietary low level of AFB_1 not only increased ($p < 0.05$) levels of hydrogen peroxide and lipid peroxidation, but also decreased ($p < 0.05$) total antioxidant capability, catalase, glutathione peroxidase, and hydroxyl radical scavenger activity in the liver and spleen of broilers. Moreover, the addition of SSGL to AFB_1-contaminated diet counteracted these negative effects, indicating that SSGL has a protective effect against aflatoxicosis.

Keywords: spores of *Ganoderma lucidum*; oxidative stress; aflatoxins; antioxidant capability

1. Introduction

Aflatoxins (AFBs) are the most common mycotoxins and are mainly produced by *Aspergillus flavus* and *Aspergillus parasiticus* [1]. Among all the AFBs, aflatoxin B_1 (AFB_1) is the most highly toxic contaminant in foods and feedstuffs, and is classified by the International Agency of Research on Cancer (IARC) as a Group 1 carcinogen. In poultry, consumption of AFB_1 can cause huge economic losses by retarding animal growth, increasing feed efficiency and the incidence of disease, and inducing damage to organs such as the liver and spleen [2,3]. Moreover, the carry-over of AFB_1 through animal-derived products into the human food chain is a potential threat to human health due to its hepatotoxicity, carcinogenicity, mutagenicity, teratogenicity, and immunosuppression [4,5]. AFB_1 is the most widespread oxidative agent of the AFBs [6], and it was reported that the toxic effects of AFB_1 were intimately linked with its pro-oxidant which could induce the generation of reactive oxygen species (ROS) and lead to the impairment of DNA, RNA, proteins, lipids, and other molecules [7,8]. Therefore, some studies suggested that the addition of antioxidants to diets could protect animals against AFB$_1$-induced toxicity by enhancing the antioxidant system and immunity [9,10].

Ganoderma lucidum (*G. lucidum*), a traditional folk medicinal mushroom, has been used as an important longevity and health-promoting herb for more than 2000 years in China [11]. The spores of *Ganoderma lucidum* (SGL), ejected from the pileus of *G. lucidum* in the mature phase, are tiny and mist-like particles of about 6.5–8.0 × 9.6–12.6 μm enwrapped with an outer bilayer of sporoderm [12]. During the past two decades, because the spores could be collected on a large scale, SGL has attracted extensive interest as studies revealed that the spores possess many bioactive substances, including polysaccharides, unsaturated fatty acids, triterpenoids, nucleotides, ergostero and other bioactive ingredients [13,14]. In vitro, the polysaccharides of SGL had strong bioactivity in scavenging 1,1-Diphenyl-2-picrylhydrazyl and oxygen radicals [15]. In addition, antioxidant activities of polysaccharides from SGL were also demonstrated in rats [16]. Furthermore, the polysaccharides from SGL had potent stimulating effects on spleen mass, lymphocyte proliferation and antibody production in mice [17,18]. Fatty acids are another kind of potential active ingredient of SGL, which showed high bioactivities in antitumor and immunomodulation in rats [19,20]. However, the bioactivities of spores are closely related to the status of the sporoderm; the intact sporoderm can inhibit the release and absorption of bioactive substances in the spores [21]. Yue et al. [22] observed that dietary sporoderm-broken spores of *Ganoderma lucidum* (SSGL) were more effective in stimulating the production of interferon-γ, interleukin-2, interleukin-4 and interleukin-6 in mice than dietary sporoderm-unbroken spores of *G. lucidum*. Therefore, we speculated that supplementation of SSGL to AFB$_1$-contaminated diets might alleviate aflatoxicosis through decreasing the oxidative status and elevating the antioxidant defense system.

Our previous study showed that different levels of SSGL (100, 200 and 500 mg/kg) improved the average daily gain, decreased the feed:gain ratio, and enhanced the antioxidant capacity in the liver and spleen of broilers. In this study, SSGL was chosen to evaluate its protective effects on growth performance, antioxidant function and serum immunoglobulins of male Arbor Acre broilers exposed to a low level of AFB$_1$.

2. Results

2.1. Effects of SSGL on Growth Performance of Male Arbor Acre Broiler Chickens Exposed to Low Level of AFB$_1$

In 0–21 d, broilers fed a diet contaminated with AFB$_1$ resulted in a significant decrease ($p < 0.05$) in average daily feed intake (ADFI) as compared to the control group (Table 1). The addition of SSGL into the diet contaminated with AFB$_1$ significantly increased ($p < 0.05$) the ADFI of broilers as compared to the AFB$_1$ group. The ADFI of broilers among control, SSGL and AFB$_1$+SSGL groups showed no significant differences ($p > 0.05$). The average daily gain (ADG) and feed:gain ratio (F:G) were not significantly affected ($p > 0.05$) by a low level of AFB$_1$ or dietary SSGL in broilers.

In 22–44 d, broilers in the AFB$_1$ group had significantly lower ($p < 0.05$) ADFI and ADG than those in the control group. However, supplementation of SSGL to the AFB$_1$-contaminated diet obviously increased ($p < 0.05$) ADFI, ADG and decreased ($p < 0.05$) F:G as compared with the AFB$_1$ group. In addition, broilers in the SSGL and AFB$_1$+SSGL groups had significantly higher ($p < 0.05$) ADG than those in the control group, but ADFI, ADG and F:G between the SSGL and AFB$_1$+SSGL groups showed no significant differences ($p > 0.05$). Similar effects of SSGL on the growth performance (ADG, ADFI and F:G) of broilers exposed to a low level of AFB$_1$ were also observed during the entire experiment (0–44 d).

Table 1. Effects of SSGL on growth performance of male Arbor Acre broiler chickens exposed to a low level of AFB$_1$.

Index [1]	Control	AFB$_1$	SSGL	AFB$_1$+SSGL	SEM	*p*-Value
			0–21 d			
ADFI (g/d)	53.0 [a]	51.0 [b]	52.2 [ab]	53.0 [a]	0.296	0.039
ADG (g/d)	33.5	32.9	33.3	33.1	0.141	0.441
F:G	1.58	1.55	1.57	1.60	0.011	0.425
			22–44 d			
ADFI (g/d)	135.7 [a]	132.0 [b]	138.0 [a]	137.9 [a]	0.776	0.008
ADG (g/d)	70.4 [b]	67.8 [c]	74.0 [a]	73.8 [a]	0.632	<0.001
F:G	1.93 [ab]	1.95 [a]	1.87 [b]	1.87 [b]	0.012	0.011
			0–44 d			
ADFI (g/d)	98.1 [a]	95.2 [b]	99.0 [a]	99.4 [a]	0.515	0.006
ADG (g/d)	53.6 [b]	51.9 [c]	55.5 [a]	55.3 [a]	0.360	<0.001
F:G	1.83 [a]	1.84 [a]	1.78 [b]	1.80 [ab]	0.008	0.069

[1] Data are expressed as group mean values; ADFI, average daily feed intake; ADG, average daily gain; F:G, feed:gain ratio, equal to ADFI/ ADG; SEM, standard error of the mean; [a-c] Means within the same row with different superscripts are significantly different ($p < 0.05$).

2.2. Effects of SSGL on Oxidative Status in Liver and Spleen of Male Arbor Acre Broiler Chickens Exposed to Low Level of AFB$_1$

The data in Table 2 showed that the hepatic levels of H$_2$O$_2$, MDA and LPO of broilers in the AFB$_1$ group were significantly higher ($p < 0.05$) than those in the control group, but these indexes were markedly decreased ($p < 0.05$) by the supplementation of SSGL into the diet contaminated with AFB$_1$. Moreover, broilers in the SSGL group had significantly lower ($p < 0.05$) hepatic levels of H$_2$O$_2$ and MDA than those in the control group. A higher ($p < 0.05$) hepatic H$_2$O$_2$ level and lower ($p < 0.05$) hepatic MDA level of broilers were observed in the AFB$_1$ group as compared to the control group, but the hepatic LPO level of broilers showed no difference ($p > 0.05$) among the control, SSGL and AFB$_1$+SSGL groups.

Table 2. Effects of SSGL on oxidative status in liver and spleen of male Arbor Acre broiler chickens exposed to a low level of AFB$_1$.

Index [1]	Control	AFB$_1$	SSGL	AFB$_1$+SSGL	SEM	*p*-Value
			Liver			
H$_2$O$_2$ (mmol/g prot)	10.74 [c]	15.06 [a]	9.60 [d]	12.38 [b]	0.540	<0.001
MDA (nmol/g prot)	2.57 [b]	3.10 [a]	2.07 [d]	2.26 [c]	0.102	<0.001
LPO (μmol/g prot)	0.82 [b]	1.02 [a]	0.82 [b]	0.85 [b]	0.024	<0.001
			Spleen			
H$_2$O$_2$ (mmol/g prot)	14.38 [b]	16.19 [a]	14.18 [b]	14.23 [b]	0.252	0.001
MDA (nmol/g prot)	1.87 [ab]	1.95 [a]	1.80 [b]	1.84 [ab]	0.019	0.046
LPO (μmol/g prot)	4.20 [b]	5.08 [a]	3.34 [c]	3.04 [d]	0.207	<0.001

[1] Data are expressed as group mean values; H$_2$O$_2$, hydrogen peroxide; MDA, malondiadehyde; LPO, lipid peroxidation. SEM, standard error of the mean; [a-d] Means within the same row with different superscripts are significantly different ($p < 0.05$).

Similar to the liver, the splenic levels of H$_2$O$_2$ and LPO of broilers in the AFB$_1$ group were significantly increased ($p < 0.05$) as compared with those in the control group (Table 2). However, the addition of SSGL in the diet contaminated with AFB$_1$ significantly reduced ($p < 0.05$) H$_2$O$_2$ and LPO levels in the spleen compared with the AFB_1 group. In addition, the splenic level of LPO in the SSGL and AFB_1+SSGL groups was significantly lower ($p < 0.05$) than that in the control group, and broilers in the SSGL group had a lower ($p < 0.05$) MDA level than those in AFB_1 group. These results showed that diets contaminated with AFB_1 could aggravate the oxidative status of broilers, which could be relieved by the supplementation of SSGL.

2.3. Effects of SSGL on Antioxidant Defense System in Liver and Spleen of Male Arbor Acre Broiler Chickens Exposed to Low Level of AFB

Broilers fed a diet contaminated with AFB_1 had significantly decreased ($p < 0.05$) T-AOC, CAT, GR, GSH-Px and HRSA levels in the liver as compared with those in the control group (Table 3), while the addition of SSGL into the diet contaminated AFB_1 significantly enhanced ($p < 0.05$) the levels of T-AOC, CAT, GR, HRSA and GSH in the liver. When compared to the control group, hepatic enzyme activities (T-AOC, CAT, GR) and the GSH level of broilers were significantly increased ($p < 0.05$) in the SSGL group, whereas a significantly lower ($p < 0.05$) hepatic activity of GSH-Px and a higher ($p < 0.05$) level of GSH were observed in broilers in the AFB_1+SSGL group. The hepatic activity of T-SOD of broilers exposed to AFB_1 was not significantly affected ($p > 0.05$) by dietary supplementation with SSGL.

Table 3. Effects of SSGL on antioxidant defense system in liver and spleen of male Arbor Acre broiler chickens exposed to a low level of AFB_1.

Index [1]	Control	AFB_1	SSGL	AFB_1+SSGL	SEM	*p*-Value
			Liver			
T-AOC (U/mgprot)	2.19 [b]	2.04 [c]	2.32 [a]	2.28 [ab]	0.032	<0.001
CAT U/mgprot)	8.13 [b]	7.35 [c]	9.43 [a]	8.05 [b]	0.205	<0.001
T-SOD (U/mgprot)	16.95	16.78	16.12	16.58	0.227	0.647
GR (U/gprot)	4.47 [b]	3.74 [c]	6.74 [a]	4.84 [b]	0.297	<0.001
GSH-Px (U)	43.25 [a]	37.50 [b]	43.49 [a]	36.63 [b]	0.853	<0.001
GSH (mg/gprot)	2.03 [b]	2.07 [b]	2.43 [a]	2.44 [a]	0.058	0.001
HRSA (U/g prot)	2.42 [a]	1.98 [b]	2.58 [a]	2.45 [a]	0.062	<0.001
			Spleen			
T-AOC (U/mgprot)	2.46 [b]	1.78 [c]	3.28 [a]	2.52 [b]	0.139	<0.001
CAT (U/mgprot)	5.35 [a]	4.33 [b]	5.42 [a]	4.48 [b]	0.131	<0.001
T-SOD (U/mgprot)	22.22 [a]	19.27 [b]	22.11 [a]	21.64 [a]	0.363	0.001
GR (U/gprot)	6.34 [c]	6.02 [c]	9.25 [a]	7.52 [b]	0.346	<0.001
GSH-Px (U)	47.73 [a]	43.39 [b]	47.59 [a]	44.94 [b]	0.542	<0.001
GSH (mg/gprot)	3.92 [b]	3.64 [c]	4.32 [a]	4.36 [a]	0.079	<0.001
HRSA (U/g prot)	1.87 [c]	1.61 [d]	2.48 [a]	2.06 [b]	0.084	<0.001

[1] Data are expressed as group mean values; T-AOC, total antioxidant capability; CAT, catalase; GR, glutathione reductase; GSH-Px, glutathione peroxidase; GSH, reduced glutathione; HRSA, Hydroxyl radical scavenger activity; SEM, standard error of the mean. [a-d] Means within the same row with different superscripts are significantly different ($p < 0.05$).

Similar negative effects of AFB_1 and ameliorative effects of dietary SSGL on the antioxidant defense system were also found in the spleen (Table 3). The splenic levels of T-AOC, CAT, T-SOD, GSH-Px, HRSA and GSH of broilers in the AFB_1 group were lower ($p < 0.05$) than those in the control group. However, supplementation of SSGL to the AFB_1-contaminated diet significantly enhanced ($p < 0.05$) splenic levels of T-AOC, T-SOD, GR, HRSA and GSH as compared to the AFB_1 group. In addition, broilers in the SSGL group had significantly higher T-AOC, GR, GSH, HRSA levels than those in the control group, and significantly lower ($p < 0.05$) activities of CAT, GSH-Px, and higher ($p < 0.05$) activities of GR, HRSA and GSH were observed in the AFB_1+SSGL group as compared to the control group. The splenic activities of T-AOC, CAT, GR, GSH-Px, HRSA in the SSGL group were higher ($p < 0.05$) than those in the AFB_1+SSGL group.

2.4. Effects of SSGL on Serum Total Protein and Immunoglobulins of Male Arbor Acre Broiler Chickens Exposed to Low Level of AFB₁

The contents of serum TP, IgA and IgG of broilers exposed to a low level of AFB_1 were significantly lower ($p < 0.05$) than those in the control group (Table 4), while these indexes were significantly improved by the addition of SSGL to the diet contaminated with AFB_1. However, the contents of serum TP, IgA and IgG of broilers among the control, SSCL and AFB_1+SSGL groups showed no significant differences ($p > 0.05$). There was no significant difference ($p > 0.05$) in the serum IgM of broilers among all the groups.

Table 4. Effects of SSGL serum immunoglobulins of male Arbor Acre broiler chickens exposed to a low level of AFB_1.

Index [1]	Control	AFB_1	SSGL	AFB_1+SSGL	SEM	*p*-Value
TP (g/L)	24.82 [a]	22.73 [b]	25.94 [a]	25.07 [a]	0.374	0.007
IgA (g/L)	0.291 [a]	0.284 [b]	0.298 [a]	0.295 [a]	0.002	0.003
IgG (g/L)	0.266 [a]	0.232 [b]	0.273 [a]	0.257 [a]	0.005	<0.001
IgM (g/L)	0.224	0.225	0.229	0.236	0.002	0.288

[1] Data are expressed as group mean values; TP, total protein; SEM, standard error of the mean; [a,b] Means within the same row with different superscripts are significantly different ($p < 0.05$).

3. Discussion

3.1. Toxicity of AFB₁ on Growth Performance

Feedstuffs are easily contaminated with AFB_1 during harvesting, transportation and storage, and it was reported that the detection rates of AFB_1 were 50% in corn, 36% in soybean meal, 94% in DDGS, respectively, in the Beijing region [23], and 15.39% in peanuts in the south of China [24]. Therefore, the presence of AFB_1 in animals' diets is hard to avoid, and it is increasingly recognized that the long-term consumption of low levels of AFBs is detrimental to animals' growth and health. In this study, diet contaminated with AFB_1 caused a significant decrease in the ADFI of broilers during the starter period (0–21 d), and not only the ADFI but also the ADG of broilers during the finisher period (22–44 d) were significantly reduced by dietary AFB_1. Similar results had been observed in the previous studies, in which an AFB_1-containing diet (50–100 µg/kg) could significantly decrease the body weight gain and feed consumption of broilers, causing economic loss [2,25]. Ducks fed diets containing 20 µg/kg of AFB_1 had a significantly lower ADG, ADF, and higher F:G [26]. These results indicated that diets containing a low level of AFB_1 (22.5–25.0 µg/kg) could inhibit broiler growth and cause great financial losses.

3.2. Toxicity of AFB₁ on Oxidative Stress of Broilers

Liver and immune system organs such as the spleen are considered to be sensitive to AFB_1 [27,28]. It had been reported that the toxic effect elicited by AFB_1 could be closely related to the generation of reactive oxygen species (ROS), mainly including superoxide (O_2^-), hydrogen peroxide (H_2O_2), and hydroxyl ($OH^·$) [29]. H_2O_2 is released from the mitochondria; hydroxyl radicals ($OH^·$) are also an extremely reactive free radical. Biosynthesis and accumulation of ROS are central to oxidative stress-related metabolism [30]. The exceeded ROS could enhance lipid peroxidation, which will impair membrane function by decreasing membrane fluidity and changing the activities of membrane-bound enzymes and receptors [31]. MDA is formed at the end of lipid peroxidation, and reflects the degree of the whole lipid oxidation in the body. In this study, the contents of H_2O_2, MDA and LPO in the liver and spleen of broilers were significantly increased by dietary AFB_1, indicating that a low level of AFB_1 (22.5–25.0 µg/kg) could increase the oxidative status in the liver and spleen of broilers. Similar toxic effects of AFB_1 on oxidative status were observed in the liver and spleen of rats [32], and broilers [33,34].

The levels of free radical molecules and lipid peroxidation are controlled by an antioxidant defense system, consisting of enzymatic components such as SOD, CAT, and GR, and non-enzymatic components such as GSH and vitamin E [35]. When oxidative stress arises as a pathologic event, the defense system will promote the regulation and expression of enzymatic and non-enzymatic components [36]. SOD can detoxicate O_2^- to H_2O_2, and then GSH-Px and CAT catalyze H_2O_2 directly to water and oxygen at the expense of GSH [37,38]. Many studies observed that dietary AFB_1 could decrease the activities of antioxidant enzymes and levels of non-enzymatic antioxidants in broilers [9,39,40]. In the present study, the antioxidant enzyme activities and GSH level in the liver and spleen of broilers fed a diet contaminated with AFB_1 were significantly lower than those in the control group, suggesting that the diet contaminated with a low level of AFB_1 (22.5–25.0 µg/kg) could suppress the antioxidant capacity of broilers.

3.3. Toxicity of AFB$_1$ on Serum Total Protein and Immunoglobulins of Broilers

Except for the toxic effects of AFB_1 on the liver and spleen, the immunosuppression in animals is also a matter of concern, because it may predispose farm animals to infectious diseases and result in economic losses [41]. Immunoglobulins, secreted by B cells, are the main antibody isotypes of the serum and extracellular fluid immune system, thereby allowing them to control the infection of body tissues. AFB_1 is known to be immunosuppressive in birds, and it has been reported that diets containing 300 µg/kg of AFB_1 significantly reduced the serum IgA, IgG and IgM of broilers [42]. In addition, chickens fed aflatoxin at a concentration of 2500 µg/kg had lower serum total protein (TP), albumin and IgG, but serum IgM was not affected by AFBs [43]. Serum TP is the indicator of protein synthesis, and the decreased serum TP induced by dietary AFB_1 may contribute to the decreased contents of immunoglobulins [44,45]. In the present study, broilers fed diets contaminated with a low level of AFB_1 (22.5–25.0 µg/kg) had significantly lower serum TP, IgA and IgG, while serum IgM was not significantly affected by dietary AFB_1. One mechanism of action of AFBs is related to the inhibition of protein synthesis, which may be responsible for the decrease in serum IgA and IgG in this study.

3.4. Effects of SSGL

Seeking effective ways to alleviate the negative effects of AFB_1 has attracted more and more attention. Nowadays, researchers found that different feed additives had the ability to relieve aflatoxicosis. Alpha-lipoic acid, known as an "ideal antioxidant", could prevent hepatic oxidative stress and down-regulate the expression of hepatic pro-inflammatory cytokines of broilers exposed to a low level of AFB_1 [46]. Selenium, playing important roles in immune function, could protect chickens from AFB_1-induced impairment of humoral and cellular immune function by reducing bursal histopathological lesions and percentages of apoptotic bursal cells [47]. Vitamin E, as an antioxidant, could ameliorate AFB_1-induced toxicity in rats [9] and in ducks [48]. Moreover, some plants or their extracts, such as Chinese cabbage powder [49] and tinospora cordifolia root extract [50], were demonstrated to counteract the detrimental toxic effects of AFBs. Although SGL, as a traditional Chinese herb, is characterized as having antioxidation, immunomodulation and hepatoprotection activities, there is no study to assess the protective effects of SGL or its extracts on AFB_1-induced toxins in broiler chickens. In the present study, the results showed that the addition of SSGL to an AFB_1-contaminated diet relieved the negative effects of AFB_1 on the growth performance of broilers.

SGL has strong bioactivities in antioxidation, and polysaccharides from SGL had been reported to have the ability to scavenge DPPH radicals, increase reducing power and inhibit lipid peroxidation in vitro [51]. Moreover, SGL was effective in protection against Cd(II)-induced hepatotoxicity through reducing the increase in serum AST, ALT and hepatic MDA of mice [52]. Supplementation with 4 or 8 g/kg of SGL significantly increased the activities of GSH and GR, and decreased the MDA content in the hippocampus of rats [15]. In type 2 diabetic rats, the serum level of MDA was 13.9% lower, and serum levels of GSH-Px and SOD were 25.9% and 38.0% higher at four weeks in the SGL group [53]. Some antioxidants such as vitamin E and alpha-lipoic acid are reported to be effective in protecting animals from oxidative damage induced by AFB_1, so SSGL may have similar abilities. Indeed, lower oxidative status and higher antioxidant capacity were observed in broilers fed on a diet supplemented with SSGL in the present study, indicating that dietary SSGL could protect broilers from oxidative damage induced by AFB_1.

In the present study, the results showed that dietary SSGL significantly increased the contents of serum TP, IgA and IgG, and decreased the AFB_1-induced toxins to serum immunity. Mohan et al. [54] observed that dietary polysaccharides of *G. lucidum* significantly increased the muscle TP of *Macrobrachium rosenbergii*. Finisher pigs fed diets that received polysaccharides of *G. lucidum* had higher contents of serum TP, globulin and IgG [55]. These results suggested that *G. lucidum* may have the ability to increase the protein synthesis of animals. Di et al. [56] found that the high-performance thin-layer chromatography fingerprint profiles of carbohydrate and acid hydrolyzates of polysaccharides from *G. lucidum* and the spores were quite similar, indicating that they may have similar bioactivities. The $(1 \rightarrow 3)$-α-D-glucan of SGL enhanced the T and B lymphocyte proliferation and antibody production of rats [17]. Thus, the mechanism of SSGL improving serum IgA and IgG in this study could be partially due to the decrease in the inhibition of protein synthesis induced by AFB_1.

4. Conclusions

The results from this study demonstrated that diets contaminated with AFB_1 (22.5–25.0 µg/kg) suppressed the growth performance, antioxidant capacity and immune action of broilers. However, the addition of SSGL significantly counteracted the adverse effects of AFB_1, effectively improving growth performance and reducing the oxidative stress and immunosuppression of broilers; so SSGL, as a feed additive for inhibiting aflatoxicosis, may have promising potential in feed industrial applications.

5. Materials and Method

5.1. Sporoderm-Broken Spores of Ganoderma lucidum

SGL in present study was supplied by Riverside Ganoderma Lucidum Planting Co. Ltd. (Xiuyan Manchu Autonomous County, Anshan, Liaoning Province, China). The spores was dried at 55 °C for 24 h, and then broken by supercritical fluid extraction device (Nantong Hua'an Co. Ltd., Nantong, China). Briefly, approximately 150 g of spores was loaded into a steel cylinder equipped with mesh filters (6.5 µm) on both ends. Liquefied CO_2 was pumped into the vessel and the pressure was raised to 35 MPa consequently. The process was last for 4 h and the temperature was set at 25 °C by the temperature controller. At the end of process, the pressure was quickly released within 1 min, and the spores was broken due to rapid depressurization. The SSGL were removed and stored at −20 °C before adding into the diets [12,57].

5.2. Birds, Diets and Management

This study was approved by the Animal Care and Use Committee of the China Agricultural University (ethical approval code: CAU20151028-2; Date: 28 October 2015). A total of 300 one-day-old male Arbor Acre broiler chickens were obtained from a commercial company (Beijing Huadu Yukou Poultry Co., Ltd., Beijing, China). Birds with similar body weights (40.0 ± 1.0 g) were randomly assigned to four treatments with five replicate pens of 15 birds per pen. The diets for treatments were: Control (the basal diets containing 20% normal peanut meal in the starter and 18% in the finisher period, without any mycotoxins); AFB_1 (moldy peanut meal naturally contaminated with 121.7 µg/kg AFB_1 substituting for all the normal peanut meal in the basal diet); SSGL (basal diet received 200 mg/kg of SSGL); AFB_1+SSGL (AFB_1 diet supplemented with 200 mg/kg of SSGL). The contents of mycotoxins (AFB_1, AFB_2, AFG_1, AFG_2, deoxynivalenol (DON), zearalenone (ZEA) and ochratoxin (OTA)) in basal diet and AFB_1-contaiminated diets were tested with HPLC (Shimadzu LC-10 AT, Shimadzu, Tokyo, Japan) method [25,58]. Briefly, 25 g of the milled samples were mixed with 80 mL of methanol-water (80:20 v/v) for aflatoxins; water for DON; acetonitrile-water (70:30 v/v) for ZEA; methanol-water (60:40 v/v) for OTA, the mixtures were shaken for 2 h. The extracts were filtered, and the filtrate was cleaned up through an immunoaffinity column (Vicam, Milford, MA, USA) before HPLC determination. The limits of detection (LOD) and quantification (LOQ) were: 0.10 and 0.50 μg/kg for AFB_1; 0.08 and 0.25 μg/kg for AFB_2; 0.15 and 0.50 μg/kg for AFG_1; 0.10 and 0.50 μg/kg for AFG_2; 1.00 and 5.00 μg/kg for ZEA; 10.00 and 50.00 μg/kg for DON; 0.21 and 0.60 μg/kg for OTA, respectively. The final contents of AFB_1 in diets received moldy peanut meal (AFB_1 and AFB_1+SSGL groups) were 25.0 ± 1.2 μg/kg in starter period and 22.5 ± 1.1 μg/kg in finisher period; the other mycotoxins were determined to be at concentrations below detection limits. Hardly any aflatoxins or other mycotoxins were found (below detection limits) in basal diets.

The formulation of the basal diets are presented in Table 5, all essential nutrients in the basal diet met NRC (1994) [59]. The feeding trial period lasted for 44 days.

Table 5. Ingredients and nutrient composition of the basal diet.

Ingredient (%)	Starter Period (0–21 d)	Finisher Period (22–44 d)
Corn	57.95	60.79
Soybean meal	14.20	13.30
Peanut meal	20.00	18.00
Limestone	1.30	1.00
Dicalcium phosphate	1.80	1.50
Salt	0.30	0.30
Soybean oil	3.00	4.00
Lysine [98.5%]	0.46	0.34
DL-Methionine	0.37	0.20
Threonine	0.19	0.14
Choline chloride	0.10	0.10
Vitamin premix [1]	0.03	0.03
Mineral premix [2]	0.30	0.30
Total	100.000	100.00
Nutrition componen		
CP	21.48	20.03
ME(Mcal/kg)	2.99	3.09
Ca	1.01	0.82
Total phosphorus	0.67	0.61
Lys	1.15	1.01
Met	0.63	0.45

[1] Provided per kilogram of diet: vitamin A, 15,000 IU; cholecalciferol, 3000 IU; vitamin E, 20 IU; vitamin K 3, 2.18 mg; thiamine, 2.15 mg; riboflavin, 8.00 mg; pyridoxine, 4.40 mg; vitamin B 12, 0.02 mg; Calcium pantothenate, 25.60 mg; nicotinic acid, 65.80 mg; folic acid, 0.96 mg; biotin, 0.20 mg. [2] Provided per kilogram of diet: Fe, 100.0 mg; Cu, 8.0 mg; Zn, 78.0 mg; Mn, 105.0 mg; I, 0.5 mg; Se, 0.3 mg.

The trial was conducted in two periods consisting of a starter period from day 1 to 21 and a finisher period from day 22 to 44. The rearing system was designed as flooring rearing. All birds were raised in weird-floored pens ($100 \times 100 \times 60$ cm) in an environmentally controlled room with continuous light. The temperature was maintained at 35–37 °C (65% relative humidity) for the first two days, and 33–35 °C for day 3–7. The room temperature was gradually decreased by 2 °C per week until to 24 °C, and then maintained unchanged. Water and diets were provided ad libitum. All birds were inoculated with Newcastle disease vaccine on day 7 and day 21 and inoculated with infectious bursa disease vaccine on day 14 and day 21.

5.3. Sample Collection

At 44 days of age, one bird close to the average weight was selected from each pen. After the birds were fasted for 12 h, blood sample was collected in tubes (without anticoagulant) from a wing vein with a 5 mL syringe. The samples were centrifuged at $1000 \times g$ for 10 min, and the serum was separated and then stored at −20 °C until analysis. The chickens were then sacrificed; the liver and spleen were removed immediately. A portion of the liver and the whole spleen were snap frozen in liquid nitrogen and stored at −70 °C.

5.4. Oxidative Stress Indices in Liver and Spleen

Liver and spleen tissues (about 1 g) were cut into small pieces and homogenized in ice-cold saline buffer (1:9, wt/v) using an Ultra-turrax (T8, IKA-labortechnik, Staufen, Germany) to form homogenate at a concentration of 0.1 g/mL for further analysis. The homogenate was centrifuged at $1000\times g$ for 10 min at 4 °C, and the supernatant was collected and stored in freezer for assays of T-AOC, CAT, GR, GSH-PX, GSH, MDA, LPO, H_2O_2, hydroxyl radical scavenger activity (HRSA) and TP. All the assays were measured with commercial assay kits (Nanjing Jiancheng Bioengineering Institute, Nanjing, China) and the procedures accordingly.

5.5. Serum Concentrations of Total Protein and Immunoglobulins

The contents of serum TP, IgA, IgM and IgG were measured using commercial kits from Nanjing Jiancheng Bioengineering Institute. The measurements were performed according to the detection kit instructions.

5.6. Statistical Analyses

All data were subjected to analysis by one-way ANOVA using SAS (Version 8e, SAS Institute, Cary, NC, USA). Duncan's multiple range test was used for multiple comparisons when a significant difference was detected. Means were considered significantly different at $p < 0.05$.

Acknowledgments: We appreciate the support of the Program for New Century Excellent Talents in the University (NCET-13-0558), the Public sector (Agriculture) scientific research of China (Grant No. 201403047), and the National Key Technology Research and Development Program of China (Grant No. 2012BAD39B00 and Grant No. 2011BAD26B04).

Author Contributions: Qiugang Ma and Lihong Zhao conceived and designed the experiments; Tao Liu, and Xinyue Wang performed the experiments; Tao Liu and Ru Jia analyzed the data; Jianyun Zhang and Cheng Ji contributed reagents and materials; Tao Liu and Lihong Zhao wrote the paper.

Conflicts of Interest: The authors declare no conflict of interest.

References

1. Diener, U.L.; Cole, R.J.; Sanders, T.H.; Payne, G.A.; Lee, L.S.; Klich, M.A. Epidemiology of aflatoxin formation by *Aspergillus flavus*. *Annu. Rev. Phytopathol.* **1987**, *25*, 249–270. [CrossRef]

2. Bintvihok, A.; Kositcharoenkul, S. Effect of dietary calcium propionate on performance, hepatic enzyme activities and aflatoxin residues in broilers fed a diet containing low levels of aflatoxin B_1. *Toxicon* **2006**, *47*, 41–46. [CrossRef] [PubMed]

3. Ortatatli, M.; Oguz, H.; Hatipoğlu, F.; Karaman, M. Evaluation of pathological changes in broilers during chronic aflatoxin (50 and 100 ppb) and clinoptilolite exposure. *Res. Vet. Sci.* **2005**, *78*, 61–68. [CrossRef] [PubMed]

4. Meissonnier, G.M.; Pinton, P.; Laffitte, J.; Cossalter, A.M.; Gong, Y.Y.; Wild, C.P.; Bertin, G.; Galtier, P.; Oswald, I.P. Immunotoxicity of aflatoxin B_1: Impairment of the cell-mediated response to vaccine antigen and modulation of cytokine expression. *Toxicol. App. Pharm.* **2008**, *231*, 142–149. [CrossRef] [PubMed]

5. Zimmermann, C.E.P.; Cruz, I.B.M.; Cadoná, F.C.; Machado, A.K.; Assmann, C.; Schlemmer, K.B.; Zanette, R.A.; Leal, D.B.R.; Santurio, J.M. Cytoprotective and genoprotective effects of b-glucans against aflatoxin B1-induced DNA damage in broiler chicken lymphocytes. *Toxicol. In Vitro* **2015**, *29*, 538–543. [CrossRef] [PubMed]

6. Alpsoy, L.; Yalvac, M.E.; Litwack, G. Key Roles of Vitamins A, C, and E in Aflatoxin B1-Induced Oxidative Stress. *Vitam. Horm.* **2011**, *86*, 287–305. [PubMed]

7.   Kezis, K.A.G.; Mulder, J.E.; Massey, T.E. In vivo treatment with aflatoxin B$_1$ increases DNA oxidation, base excision repair activity and 8-oxoguanine DNA glycosylase 1 levels in mouse lung. *Toxicology* **2014**, *321*, 21–26. [CrossRef] [PubMed]

8.   Mary, V.S.; Theumer, M.G.; Arias, S.L.; Rubinstein, H.R. Reactive oxygen species sources and biomolecular oxidative damage induced by aflatoxin B$_1$ and fumonisin B$_1$ in rat spleen mononuclear cells. *Toxicology* **2012**, *302*, 299–307. [CrossRef] [PubMed]

9.   Li, Y.; Ma, Q.G.; Zhao, L.H.; Guo, Y.Q.; Duan, G.X.; Zhang, J.Y.; Ji, C. Protective Efficacy of Alpha-lipoic Acid against AflatoxinB$_1$-induced Oxidative Damage in the Liver. *Asian Australas. J. Anim. Sci.* **2014**, *27*, 907–915.

10.   Ahmed, A.M.; Hamid, A.; Firgany, A.E.D.L. Vitamin E supplementation ameliorates aflatoxin B1-induced nephrotoxicity in rats. *Acta. Histochemica.* **2015**, *117*, 767–779.

11.   Bishop, K.S.; Kao, C.H.J.; Xu, Y.Y.; Glucina, M.P.; Paterson, R.R.M.; Ferguson, L.R. From 2000 years of *Ganoderma lucidum* to recent developments in nutraceuticals. *Phytochemistry* **2015**, *114*, 56–65. [CrossRef] [PubMed]

12.   Fu, Y.J.; Liu, W.; Zu, Y.G.; Shi, X.G.; Liu, Z.G.; Schwarz, G.; Efferth, T. Breaking the spores of the fungus *Ganoderma lucidum* by supercritical CO$_2$. *Food Chem.* **2009**, *112*, 71–76. [CrossRef]

13.   Fukuzawa, M.; Yamaguchi, R.; Hide, I.; Chen, Z.; Hirai, Y.; Sugimoto, A.; Yasuhara, T.; Nakata, Y. Possible involvement of long chain fatty acids in the spores of *Ganoderma lucidum* (Reishi Houshi) to its anti-tumor activity. *Biol. Pharm. Bull.* **2008**, *31*, 1933–1937. [CrossRef] [PubMed]

14.   Wang, S.Q.; Li, X.J.; Zhou, S.B.; Sun, D.X.; Wang, H.; Cheng, P.F.; Ma, X.R.; Liu, L.; Liu, J.X.; Wang, F.F.; et al. Intervention Effects of Spores of *Ganoderma lucidum* on Epileptiform Discharge Hippocampal Neurons and Expression of Neurotrophin-4 and N-Cadherin. *PLoS ONE* **2013**, *8*. [CrossRef] [PubMed]

15.   Heleno, S.A.; Barros, L.; Martins, A.; Queiroz, M.J.T.P.; Celestino, S.B.; Ferreira, I.C.F.R. Fruiting body, spores and in vitro produced mycelium of *Ganoderma lucidum* from Northeast Portugal: A comparative study of the antioxidant potential of phenolic and polysaccharidic extracts. *Food Res. Int.* **2012**, *46*, 135–140. [CrossRef]

16.   Zhou, Y.; Qu, Z.Q.; Zeng, Y.S.; Lin, Y.K.; Li, Y.; Chung, P.; Wong, R.; Hägg, U. Neuroprotective effect of preadministration with *Ganoderma lucidum* spore on rat hippocampus. *Exp. Toxicol. Pathol.* **2012**, *64*, 673–680. [CrossRef] [PubMed]

17.   Bao, X.F.; Duan, J.Y.; Fang, X.Y.; Fang, J.N. Chemical modifications of the $(1 \to 3)$-α-d-glucan from spores of *Ganoderma lucidum* and investigation of their physicochemical properties and immunological activity. *Carbohydr. Res.* **2001**, *336*, 127–140. [CrossRef]

18.   Bao, X.F.; Liu, C.P.; Fang, J.N.; Li, X.L. Structural and immunological studies of a major polysaccharide from spores of *Ganoderma lucidum* (Fr.) Karst. *Carbohydr. Res.* **2001**, *332*, 67–74. [CrossRef]

19.   Gao, P.; Hirano, T.; Chen, Z.Q.; Tadashi, Y.; Yoshihiro, N.; Akiko, S. Isolation and identification of C-19 fatty acids with anti-tumor activity from the spores of *Ganoderma lucidum* (reishi mushroom). *Fitoterapia* **2012**, *83*, 490–499. [CrossRef] [PubMed]

20.   Liu, X.; Yuan, J.P.; Chung, C.K.; Chen, X.J. Antitumor activity of the sporoderm-broken germinating spores of *Ganoderma lucidum*. *Cancer Lett.* **2002**, *182*, 155–161. [CrossRef]

21.   Zhu, H.S.; Yang, X.L.; Wang, L.B.; Zhao, D.X.; Chen, L. Effects of extracts from sporoderm-broken spores of *Ganoderma lucidum* on HeLa cells. *Cell Biol. Toxicol.* **2000**, *16*, 201–206. [CrossRef] [PubMed]

22.   Yue, G.G.L.; Fung, K.P.; Leung, P.C.; Lau, C.B.S. Comparative studies on the immunomodulatory and antitumor activities of the different parts of fruiting body of *Ganoderma lucidum* and Ganoderma spores. *Phytother. Res.* **2008**, *22*, 1282–1291. [CrossRef] [PubMed]

23.   Li, X.Y.; Zhao, L.H.; Fan, Y.; Jia, Y.X.; Sun, L.; Ma, S.S.; Ji, C.; Ma, Q.G.; Zhang, J.Y. Occurrence of mycotoxins in feed ingredients and complete feeds obtained from the Beijing region of China. *J. Anim. Sci. Biotechnol.* **2014**, *37*, 2–8. [CrossRef] [PubMed]

24.   Wu, L.X.; Ding, X.X.; Li, P.W.; Du, X.H.; Zhou, H.Y.; Bai, Y.Z.; Zhang, L.X. Aflatoxin contamination of peanuts at harvest in China from 2010 to 2013 and its relationship with climatic conditions. *Food Control* **2016**, *60*, 117–123. [CrossRef]

25.   Fan, Y.; Zhao, L.H.; Ma, Q.G.; Li, X.Y.; Shi, H.Q.; Zhou, T.; Zhang, J.Y.; Ji, C. Effects of Bacillus subtilis ANSB060 on growth performance, meat quality and aflatoxin residues in broilers fed moldy peanut meal naturally contaminated with aflatoxins. *Food Chem. Toxicol.* **2013**, *59*, 748–753. [CrossRef] [PubMed]

26. Han, X.Y.; Huang, Q.C.; Li, W.F.; Jiang, J.F.; Xu, Z.R. Changes in growth performance, digestive enzyme activities and nutrient digestibility of cherry valley ducks in response to aflatoxin B$_1$ levels. *Livest. Sci.* **2008**, *119*, 216–220. [CrossRef]

27. Fowler, J.; Li, W.; Bailey, C. Effects of a Calcium Bentonite Clay in Diets Containing Aflatoxin when Measuring Liver residues of Aflatoxin B 1 in Starter Broiler Chicks. *Toxins* **2015**, *7*, 3455–3464. [CrossRef] [PubMed]

28. Huff, W.E.; Kubena, L.F.; Harvey, R.B.; Corrier, D.E.; Mollenhauer, H.H. Progression of aflatoxicosis in broiler chickens. *Plout. Sci.* **1986**, *65*, 1981–1989. [CrossRef]

29. Mehrzad, J.; Klein, G.; Kamphues, J.; Wolf, P.; Grabowski, N.; Schuberth, H.J. In vitro effects of very low levels of aflatoxin B$_1$ on free radicals production and bactericidal activity of bovine blood neutrophils. *Vet. Immunol. Immunopathol.* **2011**, *141*, 16–25. [CrossRef] [PubMed]

30. Gill, S.S.; Tuteja, N. Reactive oxygen species and antioxidant machinery in abiotic stress tolerance in crop plants. *Plant Physiol. Biochem.* **2010**, *48*, 909–930. [CrossRef] [PubMed]

31. Arulselvan, P.; Subramanian, S.P. Beneficial effects of *Murraya koenigii* leaves on antioxidant defense system and ultrastructural changes of pancreatic b-cells in experimental diabetes in rats. *Chem. Biol. Interact.* **2007**, *165*, 155–164. [CrossRef] [PubMed]

32. Karabacak, M.; Eraslan, G.; Kanbur, M.; Sarica, Z.S. Effects of *Tarantula cubensis* D6 on aflatoxin-induced injury in biochemical parameters in rats. *Homeopathy* **2015**, *104*, 205–210. [CrossRef] [PubMed]

33. Shi, Y.H.; Xu, Z.R.; Feng, J.L.; Wang, C.Z. Efficacy of modified montmorillonite nanocomposite to reduce the toxicity of aflatoxin in broiler chicks. *Anim. Feed Sci. Technol.* **2006**, *129*, 138–148. [CrossRef]

34. Li, Y.; Ma, Q.G.; Zhao, L.H.; Wei, H.; Duan, G.X.; Zhang, J.Y.; Ji, C. Effects of Lipoic Acid on Immune Function, the Antioxidant Defense System, and Inflammation-related Genes Expression of Broiler Chickens Fed Aflatoxin Contaminated Diets. *Int. J. Mol. Sci.* **2014**, *15*, 5649–5662. [CrossRef] [PubMed]

35. Delles, R.M.; Xiong, Y.L.; Ture, A.D.; Ao, T.; Dawson, K.A. Dietary antioxidant supplementation enhances lipid and protein oxidative stability of chicken broiler meat through promotion of antioxidant enzyme activity. *Poult. Sci.* **2014**, *93*, 1561–1570. [CrossRef] [PubMed]

36. Mceligot, A.J.; Yang, S.; Meyskens, F.L. Redox regulation by intrinsic species and extrinsic nutrients in normal and cancer cells (abstract). *Annu. Rev. Nutr.* **2005**, *25*, 261–295. [CrossRef] [PubMed]

37. Jaeschke, H. Mechanisms of oxidant stress-induced acute tissue injury. *Proc. Soc. Exp. Biol. Med.* **1995**, *209*, 104–111. [CrossRef] [PubMed]

38. Kinnula, V.L.; Crapo, J.D. Superoxide dismutase in malignant cells and human tumors. *Free Radic. Biol. Med.* **2004**, *36*, 718–744. [CrossRef] [PubMed]

39. Zuo, R.Y.; Chang, J.; Yin, Q.Q.; Wang, P.; Yang, Y.R.; Wang, X.; Wang, G.Q.; Zheng, Q.H. Effect of the combined probiotics with aflatoxin B$_1$-degrading enzyme on aflatoxin detoxification, broiler production performance and hepatic enzyme gene expression. *Food Chem. Toxicol.* **2013**, *59*, 470–475. [CrossRef] [PubMed]

40. Fan, Y.; Zhao, L.H.; Ji, C.; Li, X.Y.; Jia, R.; Xi, L.; Zhang, J.Y.; Ma, Q.G. Protective Effects of *Bacillus subtilis* ANSB060 on Serum Biochemistry, Histopathological Changes and Antioxidant Enzyme Activities of Broilers. Fed Moldy Peanut Meal Naturally Contaminated with Aflatoxins. *Toxins* **2015**, *7*, 3330–3343. [CrossRef] [PubMed]

41. Gremmels, F.J. The role of mycotoxins in the health and performance of dairy cows. *Vet. J.* **2008**, *176*, 84–92. [CrossRef] [PubMed]

42. Chen, K.J.; Fang, J.; Peng, X.; Cui, H.M.; Chen, J.; Wang, F.Y.; Chen, Z.L.; Zuo, Z.C.; Deng, J.L.; Lai, W.M.; et al. Effect of selenium supplementation on aflatoxin B$_1$–induced histopathological lesions and apoptosis in bursa of Fabricius in broiler. *Food Chem. Toxicol.* **2014**, *74*, 91–97. [CrossRef] [PubMed]

43. Tung, H.T.; Wyatt, R.D.; Thaxon, P.; Hamilton, P.B. Concentrations of Serum Proteins during Aflatoxicosis. *Toxicol. Appl. Pharmacol.* **1975**, *34*, 320–326. [CrossRef]

44. Allameh, A.; Safamehr, A.; Mirhadi, S.A.; Shivazad, M.; Mehdi, R.A.; Azam, A.N. Evaluation of biochemical and production parameters of broiler chicks fed ammonia treated aflatoxin contaminated maize grains. *Anim. Feed Sci. Technol.* **2005**, *122*, 289–301. [CrossRef]

45. Quezada, T.; Cuellar, H.; Jaramillo, J.F.; Valdivia, A.G.; Reyes, J.L. Effects of aflatoxin B1 on the liver and kidney of broiler chickens during development. *Comp. Biochem. Physiol. C* **2000**, *125*, 265–272. [CrossRef]

46. Ma, Q.G.; Li, Y.; Fan, Y.; Zhao, L.H.; Wei, H.; Ji, C.; Zhang, J.Y. Molecular Mechanisms of Lipoic Acid Protection against Aflatoxin B 1 -Induced Liver Oxidative Damage and Inflammatory Responses in Broilers. *Toxins* **2015**, *7*, 5435–5447. [CrossRef] [PubMed]

47. Chen, K.J.; Yuan, S.H.; Chen, J.; Peng, X.; Wang, F.Y.; Cui, H.M.; Fang, J. Effects of sodium selenite on the decreased percentage of T cell subsets, contents of serum IL-2 and IFN-γ induced by aflatoxin B 1 in broilers. *Res. Vet. Sci.* **2013**, *95*, 143–145. [CrossRef] [PubMed]

48. He, J.; Zhang, K.Y.; Chen, D.W.; Ding, X.M.; Feng, G.D.; Ao, X. Effects of vitamin E and selenium yeast on growth performance and immune function in ducks fed maize naturally contaminated with aflatoxin B$_1$. *Livest. Sci.* **2013**, *152*, 200–207. [CrossRef]

49. Wang, T.Y.; Li, C.Y.; Liu, Y.; Li, T.Z.; Zhang, J.; Sun, Y.H. Inhibition effects of Chinese cabbage powder on aflatoxin B$_1$-induced liver cancer. *Food Chem.* **2015**, *186*, 13–19. [CrossRef] [PubMed]

50. Gupta, R.; Sharma, V. Ameliorative effects of *Tinospora cordifolia* root extract on histopathological and biochemical changes induced by aflatoxin B$_1$ in mice kidney. *Toxicol. Int.* **2011**, *18*, 94–98. [PubMed]

51. Kozarski, M.; Klausa, A.; Niksic, M.; Jakovljevic, D.; Helsperc, J.P.F.G.; Van Griensven, L.J.L.D. Antioxidative and immunomodulating activities of polysaccharide extracts of the medicinal mushrooms *Agaricus bisporus*, *Agaricus brasiliensis*, *Ganoderma lucidum* and *Phellinus linteus*. *Food Chem.* **2011**, *129*, 1667–1675. [CrossRef]

52. Jin, H.; Jin, F.; Jin, J.X.; Xu, J.; Tao, T.T.; Liu, J.; Huang, H.J. Protective effects of *Ganoderma lucidum* spore on cadmium hepatotoxicity in mice. *Food Chem. Toxicol.* **2013**, *52*, 171–175. [CrossRef] [PubMed]

53. Wang, F.; Zhou, Z.K.; Ren, X.C.; Wang, Y.Y.; Yang, R.; Luo, J.H.; Strappe, P. Effect of *Ganoderma lucidum* spores intervention on glucose and lipid metabolism gene expression profiles in type 2 diabetic rats. *Lipids Health* **2015**, *14*, 1–9. [CrossRef] [PubMed]

54. Mohan, K.; Padmanaban, A.M.; Uthayakumar, V.; Chandirasekar, R.; Muralisankar, T.; Santhanam, P. Effect of dietary *Ganoderma lucidum* polysaccharides on biological and physiological responses of the giant freshwater prawn *Macrobrachium rosenbergii*. *Aquaculture* **2016**, *464*, 42–49. [CrossRef]

55. Li, X.L.; He, L.P.; Yang, Y.; Liu, F.J.; Gao, Y.; Zuo, J.J. Effects of extracellular polysaccharides of *Ganoderma lucidum* supplementation on the growth performance, blood profile, and meat quality in finisher pigs. *Livest. Sci.* **2015**, *178*, 187–194. [CrossRef]

56. Di, X.; Chan, K.K.C.; Leung, H.W.; Huie, C.W. Fingerprint profiling of acid hydrolyzates of polysaccharides extracted from the fruiting bodies and spores of Lingzhi by high-performance thin-layer chromatography. *J. Chromatogr. A* **2003**, *1018*, 85–95. [CrossRef] [PubMed]

57. Li, G.L.; Shi, J.Y.; Suo, Y.R.; Sun, Z.W.; Xia, L.; Zheng, J.; You, J.M.; Liu, Y.J. Supercritical CO$_2$ cell breaking extraction of *Lycium barbarum* seed oil and determination of its chemical composition by HPLC/APCI/MS and antioxidant activity. *Food Sci. Technol.* **2011**, *44*, 1172–1178.

58. Binder, E.M.; Tan, L.M.; Chin, L.J.; Handl, J.; Richard, J. Worldwide occurrence of mycotoxins in commodities, feeds and feed ingredients. *Anim. Feed Sci. Technol.* **2007**, *137*, 265–282. [CrossRef]

59. NRC. *Nutrient Requirements of Poultry*, 9th ed.; Natl. Acad. Press: Washington, DC, USA, 1994.

toxins

MDPI

Article

Ameliorating Effects of *Bacillus subtilis* ANSB060 on Growth Performance, Antioxidant Functions, and Aflatoxin Residues in Ducks Fed Diets Contaminated with Aflatoxins

Liyuan Zhang [1,†], Qiugang Ma [1,†], Shanshan Ma [2], Jianyun Zhang [1], Ru Jia [3], Cheng Ji [1] and Lihong Zhao [1,*]

[1] State Key Laboratory of Animal Nutrition, College of Animal Science and Technology, China Agricultural University, Beijing 100193, China; zhangliyuan619@163.com (L.Z.); maqiugang@cau.edu.cn (Q.M.); jyzhang@cau.edu.cn (J.Z.); jicheng@cau.edu.cn (C.J.)
[2] Heilongjiang Animal Science Institute, Qiqihar 161005, China; mashanshan33@foxmail.com
[3] College of Life Science, Shanxi University, Taiyuan 030006, China; jiaru.bjdk071@163.com
* Correspondence: zhaolihongcau@cau.edu.cn; Tel./Fax: +86-10-6273-2774
† These authors contributed equally to this work.

Academic Editor: Ting Zhou
Received: 22 August 2016; Accepted: 15 December 2016; Published: 22 December 2016

Abstract: *Bacillus subtilis* ANSB060 isolated from fish gut is very effective in detoxifying aflatoxins in feed and feed ingredients. The purpose of this research was to investigate the effects of *B. subtilis* ANSB060 on growth performance, body antioxidant functions, and aflatoxin residues in ducks fed moldy maize naturally contaminated with aflatoxins. A total of 1500 18-d-old male Cherry Valley ducks with similar body weight were randomly assigned to five treatments with six replicates of 50 ducks per repeat. The experiment design consisted of five dietary treatments labeled as C0 (basal diet containing 60% normal maize), M0 (basal diet containing 60% moldy maize contaminated with aflatoxins substituted for normal maize), M500, M1000, and M2000 (M0 +500, 1000 or 2000 g/t aflatoxin biodegradation preparation mainly consisted of *B. subtilis* ANSB060). The results showed that ducks fed 22.44 ± 2.46 µg/kg of AFB_1 (M0) exhibited a decreasing tendency in average daily gain (ADG) and total superoxide dismutase (T-SOD) activity in serum, and T-SOD and glutathione peroxidase (GSH-Px) activities in the liver significantly decreased along with the appearance of AFB_1 and AFM_1 compared with those in Group C0. The supplementation of *B. subtilis* ANSB060 into aflatoxin-contaminated diets increased the ADG of ducks ($p > 0.05$), significantly improved antioxidant enzyme activities, and reduced aflatoxin accumulation in duck liver. In conclusion, *Bacillus subtilis* ANSB060 in diets showed an ameliorating effect to duck aflatoxicosis and may be a promising feed additive.

Keywords: aflatoxin B_1; aflatoxin biodegradation preparation; *Bacillus subtilis* ANSB060; ameliorating effects; growth performance; antioxidant function; residue

1. Introduction

Aflatoxins (AF) are secondary toxic metabolites produced by certain fungi belonging to the genus *Aspergillus* and can occur as natural contaminants of poultry food [1–5]. The principal aflatoxins commonly found in feedstuffs are aflatoxin B_1 (AFB_1), aflatoxin B_2 (AFB_2), aflatoxin G_1 (AFG_1), and aflatoxin G_2 (AFG_2) [6]. AFB_1 has the highest toxic potency and is classified as a Group I carcinogen by the International Agency for Research on Cancer (IARC, International Agency for Research on Cancer, 1993). After being ingested by animals, AFB_1 is converted to AFB_1-8, 9-epoxide in the liver with a reaction catalyzed by cytochrome P-450 isozymes, forming adducts with the guanine base of DNA, thus resulting in the toxicity of AFB_1 [7,8].

Feeds contaminated with AFB_1 can result in aflatoxicosis in poultry [9,10]. Many studies have found that feeding grains contaminated with AFB_1, either naturally or purified, can reduce growth performance and immunity ability, alter intestinal morphology and blood biochemistry parameters, and damage liver and kidney tissues in broilers [11,12]. Additionally, residues of aflatoxins and their metabolites may be present in meat and other products of animals fed rations contaminated with aflatoxins and potentially result in health problems in human.

It is known that aflatoxins are difficult or impossible to be removed completely from plants and living organisms, as they can alter the chemical structure of mycotoxins as part of their defense against xenobiotics [13]. Researchers have tried for decades to find effective strategies regarding mycotoxin detoxification. The multifaceted approaches in previous reports could simply be categorized into physical, chemical, and biological methods. A physical method is using non-nutritive absorptive materials in animal diets to reduce the toxin absorption from the gastrointestinal tract of chickens [14,15]. A study showed that Blasticidin A (BcA), an antibiotic produced by Streptomyces, can control aflatoxin production by inhibiting protein synthesis [16]. Recently, the interest of researchers has turned to the biological detoxification of mycotoxins using microorganisms or enzymatic preparation for their specificity, efficiency, and simplicity [17–21]. Some microbes such as *Armillariella tabescens* [22], *Rhodococcus erythropolis* [23], and *Myxococcus fulvus* [24] have been reported to possess the ability to degrade aflatoxins. However, most of the microorganisms are not allowed to be applied in animal feed under the supervision of the U.S. Food and Drug Administration (FDA), the Association of American Feed Control Officials (AAFCO), and the Direct of Ministry of Agriculture, People's Republic of China. Based on this concern, we have isolated a strain of *Bacillus subtilis* ANSB060 from fish gut that possesses a strong ability to detoxify aflatoxins. We also have confirmed that this strain can inhibit the growth of pathogens and resist unfavorable conditions within simulated gut environments [25]. Since the nutritional and pharmaceutical use of *Bacillus subtilis* is generally recognized as safe (GRAS), the Ministry of Agriculture of the People's Republic of China has permitted *B. subtilis* for use in animal feed. Ma et al. (2012) have shown that the addition of *B. subtilis* ANSB060 can enhance the activity of antioxidant enzymes and recover the protein synthesis in the livers of laying hens [26], and Fan et al. (2013, 2015) verified its positive effects of detoxifying aflatoxins on broilers [27,28].

Ducks are considered to be highly susceptible because they are unable to efficiently metabolize aflatoxins [10]. A study examining the effects of AFB_1 on liver lesions in poultry reported that ducks generated hepatic lesions on alternate days per os 15 µg AFB_1 [29]. Other literature on ducks fed diets naturally contaminated with AFB_1 is also available [30,31]. However, to the best of our knowledge, little information on biological detoxification methods dealing with duck diets contaminated with AFB_1 has been found. China has a large consumer of duck products in the world (FAO, 2003), so it is of great importance to study the effects of aflatoxins on ducks and develop solutions to detoxify aflatoxins to ensure the qualities of duck products. We have investigated the toxic effects of aflatoxins and the efficacy of *Bacillus subtilis* ANSB060 for the amelioration of aflatoxicosis in broiler chickens. Whether low levels of aflatoxins can affect the growth and antioxidant functions of ducks and which level of aflatoxins in diets can result in residues in duck tissues is still under consideration. Thus, certifying the effects of low levels of aflatoxins on ducks and finding a reasonable method to degrade aflatoxins in diets is important. Therefore, the objective of this study was to investigate the protective effects of aflatoxin biodegradation preparation (*B. subtilis* ANSB060) on growth performance, body antioxidant functions, and aflatoxin residues in ducks fed maize naturally contaminated with AFB_1.

2. Results

2.1. Growth Performance

The effects of *B. subtilis* ANSB060 on growth performance of Cherry Valley ducks fed moldy maize naturally contaminated with aflatoxins are presented in Table 1. There were no significant difference on ADG and ADFI among ducks fed basal diet without AFB_1 (C0), a moldy diet naturally contaminated with AFB_1 (M0), and diets (M500, M1000, and M2000) naturally contaminated with AFB_1 mixed with *B. subtilis* ANSB060 during the experimental phase ($p > 0.05$). However, birds fed a diet with 22.44 ± 2.46 µg/kg of AFB_1 (M0) had a relatively smaller ADG ($p > 0.05$) compared to those fed the basal diet without AFB_1 (C0), while ducks fed 22.44 ± 2.46 µg/kg of AFB_1 with *B. subtilis* ANSB060 exhibited relatively higher ADG ($p > 0.05$) compared to those in Group M0. The feed/gain ratio (F:G) of ducks was significantly affected by the dietary aflatoxins and *B. subtilis* ANSB060 during the experimental phase ($p < 0.05$). The F:G of birds in Group M1000 was markedly decreased compared with that in Group M0 ($p < 0.05$), implying that *B. subtilis* ANSB060 in diets might relieve the negative effects of aflatoxion on birds and improve the feed efficiency of ducks.

Table 1. Effects of *B. subtilis* ANSB060 on the growth performance of ducks fed moldy maize naturally contaminated with aflatoxins during the experiment.

Item	Initial BW, g	Final BW, g	ADG, g/bird d	ADFI, g/bird d	F:G, g/g
C0	759.75 ± 7.13	3159.73 ± 19.62	109.09 ± 0.78	235.28 ± 1.79	2.17 ± 0.02 [a]
M0	758.15 ± 6.02	3120.93 ± 8.43	107.40 ± 0.53	234.28 ± 0.61	2.18 ± 0.01 [a]
M500	754.92 ± 3.61	3193.96 ± 22.20	110.86 ± 1.05	237.23 ± 3.08	2.14 ± 0.02 [a]
M1000	758.47 ± 5.26	3195.05 ± 11.78	110.75 ± 0.46	230.38 ± 2.21	2.08 ± 0.02 [b]
M2000	759.60 ± 7.01	3186.97 ± 23.95	110.33 ± 1.31	237.32 ± 2.44	2.17 ± 0.03 [a]
p Value	0.967	0.067	0.075	0.332	0.014

Each value is a mean $\pm$ SD from 6 replicates of 50 ducks each; [a,b] Means values with different superscripts in each column are significantly different ($p < 0.05$).

2.2. Serum Antioxidant Indices

The effects of *B. subtilis* ANSB060 on the serum antioxidant indices of ducks exposed to aflatoxins are shown in Table 2. The total superoxide dismutase (T-SOD) activity of the M0 group was decreased by the dietary aflatoxins compared to that in Group C0 during the experimental phase ($p < 0.05$). However, the addition of *B. subtilis* ANSB060 into moldy diets significantly improved the activity of serum T-SOD in the Group M500, M1000 and M2000 ($p < 0.05$), maintaining it in a normal status showed in Group C0. Though the glutathione peroxidase (GSH-Px) activity and malondiadehyde (MDA) content were not different among all groups ($p > 0.05$), Group M0 had the lowest GSH-Px activity and the highest MDA content, and the dietary *B. subtilis* ANSB060 could raise the activity of GSH-Px and decrease MDA concentration, keeping them to the normal level.

Table 2. The effects of *B. subtilis* ANSB060 on the serum antioxidant indexes of ducks fed moldy maize naturally contaminated with aflatoxins during the experiment.

Item	T-SOD, U/mL	GSH-Px, U/mL	MDA, nmol/mL
C0	112.41 ± 4.32 [a,b]	1718.89 ± 141.66	2.55 ± 0.16
M0	107.36 ± 2.70 [a]	1508.71 ± 184.24	2.83 ± 0.32
M500	121.50 ± 2.13 [b]	1601.39 ± 81.31	2.62 ± 0.16
M1000	121.34 ± 4.12 [b]	1818.50 ± 96.85	2.58 ± 0.28
M2000	120.30 ± 3.99 [b]	1818.50 ± 264.37	2.45 ± 0.20
p Value	0.029	0.581	0.810

Each value is a mean $\pm$ SD from 6 replicates of 2 ducks each; [a,b] Means values with different superscripts in each column are significantly different ($p < 0.05$).

2.3. Liver Antioxidant Indices

The effects of *B. subtilis* ANSB060 on liver antioxidant indices of ducks exposed to aflatoxins are presented in Table 3. The result indicated that a low dose of aflatoxins in diets (Group M0) can decrease the activities of liver T-SOD and GSH-Px in ducks during the experimental period ($p < 0.05$); however, the presence of *B. subtilis* ANSB060 in some diets (Group M500, M1000 and M2000) exhibited a powerful capability of recovering T-SOD and GSH-Px activities in duck livers. The MDA level in liver was detected as the index of lipid peroxidation, and there were no significant effects of dietary aflatoxins and *B. subtilis* ANSB060 on MDA levels in duck livers in all groups ($p > 0.05$).

Table 3. The effects of *B. subtilis* ANSB060 on the liver antioxidant indices of ducks fed moldy maize naturally contaminated with aflatoxins during the experiment.

Item	T-SOD, U/mgprot	GSH-Px, U/mgprot	MDA, nmol/mgprot
C0	111.10 ± 5.73 [c]	116.45 ± 6.12 [a]	0.55 ± 0.05
M0	85.21 ± 3.44 [a]	87.11 ± 5.28 [b]	0.90 ± 0.29
M500	104.83 ± 3.50 [b,c]	109.96 ± 3.31 [a]	0.79 ± 0.08
M1000	98.37 ± 3.00 [b]	106.58 ± 3.82 [a]	0.56 ± 0.05
M2000	105.80 ± 3.92 [b,c]	104.44 ± 5.86 [a]	0.64 ± 0.06
p-value	0.001	0.007	0.398

Each value is a mean ± SD from 6 replicates of 2 ducks each; [a–c] Means values with different superscripts in each column are significantly different ($p < 0.05$).

2.4. Aflatoxin Residues in Liver

The aflatoxin residues in the livers of ducks fed moldy diet naturally contaminated with 22.44 ± 2.46 µg/kg AFB_1 with or without *B. subtilis* ANSB060 are given in Table 4. Aflatoxins were not detected in the livers of ducks consuming a normal diet (Group C0). The AFB_1 and AFM_1 residues in Group M0 were the highest, being 0.12 and 0.10 ng/g, respectively. *B. subtilis* ANSB060 in diets could degrade aflatoxins into nontoxic products in the duck intestinal tract such that AFB_1 and AFM_1 residues in the liver were decreased. The AFB_1 residues was significantly decreased by 41.7%, 50.0%, and 58.3% respectively in Groups M500, M1000, and M2000 compared to those in Group M0 ($p < 0.05$). Similarly, the AFM_1 residues were significantly reduced by 40.0% in Groups M500, M1000, and M2000, respectively. The AFB_2, AFG_1, and AFG_2 residues in liver were not detected in all treatments.

Table 4. The effects of *B. subtilis* ANSB060 on aflatoxin residues (ng/g) in livers from ducks fed moldy maize naturally contaminated with aflatoxins during the experiment.

Item	AFB_1	AFB_2	AFG_1	AFG_2	AFM_1
C0	ND *	ND	ND	ND	ND
M0	0.12 ± 0.01 [a]	ND	ND	ND	0.10 ± 0.02 [a]
M500	0.07 ± 0.01 [b]	ND	ND	ND	0.06 ± 0.00 [b]
M1000	0.06 ± 0.01 [c]	ND	ND	ND	0.06 ± 0.01 [b]
M2000	0.05 ± 0.01 [d]	ND	ND	ND	0.05 ± 0.01 [b]

Each value is a mean ± SD from 6 replicates of 1 duck each; [a–c] Means values with different superscripts in each column are significantly different ($p < 0.05$); * ND = Not detected.

3. Discussion

3.1. Growth Performance

Aflatoxins are mycotoxins produced by *Aspergillus* spp. that contaminate animal feed and human food. Aflatoxins can lead to growth reduction, immunity suppression, and oxidative damage, along with residues in animal tissues [9,32–34], and it has become a threat to human health. *B. subtilis* ANSB060 has shown a strong ability to detoxify aflatoxins in vitro; the degradation percentages of aflatoxins B_1, M_1, and G_1 in this strain are 81.5%, 60%, and 80.7%, respectively [25]; meanwhile, their protective effects on the performance and antioxidant function in layers [26] and the reduced accumulation of aflatoxin residues in livers of broilers [27] have also been verified.

In this study, ducks fed diets contaminated with 22.44 ± 2.46 μg/kg of AFB_1 show a tendency to decline in body weight, ADG, and feed efficiency compared with ones in the control group, illustrating a toxic effect of aflatoxins on the growth performance of ducks, which is consistent with previous research. A significant decrease in ADG and ADFI in ducks fed a diet containing 200 ppm AFB_1 for three weeks were observed in the research conducted by Cheng et al. (2001) [35]. Meanwhile, Fowler et al. (2015) reported a reduced body weight (BW) from the first week to the third week in broilers fed aflatoxin-contaminated diets, and this reduction grew greater along with the increase of aflatoxin concentrations in diets from 0 to 1800 μg/kg; the percentages of body weight reduction ranged from 8.8% to 24.5% during Week 2, and from 15.6% to 41.3% during Week 3. Additionally, a negative effect of aflatoxins on the cumulative F:G ratio of broiler chicks was also shown [36]. However, the deleterious effect on the ADG of ducks in our study was not significant during the trial phase. This may be due to lower concentration of aflatoxins and the inadequate duration of the experiment. Previous results indicate that a decrease in ADG was due to anorexia, reluctance, and an inhibition of protein synthesis and lipogenesis [37–39]. This growth inhibition in ducks suggests that diets naturally contaminated with aflatoxins at a low level (22.44 ± 2.46 μg/kg) appear to have a toxic effect on the growth performance of ducks in vivo. This result indicates that aflatoxin degradation preparation mainly consisting of *B. subtilis* ANSB060 has a great ability to ameliorate the toxic effects of aflatoxins on the growth performance of ducks. The basic mechanism seems to be one in which the spores of *B. subtilis* ANSB060 can survive and colonize in an animal's intestinal tract and secrete active substances to degrade aflatoxins in diets so that the adsorption of aflatoxins declines [27]. Some studies have suggested that microbial enzymes cleave the lactone ring of the AFB_1 molecule in vitro to lower its toxicity [40,41]. Nevertheless, the specific biotransformation mechanism of *B. subtilis* ANSB060 detoxifying aflatoxins in the animal's intestinal tract is unclear, and further studies should be performed with this strain [28].

3.2. Serum and Liver Antioxidant Indices

The inability to protect against reactive oxygen species (ROS) is the origin of oxidative stress. ROS are produced during the procedure of normal metabolism, and do much harm to the animal body. ROS mainly contain hydrogen peroxide (H_2O_2), a superoxide anion radical ($O_2{}^{\bullet-}$), and a hydroxyl radical ($^\bullet OH$). $^\bullet OH$ is one of the most toxic oxygen-based radicals and wreaks havoc within cells, particularly with macromolecules. There is a dynamic equilibrium between oxidant and antioxidant systems and is vital for cell function, regulation, and adaptation to diverse growth conditions to remain healthy in animal bodies [42], and oxidant stress can be induced subsequently if the balance is broken down and the generation of ROS in a system exceed the system's ability to neutralize and eliminate them [43]. Oxidant stress can cause biological membrane lipid oxidation, denaturation of intracellular protein and enzymes, DNA damage, pain, and ultimately various diseases. Enzymes such as SOD and GSH-Px are crucial components of the antioxidant system and play a key role in removing ROS and relieving oxidative damage [42]. MDA is the main product of lipid peroxidation that can bring about protein damage and inactivation of membrane-bound enzymes [44]; thus, the level of MDA can be an important index reflecting the body's antioxidant ability [45].

One of our aims is to study the effects of adding *B. subtilis* ANSB060 to diets contaminated with a concentration of 22.44 ± 2.46 μg/kg aflatoxins on the antioxidant function in ducks. Our findings related to body antioxidant ability indicate that feeding diets contaminated with 22.44 ± 2.46 μg/kg aflatoxins can result in a significant decrease in T-SOD activity and an increase in the MDA level both in serum and the liver, while the activity of GSH-Px in duck livers is markedly decreased compared with that in the control group from Day 18 to Day 39. These results are in accordance with previous studies [46–49], demonstrating that lipid peroxidation took place in the ducks fed diet contaminated with aflatoxins.

While the activity of T-SOD is significantly increased and the MDA level is relatively decreased both in serum and liver, the activity of GSH-Px has a tendency to increase in serum but markedly increase in liver when ducks ingest aflatoxin-contaminated diets containing *B. subtilis* ANSB060 during the experiment compared with those fed no *B. subtilis* ANSB060 (M0). SOD and GSH-Px take part in lipid peroxidation by catalyzing the conversion of lipid hydroperoxide to hydroxy acids in the presence of GSH [50]. Liver is shown to be the major target organ for aflatoxins [51], and the increase in SOD and GSH-Px activities as well as the decrease in MDA level in serum and liver reflect that *B. subtilis* ANSB060 contributes to the inhibition of lipid peroxidation and relieve the harm produced by lipid peroxidation in ducks.

As a cause of liver degeneration and cancer, the elimination of these toxins from feedstuffs and foodstuffs bears great importance. In poultry, several nonnutritive adsorbents have been tested and have made some progress [46,52]; now, *B. subtilis* ANSB060 has been shown to have the capacity to inhibit lipid peroxidation, providing another biological method to alleviating oxidative damage to ducks caused by aflatoxins.

3.3. Aflatoxin Residues in Liver

The residue of aflatoxins and their metabolites in animal products have caused wide public concern. Liver is the main detoxifying organ that removes wastes and xenobiotics by metabolic conversion and biliary excretion [53]. Among the metabolites of AFB_1, the toxicity of AFM_1 in animals seems to be comparable to or slightly less than that of AFB_1 [12]. In our study, low levels (0.12 and 0.10 ng/g) of AFB_1 and AFM_1 are retained in the livers of ducks fed the aflatoxin-contaminated diet for 21 days. Residue levels of AFB_1 (0.05 and 0.13 μg/kg) and AFM_1 (0.10 and 0.32 μg/kg) were also observed in the livers of broilers given diets contain 50 and 100 μg of AFB_1/kg for 42 days [15]. AFB_1 residue was found in the livers of laying hens fed 2.5 mg/kg AFB_1 diet for four weeks [54]. Residue levels may be different because of the type of bird and diet, the concentrations of AFB_1, and the duration, and enhanced tolerance to aflatoxins. In many countries, the maximum tolerance level of AFB_1 in human food products is 2 ng/g. AFB_1 residues in liver not only affect the performance and health of poultry, but also impair the health of the poultry product consumers as aflatoxins accumulate in edible parts of poultry, so it is necessary to control the quality of poultry products and analyze aflatoxin residues in different tissues of birds considering public health and safety. The results of our study show significantly decreased residue levels of AFB_1- and AFM_1-incorporated *B. subtilis* ANSB060 in ducks exposed to aflatoxins. The protective effects of *B. subtilis* ANSB060 from aflatoxins may be due to their specific biotransformation of aflatoxins in the intestinal tract, which leads to the reduction of aflatoxins absorbed by the intestinal tract and, consequently, a decrease in aflatoxin residues in the liver. However, the specific mechanism has not been tested, and more research should be done to determine this.

4. Conclusions

In conclusion, this study clearly indicates that aflatoxins in diets at a level of 22.44 μg/kg resulted in depressed growth performance and antioxidant capacity in liver and serum, as well as aflatoxin liver residues in ducks. Adding aflatoxin biodegradation preparation *B. subtilis* ANSB060 into moldy diets significantly recovers the growth performance of ducks and reduces the accumulation of aflatoxin residues in liver, effectively improving the antioxidant capacity to some extent. These results suggest that *B. subtilis* ANSB060 counteract the adverse effects of aflatoxins in animal diets, and thus is believed to be a very promising feed additive to detoxify aflatoxins in feed.

5. Materials and Methods

5.1. Ducks, Diets, and Management

One-day-old Cherry Valley ducks were obtained from a commercial hatchery and fed a commercial rearing diet for 17 days to adapt to the surroundings. On Day 18, 1500 male ducks were randomly assigned into 5 treatment groups with 6 replicates of 50 birds each. Five dietary treatment groups were composed as follows: Group C0 (the negative control), consisting of the basal diet with 60% normal maize; Group M0 (the positive control), containing 60% moldy maize taking the place of normal maize; and Groups M500, M1000, and M2000, containing an added 500 g/t, 1000 g/t and 2000 g/t of aflatoxin biodegradation preparation in the diet of Group M0, respectively. In this experiment, aflatoxin biodegradation preparation was mainly composed of industrially fermented and dried *B. subtilis* ANSB060 products via certain processing technology. The viable count of *B. subtilis* ANSB060 was more than 1×10^9 CFU/g in the aflatoxin biodegradation preparation. The composition of the basal diet was presented in Table 5.

Table 5. Basal diet formulations and nutritional contents.

Ingredients	Percentage (%)	Nutrition Component	Content
Maize	60.00	Crude protein, %	16.50
Soybean meal	13.64	Metabolizable energy, MJ/kg	12.39
Wheat	4.75	Calcium, %	1.00
Flour	3.50	Total phosphorus, %	0.67
Peanut meal	4.20	Available phosphorus, %	0.42
Zein meal	3.50	Methionine, %	0.46
Calcium hydrophosphate	1.36	Methionine + Cystine, %	0.76
Limestone	1.36	Lysine, %	0.96
Salt	0.32		
Soybean oil	1.77		
Rice bran	2.30		
Rice bran meal	2.00		
Premix [1]	1.30		
Total	100.00		

[1] Provided per kilogram of diet: vitamin A: 15,000 IU; cholecalciferol: 4000 IU; vitamin E: 40 IU; vitamin K3: 5.0 mg; thiamine: 3.0 mg; riboflavin: 3.0 mg; pyridoxine: 4.6 mg; vitamin B12: 0.08 mg; folic acid: 1.6 mg; niacin: 50 mg; pantothenic acid: 15 mg; choline chloride: 1000 mg; biotin: 0.25 mg; manganese: 100 mg; zinc: 100 mg; selenium: 0.3 mg; iodine: 0.3 mg; iron: 50 mg; copper: 15 mg.

This study was approved by the Animal Care and Use Committee of the China Agricultural University (ethical approval code: CAU20131028-2; Date: 28 October 2013). All ducks were reared on floor in an environmentally controlled house equipped with central heating and temperature controllers. The size of barrier was 8 m × 2 m. Animals were cared for in accordance with the guidelines for the care and use of laboratory animals presented in the guide issued by the National Institute of Health and China's Ministry of Agriculture. Body weight (BW), average daily gain (ADG), average daily feed intake (ADFI), and feed/gain ratio (F:G) were measured on a replicate basis at the end of the experiment (39 day old).

5.2. Analysis of Dietary Mycotoxin

The contents of mycotoxins (AFB_1, AFB_2, AFG_1, AFG_2, deoxynivalenol, zearalenone, and ochratoxin A) in feed ingredients and formulated diets in the study were determined using high performance liquid chromatography (HPLC) according to appropriate methods [55]. After these five diets were prepared, the final contents of mycotoxins for each diet were determined. The concentration of AFB_1, AFB_2, AFG_1, AFG_2, and zearalenone of the diets fluctuated at 22.44 ± 2.46 µg/kg, 6.69 ± 1.32 µg/kg, 1.65 ± 0.65 µg/kg, 0.00 µg/kg, and 5.20 ± 0.68 µg/kg in Groups M0, M500,

M1000, and M2000, respectively. Only AFB_1 was found in diet of Group C0, but the concentration of AFB_1 was less than 0.20 µg/kg in this group.

5.3. Sampling of Antioxidant Indices and Measurements

At 39 day of age, 12 ducks (2 birds for each replicate) close to the average weight of treatment were slaughtered by dislocation of the neck vertebrae and bleeding after fasting for 12 h. Blood samples were collected by puncture of wing vein; after being centrifuged at $1000 \times g$ at 4 °C for 10 min, the serum was separated and transferred into 1.5 mL plastic centrifuge tubes for analysis of antioxidant indices including superoxide dismutase (SOD), glutathione peroxidase (GSH-Px) activity, and malondiadehyde (MDA) content.

The liver tissue samples were washed with ice-cold sterilized saline (0.85%), snap frozen in liquid nitrogen, and stored at -80 °C for further analysis of liver antioxidant indices. Liver tissue (1 g) was cut into small pieces and homogenized in ice-cold saline buffer (0.85%, pH = 7.4) (1:9, w/v) with an Ultra-Turrax (T8, IKA-labortechnik, Staufen, Germany) to form homogenates at a concentration of 0.1 g/mL for further analysis. Liver homogenates were centrifuged at $1000 g$ for 15 min at 4 °C, and the supernatants were collected. The supernatants were used for the assays of SOD and GSH-Px activity, and MDA content.

These parameters were determined with the clinical chemistry analyzer (Commercial Kit, Nanjing Jiancheng Bioengineering Institute, Nanjing, China) according to the manufacturer's recommended procedure.

5.4. Analysis of Aflatoxin Residues in Liver

Livers samples from the pectoralis major of the right side of the duck were excised and frozen at -20 °C to analyze the residues of the aflatoxins. For this analysis, liver samples of six birds from each treatment (one for each replicate) were selected. Analysis of AFB_1, AFB_2, AFG_1, AFG_2, and AFM_1 residues in the tissues was performed according to Fan et al. [27].

The ground defrosted liver sample (25 g), when 5 g of NaCl was added, was blended in 100 mL of methanol–water (80:20) for 3 min. Through a paper filter, an aliquot of 10 mL of filtrate was diluted with 40 mL of PBS/0.1% Tween-20 Wash Buffer and applied to an immunoaffinity column. Aflatoxins were eluted with 1.0 mL of methanol in a glass vial and dried near to dryness under a gentle stream of nitrogen and dissolved in a HPLC mobile phase. AFB_1, AFB_2, AFG_1, and AFG_2 were determined by the HPLC system (Shimadzu LC-10 AT) equipped with reverse phase column (Diamonsil, C18, 5 µm, 15 cm $\times$ 4.6 cm ID) and post-column photochemical derivatization (AURA, New York, NY, USA) and fluorescence monitor (Shimadzu RF-20A), with excitation at 360 nm and emission at 440 nm, with methanol/water (45:55) as the mobile phase, at a flow rate of 1 mL/min. For the determination of AFM_1, HPLC was performed with a fluorescence monitor at 365 nm for excitation and 425 nm for emission, and water/acetonitrile/methanol (68:24:8) as the mobile phase at a flow rate of 1 mL/min.

5.5. Statistical Analyses

Data were analyzed using the GLM procedure of SAS software (version 9; SAS Institute, Inc., Cary, NC, USA). Duncan's multiple range tests were used for multiple comparisons when the analysis indicated significant differences among treatments. All statements of statistical significance were based on probability ($p < 0.05$). All data were expressed as means $\pm$ SD.

Acknowledgments: This study was supported by the National Natural Science Foundation of China (Grant No. 31301981), a Special Fund for Agro-scientific Research in the Public Interest (201403047), and the Program of State Key Laboratory of Animal Nutrition of China.

Author Contributions: Liyuan Zhang and Lihong Zhao conceived the study design, collected and analyzed data, and wrote the manuscript. Shanshan Ma and Ru Jia analyzed samples and supervised the analysis of data. Qiugang Ma, Jianyun Zhang, and Cheng Ji contributed to the supervision and guidance of the present study.

Conflicts of Interest: The authors declare no conflict of interest.

References

1. Diener, U.L.; Cole, R.J.; Sanders, T.H.; Payne, G.A.; Lee, L.S.; Klich, M.A. Epidemiology of Aflatoxin Formation by Aspergillus-flavus. *Ann. Rev. Phytopathol.* **1987**, *25*, 249–270. [CrossRef]

2. Goto, T.; Ito, Y.; Peterson, S.W.; Wicklow, D.T. Mycotoxin producing ability of *Aspergillus tamarii*. *Mycotoxins* **1997**, *1997*, 17–20. [CrossRef]

3. Ito, Y.; Peterson, S.; Wicklow, D.; Goto, T. *Aspergillus pseudotamarii*, a new aflatoxin producing species in *Aspergillus* section *Flavi*. *Mycol. Res.* **2001**, *105*, 233–239. [CrossRef]

4. Kurtzman, C.P.; Horn, B.W.; Hesseltine, C.W. *Aspergillus-nomius*, a New Aflatoxin-producing Species Related to *Aspergillus-flavus* and *Aspergillus-tamarii*. *J. Microbiol.* **1987**, *53*, 147–158. [CrossRef]

5. Yiannikouris, A.; Jouany, J. Mycotoxins in feeds and their fate in animals: A review. *Anim. Res.* **2002**, *51*, 81–99. [CrossRef]

6. Hussein, H.S.; Brasel, J.M. Toxicity, metabolism, and impact of mycotoxins on humans and animals. *Toxicology* **2001**, *167*, 101–134. [CrossRef]

7. Mishra, H.; Das, C. A review on biological control and metabolism of aflatoxin. *Crit. Rev. Food Sci.* **2003**, *43*, 245–264. [CrossRef] [PubMed]

8. Groopman, J.D.; Hasler, J.A.; Trudel, L.J.; Pikul, A.; Donahue, P.R.; Wogan, G.N. Molecular dosimetry in rat urine of aflatoxin-N7-guanine and other aflatoxin metabolites by multiple monoclonal antibody affinity chromatography and immunoaffinity/high performance liquid chromatography. *Cancer Res.* **1992**, *52*, 267–274. [PubMed]

9. Rustemeyer, S.M.; Lamberson, W.R.; Ledoux, D.R.; Rottinghaus, G.E.; Shaw, D.P.; Cockrum, R.R.; Kessler, K.L.; Austin, K.J.; Cammack, K.M. Effects of dietary aflatoxin on the health and performance of growing barrows. *J. Anim. Sci.* **2010**, *88*, 3624–3630. [CrossRef] [PubMed]

10. Dalvi, R. An overview of aflatoxicosis of poultry: Its characteristics, prevention and reduction. *Vet. Res. Commun.* **1986**, *10*, 429–443. [CrossRef] [PubMed]

11. Kermanshahi, H.; Hazegh, A.R.; Afzali, N. Effect of sodium bentonite in broiler chickens fed diets contaminated with aflatoxin B1. *J. Anim. Vet. Adv.* **2009**, *8*, 1631–1636.

12. Magnoli, A.P.; Monge, M.P.; Miazzo, R.D.; Cavaglieri, L.R.; Magnoli, C.E.; Merkis, C.I.; Cristofolini, A.L.; Dalcero, A.M.; Chiacchiera, S.M. Effect of low levels of aflatoxin B(1) on performance, biochemical parameters, and aflatoxin B(1) in broiler liver tissues in the presence of monensin and sodium bentonite. *Poult. Sci.* **2011**, *90*, 48–58. [CrossRef] [PubMed]

13. Berthiller, F.; Crews, C.; Dall'Asta, C.; Saeger, S.D.; Haesaert, G.; Karlovsky, P.; Oswald, I.P.; Seefelder, W.; Speijers, G.; Stroka, J. Masked mycotoxins: A review. *Mol. Nutr. Food Res.* **2013**, *57*, 165–186. [CrossRef] [PubMed]

14. Ledoux, D.R.; Rottinghaus, G.E.; Bermudez, A.J.; Alonso-Debolt, M. Efficacy of a hydrated sodium calcium aluminosilicate to ameliorate the toxic effects of aflatoxin in broiler chicks. *Poult. Sci.* **1999**, *78*, 204–210. [CrossRef] [PubMed]

15. Bintvihok, A.; Kositcharoenkul, S. Effect of dietary calcium propionate on performance, hepatic enzyme activities and aflatoxin residues in broilers fed a diet containing low levels of aflatoxin B1. *Toxicon* **2006**, *47*, 41–46. [CrossRef] [PubMed]

16. Yoshinari, T.; Noda, Y.; Yoda, K.; Sezaki, H.; Nagasawa, H.; Sakuda, S. Inhibitory activity of blasticidin A, a strong aflatoxin production inhibitor, on protein synthesis of yeast: Selective inhibition of aflatoxin production by protein synthesis inhibitors. *J. Antibiot.* **2010**, *63*, 309–314. [CrossRef] [PubMed]

17. Karlovsky, P. Biological detoxification of fungal toxins and its use in plant breeding, feed and food production. *Nat. Toxins* **1999**, *7*, 1–23. [CrossRef]

18. Bata, A.; Lasztity, R. Detoxification of mycotoxin-contaminated food and feed by microorganisms. *Trends Food Sci. Technol.* **1999**, *10*, 223–228. [CrossRef]

19. Wu, Q.; Jezkova, A.; Yuan, Z.; Pavlikova, L.; Dohnal, V.; Kuca, K. Biological degradation of aflatoxins. *Drug Metab. Rev.* **2009**, *41*, 1–7. [CrossRef] [PubMed]

20. Kabak, B.; Dobson, A.D.W.; Var, I. Strategies to prevent mycotoxin contamination of food and animal feed: A review. *Crit. Rev. Food Sci.* **2006**, *46*, 593–619. [CrossRef] [PubMed]

21. Taylor, W.J.; Draughon, F.A. Nannocystis exedens: A potential biocompetitive agent against *Aspergillus flavus* and *Aspergillus parasiticus*. *J. Food Protect.* **2001**, *64*, 1030–1034. [CrossRef]

22. Liu, D.L.; Yao, D.S.; Liang, R.; Ma, L.; Cheng, W.Q.; Gu, L.Q. Detoxification of aflatoxin B-1 by Enzymes Isolated from *Armillariella tabescens*. *Food Chem. Toxicol.* **1998**, *36*, 563–574. [CrossRef]
23. Teniola, O.D.; Addo, P.A.; Brost, I.M.; Farber, P.; Jany, K.D.; Alberts, J.F.; van Zyl, W.H.; Steyn, P.S.; Holzapfel, W.H. Degradation of aflatoxin B-1 by cell-free extracts of *Rhodococcus erythropolis* and *Mycobacterium fluoranthenivorans* sp. nov DSM44556(T). *Int. J. Food Microbiol.* **2005**, *105*, 111–117. [CrossRef] [PubMed]
24. Zhao, L.H.; Guan, S.; Gao, X.; Ma, Q.G.; Lei, Y.P.; Bai, X.M.; Ji, C. Preparation, purification and characteristics of an aflatoxin degradation enzyme from *Myxococcus fulvus* ANSM068. *J. Appl. Mocrpbiol.* **2011**, *110*, 147–155. [CrossRef] [PubMed]
25. Gao, X.; Ma, Q.; Zhao, L.; Lei, Y.; Shan, Y.; Ji, C. Isolation of *Bacillus subtilis*: Screening for aflatoxins B1, M1, and G1 detoxification. *Eur. Food Res. Technol.* **2011**, *232*, 957–962. [CrossRef]
26. Ma, Q.G.; Gao, X.; Zhou, T.; Zhao, L.H.; Fan, Y.; Li, X.Y.; Lei, Y.P.; Ji, C.; Zhang, J.Y. Protective effect of *Bacillus subtilis* ANSB060 on egg quality, biochemical and histopathological changes in layers exposed to aflatoxin B1. *Poult. Sci.* **2012**, *91*, 2852–2857. [CrossRef] [PubMed]
27. Fan, Y.; Zhao, L.; Ma, Q.; Li, X.; Shi, H.; Zhou, T.; Zhang, J.; Ji, C. Effects of *Bacillus subtilis* ANSB060 on growth performance, meat quality and aflatoxin residues in broilers fed moldy peanut meal naturally contaminated with aflatoxins. *Food Chem. Toxicol.* **2013**, *59*, 748–753. [CrossRef] [PubMed]
28. Fan, Y.; Zhao, L.; Ji, C.; Li, X.; Jia, R.; Xi, L.; Zhang, J.; Ma, Q. Protective Effects of *Bacillus subtilis* ANSB060 on Serum Biochemistry, Histopathological Changes and Antioxidant Enzyme Activities of Broilers Fed Moldy Peanut Meal Naturally Contaminated with Aflatoxins. *Toxins* **2015**, *7*, 3330–3343. [CrossRef] [PubMed]
29. Anbiah, S.; Mohan, C.; Manohar, B.; Balachandram, C. Chronic aflatoxin B-1 induced hepatopathy in ducks. *Indian Vet. J.* **2004**, *81*, 1210–1212.
30. He, J.; Zhang, K.Y.; Chen, D.W.; Ding, X.M.; Feng, G.D.; Ao, X. Effects of maize naturally contaminated with aflatoxin B1 on growth performance, blood profiles and hepatic histopathology in ducks. *Liver Sci.* **2013**, *152*, 192–199. [CrossRef]
31. Han, X.; Huang, Q.; Li, W. Changes in growth performance, digestive enzyme activities and nutrient digestibility of cherry valley ducks in response to aflatoxin B-1 levels. *Liver Sci.* **2008**, *119*, 216–220. [CrossRef]
32. Yarru, L.P.; Settivari, R.S.; Gowda, N.K.; Antoniou, E.; Ledoux, D.R.; Rottinghaus, G.E. Effects of turmeric (*Curcuma longa*) on the expression of hepatic genes associated with biotransformation, antioxidant, and immune systems in broiler chicks fed aflatoxin. *Poult. Sci.* **2009**, *88*, 2620–2627. [CrossRef] [PubMed]
33. Matur, E.; Ergul, E.; Akyazi, I.; Eraslan, E.; Inal, G.; Bilgic, S.; Demircan, H. Effects of Saccharomyces cerevisiae extract on haematological parameters, immune function and the antioxidant defence system in breeder hens fed aflatoxin contaminated diets. *Br. Poult. Sci.* **2011**, *21*, 806–815.
34. Li, Y.; Liu, Y.H.; Yang, Z.B.; Wan, X.L.; Chi, F. The efficacy of clay enterosorbent to ameliorate the toxicity of aflatoxin B1 from contaminated corn (Zea mays) on hematology, serum biochemistry, and oxidative stress in ducklings. *J. Appl. Anim. Res.* **2012**, *21*, 806–815. [CrossRef]
35. Cheng, Y.H.; Shen, T.F.; Pang, V.F.; Chen, B.J. Effects of aflatoxin and carotenoids on growth performance and immune response in mule ducklings. *Comp. Biochem. Physiol.* **2001**, *128*, 19–26. [CrossRef]
36. Fowler, J.; Li, W.; Bailey, C. Effects of a Calcium Bentonite Clay in Diets Containing Aflatoxin when Measuring Liver Residues of Aflatoxin B1 in Starter Broiler Chicks. *Toxins* **2015**, *7*, 3455–3464. [CrossRef] [PubMed]
37. Ghosh, R.C.; Chauhan, H.V.S.; Roy, S. Immunosuppression in broilers under experimental aflatoxicosis. *Br. Vet. J.* **1990**, *146*, 457–462. [CrossRef]
38. Oguz, H.; Kurtoglu, V. Effect of clinoptilolite on fattening performance of broiler chickens during experimental aflatoxicosis. *Br. Poult. Sci.* **2000**, *41*, 512–517. [CrossRef] [PubMed]
39. Oguz, H.; Kececi, T.; Birdane, Y.O.; Onder, F.; Kurtoglu, V. Effect of clinoptilolite on serum biochemical and haematological characters of broiler chickens during experimental aflatoxicosis. *Res. Vet. Sci.* **2000**, *69*, 89–93. [CrossRef] [PubMed]
40. Guan, S.; Zhao, L.; Ma, Q.; Zhou, T.; Wang, N.; Hu, X.; Ji, C. In Vitro Efficacy of *Myxococcus fulvus* ANSM068 to Biotransform Aflatoxin B1. *Int. J. Mol. Sci.* **2010**, *11*, 4063–4079. [CrossRef] [PubMed]
41. Liu, D.L.; Yao, D.S.; Liang, Y.Q.; Zhou, T.H.; Song, Y.P.; Zhao, L.; Ma, L. Production, purification, and characterization of an intracellular aflatoxin-detoxifizyme from *Armillariella tabescens* (E-20). *Food Chem. Toxicol.* **2001**, *39*, 461–466. [CrossRef]

42. Yang, J.; Bai, F.; Zhang, K.; Bai, S.; Peng, X.; Ding, X.; Li, Y.; Zhang, J.; Zhao, L. Effects of feeding corn naturally contaminated with aflatoxin B1 and B2 on hepatic functions of broilers. *Poult. Sci.* **2012**, *91*, 2792–2801. [CrossRef] [PubMed]

43. Fiers, W.; Beyaert, R.; Declercq, W. More than one way to die: Apoptosis, necrosis and reactive oxygen damage. *Oncogene* **1999**, *18*, 7719–7730. [CrossRef] [PubMed]

44. Coskun, O.; Kanter, M.; Korkmaz, A. Quercetin, a flavonoid antioxidant, prevents and protects streptozotocin-induced oxidative stress and beta-cell damage in rat pancreas. *Pharmacol. Res.* **2005**, *51*, 117–123. [CrossRef] [PubMed]

45. Wills, E.D. Mechanisms of Lipid Peroxide Formation in Animal Tissues. *Biochem. J.* **1966**, *99*, 667–676. [CrossRef] [PubMed]

46. Essiz, D.; Altintas, L.; Das, Y.K. Effects of aflatoxin and various adsorbents on plasma malondialdehyde levels in quails. *Bull. Vet. Inst. Pulway* **2006**, *50*, 585–588.

47. Eraslan, G.; Akdogan, M.; Yarsan, E.; Sahindokuyucu, F.; Essiz, D.; Altintas, L. The effects of aflatoxins on oxidative stress in broiler chickens. *Turk. J. Vet. Anim. Sci.* **2005**, *29*, 701–707.

48. Guerre, P.; Larrieu, G.; Burgat, V. Cytochrome P450 decreases are correlated to increased microsomal oxidative damage in rabbit liver and primary cultures of rabbit hepatocytes exposed to AFB1. *Toxicol. Lett.* **1999**, *104*, 117–125. [CrossRef]

49. Meki, A.; Esmail, E.; Hussein, A. Caspase-3 and heat shock protein-70 in rat liver treated with aflatoxin B1: effect of melatonin. *Toxincon* **2004**, *43*, 93–100. [CrossRef] [PubMed]

50. Preetha, S.P.; Kanniappan, M.; Selvakumar, E.; Nagaraj, M.; Varalakshmi, P. Lupeol ameliorates aflatoxin B1-induced peroxidative hepatic damage in rats. *Comp. Biochem. Physiol.* **2006**, *143*, 333–339. [CrossRef] [PubMed]

51. Marin, S.; Ramos, A.J.; Cano-Sancho, G.; Sanchis, V. Mycotoxins: Occurrence, toxicology, and exposure assessment. *Food Chem. Toxicol.* **2013**, *60*, 218–237. [CrossRef] [PubMed]

52. Ortatatli, M.; Oguz, H.; Hatipoglu, F.; Karaman, M. Evaluation of pathological changes in broilers during chronic aflatoxin (50 and 100 ppb) and clinoptilolite exposure. *Res. Vet. Sci.* **2005**, *78*, 61–68. [CrossRef] [PubMed]

53. Taub, R. Liver regeneration: From myth to mechanism. *Nat. Rev. Mol. Cell Biol.* **2004**, *5*, 836–847. [CrossRef] [PubMed]

54. Zaghini, A.; Martelli, G.; Roncada, P. Mannanoligosaccharides and Aflatoxin B1 in Feed for Laying Hens: Effects on Egg Quality, Aflatoxins B1 and M1 Residues in Eggs, and Aflatoxin B1 Levels in Liver. *Poult. Sci.* **2005**, *84*, 825–832. [CrossRef] [PubMed]

55. Binder, E.M.; Tan, L.M.; Chin, L.J.; Handl, J.; Richard, J. Worldwide occurrence of mycotoxins in commodities, feeds and feed ingredients. *Anim. Feed Sci. Technol.* **2007**, *137*, 265–282. [CrossRef]

toxins

MDPI

Article

Curcumin Prevents Aflatoxin B₁ Hepatoxicity by Inhibition of Cytochrome P450 Isozymes in Chick Liver

Ni-Ya Zhang [1], Ming Qi [1], Ling Zhao [1], Ming-Kun Zhu [1], Jiao Guo [1], Jie Liu [1], Chang-Qin Gu [2], Shahid Ali Rajput [1], Christopher Steven Krumm [3], De-Sheng Qi [1] and Lv-Hui Sun [1,*]

[1] Department of Animal Nutrition and Feed Science, College of Animal Science and Technology, Huazhong Agricultural University, Wuhan 430070, China; zhangniya@mail.hzau.edu.cn (N.-Y.Z.); qi_ming1993@126.com (M.Q.); ling930910@126.com (L.Z.); zhumingkun1989@126.com (M.-K.Z.); guojiao1991@sina.com (J.G.); cherishlj@webmail.hzau.edu.cn (J.L.); dr.shahidali@hotmail.com (S.A.R.); qds@mail.hzau.edu.cn (D.-S.Q.)

[2] College of Veterinary Medicine, Huazhong Agricultural University, Wuhan 430070, China; guchangqin@mail.hzau.edu.cn

[3] Department of Animal Science, Cornell University, Ithaca, NY 14853, USA; csk97@cornell.edu

* Correspondence: lvhuisun@mail.hzau.edu.cn; Tel.: +86-27-8728-1793; Fax: +86-27-8728-1033

Academic Editor: Ting Zhou
Received: 31 August 2016; Accepted: 7 November 2016; Published: 10 November 2016

Abstract: This study was designed to establish if Curcumin (CM) alleviates Aflatoxin B₁ (AFB₁)-induced hepatotoxic effects and to determine whether alteration of the expression of cytochrome P450 (CYP450) isozymes is involved in the regulation of these effects in chick liver. One-day-old male broilers ($n = 120$) were divided into four groups and used in a two by two factorial trial in which the main factors included supplementing AFB₁ (< 5 vs. 100 µg/kg) and CM (0 vs. 150 mg/kg) in a corn/soybean-based diet. Administration of AFB₁ induced liver injury, significantly decreasing albumin and total protein concentrations and increasing alanine aminotransferase and aspartate aminotransferase activities in serum, and induced hepatic histological lesions at week 2. AFB₁ also significantly decreased hepatic glutathione peroxidase, catalase, and glutathione levels, while increasing malondialdehyde, 8-hydroxydeoxyguanosine, and exo-AFB₁-8,9-epoxide (AFBO)-DNA concentrations. In addition, the mRNA and/or activity of enzymes responsible for the bioactivation of AFB₁ into AFBO—including CYP1A1, CYP1A2, CYP2A6, and CYP3A4—were significantly induced in liver microsomes after 2-week exposure to AFB₁. These alterations induced by AFB₁ were prevented by CM supplementation. Conclusively, dietary CM protected chicks from AFB₁-induced liver injury, potentially through the synergistic actions of increased antioxidant capacities and inhibition of the pivotal CYP450 isozyme-mediated activation of AFB₁ to toxic AFBO.

Keywords: curcumin; aflatoxin B₁; CYP450; AFBO–DNA; chicks

---

1. Introduction

Aflatoxins (AF) are secondary fungal metabolites that are largely produced by the fungi *Aspergillus flavus* and *Aspergillus parasiticus* [1,2]. Among the various dangerous AF and their metabolites, aflatoxin B₁ (AFB₁) is the most toxic, exhibiting harmful hepatotoxic, teratogenic, mutagenic, and carcinogenic effects on humans and many species of livestock [3–6]. It is also classified as a Group I carcinogen [7]. Human or animal consumption of the food or feed contaminated by AFB₁ can pose serious problems to their health and productivity, and thus result in significant economic losses [8,9]. The toxic effects of AFB₁ are associated with its toxification and detoxification biotransformation pathways. Upon being delivered to the liver, AFB₁ is bioactivated by cytochrome P450 (CYP450)—a member of the phase I metabolizing enzymes—into the highly reactive exo-AFB1-8,9-epoxide (AFBO) [3,10]. AFBO can form adducts with DNA and other critical macromolecules, causing toxicity, mutations, and cancer [10]. Meanwhile, AFB_1 can induce the generation of reactive oxygen species (ROS), which can lead to oxidative stress, potentially mediated via CYP450 activity [11,12]. On the other hand, AFBO can be detoxified via conjugation with glutathione (GSH) to form a non-toxic adduct, which can be catalyzed by glutathione-S transferases (GSTs), the phase II detoxification enzymes [10].

Curcumin (CM) is a natural polyphenolic compound extracted from rhizomes of *Curcuma longa* Linn (turmeric), widely used as household spice, natural food colorant, and herbal medicine in many Asian countries for thousands of years [13]. It possesses antioxidant, anti-inflammatory, radio-protective, chemotherapeutic, anti-cancer, and detoxification abilities in laboratory animals and humans [14–17]. Previous publications have described that CM can effectively mitigate AFB_1-induced adverse effects in several animal species [6,15,18–20]. Moreover, the protective action of CM against AFB_1-induecd adverse effects was further demonstrated by improving the antioxidant capacity [18,19,21,22]. Notably, as different CYP450 isozymes catalyze AFB_1 to various metabolites, including the highly toxic AFBO and the less- or non-toxic aflatoxicol, AFM_1, AFP_1, and AFQ_1, examination of the regulation of the proportions of CYP450 isozymes by CM would be valuable [10,23]. Previous studies have described that CM can alter various CYP450 isozymes in vivo and in vitro [24–27]. Moreover, CM inhibition of AFB_1 toxicity has been reported by modulating CYP450 function [14], while the knowledge of which crucial CYP450 isozymes are involved in this process remains unknown. Chicken orthologs of human CYP1A, 2A, and 3A families are the main CYP450 enzymes responsible for the bioactivation of AFB_1 into the highly toxic AFBO in chicken [23]. However, there is limited information on the effect of CM on these pivotal CYP450 isozymes that are involved in AFB_1 metabolism. Therefore, we selected chickens to investigate whether dietary supplementation of CM mitigated AFB_1-induced hepatotoxic effects though the regulation of these key CYP450 isozymes.

2. Results

2.1. Growth Performance, Serum Biochemistry, and Liver Histology

Non-significant differences in average daily feed intake, average daily gain, and feed/gain ratio were observed among the four groups throughout the experiment (Table S1). Serum biochemical parameters were significantly affected by either supplementation of AFB_1 or CM at week 2 (Table 1). Compared to the control, the activities of alanine aminotransferase (ALT) and aspartate aminotransferase (AST) increased ($p < 0.05$) by 33.3% and 43.8% respectively, while the concentrations of albumin (ALB) and total protein (TP) decreased ($p < 0.05$) by 33.8% and 26.0%, respectively, in the serum of chicks by AFB_1 supplementation. Notably, the serum biochemical parameter changes observed in the AFB_1 group were prevented in the AFB_1 + CM group. However, no significant differences in these serum biochemical parameters were observed among the four groups at week 4 (Table 1). Since the serological results indicated AFB_1 only induced liver injury at week 2, we selected samples from week 2 to explore the mechanism. Furthermore, dietary AFB_1 exposure induced liver injury as shown through bile duct hyperplasia and necrosis at week 2. Strikingly, the AFB_1 + CM group prevented the hepatic injury observed in the AFB_1 group (Figure 1).

2.2. Hepatic Antioxidant Parameters and CYP450 Isozyme Activities

After 2 weeks of experimental treatments, the antioxidant parameters and CYP450 isozyme activities were significantly altered by either supplementation of AFB_1 or CM (Tables 2 and 3). Compared to the control, supplementation of dietary AFB_1 led to a decrease (0.05) in the activities of glutathione peroxidase (GPX, 13.1%), catalase (CAT, 16.2%), and GSH concentration (30.9%), along with increase ($p < 0.05$) in concentrations of malondialdehyde (MDA, 100.0%) and 8-hydroxydeoxyguanosine (8-OHdG, 17.9%) in the liver of chicks at week 2, respectively. Interestingly, the antioxidant parameter changes observed in the AFB_1 group were prevented in the AFB_1 + CM group (Table 2). In addition, supplementation of CM alone increased ($p < 0.05$) activity of GPX (25%), while it did not affect the other antioxidant parameters, when compared with the control. Meanwhile, the dietary AFB_1 supplementation led to an increase ($p < 0.05$) in the activity of CYP1A1 (270.6%), CYP1A2 (99.4%), CYP2A6 (184.5%), and CYP3A4 (29.2%) in the liver microsomes of chickens, respectively (Table 3). Strikingly, the increased CYP450 isozyme activities observed in the AFB_1 group were reduced in the AFB_1 + CM group.

Table 1. Effects of dietary AFB_1 and CM concentrations on serum biochemical parameters in chicks [1].

Item	Control	AFB1	CM	AFB_1 + CM
		Week 2		
ALT, U/L	1.2 ± 0.1 [a]	1.6 ± 0.4 [b]	1.2 ± 0.3 [a,b]	1.3 ± 0.2 [a,b]
AST, U/L	176.7 ± 27.0 [a]	254.2 ± 53.9 [b]	178.4 ± 38.1 [a]	193.5 ± 39.4 [a,b]
TP, g/L	17.7 ± 1.1 [b]	13.1 ± 2.3 [a]	17.3 ± 1.8 [b]	19.1 ± 2.0 [b]
ALB, g/L	7.4 ± 0.3 [b]	4.9 ± 1.0 [a]	7.1 ± 1.2 [b]	8.1 ± 1.0 [b]
		Week 4		
ALT, U/L	1.0 ± 0.1	1.2 ± 0.3	1.4 ± 0.4	1.6 ± 0.5
AST, U/L	217.4 ± 26.8	211.2 ± 22.0	223.6 ± 37.5	245.9 ± 83.4
TP, g/L	22.8 ± 3.9	23.3 ± 2.7	21.8 ± 4.4	26.3 ± 7.8
ALB, g/L	9.9 ± 2.1	9.8 ± 1.9	9.3 ± 2.4	11.6 ± 3.7

[1] Values are expressed as means ± SD (n = 5), and means with different superscript letters differ ($p < 0.05$). AFB_1, aflatoxin B_1; ALB, albumin; ALT, alanine aminotransferase; AST, aspartate aminotransferase; CM, curcumin; TP, total protein. Experimental details of Control and AFB_1 groups are given in Sun et al. (2016) [12].

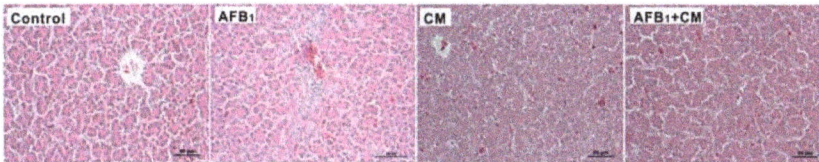

Figure 1. Photomicrographs of hepatic sections stained with hematoxylin and eosin (400× magnification) of chicks from different treatment groups at week 2. AFB_1, aflatoxin B_1; CM, curcumin. Experimental details of Control and AFB_1 groups are given in Sun et al. (2016) [12].

Table 2. Effects of dietary AFB_1 and CM concentrations on hepatic antioxidant parameters in chicks at week 2 [1].

Item	Control	AFB_1	CM	AFB_1 + CM
GPX, U/mg	127.8 ± 5.1 [b]	111.1 ± 10.3 [a]	159.9 ± 15.1 [c]	159.7 ± 8.9 [c]
SOD, U/mg	156.8 ± 5.2 [a,b]	149.9 ± 9.7 [a]	170.2 ± 6.7 [b]	162.5 ± 8.7 [a,b]
CAT, U/mg	13.6 ± 0.9 [b]	11.4 ± 1.6 [a]	15.4 ± 1.2 [b,c]	16.7 ± 1.0 [c]
GST, U/mg	61.5 ± 1.1	60.6 ± 1.9	62.3 ± 5.1	62.0 ± 2.4
GSH, μmol/g	48.5 ± 10.1 [b]	33.5 ± 3.9 [a]	62.8 ± 17.6 [b]	53.4 ± 15.6 [b]
MDA, μmol/g	3.2 ± 0.4 [a]	6.4 ± 1.3 [b]	2.9 ± 0.5 [a]	3.2 ± 0.6 [a]
8-OHdG, nmol/mg	152.2 ± 8.1 [a]	179.5 ± 5.4 [b]	157.9 ± 2.9 [a]	156.8 ± 4.7 [a]

[1] Values are expressed as means ± SD (n = 5), and means with different superscript letters differ ($p < 0.05$). AFB_1, aflatoxin B_1; CAT, catalase; CM, curcumin; GPX, glutathione peroxidase; GSH, glutathione; GST, glutathione-S transferases; MDA, malondialdehyde; SOD, superoxide dismutase; 8-OHdG, 8-hydroxydeoxyguanosine. Experimental details of Control and AFB_1 groups are given in Sun et al. (2016) [12].

Table 3. Effects of dietary AFB$_1$ and CM concentrations on the activities of chicken cytochrome P450 (CYP450) orthologs in the liver at week 2 [1].

Item	Control	AFB$_1$	CM	AFB$_1$ + CM
CYP1A1, nmol/(mgprotein·min)	0.34 ± 0.13 [a]	1.26 ± 0.17 [c]	0.83 ± 0.15 [b]	0.79 ± 0.16 [b]
CYP1A2, nmol/(mgprotein·min)	1.71 ± 0.45 [a]	3.41 ± 0.49 [b]	2.45 ± 0.63 [a,b]	1.95 ± 0.55 [a]
CYP2A6, nmol/(mgprotein·min)	2.07 ± 0.30 [a]	5.89 ± 1.37 [b]	3.00 ± 1.18 [a]	3.08 ± 0.63 [a]
CYP3A4, nmol/(mgprotein·min)	26.93 ± 2.22 [a]	34.79 ± 3.06 [b]	24.86 ± 1.51 [a]	22.60 ± 2.26 [a]

[1] Values are expressed as means ± SD (n = 5), and means with different superscript letters differ ($p < 0.05$). AFB$_1$, aflatoxin B$_1$; CM, curcumin; CYP1A1, Cytochrome P450 1A1; CYP1A2, Cytochrome P450 1A2; CYP2A6, Cytochrome P450 2A6; CYP3A4, Cytochrome P450 3A4. Experimental details of Control and AFB$_1$ groups are given in Sun et al. (2016) [12].

2.3. Hepatic AFBO–DNA Adduct Contents

The concentrations of AFBO–DNA adducts in the liver were significantly affected by either supplementation of AFB$_1$ or CM at week 2 (Figure 2). Compared to the control, the hepatic AFBO–DNA adduct content was increased ($p < 0.05$) 12 times by AFB$_1$ supplementation. Interestingly, the AFB$_1$ + CM group decreased ($p < 0.05$) the concentration of AFBO–DNA adduct (63.7%) in the liver when compared to the AFB$_1$ group.

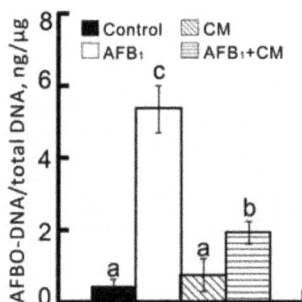

Figure 2. Effects of dietary AFB$_1$ and CM concentrations on the contents of AFBO–DNA adducts in the liver of chicks at week 2. Values are expressed as means ± SD (n = 5), and means with different superscript letters differ ($p < 0.05$). AFB$_1$, aflatoxin B$_1$; AFBO, exo-AFB1-8,9-epoxide; CM, curcumin. Experimental details of Control and AFB1 groups are given in Sun et al. (2016) [12].

2.4. Hepatic CYP450 Isozyme Activities and Gene Expression

The mRNA levels of *CYP1A1*, *CYP1A2*, and *CYP3A4* in the liver were significantly altered by either supplementation of AFB$_1$ or CM (Figure 3). Specifically, dietary AFB$_1$ supplementation led to upregulated ($p < 0.05$) mRNA levels of *CYP1A1*, *CYP1A2*, and *CYP3A4* in liver microsomes. Strikingly, the increased hepatic CYP450 isozyme mRNA levels observed in the AFB$_1$ group were suppressed in the AFB$_1$ + CM group. It is fascinating to find that the effects of AFB$_1$ and CM on changes in hepatic CYP450 isozyme mRNA levels were in parallel with their activities.

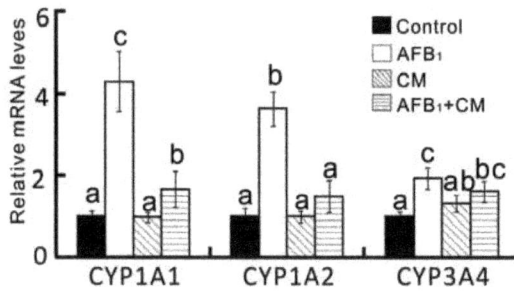

Figure 3. Effects of dietary AFB$_1$ and CM concentrations on relative mRNA abundance of CYP450 isozyme genes in liver of chicks at week 2. Values are expressed as means $\pm$ SD (n = 5), and means with different superscript letters differ ($p < 0.05$). AFB$_1$, aflatoxin B$_1$; CM, curcumin; *CYP1A1*, Cytochrome P450 1A1; *CYP1A2*, Cytochrome P450 1A2; *CYP3A4*, Cytochrome P450 3A4. Experimental details of Control and AFB1 groups are given in Sun et al. (2016) [12].

3. Discussion

Protection against AFB$_1$-induced hepatotoxic effects was successfully replicated in broiler chicks fed an AFB$_1$-contaminated corn–soybean diet with CM supplementation. Although the dietary AFB$_1$ had no significant effect on growth performance in chicks, it induced the typical clinical signs of hepatic injury, including increased activities of AST and ALT, decreased concentrations of ALB and TP in serum, as well as bile duct hyperplasia and necrosis in the liver of chicks at week 2 [28,29]. However, the serological results indicated that AFB$_1$-induced liver injury disappeared at week 4. The reasons for this might be that older poultry was more resistant to aflatoxicosis than young poultry [30], and the AFBO–DNA adducts could be repaired by the nucleotide excision repair system in liver [31]. Intriguingly, dietary supplementation of CM mitigated serum and histopathological parameter alterations that were induced by dietary supplementation of AFB$_1$ at week 2. These outcomes were consistent with previous studies, which provided evidence that hepatic injury was induced by dietary AFB$_1$ as well as that dietary CM supplementation displayed protective effects against its negative effects [15,19–21]. Moreover, the present study displayed AFB$_1$-induced oxidative stress in chickens as evidenced by decreased antioxidant ability (GPX, CAT, and GSH), increased lipid peroxidation (MDA), and DNA damage (8-OHdG). On the other hand, dietary CM supplementation inhibited these changes. Meanwhile, dietary supplementation of CM alone improved the hepatic GPX activity, which was consistent with previous studies in which CM increased GPX activity, probably by activating the Nrf2–keap1 pathway [32,33]. A previous study showed that dietary supplementation of 5000 mg/kg CM increased hepatic phase II detoxification enzyme (GST) activity in rats that were exposed to AFB$_1$ [34], while GST activity was not affected by CM in this study. The divergence between these reports could be attributed to the different animal species and ingestion dose. Taken together, these outcomes are similar to former studies, which reported that oxidative stress could be due to the direct effects of AFB$_1$, its metabolites, and/or the generation of free radicals [11,35]. Dietary supplementation of CM, however, showed protective actions against AFB$_1$-induced hepatic injury, which were associated with the enhancement of antioxidant capacities [18,19,21,22].

The most interesting finding from the present study was that the four major CYP450 isozymes were significantly inhibited to a large extent by dietary supplementation of CM upon exposure to dietary AFB$_1$. The hepatic mRNA levels and/or enzyme activities of CYP1A1, CYP1A2, CYP2A6, and CYP3A4 were significantly increased when chicks were exposed to dietary AFB$_1$, while dietary supplementation of CM inhibited these changes. Because a previous study reported that CYP2A6 and (to a lesser extent) CYP1A1 are responsible for the bioactivation of AFB$_1$ into AFBO in chicken hepatic microsomes, and that CYP1A2 and CYP3A4 are the most important enzymes capable of bioactivating AFB$_1$ into AFBO in mammals [23,36], inhibition of the activities of these enzymes could decrease the production of AFBO. Indeed, as a major toxic adduct of AFBO [10,36], the AFBO–DNA was sharply decreased by the dietary supplementation of CM when chicks were exposed to dietary AFB_1. These findings suggest that the protective actions of CM may be mediated through inhibited activities of these crucial CYP450 isozymes, which could decrease the production of the highly toxic AFBO. These outcomes are in agreement with previous studies, which showed that CM-mediated inhibition of AFB_1 toxicity was associated with a reduction of the formation of AFBO–DNA by modulating CYP450 function, including CYP1A1 activity [14,20], while the activities of CYP1A2, CYP2A6, and CYP3A4 inhibited by CM in this process are described, to our knowledge, for the first time in the current study. Similarly, CM has been shown to inhibit tumorigenesis induced by benzo(a)pyrene and 2,3,7,8-tetrachlorodibenzo-p-dioxin through the inhibition of CYP1A1, CYP1B1, and/or CYP1A2 enzyme on mRNA, protein, and/or activities levels [25,37].

The CYP450s are important metabolizing enzymes, most of which are mainly expressed in the liver. Clinically, these enzymes play vital roles in drug metabolism and are required for the efficient clearance of xenobiotics from the body [38,39]. On the other hand, these enzymes can also bioactivate biologically inert carcinogens and toxins, such as 4-(methylnitrosamino)-1-(3-pyridyl)-1-butanone [40], *N*-nitrosonornicotine [41], hexamethylphosphoramide [42], benzo(a)pyrene [25], AFB_1 [10], and 2,6-dichlorobenzonitrile [43] to electrophilic metabolites that can cause toxicity, cell death, and sometimes cellular transformation that results in cancer. Given that CM is a strong inhibitor of CYP450 isozymes, it could be a potentially promising chemopreventive agent for many carcinogens and toxins, even though its mechanism of action remains to be clarified.

In summary, the present study successfully confirmed that dietary CM supplementation could alleviate AFB_1-induced liver injury, with regard to the suppression of serum biochemistry changes and histopathological lesions in the liver of broilers. The protective mechanism of CM against AFB_1-induced adverse effects may be associated with: (1) the common mechanism that CM could reduce AFB_1-induced oxidative stress by increasing antioxidant capacities; and (2) the novel finding that CM could effectively inhibit the regulatory role of CYP450 isozymes that are crucial for the activation of AFB_1 to highly toxic AFBO. Further validation of the potential mechanisms of the interactions between CM and the CYP450 isozymes would be beneficial to gain a better understanding of the detoxification mechanism of CM and applications of CM to control chemical carcinogenesis.

4. Materials and Methods

4.1. Chicks, Treatments, and Samples Collection

Our animal protocol was approved by the Scientific Ethic Committee of Huazhong Agricultural University on 8 August 2015. The project identification code is HZAUCH-2015-007. In total, 120 one-day-old male avian broilers were randomly divided into four treatment groups with five replicates of six chicks per pen. All chicks were allowed free access to a corn/soybean-based diet (BD) formulated to meet the nutritional requirements of broilers' diets (Table S2) as reported previously [12], and distilled water. After three days of acclimation, the four experimental groups were arranged in a two by two factorial design trial that included BD diet supplemented with AFB_1 (Sigma-Aldrich, St. Louis, MO, USA) and CM (China National Medicines Corp. Ltd., Beijing, China), as follows: Control; AFB_1 (100 μg AFB_1/kg); CM (150 mg CM/kg); and AFB_1 + CM (100 μg AFB_1/kg with 150 mg CM/kg). The experiment lasted for four weeks. The doses were chosen based on previous studies, which reported that dietary consumption of approximately 100 μg AFB_1/kg induced adverse effects [44], while dietary supplementation of 74–222 mg CM/kg displayed a protective effect on AFB_1 in broilers [19]. Individual body weights and feed intake of broilers were measured biweekly. Meanwhile, chicks ($n = 5$/group) were euthanized by decapitation to collect blood and livers for the preparation of serum and liver histological tissue samples as previously described [29].

4.2. Dietary AFB₁ Analysis, and Feed Preparation

Twenty grams of moldy corn was extracted with 100 mL of methanol (Fisher, Pittsburgh, PA, USA):water (70:30, *v/v*) for AFB_1 detection. After shaking for 3 min, the supernatant of the extract was filtered through a Whatman filter (Whatman, Clifton, NJ, USA), and the filtrate was collected. Then, the concentration of AFB_1 in filtrate was measured followed the protocol of the ELISA kit (AgraQuant® Aflatoxin B₁ Assay, Romer, Singapore). The powdered feed was mixed with a vertical mixer.

4.3. Serum Biochemical and Histological Analysis

The serum activities of aspartate aminotransferase (AST) and alanine aminotransferase (ALT), along with concentrations of albumin (ALB) and total protein (TP) were determined in serum samples. Analysis of the serum samples was measured by an automatic biochemistry analyzer (Beckman Synchron CX4 PRO, Fullerton, CA, USA). The liver tissues were fixed in 10% neutral buffered formalin and processed for paraffin embedding, sectioned at 5 μm, and then stained with hematoxylin and eosin, by standard procedure [29]. Liver sections from all broilers were microscopically examined.

4.4. Antioxidant Enzyme Activities Analysis

Liver samples (0.5 g) were thawed in 4.5 mL isotonic saline on ice and homogenized as previously described [33]. The supernatants were then prepared by centrifugation at $12,000 \times g$ for 15 min at 4 °C. Activities of superoxide dismutase (SOD), glutathione peroxidase (GPX), catalase (CAT), and GST, as well as concentrations of GSH and malondialdehyde (MDA) were determined using the colorimetric method through the specific assay kits (A001, A005, A007-1, A004 A006-1 and A003), which were purchased from the Nanjing Jiancheng Bioengineering Institute of China. The concentration of 8-hydroxydeoxyguanosine (8-OHdG) in serum was measured using the ELISA kit (H165, Nanjing Jiancheng Bioengineering Institute, Nanjing, China). Concentrations of protein were determined using the bicinchoninic acid assay [45].

4.5. Hepatic AFBO–DNA Adduct Analysis

Liver genomic DNA was extracted using the DNA extraction kit following the manufacturer's instructions (Qiagen, Shanghai, China). The DNA concentrations were quantified by the 260/280 nm absorbance ratio by an Agilent Bioanalyzer 2100 (Agilent Technologies, Amstelveen, The Netherlands). Genomic DNA (15 μg) was used to determine the AFBO–DNA adduct amount using a competitive ELISA method, according to the manufacturer's instructions (Cell Biolabs, Inc., San Diego, CA, USA).

4.6. Hepatic Microsomal CYP450 Isozyme Activities Analysis

The liver microsomes and cytosolic fractions were prepared as described previously [46]. The microsomal activities of 7-ethoxyresorufin-*O*-deethylase, methoxyresorufin-*O*-demethylase, coumarin 7-hydroxylase, and nifedipine oxidation were determined to assess chicken orthologs of human CYP1A1, CYP1A2, CYP2A6, and CYP3A4 activities, respectively [23]. The concentrations of protein were measured as described above [45].

4.7. Real-Time Quantitative Polymerase Chain Reaction (qPCR)

Total RNA was extracted from liver samples by using Trizol (Invitrogen, Carlsbad, CA, USA) following the manufacturer's instructions. The quality and quantity of the RNA samples were analyzed by an Agilent Bioanalyzer 2100 using an RNA 6000 Labchip kit (Agilent Technologies, Amstelveen, The Netherlands). The cDNA was synthesized from 1.0 μg total RNA by using Super Script III reverse transcriptase (Invitrogen). The mRNA levels of CYP450 isozyme genes were determined by qPCR (7900 HT; Applied Biosystems, Foster City, CA, USA). The primers of CYP450 isozymes and reference gene glyceraldehyde-3-phosphate dehydrogenase were reported previously [12]. The 2^{-ddCt} method was used for the quantification, with glyceraldehyde 3-phosphate dehydrogenase (*GAPDH*) as a housekeeping gene and the relative abundance normalized to the control (as 1) [47].

4.8. Statistical Analysis

One-way ANOVA followed by Fisher's least significant difference (LSD) test was used to compare the effects between groups on each variable within the same tissue. Data are presented as means ± SD, and significance level was set at $p < 0.05$. All analyses were conducted using SAS 8.2 (SAS Institute, Cary, NC, USA).

Supplementary Materials: The following are available online at www.mdpi.com/2072-6651/8/11/327/s1, Table S1: Effects of dietary AFB_1 and CM concentrations on growth performance in chicks, Table S2: Basal diet formulations and nutritional contents.

Acknowledgments: This project was supported by the Chinese Natural Science Foundation projects (31501987 and 31272479), National Key Research and Development Program of China (2016YFD0501207). We thank the Journal of Nutrition for the use of data of Control and AFB1 groups that were originally published therein.

Author Contributions: L.-H.S. designed the research; N.-Y.Z., M.Q., L.Z., M.-K.Z., J.G., J.L. and C.-Q.G. conducted the experiments and analyzed the data; L.-H.S., N.-Y.Z., S.A.R., C.S.K. and D.-S.Q. wrote the paper; and L.-H.S. had primary responsibility for the final content.

Conflicts of Interest: The authors declare no conflict of interest.

References

1. Koehler, P.E.; Hanlin, R.T.; Beraha, L. Production of aflatoxins B1 and G1 by Aspergillus flavus and Aspergillus parasiticus isolated from market pecans. *Appl. Microbiol.* **1975**, *30*, 581–583. [PubMed]
2. Sun, L.H.; Lei, M.Y.; Zhang, N.Y.; Zhao, L.; Krumm, C.S.; Qi, D.S. Hepatotoxic effects of mycotoxin combinations in mice. *Food Chem. Toxicol.* **2014**, *74*, 289–293. [CrossRef] [PubMed]
3. Hussein, H.S.; Brasel, J.M. Toxicity, metabolism, and impact of mycotoxins on humans and animals. *Toxicology* **2001**, *167*, 101–134. [CrossRef]
4. Monson, M.S.; Settlage, R.E.; McMahon, K.W.; Mendoza, K.M.; Rawal, S.; El-Nezami, H.S.; Coulombe, R.A.; Reed, K.M. Response of the hepatic transcriptome to aflatoxin B_1 in domestic turkey (Meleagris gallopavo). *PLoS ONE* **2014**, *9*, e100930. [CrossRef] [PubMed]
5. Rustemeyer, S.M.; Lamberson, W.R.; Ledoux, D.R.; Wells, K.; Austin, K.J.; Cammack, K.M. Effects of dietary aflatoxin on the hepatic expression of apoptosis genes in growing barrows. *J. Anim. Sci.* **2011**, *89*, 916–925. [CrossRef] [PubMed]
6. Yarru, L.P.; Settivari, R.S.; Gowda, N.K.; Antoniou, E.; Ledoux, D.R.; Rottinghaus, G.E. Effects of turmeric (Curcuma longa) on the expression of hepatic genes associated with biotransformation, antioxidant, and immune systems in broiler chicks fed aflatoxin. *Poult. Sci.* **2009**, *88*, 2620–2627. [CrossRef] [PubMed]
7. International Agency for Research on Cancer (IARC). *IARC Monographs on the Evaluation of Carcinogenic Risk of Chemicals to Humans—Overall Evaluation of Carcinogenicity: An Updating of IARC Monographs, Volumes 1–42;* IARC: Lyon, France, 1987.
8. Wu, F.; Guclu, H. Aflatoxin regulations in a network of global maize trade. *PLoS ONE* **2012**, *7*, e45151. [CrossRef] [PubMed]
9. Zain, M.E. Impact of mycotoxins on humans and animals. *J. Saudi Chem. Soc.* **2011**, *15*, 129–144. [CrossRef]
10. Yunus, A.W.; Razzazi-Fazeli, E.; Bohm, J. Aflatoxin B_1 in affecting broiler's performance, immunity, and gastrointestinal tract: A review of history and contemporary issues. *Toxins (Basel)* **2011**, *3*, 566–590. [CrossRef] [PubMed]
11. Mary, V.S.; Theumer, M.G.; Arias, S.L.; Rubinstein, H.R. Reactive oxygen species sources and biomolecular oxidative damage induced by aflatoxin B_1 and fumonisin B1 in rat spleen mononuclear cells. *Toxicology* **2012**, *302*, 299–307. [CrossRef] [PubMed]
12. Sun, L.H.; Zhang, N.Y.; Zhu, M.K.; Zhao, L.; Zhou, J.C.; Qi, D.S. Prevention of Alfatoxin B1 hepatoxicity by dietary selenium is associated with inhibition of cytochrome P450 isozymes and up-regulation of six selenoprotein genes in chick liver. *J. Nutr.* **2016**, *143*, 1115–1122. [CrossRef] [PubMed]

13. Aggarwal, B.B.; Sundaram, C.; Malani, N.; Ichikawa, H. Curcumin: The Indian solid gold. *Adv. Exp. Med. Biol.* **2007**, *595*, 1–75. [PubMed]

14. Firozi, P.F.; Aboobaker, V.S.; Bhattacharya, R.K. Action of curcumin on the cytochrome P450-system catalyzing the activation of aflatoxin B_1. *Chem. Biol. Interact.* **1996**, *100*, 41–51. [CrossRef]

15. Poapolathep, S.; Imsilp, K.; Machii, K.; Kumagai, S.; Poapolathep, A. The effects of curcumin on aflatoxin B_1-induced toxicity in rats. *Biocontrol. Sci.* **2015**, *20*, 171–177. [CrossRef] [PubMed]

16. Satoskar, R.R.; Shah, S.J.; Shenoy, S.G. Evaluation of anti-inflammatory property of curcumin (diferuloyl methane) in patients with postoperative inflammation. *Int. J. Clin. Pharmacol. Ther. Toxicol.* **1986**, *24*, 651–654. [PubMed]

17. Smith, W.A.; Freeman, J.W.; Gupta, R.C. Effect of chemopreventive agents on DNA adduction induced by the potent mammary carcinogen dibenzo[a,l]pyrene in the human breast cells MCF-7. *Mutat. Res.* **2001**, *480–481*, 97–108. [CrossRef]

18. El-Bahr, S.M. Effect of curcumin on hepatic antioxidant enzymes activities and gene expressions in rats intoxicated with aflatoxin B_1. *Phytother. Res.* **2015**, *29*, 134–140. [CrossRef] [PubMed]

19. Gowda, N.K.; Ledoux, D.R.; Rottinghaus, G.E.; Bermudez, A.J.; Chen, Y.C. Antioxidant efficacy of curcuminoids from turmeric (*Curcuma longa* L.) powder in broiler chickens fed diets containing aflatoxin B_1. *Br. J. Nutr.* **2009**, *102*, 1629–1634. [CrossRef] [PubMed]

20. Nayak, S.; Sashidhar, R.B. Metabolic intervention of aflatoxin B_1 toxicity by curcumin. *J. Ethnopharmacol.* **2010**, *127*, 641–644. [CrossRef] [PubMed]

21. El-Agamy, D.S. Comparative effects of curcumin and resveratrol on aflatoxin B_1-induced liver injury in rats. *Arch. Toxicol.* **2010**, *84*, 389–396. [CrossRef] [PubMed]

22. Verma, R.J.; Mathuria, N. Curcumin ameliorates aflatoxin-induced lipid-peroxidation in liver and kidney of mice. *Acta Pol. Pharm.* **2008**, *65*, 195–202. [PubMed]

23. Diaz, G.J.; Murcia, H.W.; Cepeda, S.M. Cytochrome P450 enzymes involved in the metabolism of aflatoxin B_1 in chickens and quail. *Poult. Sci.* **2010**, *89*, 2461–2469. [CrossRef] [PubMed]

24. Appiah-Opong, R.; Commandeur, J.N.; van Vugt-Lussenburg, B.; Vermeulen, N.P. Inhibition of human recombinant cytochrome P450s by curcumin and curcumin decomposition products. *Toxicology* **2007**, *235*, 83–91. [CrossRef] [PubMed]

25. Garg, R.; Gupta, S.; Maru, G.B. Dietary curcumin modulates transcriptional regulators of phase I and phase II enzymes in benzo[a]pyrene-treated mice: Mechanism of its anti-initiating action. *Carcinogenesis* **2008**, *29*, 1022–1032. [CrossRef] [PubMed]

26. Hsieh, Y.W.; Huang, C.Y.; Yang, S.Y.; Peng, Y.H.; Yu, C.P.; Chao, P.D.; Hou, Y.C. Oral intake of curcumin markedly activated CYP 3A4: In vivo and ex vivo studies. *Sci. Rep.* **2014**, *4*, 6587. [CrossRef] [PubMed]

27. Thapliyal, R.; Maru, G.B. Inhibition of cytochrome P450 isozymes by curcumins in vitro and in vivo. *Food Chem. Toxicol.* **2001**, *39*, 541–547. [CrossRef]

28. Bagherzadeh Kasmani, F.; Karimi Torshizi, M.A.; Allameh, A.; Shariatmadari, F. A novel aflatoxin-binding Bacillus probiotic: Performance, serum biochemistry, and immunological parameters in Japanese quail. *Poult. Sci.* **2012**, *91*, 1846–1853. [CrossRef] [PubMed]

29. Sun, L.H.; Zhang, N.Y.; Sun, R.R.; Gao, X.; Gu, C.; Krumm, C.S.; Qi, D.S. A novel strain of Cellulosimicrobium funkei can biologically detoxify aflatoxin B_1 in ducklings. *Microb. Biotechnol.* **2015**, *8*, 490–498. [CrossRef] [PubMed]

30. Klein, P.J.; Van Vleet, T.R.; Hall, J.O.; Coulombe, R.A., Jr. Biochemical factors underlying the age-related sensitivity of turkeys to aflatoxin B_1. *Comp. Biochem. Physiol. C Toxicol. Pharmacol.* **2002**, *132*, 193–201. [CrossRef]

31. Bedard, L.L.; Massey, T.E. Aflatoxin B_1-induced DNA damage and its repair. *Cancer Lett.* **2006**, *241*, 174–183. [CrossRef] [PubMed]

32. Piper, J.T.; Singhal, S.S.; Salameh, M.S.; Torman, R.T.; Awasthi, Y.C.; Awasthi, S. Mechanisms of anticarcinogenic properties of curcumin: The effect of curcumin on glutathione linked detoxification enzymes in rat liver. *Int. J. Biochem. Cell Biol.* **1998**, *30*, 445–456. [CrossRef]

33. Soetikno, V.; Sari, F.R.; Lakshmanan, A.P.; Arumugam, S.; Harima, M.; Suzuki, K.; Kawachi, H.; Watanabe, K. Curcumin alleviates oxidative stress, inflammation, and renal fibrosis in remnant kidney through the Nrf2-keap1 pathway. *Mol. Nutr. Food Res.* **2013**, *57*, 1649–1659. [CrossRef] [PubMed]

34. Kelly, V.P.; Ellis, E.M.; Manson, M.M.; Chanas, S.A.; Moffat, G.J.; McLeod, R.; Judah, D.J.; Neal, G.E.; Hayes, J.D. Chemoprevention of aflatoxin B_1 hepatocarcinogenesis by coumarin, a natural benzopyrone that is a potent inducer of aflatoxin B_1-aldehyde reductase, the glutathione *S*-transferase A5 and P1 subunits, and NAD(P)H: Quinone oxidoreductase in rat liver. *Cancer Res.* **2000**, *60*, 957–969. [PubMed]

35. Kanbur, M.; Eraslan, G.; Sarıca, Z.S.; Aslan, O. The effects of evening primrose oil on lipid peroxidation induced by subacute aflatoxin exposure in mice. *Food Chem. Toxicol.* **2011**, *49*, 1960–1964. [CrossRef] [PubMed]

36. Guengerich, F.P.; Johnson, W.W.; Shimada, T.; Ueng, Y.F.; Yamazaki, H.; Langouët, S. Activation and detoxication of aflatoxin B_1. *Mutat. Res.* **1998**, *402*, 121–128. [CrossRef]

37. Choi, H.; Chun, Y.S.; Shin, Y.J.; Ye, S.K.; Kim, M.S.; Park, J.W. Curcumin attenuates cytochrome P450 induction in response to 2,3,7,8-tetrachlorodibenzo-p-dioxin by ROS-dependently degrading AhR and ARNT. *Cancer Sci.* **2008**, *99*, 2518–2524. [CrossRef] [PubMed]

38. Anzenbacher, P.; Anzenbacherová, E. Cytochromes P450 and metabolism of xenobiotics. *Cell Mol. Life Sci.* **2001**, *58*, 737–747. [CrossRef] [PubMed]

39. Zanger, U.M.; Schwab, M. Cytochrome P450 enzymes in drug metabolism: Regulation of gene expression, enzyme activities, and impact of genetic variation. *Pharmacol. Ther.* **2013**, *138*, 103–141. [CrossRef] [PubMed]

40. Smith, T.J.; Stoner, G.D.; Yang, C.S. Activation of 4-(methylnitrosamino)-1-(3-pyridyl)-1-butanone (NNK) in human lung microsomes by cytochromes P450, lipoxygenase, and hydroperoxides. *Cancer Res.* **1995**, *55*, 5566–5573. [PubMed]

41. Patten, C.J.; Smith, T.J.; Friesen, M.J.; Tynes, R.E.; Yang, C.S.; Murphy, S.E. Evidence for cytochrome P450 2A6 and 3A4 as major catalysts for N9-nitrosonornicotine a-hydroxylation by human liver microsomes. *Carcinogenesis (Lond.)* **1997**, *18*, 1623–1630. [CrossRef]

42. Su, T.; Bao, Z.; Zhang, Q.Y.; Smith, T.J.; Hong, J.Y.; Ding, X. Human cytochrome P450 CYP2A13: Predominant expression in the respiratory tract and its high efficiency metabolic activation of a tobacco-specific carcinogen, 4-(methylnitrosamino)-1-(3-pyridyl)-1-butanone. *Cancer Res.* **2000**, *60*, 5074–5079. [PubMed]

43. Xie, F.; D'Agostino, J.; Zhou, X.; Ding, X. Bioactivation of the nasal toxicant 2,6-dichlorobenzonitrile: An assessment of metabolic activity in human nasal mucosa and identification of indicators of exposure and potential toxicity. *Chem. Res. Toxicol.* **2013**, *26*, 388–398. [CrossRef] [PubMed]

44. Yang, J.; Bai, F.; Zhang, K.; Bai, S.; Peng, X.; Ding, X.; Li, Y.; Zhang, J.; Zhao, L. Effects of feeding corn naturally contaminated with aflatoxin B_1 and B2 on hepatic functions of broilers. *Poult. Sci.* **2012**, *91*, 2792–2801. [CrossRef] [PubMed]

45. Liu, Y.; Zhao, H.; Zhang, Q.; Tang, J.; Li, K.; Xia, X.J.; Wang, K.N.; Li, K.; Lei, X.G. Prolonged dietary selenium deficiency or excess does not globally affect selenoprotein gene expression and/or protein production in various tissues of pigs. *J. Nutr.* **2012**, *142*, 1410–1416. [CrossRef] [PubMed]

46. Klein, P.J.; Buckner, R.; Kelly, J.; Coulombe, R.A., Jr. Biochemical basis for the extreme sensitivity of turkeys to aflatoxin B_1. *Toxicol. Appl. Pharmacol.* **2000**, *165*, 45–52. [CrossRef] [PubMed]

47. Sun, L.H.; Li, J.G.; Zhao, H.; Shi, J.; Huang, J.Q.; Wang, K.N.; Xia, X.J.; Li, L.; Lei, X.G. Porcine serum can be biofortified with selenium to inhibit proliferation of three types of human cancer cells. *J. Nutr.* **2013**, *143*, 1115–1122. [CrossRef] [PubMed]

MDPI

St. Alban-Anlage 66

4052 Basel

Switzerland

Tel. +41 61 683 77 34

Fax +41 61 302 89 18

www.mdpi.com

Toxins Editorial Office

E-mail: toxins@mdpi.com

www.mdpi.com/journal/toxins